Eighth Edition

RESEARCH METHODS IN PHYSICAL ACTIVITY

Jerry R. Thomas
Philip E. Martin
Jennifer L. Etnier
Stephen J. Silverman

Loose-leaf version up to 35% off!

HUMAN KINETICS

EIGHTH EDITION

Research Methods in Physical Activity

Jerry R. Thomas, EdD

Philip E. Martin, PhD

Jennifer L. Etnier, PhD

Stephen J. Silverman, EdD

HUMAN KINETICS

Library of Congress Cataloging-in-Publication Data

Names: Thomas, Jerry R., author. | Martin, Philip E. (Philip Edward), 1955-
  author. | Etnier, Jennifer L., author. | Silverman, Stephen J., author.
Title: Research methods in physical activity / Jerry R. Thomas, Philip E.
  Martin, Jenny L. Etnier, Stephen J. Silverman .
Description: Eighth edition. | Champaign, IL : Human Kinetics, Inc., 2023.
  | Includes bibliographical references and indexes.
Identifiers: LCCN 2021042222 (print) | LCCN 2021042223 (ebook) | ISBN
  9781718201026 (paperback) | ISBN 9781718201033 (epub) | ISBN
  9781718201040 (pdf)
Subjects: LCSH: Physical education and training--Research. |
  Health--Research. | Recreation--Research.
Classification: LCC GV361 .T47 2023 (print) | LCC GV361 (ebook) | DDC
  613.7/1072--dc23
LC record available at https://lccn.loc.gov/2021042222
LC ebook record available at https://lccn.loc.gov/2021042223
ISBN: 978-1-7182-0102-6 (paperback)
ISBN: 978-1-7182-1304-3 (loose-leaf)

The web addresses cited in this text were current as of October 2021, unless otherwise noted.

Acquisitions Editor: Diana Vincer
Developmental Editor: Melissa J. Zavala
Copyeditor: Amy Pavelich
Proofreader: Leigh Keylock
Indexer: Nan N. Badgett
Permissions Manager: Dalene Reeder
Senior Graphic Designer: Nancy Rasmus
Cover Designer: Keri Evans
Cover Design Specialist: Susan Rothermel Allen
Photographs (interior): © Human Kinetics, unless otherwise noted
Photo Asset Manager: Laura Fitch
Photo Production Manager: Jason Allen
Senior Art Manager: Kelly Hendren
Illustrations: © Human Kinetics, unless otherwise noted
Printer: Walsworth

Printed in the United States of America   10 9 8 7 6 5 4 3 2 1

The paper in this book was manufactured using responsible forestry methods.

Human Kinetics
1607 N. Market Street
Champaign, IL 61820
USA

United States and International
Website: **US.HumanKinetics.com**
Email: info@hkusa.com
Phone: 1-800-747-4457

Canada
Website: **Canada.HumanKinetics.com**
Email: info@hkcanada.com

E8222 (paperback) /
E8605 (loose-leaf)

Tell us what you think!
Human Kinetics would love to hear what we can do to improve the customer experience. Use this QR code to take our brief survey.

Jack Kimberly Nelson
(September 14, 1932-January 12, 2018)

Dr. Jack K. Nelson grew up in Valier, Montana, where he worked on a ranch, survived polio, and enlisted in the Air Force, where he was a pilot. He graduated from the University of Montana at Missoula, then matriculated to the University of Oregon, where he earned his PhD working with H. Harrison Clark. He was a professor at Louisiana State University (1962-1990) and University of Idaho (1990-1996).

Jack Nelson and Jerry Thomas taught, as a team, research methods at Louisiana State University and, as a result, wrote the first edition of this textbook. This was the beginning of a long, professional partnership and a deep, personal friendship. Jack had a quick and dry wit, often offered a helping hand, and was always a trusted friend. Among his numerous accomplishments and contributions to the field of research methods, he taught research methods for 35 years; conducted research; had over 80 publications, including authoring multiple textbooks; was an adviser on more than 100 doctoral dissertations and master's theses combined; and served as editor of research publications. A fellow in the Research Consortium, he was also a member of American Alliance for Health, Physical Education, Recreation and Dance (now SHAPE America), American Educational Research Association (AERA), and American College of Sports Medicine (ACSM). He also served as president of the Association for Research, Administration, Professional Councils and Societies (now AAALF) and as vice president of AAHPERD.

Jack was a remarkable kind of friend who was easygoing and could make you laugh until it hurt. He was very proud of his children and grandchildren and was loved by his family, friends, colleagues, and students. Jack had a spirit of adventure and was keen to taking a spur-of-the-moment trip on a houseboat or train and yet perfectly content with sitting on the screened-in porch, completing a *New York Times* crossword. Professionally, he brought to each project a combination of superior intellect, strong work ethic, integrity, abundant kindness, and humor. As our careers diverged, we found ourselves looking forward to subsequent editions of this book because the work brought us together again. It was an honor to have him in our lives; we are so thankful for all the wonderful and comforting memories of him. And so, nearly 40 years after its original publication, we've developed this latest edition without Jack. It is our sincerest hope that he would be proud of our work.

CONTENTS

The first edition was published in 1985 and was titled *Introduction to Research in Health, Physical Education, Recreation and Dance*. Publishing the eighth edition is rewarding and surprising. In 1985, Human Kinetics was a new publisher. We could not have guessed that we would do this many editions nor did we expect the field to evolve so dramatically. The second edition recognized the evolving field with a title to represent its breadth, *Research in Physical Activity*. The title was inclusive of the field: sport science, exercise science, kinesiology, physical education, and so forth. In 1985, we could not have predicted that physical activity would become a key factor in public health. By 1990, however, it was clear that physical activity was becoming important and of interest to those outside sports, exercise, dance, and physical education. Research in physical activity was published in the parent journals—for example, *Developmental Psychology*, *Physiology*—and in other highly regarded journals such as *Journal of the American Medical Association*. Scholars in physical activity were recognized for their high-quality work. Further, Human Kinetics was growing.

We take this opportunity to thank all the people who have used this book over the years. Once again, we hope that you have learned about research methods in the study of physical activity such that you will be an informed consumer of research and a knowledgeable scholar. Maybe you have even enjoyed the humorous stories, jokes, and pictures that we have included to enliven the reading. We also thank the reviewers for their helpful comments and suggestions, which we have tried to address in this edition. When we read reviews, we feel as Day (1983, p. xi) did when he read that a reviewer described his book as both good and original, but then went on to say that "the part that is good is not original, and the part that is original is not good." We are also delighted that many of you in other English-speaking countries have also used this book. In addition, we appreciate that earlier editions have been translated into Chinese (twice), Greek, Korean, Italian, Japanese, Spanish, and Portuguese.

Dr. Stephen Silverman joined us as a coauthor on the fifth edition and, in spite of our sense of humor, agreed to continue on the subsequent editions. Dr. Silverman is a well-known scholar and methodologist in physical education pedagogy and is a former editor-in-chief of *Research Quarterly for Exercise and Sport*. Joining the team on this edition are Dr. Philip E. Martin and Dr. Jennifer L. Etnier. Dr. Martin is a professor and the chair emeritus at Iowa State University and a biomechanist with an impressive scholarly record. Dr. Etnier is a distinguished professor and department chair at the University of North Carolina–Greensboro. She is an exercise psychologist who has led research on Alzheimer's disease as well as national professional organizations in sport psychology and kinesiology, all while producing a stellar scholarly record. All members of this team have taught research methods.

The main use of this text still appears to be in the first graduate-level research methods courses, although it is also being used in undergraduate research methods courses and as a resource for those engaged in research planning and analysis. Our use of the term *physical activity* in the book title is meant to convey the broadly conceived field of study often labeled kinesiology, exercise science, exercise and sport science, human movement, sport studies, or physical education, as well as related fields such as physical therapy, rehabilitation, and occupational therapy. We hope that everyone who reads, understands, plans, carries out, writes, or presents research will find the book a useful tool to enhance their efforts.

This eighth edition retains the basic organization of the seventh edition, as follows:

- *Part I* is an overview of the research process, including developing the problem, using the literature, preparing a research plan, and understanding ethical issues in research and writing.

- *Part II* introduces statistical and measurement issues in research, including statistical descriptions, power, interrelationships of variables, differences between groups, nonparametric procedures, and measurement issues in research.

- *Part III* presents the types of, or approaches to, research, including historical, philosophical, research synthesis, survey, descriptive, epidemiological, experimental, qualitative, and mixed methods.

- *Part IV* will help you complete the research process, which includes writing the results and discussion, organizing the research paper, developing good figures and tables, and presenting research in written and oral forms.

- The appendix includes statistical tables.

Instructors using this text in their courses will find an instructor guide, test package, chapter quizzes, presentation package, and image bank in HK*Propel*. The image bank includes most of the art, tables, and example elements from the text, which can be used to create custom presentations. The instructor guide includes chapter overviews, sample course syllabuses, supplemental class activities, and student handouts. The test bank includes over 600 questions. The chapter quizzes contain ready-made quizzes, with about 10 questions per chapter drawn from the test package, to assess student comprehension of the most important concepts in each chapter.

Although the format of the book remains similar to that of the seventh edition, we have made a number of changes that we hope improve and update the text. Following is a short review of the changes in this eighth edition:

- *Part I*: *Overview of the Research Process.* Each chapter includes minor revisions that reflect updated information and more recent reports. Chapter 1 includes examples that more broadly represent physical activity research and the inclusion of cases studies. We have again made a significant revision to chapter 2 about using library techniques by adding much more on electronic searches. In addition, chapter 5 on ethical issues has been updated with particular attention to procedures for the use of human and animal subjects with expanded focus on security.

- *Part II*: *Statistical and Measurement Concepts in Research.* We strive in each edition to increase the relevance of the examples and provide easy-to-understand calculations for basic statistics. We have reduced the examples of hand calculations and formulas and replaced them with sample output from the Statistical Package for the Social Sciences. We have included 2019 player performance data for outcome and skill variables from the Professional Golfers Association website as examples for analysis in the statistical chapters. In chapter 6, we have added a more in-depth introduction to the need for and types and uses of statistics. More information on sampling, as well as greater detail on the stem-and-leaf technique, has been included. Chapters 7, 8, and 9 have been reorganized with some information shifted among the chapters in this section. Along with the chapter examples, the learning activities in the instructor guide should help students grasp the fundamentals of statistical techniques. We continue to use a unified approach to parametric and nonparametric techniques.

- *Part III*: *Types of Research.* We have continued our use of expert authors to present coherent views of sociohistorical research (David Wiggins and Daniel Mason), philosophical research (Tim Elcombe and R. Scott Kretchmar), and epidemiological research in physical activity (Duck-chul Lee and Angelique Brellenthin). These three types of research are outside our expertise, and we wanted them presented by expert scholars. Chapter 14 has been expanded to include systematic reviews. In addition, we have made minor revisions and updates to all the other chapters in this part.

- *Part IV: Writing the Research Report.* The two chapters in this section remain essentially the same, with changes and updates. Chapter 21 includes these headings: Thesis and Dissertation Proposals, Advisor and Dissertation Committee, The Good Scholar Must Research and Write, Scientific Writing, and First Things Are Sometimes Best Done Last. Chapter 22 has greater focus on all aspects of presenting research results.

As we have said in each edition, we are grateful for the help of our friends, both for help that we acknowledge in various places in the book and for help in other places where we have inadvertently taken an idea without giving credit.

> **After the passage of time, one can no longer remember who originated what idea. After the passage of even more time, it seems to me that all of the really good ideas originated with me, a proposition which I know is indefensible. (Day, 1983, p. xv)**

We believe that this book provides the necessary information for both the consumer and the producer of research. Although no amount of knowledge about the tools of research can replace expertise in the content area, good scholars of physical activity cannot function apart from the effective use of research tools. Researchers, teachers, clinicians, technicians, health workers, exercise leaders, sport managers, athletic counselors, and coaches need to understand the research process. If they do not, they are forced to accept information at face value or on the recommendation of others. Neither is necessarily bad, but the ability to evaluate and reach a valid conclusion based on data, method, and logic is the mark of a professional.

Inserted into some chapters are humorous stories, anecdotes, sketches, laws, and corollaries. These are intended to make a point and enliven the reading without distracting from the content. Research processes are not mysterious events that graduate students should fear. To the contrary, they are useful tools that every academic and professional should have access to; they are, in fact, the very basis on which we make competent decisions.

Jerry R. Thomas

Philip E. Martin

Jennifer L. Etnier

Stephen J. Silverman

Dear Student of Research Methods:

We want you to learn the material here, and most of you are learning it in a classroom setting as well as by reading the textbook. From many years of teaching research methods, we have arrived at the following recommendations:

1. Attend and participate in class—90% of life is showing up!
2. Take notes in class—writing it down is an effective way to learn.
3. Read the assigned materials before class—duh!
4. Plan for and ask at least one question in every class.
5. At the conclusion of class, recall everything you can about the class content—research shows this enhances learning.
6. Develop and work with a study group.
7. Prepare for exams and tests—do not cram; study over several days.
8. Use campus resources to improve learning—library, computer, the Internet.
9. Visit often with your professor—those of us teaching research methods are likable folks!

The following list will help you determine your readiness to be a student of research methods.

Score one point for each of the following statements that describes you:

Your library carrel is better decorated than is your apartment.

You have taken a scholarly article to a bar or coffee shop.

You rate coffee shops on the availability of Wi-Fi and outlets for your electronic devices.

You have discussed academic matters at a sport event.

You actually have a preference between microfilm and microfiche.

You always read the reference lists in research articles.

You think that the sorority sweatshirt Greek letters are a statistical formula.

You need to explain to children why you are in the 20th grade.

You refer to stories as "Snow White et al."

You wonder how to cite talking to yourself in APA style.

Scoring Scale

5 or 6—definitely ready to be a student in research methods

7 or 8—probably a master's student

9 or 10—probably a doctoral student

Humorously yours,

Professors of Research Methods

ACKNOWLEDGMENTS

As with any work, numerous people contributed to this book, and we want to recognize them. Many are former students and colleagues who have said or done things that better developed our ideas as expressed in these pages. Also, a number of faculty members who have used previous editions have either written reviews or made suggestions that have improved the book. Although we cannot list or even recall all these contributions, we do know that you made them, and we thank all of you.

In particular, Scott Kretchmar, Tim Elcombe, David Wiggins, Daniel Mason, Duck-chul Lee and Angelique G. Brellenthin made invaluable contributions with their chapters on research methods in the areas of philosophy, history, and exercise epidemiology, which are areas we simply could not write about effectively.

Finally, we thank the staff at Human Kinetics—in particular Diana Vincer and Melissa Zavala, for their support and contributions. They have sharpened our thinking and improved our writing.

Jerry R. Thomas

Philip E. Martin

Jennifer L. Etnier

Stephen J. Silverman

PART I

Overview of the Research Process

> The researches of many have thrown much darkness on the subject and if they continue, soon we shall know nothing at all about it.
>
> —Attributed to Mark Twain

Part I provides an overall perspective of the research process. The introductory chapter defines and reviews the types of research done in physical activity and provides some examples. We define *science* as "systematic inquiry," and we discuss the steps in the scientific method. This logical method answers the following four questions (Day, 1983, p. 4), which constitute the parts of a typical thesis, dissertation, or research report:

1. *What was the problem?* Your answer is the introduction.
2. *How did you study the problem?* Your answer is the materials and methods.
3. *What did you find?* Your answer is the results.
4. *What do these findings mean?* Your answer is the discussion.

We also present alternative approaches for doing research relative to a more philosophical discussion of science and ways of knowing. In particular, we address qualitative research, the use of field studies, and methods of introspection as strategies for answering research questions instead of relying on the traditional scientific paradigm as the only approach to research problems.

Chapter 2 suggests ways of developing a problem and using the literature to clarify the research problem, specify hypotheses, and develop the methodology. In particular, we emphasize the use of new electronic technology for searching, reading, analyzing, synthesizing, organizing, and writing literature reviews.

The next two chapters in part I present the format of the research proposal with examples. This information is typically required of the master's or doctoral student before collecting data for the thesis or dissertation. Chapter 3 addresses defining and delimiting the research problem, including the introduction, statement of the problem, research hypotheses, operational definitions, assumptions and limitations, and significance. Information is provided for both quantitative and qualitative approaches. Chapter 4 covers methodology, or how to do the research, using either quantitative or qualitative methods. Included are the topics of participant selection, instrumentation or apparatuses, procedures, and design and analysis. We emphasize the value of pilot work conducted before the research and how cause and effect may be established.

Chapter 5 discusses ethical issues in research and scholarship. We include information on misconduct in science; security of data, ethical considerations in research writing, collaborative work with advisors, and copyright; and the use of humans and animals in research.

When you have completed part I, you should have a better understanding of the research process. Then comes the tricky part: learning all the details. We consider these details in part II (Statistical and Measurement Concepts in Research), part III (Types of Research), and part IV (Writing the Research Report).

1

Introduction to Research in Physical Activity

> **Everything that can be invented has been invented.**
>
> —Charles H. Duell, Commissioner, U.S. Office of Patents, 1899

To each person, the word *research* conjures up a different picture. One might think of searching the Internet or going to the library; another might visualize a lab filled with test tubes, vials, and perhaps little, white rats. Therefore, as we begin a text on the subject, we must establish a common understanding of research. In this chapter, we introduce you to the nature of research. We do this by discussing methods of problem-solving and types of research. We explain the research process and relate it to the parts of a thesis. By the time you reach the end of chapter 1, you should understand what research really involves.

The Nature of Research

The object of research is to determine how things are as compared to how they might be. To achieve this, research implies a careful and systematic means of solving problems and involves the following five characteristics (Tuckman, 1978):

- *Systematic.* Problem-solving begins with and is accomplished by identifying and labeling variables. Research is then designed to test the relationships of these variables. Data are collected that, when related to the variables, allow the evaluation of the problem and hypotheses.
- *Logical.* Examination of the procedures used in the research process allows researchers to evaluate the conclusions they've drawn.
- *Empirical.* Researchers collect data on which to base decisions.
- *Reductive.* Researchers take individual events (data) and use them to establish general relationships.
- *Replicable.* The research process is recorded, enabling others to test the findings by repeating the research or to build future research on previous results.

Problems to be solved come from many sources and can entail resolving controversial issues, testing theories, and trying to improve present practices. For example, a popular topic of concern is obesity and the methods for losing weight. Suppose we want to investigate this issue by comparing the effectiveness of two exercise programs in reducing body fat in people who are overweight. Since we know that caloric expenditure can contribute to a reduction in body fat, we will try to find out which program does this better under specified conditions. *Note:* Our approach here is to give a simple, concise overview of a research study. We do not intend it to be a model of originality or sophistication.

This study is an example of **applied research**. Rather than try to measure the calories expended and so on, we approach it strictly from a programmatic standpoint. Let's say we operate a health club and offer aerobic dance and jogging classes for people who want to lose weight. Our research question would be: *Which program is more effective in reducing fat?*

Suppose we have a pool of participants to draw from and can randomly assign two-thirds of them to the two exercise programs and one-third to a control group. We have their scout's honor that no one is drastically dieting or engaging in any other strenuous activities for the duration of the study. Both classes are one hour long and held five times a week for 10 weeks. The same enthusiastic and immensely qualified instructor teaches both classes.

Our method for measuring body composition is by using a portable **bioelectric impedance** system. Of course, we could use other measures, such as **hydrostatic weighing** or **DXA scanning** (or some other estimate of adiposity). In any case, we can defend our measures as valid and reliable indicators of adiposity, and bioelectric impedance is a functional field measure. We measure all the participants, including those in the control group, at the beginning and the end of the 10-week period. During the study, we try to ensure that the two programs are similar in procedural aspects, such as motivational techniques and the aesthetics of the surroundings. In other words, we do not favor one group by cheering them on and not encouraging the other, nor do we have one group exercise in an air-conditioned, cheerful, and healthful facility while the other has to sweat it out in a dingy room or a parking lot. We make the programs as similar as possible in every respect except the experimental treatments: The control group does not engage in any regular exercise.

After we have measured all the participants on our criterion of adiposity at the end of the 10-week program, we are ready to analyze our data. We want to see how much change in fat mass has occurred and whether differences have occurred between the two types of exercise. Because we are dealing with samples of people (from a whole universe of similar people), we need to use some type of statistics to establish how confident we can be in our results. In other words, we need to determine the significance of our results. Suppose the mean (average) scores for the groups are as follows:

- Aerobic dance: −3.1%
- Jogging: −3.7%
- Control: +1.1%

These hypothetical values represent the average changes in percent body fat for each group. The two experimental groups lost fat, but the control group actually showed an increase over the 10-week period.

We decide to use the statistical technique of analysis of variance. We find a significant F ratio, indicating that significant differences exist between the three groups. Using a follow-up test procedure, we discover that both exercise groups significantly differ from the control group. But we find only a slight difference between the aerobic dance and the jogging groups. (Many of you may not have the foggiest idea what we are talking about with the statistical terms F *ratio* and *significance*, but do not worry about it! All that is explained later. This book is directly concerned with those kinds of concepts.)

We conclude from our study that over a course of 10 weeks, both aerobic dance and jogging are effective (and, apparently, equally so) in reducing adiposity in people (such as those in our study). Although these results are reasonable, remember that this is a hypothetical scenario. We could also pretend that this study was published in a prestigious journal and that we won the Nobel Prize.

Research Continuum

Research in our field can be placed on a continuum that has **basic research** at one extreme and applied research at the opposite extreme. The research extremes are generally associated with certain characteristics. Basic research usually deals with theoretical problems. It uses the laboratory as the setting, sometimes uses animals as subjects, carefully controls conditions, and produces results that have limited direct application. At the other extreme, applied research tends to address immediate problems. It is conducted in real-world settings, uses human participants, and involves limited control over the research setting. Applied research provides practitioners with results that have direct value.

Christina (1989) suggested that basic and applied forms of research are useful in informing each other as to future research directions. Table 1.1 demonstrates how research problems in motor learning might vary along a continuum from basic to applied depending on their goals and approaches.

To some extent, the strengths of applied research are the weaknesses of basic research and vice versa. Considerable controversy exists in the literature on social science (e.g., Creswell, 2009; Jewczyn, 2013) and physical activity (e.g., Christina, 1989) about whether research should be more basic or more applied. This issue, labeled **ecological validity**, deals with two concerns: Is the research setting perceived by the research participant in the way intended by the experimenter? Does the setting have enough of the real-world characteristics to allow generalizing to reality?

Most research incorporates elements of both basic and applied research to some degree. We believe that systematic efforts are needed in the study of physical activity to produce research that moves back and forth across Christina's (1989) levels of research (table 1.1). Excellent summaries of this type of research and the accumulated knowledge are provided in three

basic research—A type of research that may have limited direct application but allows researchers to have careful control of the conditions.

ecological validity—The extent to which research emulates the real world.

TABLE 1.1
Levels of Relevance of Motor Learning Research for Finding Solutions to Practical Problems in Sport

	Level 1 Least direct relevance Basic research	Level 2 Moderate direct relevance Applied research	Level 3 Most direct relevance Applied research
Ultimate goal	Develop theory-based knowledge appropriate for understanding motor learning in general with no requirement to demonstrate its relevance for solving practical problems	Develop theory-based knowledge appropriate for understanding the learning of sport skills in sport settings with no requirement to find immediate solutions to learning problems in sport	Find immediate solutions to learning problems in sport with no requirement to demonstrate, or develop theory-based knowledge at either level 1 or level 2
Main approach	Test hypotheses in a laboratory setting using experimenter-designed motor tasks	Test hypotheses in a sport setting or in an applied laboratory setting using sport skills or motor tasks that have properties of those skills	Test solutions to specific learning problems in sport in the settings described under the applied research at level 2

From R.W. Christina, "Whatever happened to applied research in motor learning?" in *Future Directions in Exercise and Sport Science Research*, edited by J.S. Skinner et al. (Champaign, IL: Human Kinetics, 1989). By permission of Robert W. Christina.

edited volumes representing exercise physiology, exercise and sport psychology, and motor control: *Physical Activity and Health* (Bouchard, Blair, & Haskell, 2012), *Psychobiology of Physical Activity* (Acevedo & Ekkekakis, 2006), and *Motor Control: Theories, Experiments, and Applications* (Danion & Latash, 2011). Experts prepared each chapter in these books to summarize theories as well as to present basic and applied research about areas related to exercise physiology, exercise and sport psychology, and motor control. The novice researcher would do well to read several of these chapters as examples of how knowledge is developed and accumulated in the study of physical activity. We need more efforts to produce a related body of knowledge in the study of physical activity. Although the research base has grown tremendously in our field over the past 40 years, much remains to be done.

There is a great need to prepare proficient consumers and producers of research in kinesiology. To be proficient, people must thoroughly understand the appropriate knowledge base (biomechanics, exercise physiology, exercise psychology, motor control, pedagogy, and the social and biological sciences) as well as research methods (qualitative, quantitative, and mixed methods). In this book, we attempt to explain the tools necessary for consuming and producing research. Many of the same methods are used in the various areas of kinesiology (as well as in the fields of psychology, sociology, education, and physiology). Quality research efforts always involve some or all of the following actions:

- Identifying and delimiting a problem
- Searching, reviewing, critically analyzing, integrating, and effectively summarizing relevant literature
- Specifying and defining testable hypotheses
- Designing the research to test the hypotheses
- Selecting, describing, testing, and treating the participants
- Analyzing and reporting the results
- Discussing the meaning and implications of the findings

Practicality and Accessibility

We recognize that not everyone is a researcher. Many people in kinesiology have little interest in research per se. In fact, some have a decided aversion to it. The public at large sometimes may view researchers as people with eccentricities who deal with "insignificant" problems and are out of touch with the real world. In an informative yet entertaining book on writing scientific papers, Gastel and Day (2016, pp. 213) related the story about two men who, while riding in a hot-air balloon, encountered some cloud coverage and lost their way. When they finally descended, they did not recognize the terrain and had not the faintest idea where they were. It so happened that they were drifting over the grounds of one of our more famous scientific research institutes. When the balloonists saw a man walking alongside a road, one of them called out, "Hey, mister, where are we?" The man looked up, took in the situation, and after a few moments of reflection said, "You're in a hot-air balloon." One balloonist turned to the other and said, "I'll bet that man is a researcher." The other balloonist asked, "What makes you think so?" The first replied, "His answer is perfectly accurate—and totally useless."

All kidding aside, the need for research in any profession cannot be denied. After all, one of the primary distinctions between a discipline or profession and a trade is that the trade deals only with how to do something, whereas the discipline or profession concerns itself not only with how but also with why something should be done in a certain manner (and why it should even be done at all). But although most people in a discipline or profession recognize the need for research, most do not read research results. This situation is not unique to our field. It has been reported that only 1% of chemists read research publications, that fewer than 7% of psychologists read psychological research journals, and so on. The big question is *why*? Our best guess is that most professionals believe that the findings either are not practical enough or do not directly pertain to their work, rendering the act of reading such publications unnecessary. Another reason practitioners give for not reading research publications is the literature is indecipherable: The language is too technical, and the terminology is unfamiliar and confusing. This complaint is valid, but we could argue that if the professional preparation programs were more scientifically oriented, the problem would diminish. Nevertheless, the research literature is extremely difficult for most people to understand. We must continue efforts to decrease this communication gap.

Reading Research

Someone once said (facetiously) that scientific papers are meant not to be read but to be published. Unfortunately, we find considerable truth in this observation. We writers are often guilty of trying to use language to dazzle the reader and perhaps to give the impression that our subject matter is more esoteric than it really is. We tend to write for the benefit of a rather small audience of readers—that is, other researchers in our field.

We have the problem of jargon, of course (Plaven-Sigray et al., 2017). In any field, whether it is physics, football, or cake baking, jargon confounds the outsider. The use of jargon serves as a kind of shorthand. It provides meaning to the people within the field because everyone uses those truncations in the same context. Research literature is famous for using a three-dollar word when a nickel word would do. As Gastel and Day (2016) asked, "who would use the three-letter word *now* instead of the elegant expression *at this point in time*" (p. 209)? Researchers never *do* anything, they perform it; they never *start*, they *initiate*; and they *terminate* instead of *end*. Gastel and Day further remarked that "an occasional author will slip and use the word *drug*, but most will salivate like Pavlov's dogs in anticipation of using *chemotherapeutic agent*" (p. 209).

The need to bridge the gap between the researcher and the practitioner has been recognized for years. For example, the *Translational Journal of the American College of Sports Medicine* was created to communicate implications of basic, clinical, and policy research to practitioners. The website for the American Kinesiology Association (www.americankinesiology .org) regularly has a section on applied research. Yet despite these and other attempts to bridge the gap between researchers and practitioners, the gap is still imposing.

It goes without saying that if you are not knowledgeable about the subject matter, you cannot read the research literature. Conversely, if you know the subject matter, you can probably wade through the researcher's jargon more effectively. For example, if you know baseball and the researcher is recommending that by shortening the radius, the hitter can increase the angular velocity, you can figure out that the researcher means to choke up on the bat.

One of the big stumbling blocks is the statistical analysis part of research reports. Even the most ardent seeker of knowledge can be turned off by such descriptions as this: "The tetrachoric correlations among the test variables were subjected to a centroid factor analysis, and orthogonal rotations of the primary axes were accomplished by Zimmerman's graphical method until simple structure and positive manifold were closely approximated." Please note that we are not criticizing the authors for such descriptions, because reviewers and editors usually require them. We are just acknowledging that statistical analysis is frightening to someone who is trying to read a research article and does not know a factor analysis from a plank exercise.

How to Read Research

Despite all the hurdles that loom in a practitioner's path when reading research, we contend that you can read and profit from the research literature even if you are not well grounded in research techniques and statistical analysis. We offer the following suggestions on reading the research literature:

- Become familiar with a few publications that contain pertinent research in your field. You might get some help on choosing the publications from a professor or librarian.

- Learn to use search tools (e.g., Web of Science, Pubmed, Scopus) for identifying research literature relevant to your interests.

- Read only studies that are of interest to you. This point may sound too trite to mention, but some people feel obligated to wade through every article.

- Read as a practitioner would. Do not look for eternal truths. Look for ideas and indications. No study is proof of anything. Only when it has been verified repeatedly does it constitute knowledge.

- Read the abstract first. This saves time by helping you determine whether you wish to read the whole thing. If you are still interested, then you can read the study to gain a better understanding of the objectives, hypotheses, methods, and interpretations, but do not get bogged down with details.

- Do not be too concerned about statistical significance. Understanding the concept of *significance* certainly helps, but a little common sense serves you about as well as knowing the difference between the 0.02 and the 0.01 levels, or a one-tailed test versus a two-tailed test. Think in terms of meaningfulness. For example, if two regimens of resistance training result in an average difference in strength improvement of 0.5%, does it matter whether the difference is significant? On the other hand, if a big difference is present but not significant, further investigation is warranted, especially if the study involved a small number of participants. Knowing the concepts of the types of statistical analysis is certainly helpful, but it is not crucial to being able to read a study. Just skip that part.

- Be critical but objective. You can usually assume that a national research journal selects studies for publication by the jury method. Two or three qualified people read and judge the relevance of the problem, the validity and reliability of the procedures, the efficacy of the experimental design, and the appropriateness of the statistical analysis. Certainly, some studies are published that should not be. Yet if you are not an expert in research, you do not need to be suspicious about the scientific worth of a study that appears in a recognized journal. If it is too far removed from any practical application to your situation, do not read it.

You will find that the more you read, the more you will understand, simply because you enhance your familiarity with the language and the methodology, like the person who was thrilled to learn they had been speaking prose all their life.

An Example of Practical Research

To illustrate our research consumer suggestions, consider the following fictional account of Sonjia Roundball, a newly trained physical education teacher and coach (Nelson, 1988).

In a moment of curiosity, Coach Roundball began browsing the *Research Quarterly for Exercise and Sport*, which had been left in her car by a graduate student friend. An article titled "The Effects of a Season of Basketball on the Cardiorespiratory Responses of High School Girls" immediately caught her attention. In its introductory passages, the article stated that only a negligible amount of information was available on the specific physiological changes in girls that result from sport participation. The article cited a few studies on swimmers and other sport participants, and the rather broad takeaway of these studies was that the female athletes possessed higher levels of cardiorespiratory fitness than nonathletes. The author emphasized that no studies had tried to detect changes in girls' fitness during a season of basketball.

The article's next section described methods used in one particular study as well as noting the length of the season, numbers of games and practices (including their lengths), and a breakdown of time devoted to drills, scrimmages, and individual practice. Participants were placed into two groups. The first group comprised 12 girls who played on a high school basketball team. The control group was made up of 14 nonplayers who took physical education classes and had academic and activity schedules similar to the basketball players. All participants in the study were tested at the start and end of the season for maximal oxygen consumption and various other physiological measurements dealing with ventilation, heart rate, and blood pressure. Since these were concepts Coach Roundball remembered from her exercise physiology course several years earlier, she was willing to accept them as appropriate indicators of cardiorespiratory fitness.

Coach Roundball was inexperienced with interpreting the kinds of results that were presented in tables, so she was inclined to trust the author's claims. No significant increases in any of the cardiorespiratory measures from the pre- and postseason tests for either group were detected, which raised the first red flag. Surely, a strenuous sport such as basketball should produce improvements in fitness. The coach continued to read with more skepticism. The author reported that the basketball players had higher values of maximal oxygen consumption than the control group did at both the beginning and the end of the season. The **discussion** mentioned that the values were actually higher than similar values in other studies. *So what?* Coach Roundball thought. Additionally, the author stated that the number of participants was small and some changes may not have been detected. Despite these and other potentially flawed aspects of the study, the author still concluded that the training program used in this study was not strenuous enough to induce significant improvement in cardiorespiratory fitness.

> **discussion**—The chapter or section of a research report that explains what the results mean.

Coach Roundball understood the limitations of the single study. Nevertheless, the practice schedule and general practice routines used in the study were similar to her own. The article's references section listed three studies from a journal she had never read called *Medicine & Science in Sports & Exercise*. The following weekend, she visited the university library and found the journal's latest issue, which happened to have an article on the conditioning effects of swimming on college-age women. Although a different sport and age group, Coach Roundball reasoned that this article could contain useful information that might resolve some questions about the earlier article she had read. The *Medicine & Science in Sports & Exercise* article cited, of all things, a recent study on aerobic capacity, heart rate, and energy cost during a season of girls' basketball. The coach quickly located that study and was pleasantly surprised at how she could read this study with ease now that she was more familiar with the terminology and general organization of the research literature.

Coincidentally, this study also reported no improvement in aerobic capacity during the season. While monitoring heart rates during games by telemetry, researchers rarely observed heart rates above 170 beats per minute (bpm). They concluded that the practice sessions were apparently too moderate in intensity and that the training should be structured to meet both the skill and the fitness demands of the sport.

Coach Roundball returned to her school determined to take a more scientific approach to her basketball program. To start, she had one of her managers chart the number of minutes players were actually engaged in movement in the practice sessions. She also had the players take their pulses at various intervals throughout the sessions. She was surprised to find that the heart rates rarely surpassed 130 bpm. As an outgrowth of her recent literature search, she recalled that an intensity threshold would be necessary to bring about improvement in cardiorespiratory fitness. She knew that for this age group, a heart rate of about 160 bpm was needed to provide a significant training effect. By adjusting practice sessions to include more conditioning drills and make the scrimmages more intensive and game-like, Coach Roundball's team would go on to have an overall stronger competitive advantage in district and state championships.

Summarizing the Nature of Research

science—A process of careful and systematic inquiry.

Thomas Huxley, the famous British scientist who promoted Darwin's theory of evolution, wrote that **science** is simply common sense at its best. However, the status science holds is based on findings being correct most of the time as well as finding the instances in which reported findings are not correct. Science is systematic; if you and I do the same experiment at the same time or two years apart, we should get the same answer. Unfortunately, several recently reported attempts to replicate earlier studies have not been successful. Of the more than two million scholarly papers published in journals each year, an important question might be, *How many reported findings are wrong?*

Having said that about science, discovery can be rewarding, whether that discovery is research that applies to and can improve your situation or is simply new knowledge obtained while researching your thesis or dissertation. We must work against the common misperception that research is some dark and mysterious realm inhabited by impractical people who speak and write in baffling terms. In general, research should be viewed for what it is: a methodic approach for solving problems. We firmly believe that practitioners can read research literature. Our intent with this book is to help facilitate that process of turning our readers into research consumers.

Unscientific Versus Scientific Methods of Problem-Solving

Although the term *research* has many definitions, nearly all characterize research activity as some type of structured problem-solving. The word *structured* refers to the fact that a number of research techniques can be used as long as they are considered acceptable by scholars in the field. Thus, research is concerned with problem-solving, which then may lead to new knowledge.

scientific method of problem-solving—A method of solving problems that uses the following steps in this order: (a) define and delimit the problem; (b) form a hypothesis; (c) gather data; (d) analyze the data; and (e) interpret the results.

The problem-solving process involves several steps whereby the problem is developed, defined, and delimited; hypotheses are formulated; methods are planned and employed to gather and analyze data; and the results are interpreted with regard to the acceptance or rejection of the hypotheses. These steps are often referred to as the **scientific method of problem-solving**. The steps also constitute the chapters, or sections, of the research paper, thesis, or dissertation. Consequently, we devote much of this text to the specific ways these steps are accomplished.

Some Unscientific Methods of Problem-Solving

Before we go into more detail concerning the scientific method of problem-solving, we should recognize some other ways by which humankind has acquired knowledge. All of us have used these methods, so they are recognizable. Helmstadter (1970) labeled the methods *tenacity, intuition, authority, the rationalistic method,* and the *empirical method*.

tenacity—An unscientific method of problem-solving in which people cling to certain beliefs regardless of a lack of supporting evidence.

Tenacity

People are prone to clinging to certain beliefs despite a lack of supporting evidence. Our superstitions are good examples of the method called **tenacity**. Coaches and athletes are notoriously superstitious. A coach may wear a particular sport coat, hat, tie, or pair of shoes because the team won the last time he wore it. Athletes frequently have a set pattern that they consider lucky for dressing, warming up, or entering the stadium. Although they acknowledge no logical relationship between the game's outcome and their particular routine, they are afraid to break the pattern.

For example, take the man who believed that black cats bring bad luck. One night while he was returning to his ranch, a black cat started to cross the road. The man swerved off the road to keep the cat from crossing in front of him and hit a hard bump that caused the headlights to turn off. Unable to see the black cat in the dark, he sped frantically over rocks, mounds, and holes until he came to a sudden stop in a ravine and wrecked his car. Of course, this episode just confirmed his staunch belief that black cats do indeed bring bad luck. Obviously, tenacity has no place in science. It is the least reliable source of knowledge.

Intuition

Intuitive knowledge is sometimes considered common sense or self-evident. Many self-evident truths, however, often turn out to be false. That the earth is flat is a classic example of the intuitively obvious; that the sun revolves around the earth was once self-evident; that no one could run a mile (1.6 km) in less than four minutes once was self-evident. Furthermore, for anyone to shot-put more than 70 ft (21 m) or pole-vault more than 18 ft (5.5 m), or for a woman to run distances over 0.5 mi (0.8 km), such feats were impossible. One fundamental tenet of science is that we must be ever cognizant of the importance of substantiating our convictions with factual evidence.

Authority

Reference to some authority has long been used as a source of knowledge. Although this approach is not necessarily invalid, it does depend on the authority and the rigidity of adherence. But the appeal to authority has been carried to absurd lengths. Even personal observations and experiences have been deemed unacceptable when they dispute authority. For example, people purportedly refused to look through Galileo's telescope when he disputed Ptolemy's explanation of the world and the heavens. Galileo was later jailed and forced to recant his beliefs. Bruno also rejected Ptolemy's theory and was burned at the stake. (Scholars read and believed Ptolemy's book on astrology and astronomy for 1,200 years after his death!) In 1543, Vesalius wrote a book on anatomy, much of which is still considered correct today. But because his work clashed with Galen's theories, he met with such ridicule that he gave up his study of anatomy.

Perhaps the most crucial aspect of the appeal to authority as a means of obtaining knowledge is the right to question and to accept or reject the information. Furthermore, the authority's qualifications and the methods by which the authority acquired the knowledge also determine the validity of this source of information.

Rationalistic Method

In the rationalistic method, we derive knowledge through reasoning. A good example is the following classic syllogism:

All men are mortal (major premise).

The emperor is a man (minor premise).

Therefore, the emperor is mortal (conclusion).

Although you probably would not argue with this reasoning, the key to this method is the truth of the premises and their relationship to each other, as shown in the following example:

Basketball players are tall.

Tom Thumb is a basketball player.

Therefore, Tom Thumb is tall.

In this case, however, Tom is very short. The conclusion is trustworthy only if it is derived from premises (assumptions) that are true. Also, the premises may not in fact be premises but rather descriptions of events or statements of fact. The statements are not connected in a cause-and-effect manner. Consider the following example:

There is a positive correlation between shoe size and mathematics performance among elementary school children. In other words, children with large shoe sizes do well in math.

Herman is in elementary school and wears large shoes.

Therefore, Herman is good in mathematics.

Of course, in the first statement, the factor common to both mathematics achievement and shoe size is age. Older children tend to be bigger and thus have bigger feet than younger children. Older children also have higher achievement scores in mathematics, but there is no cause-and-effect relationship. You must always be aware of this when dealing with correlation. Reasoning is fundamental in the scientific method of problem-solving, but it cannot be used by itself to arrive at knowledge.

Empirical Method

empirical—A description of data or a study that is based on objective observations.

The word **empirical** denotes experience and the gathering of data. Certainly, data gathering is part of the scientific method of solving problems. But relying too much on your own experience (or data) has drawbacks. First, your own experience is limited. Furthermore, your retention depends substantially on how the events agree with your experience and beliefs, on whether things "make sense," and on what your state of motivation to remember is. Nevertheless, the use of data (and the empirical method) is high on the continuum of methods of obtaining knowledge as long as you are aware of the limitations of relying too heavily on this method.

Scientific Method of Problem-Solving

The methods of acquiring knowledge previously discussed lack the objectivity and control that characterize the scientific approach to problem-solving. The scientific method involves several basic steps. Some authors list seven or eight steps, and others condense these steps into three or four. Regardless, all the authors are in general agreement as to the sequence and processes involved. The steps are briefly described next. The basic processes are covered in detail in other chapters.

Step 1: Developing the Problem (Defining and Delimiting It)

This step may sound contradictory, because how could the development of the problem be part of solving it? Actually, the discussion here is not about finding a problem to study (ways of identifying a problem are discussed in chapter 2); the assumption is that the researcher has already selected a topic. But to design and execute a sound investigation, the researcher must be specific about what is to be studied and to what extent it will be studied.

independent variable—The part of the experiment that the researcher is manipulating; also called the *experimental variable* or *treatment variable.*

Many ramifications constitute this step, an important one being the identification of the independent and dependent variables. The **independent variable** is what the researcher is manipulating. If, for example, two training programs for improving balance in older adults are being compared, then type of training program is the independent variable; this item is sometimes called the *experimental*, or *treatment*, *variable*.

dependent variable—The effect of the independent variable; also called the *yield.*

The **dependent variable** is the effect of the independent variable. In the comparison of balance training programs, the measure of balance is the dependent variable. If you think of an experiment as a cause-and-effect proposition, the cause is the independent variable and the effect is the dependent variable. The latter is sometimes referred to as the *yield*. Thus, the researcher must define exactly what will be studied and what will be the measured effect. When this question is resolved, the experimental design can be determined.

hypothesis—The anticipated outcome of a study or experiment.

Step 2: Formulating the Hypothesis

The **hypothesis** is the expected result. A person setting out to conduct a study generally has an idea as to what the outcome will be. This anticipated solution to the problem may be based on some theoretical construct, on the results of previous studies, or perhaps on the experimenter's own experiences and observations. (Remember: The last source is least likely and least defensible because of the weaknesses of the unscientific methods of acquiring knowledge discussed previously.) Regardless, the research should have some experimental hypothesis about each subproblem in the study.

One of the essential features about the hypothesis is that it be testable. The study must be designed in such a way that the hypothesis can be either supported or refuted. Obviously, then, the hypothesis cannot be a type of value judgment or an abstract phenomenon that cannot be observed.

For example, you might hypothesize that success in athletics depends solely on fate. In other words, if a team wins, it is because it was meant to be; similarly, if a team loses, a victory was just not meant to be. Refuting this hypothesis is impossible because no evidence could be obtained to test it.

Step 3: Gathering the Data

Next, the researcher must decide on the proper methods of acquiring the necessary data to be used in testing the research hypothesis. The reliability of the measuring instruments, the controls that are employed, and the overall objectivity and precision of the data-gathering process are crucial to solving the problem.

In terms of difficulty, gathering data may be the easiest step because in many cases, it is routine. Planning the method, however, is one of the most difficult steps. Good methods attempt to maximize both the **internal validity** and the **external validity** of the study.

internal validity—The extent to which the results of a study can be attributed to the treatments used in the study.

external validity—The generalizability of the results of a study.

Internal validity and external validity relate to the research design and controls that are used. Internal validity refers to the extent to which the results can be attributed to the treatments used in the study. In other words, the researcher must try to control all other variables that could influence the results. For example, Jim Nasium wants to assess the effectiveness of his exercise program in developing physical fitness in young boys. He tests his participants first at the beginning and then at the end of a nine-month training program and concludes that the program brought about significant improvement in fitness. What is wrong with Jim's conclusion? His study contains several flaws. The first is that Jim did not consider maturity. Nine months of maturation produced significant changes in size and in accompanying strength and endurance. Also, what else were the participants doing during this time? How do we know that other activities were not responsible, or partly so, for the changes in their fitness levels? Chapter 18 deals with these threats to internal validity.

External validity pertains to the generalizability of the results. To what extent can the results apply to the real world? A paradox often occurs for research in the behavioral sciences because of the controls required for internal validity. In motor-learning studies, for example, the task is often something novel so that it provides a control for experience. Furthermore, being able to measure the performance objectively and reliably is desirable. Consequently, the learning task is frequently a maze, a rotary pursuit meter, or a linear position task, all of which may meet the demands for control with regard to internal validity. But then you face the question of external validity: How does performance in a laboratory setting with a novel, irrelevant task apply to learning a real-world motor skill? These questions are important and sometimes vexing, but they are not insurmountable. (They are discussed later.)

Possible Misinterpretations of Results

- We will never run out of math professors because they always multiply.
- When the body is fully immersed in water, the telephone rings.
- If there are only two people in a locker room, they will have adjacent lockers.
- The ocean would be much deeper without sponges.

Step 4: Analyzing and Interpreting Results

Any researcher new to the field finds this step to be the most formidable for several reasons. First, this step usually involves some statistical analysis, and the novice researcher (particularly a graduate student) often has a limited background in and a fear of statistics. Second, analysis and interpretation require considerable knowledge, experience, and insight, which the novice may lack.

It goes without question that analyzing and interpreting results is the most crucial and challenging of all the steps in the scientific method of problem-solving. It is here that the researcher must provide evidence for the support or rejection of the research hypothesis. In doing this, the researcher also compares the results with those of others (the related literature) and perhaps attempts to relate and integrate the results into some theoretical model. Inductive reasoning is employed in this step (whereas deductive reasoning is primarily used in the statement of the problem; we'll more thoroughly address inductive and deductive reasoning in chapter 2). The researcher attempts to synthesize the data from their study along with the results of other studies to contribute to the development or substantiation of a theory.

Alternative Models of Research

In the preceding section, we summarized the basic steps in the scientific method of problem-solving. Science is a way of knowing and is often defined as structured inquiry. One basic goal of science is to explain things or to generalize and build a theory. When a scientist develops a useful model to explain behavior, scholars often test predictions from this model using the steps of the scientific method. The model and the approaches used to test the model are called a *paradigm*.

Normal Science

normal science—An objective manner of study grounded in the natural sciences that is systematic, logical, empirical, reductive, and replicable.

For centuries, the scientific approaches used in studying problems in both the natural and the social sciences have been what Thomas Kuhn (1970), a noted science historian, termed **normal science**. This manner of study is characterized by the elements we listed at the beginning of this chapter: systematic, logical, empirical, reductive, and replicable. Its basic doctrine is objectivity. It is quantitative in nature; that is, phenomena are described or measured numerically. Normal science is grounded in the natural sciences, which have long adhered to the idea of the orderliness and reality of matter—that is, that nature's laws are absolute and discoverable by objective, systematic observations and investigations that are not influenced by (i.e., independent of) humans. The experiments are theory driven and have testable hypotheses.

Normal science received a terrific jolt with Einstein's theory of relativity and the quantum theory, which indicated that nature's laws could be influenced by humans (that is, that reality depends to a great extent on how one perceives it). Moreover, some things, such as the decay of a radioactive nucleus, happen for no reason at all. The fundamental laws that had been believed to be absolute were now considered statistical rather than deterministic. Phenomena could be predicted statistically but not explained deterministically (Jones, 1988).

Challenges to Normal Science

Relatively recently (since about 1960), serious challenges have arisen regarding normal science's concept of objectivity (i.e., that the researcher can be detached from the instruments and conduct of the experiment). Two of the most powerful challengers to the idea of objective knowledge were Michael Polanyi (1958) and Thomas Kuhn (1970). They contended that objectivity is a myth and has no basis in reality. From the first inception of the idea for the hypothesis through the selection of apparatus to the analysis of the results, the observer is involved. The conduct of the experiment and the results can be considered expressions of the researcher's point of view. Polanyi was especially opposed to the adoption of normal science for the study of human behavior.

Kuhn (1970) maintained that normal science does not really evolve in systematic steps the way that scientific writers describe it. Kuhn discussed the **paradigm crisis phenomenon**, in which researchers who have been following a particular paradigm begin to find discrepancies in it. The findings no longer agree with the predictions, and a new paradigm is advanced. Interestingly, the old paradigm does not die completely; rather, it, metaphorically speaking, just develops varicose veins and fades away. Many researchers with a great deal of time and effort invested in the old paradigm are reluctant to change, so it is usually a new group of researchers who propose the new paradigm. Thus, normal science progresses by revolution, with a new group of scientists breaking away and replacing the old. Nevertheless, normal science has been and will continue to be successful in the natural sciences and in certain aspects of the study of humans. Martens (1987), however, contended that it has failed miserably in the study of human behavior, especially in human behavior's more complex functions.

As a sport psychologist, Martens asserted that laboratory experiments have limited use in answering questions about complex human behavior in sport. He considered his role as a practicing sport psychologist to have been far more productive in gaining knowledge about athletes and coaches and the solutions to their problems. Other workers in the so-called helping professions have made similar observations about both the limitations of normal science and the importance of alternative sources of knowledge in forming and shaping professional beliefs. Schein (1987), a noted scholar of social psychology, related an interesting (some might call it shocking) revelation concerning the relative influence of published research results as opposed to practical experience. At a conference, he and a number of his colleagues were discussing what they relied on most for their classroom teaching. These professors seemed to agree that the data they really believed in and used in the classroom came from personal experience and information learned in the field. Schein was making the point that different categories of knowledge can be obtained by different methods. In effect, some people are more influenced by sociological and anthropological research models than by the normal science approach.

For some time, many scholars in education, psychology, sociology, anthropology, sport psychology, physical education, and other disciplines have proposed methods of studying human behavior other than those of conventional normal science. Anthropologists, sociologists, and clinical psychologists have used in-depth observation, description, and analysis of human behavior for nearly three-quarters of a century. For over 60 years, researchers in education have used participant and nonparticipant observation to obtain comprehensive, firsthand accounts of teacher and student behaviors as they occur in real-world settings. More recently, physical educators, sport psychologists, and exercise specialists have also become engaged in this type of field research. This general form of research is referred to by several names: *ethnographic, qualitative, grounded, naturalistic,* and *participant observational research*. Regardless of the names and the commitments to and beliefs of the researchers, this type of research was not well received initially by the adherents of normal science and the scientific method. In fact, this form of research (we include all its forms under the name **qualitative research**) has often been labeled by normal scientists as superficial, lacking in rigor, and just plain unscientific. As qualitative research methods have evolved, so has the thinking of many of these people. As you will see in chapters 19 and 20, many of the research tenets listed by Kuhn (1970) are found in contemporary qualitative research.

Martens (1987) referred to such adherents of normal science as the gatekeepers of knowledge because they are the research journal editors and reviewers who decide which research gets published, who serves on the editorial boards, and whose papers are presented at conferences. Studies without internal validity are not published, yet studies without external

paradigm crisis phenomenon—The development of discrepancies in a paradigm leading to proposals of a new paradigm that better explains the data.

qualitative research—A research method that often involves intensive, long-term observation in a natural setting; precise and detailed recording of what happens in the setting; and the interpretation and analysis of the data using description, narratives, quotes, charts, and tables. Also called *ethnographic, naturalistic, interpretive, grounded, phenomenological, subjective,* and *participant observational research*.

quantitative research— Research involving measurement of phenomena or outcomes numerically using reliable and objective data gathering and analysis methods consistent with normal science and the scientific method of problem solving.

reductionism— A characteristic of normal science that assumes that complex behavior can be reduced, analyzed, and explained as parts that can then be put back together to understand the whole.

validity lack practical significance. Martens charged that normal science (in psychology) prefers publication to practical significance.

The debates over qualitative and normal (often classified as **quantitative) research** methods have been heated and prolonged. The qualitative proponents have gained momentum as well as more researchers' confidence in recent years and qualitative research is now recognized as a viable method of addressing problems in the behavioral sciences. Credibility is established by systematically categorizing and analyzing causal and consequential factors. The naturalistic setting of qualitative research both facilitates analysis and precludes precise control of so-called extraneous factors, as does much other research occurring in field settings. The holistic interrelationship of observations and the complexity and dynamic processes of human interaction make it impossible to limit the study of human behavior to the sterile, reductionistic approach of normal science. **Reductionism**, a characteristic of normal science, assumes that complex behavior can be reduced, analyzed, and explained as parts that can then be put back together as a whole and understood. Critics of the conventional approach to research believe that the central issue is the unjustified belief that normal science is the only source of true knowledge.

Implications of Challenges to Normal Science

The challenges to normal science involve many implications. For example, when we study simple movements, such as linear positioning in a laboratory to reflect cognitive processing of information, do we learn anything about movements and performance of sport skills in real-world settings? When we evaluate EMG activity in specific muscle groups during a simple movement, does the result really tell us anything about the way the nervous system controls complex movements in athletics in natural settings? Can we study the association of psychological processes related to movement in laboratory settings and expect the results to apply in sport and exercise situations? When we conduct these types of experiments, are we studying nature's phenomena or laboratory phenomena?

Do not misinterpret the intent of these questions. They do not mean that nothing important can be discovered about physical activity from laboratory research. What they suggest is that these findings do not necessarily accurately model the way humans plan, control, and execute movements in natural settings associated with exercise and sport.

Kuhn's (1970) explanations of how science advances and of the limitations of applying normal science to natural settings demonstrate that scientists need to consider the various ways of obtaining knowledge and that the strict application of the normal scientific method of problem-solving may sometimes hinder rather than advance science. If the reductionistic approach of the scientific method has not well served the natural scientists who developed it, then certainly human behavior researchers need to assess the relative strengths and weaknesses of conventional and alternative research paradigms for their particular research questions.

Alternative Forms of Scientific Inquiry

Martens (1987, p. 52) suggested that we view knowledge not as being either scientific or unscientific or as being either reliable or unreliable but rather as existing on a continuum, as illustrated in figure 1.1. This continuum, labeled *DK* for *degrees of knowledge*, ranges from "don't know" to "damn konfident." Considered in this way, varying approaches to disciplined inquiry are useful in accumulating knowledge. As examples, Martens (1979, 1987) urged sport psychologists to consider the idiographic approach, introspective methods, and field studies instead of relying on the paradigm of normal science as the only answer to research questions in sport psychology. Thomas, French, and Humphries (1986) detailed how to study children's sport knowledge and skills in games and sports. Costill (1985) discussed the study of physiological responses in practical exercise and sport settings. Locke (1989) presented a tutorial on the use of qualitative research in physical education and sport. In later chapters, we provide greater detail about some of these alternative strategies for research, particularly the historical, philosophical, qualitative, and mixed methods.

What we hope you gain from this section is that science is disciplined inquiry, not a set of specific procedures. Although advocates of alternative methods of research are often persuasive, we do not want you to conclude that the study of physical activity should abandon the traditional methods of normal science. We have learned much from these techniques and will continue to do so. Furthermore, we certainly do not want you to toss away this book as being pointless. We have not even begun to tell you all the fascinating things that we have learned over the years—it is hard to tell whether some of these things should be classified as normal or abnormal science. In addition, we have many stories of the abnormal humor variety yet to tell. Aside from these compelling reasons for continuing with the book, we want you to realize and appreciate that so-called normal science is not the solution to all questions raised in our field. Furthermore, none of the alternative methods of research denounces the scientific method of problem-solving.

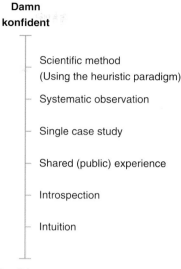

Figure 1.1 The degrees of knowledge theory with examples of methods varying in degree of reliability.

Reprinted by permission from R. Martens, "Science, Knowledge, and Sport Psychology," *The Sport Psychologist* 1, no. 1 (1987): 46.

The bottom line is that different problems require different solutions. As we said before, science is disciplined inquiry, not a set of specific procedures. We need to embrace all systematic forms of inquiry. Rather than argue about the differences, we should capitalize on the strengths of all scholarly methods to provide useful knowledge about human movement. The nature of the research questions and setting should drive the selection of approaches to acquiring knowledge. In fact, just as Christina (1989) suggested, researchers might move among levels of research (basic to applied), and so researchers might move among paradigms (quantitative to qualitative to mixed methods) to acquire knowledge. In addressing this issue, we highlight qualitative and mixed-method research in chapters 19 and 20, which focus on using varying types of research approaches. Of course, we do not want to be perceived like Danae in Wiley's comic strip *Non Sequitur*. Danae, a young girl, says to her horse that she wants to grow up to be a preconceptual scientist. Her horse asks, "What is that?" to which Danae responds "The new science of reaching a conclusion before doing any research and then simply dismissing anything contrary to your preconceived notions."

Types of Research

Research is a structured way of solving problems. Different kinds of problems attend the study of physical activity; thus, different types of research are used to solve these problems. This text concentrates on five types of research: analytical, descriptive, experimental, qualitative, and mixed methods. A brief description of each follows.

Analytical Research

As the name implies, **analytical research** involves in-depth study and the evaluation of available information in an attempt to explain complex phenomena. The types of analytical research are historical, philosophical, reviews, and research synthesis.

analytical research— A type of research that involves in-depth study and the evaluation of available information in an attempt to explain complex phenomena; can be categorized in the following way: historical, philosophical, reviews, and research synthesis.

My research methods teacher will love this idea.

© Creativa/fotolia

Historical Research

As its name indicates, historical research deals with events that have already occurred. Historical research focuses on events, organizations, institutions, and people. In some studies, the researcher is interested mostly in preserving the record of events and accomplishments. In other investigations, the researcher attempts to discover facts that will provide more meaning and understanding of past events to explain the present state of affairs. Some historians have even attempted to use information from the past to predict the future. The research procedures associated with historical studies are addressed in considerable detail in chapter 12.

Philosophical Research

Critical inquiry characterizes philosophical research. The researcher establishes hypotheses, examines and analyzes facts, and synthesizes the evidence into a workable theoretical model. Many of the most important problem areas must be dealt with by the philosophical method. Problems involving objectives, curricula, course content, requirements, and methodology are but a few of the important issues that can be resolved only through the philosophical method of problem-solving.

Although some authors emphasize the differences between science and philosophy, the philosophical method of research follows essentially the same steps as other methods of scientific problem-solving. The philosophical approach uses scientific facts as the basis for formulating and testing research hypotheses.

An example of such philosophical research is Morland's 1958 study in which he analyzed the educational views held by leaders in American physical education and categorized them into educational philosophies of reconstructionism, progressivism, essentialism, and perennialism.

Having an opinion is not the same as having a philosophy. In philosophical research, beliefs must be subjected to rigorous criticism in light of the fundamental assumptions. Academic preparation in philosophy and a solid background in the fields from which the facts are derived are necessary. Other examples and a more detailed explanation of philosophical research are given in chapter 13.

Reviews

A **review** is a critical evaluation of recent research on a particular topic. The author must be extremely knowledgeable about the available literature as well as the research topic and procedures. A review involves an analysis, evaluation, and integration of the published literature, often leading to important conclusions concerning the research findings up to that time. For good examples of reviews, see Blair (1993) and Silverman and Subramaniam (1999).

Certain publications consist entirely of reviews, such as *Psychological Review*, *Annual Review of Physiology*, *Exercise and Sports Sciences Reviews*, *Review of Educational Research*, *Sports Medicine*, and *Kinesiology Review*. A number of journals publish reviews periodically, and some occasionally devote entire issues to reviews. For example, the 75th anniversary issue of *Research Quarterly for Exercise and Sport* (Silverman, 2005) contains some excellent reviews on various topics.

> **review**—A critical evaluation of research on a particular topic.

Research Synthesis

Reviews of literature are difficult to write because they require the synthesis of a large number of studies to determine common underlying findings, agreements, or disagreements. To some extent, this is like trying to make sense of data collected on a large number of participants by simply looking at the data. Glass (1977) and Glass, McGaw, and Smith (1981) proposed a quantitative means of analyzing the findings from numerous studies; this method is called *meta-analysis*. Findings between studies are compared by changing results within studies to a common metric called *effect size*. Over the years, many meta-analyses have been reported in the physical activity literature (e.g., Lee, Folsum & Blair, 2003; Rawdon, Sharp, Shelley, & Thomas, 2012; Sibley & Etnier, 2003; Schieffer & Thomas, 2012; Vazou, Pesce, Lakes, & Smiley-Oyen, 2019). This technique is discussed in more detail in chapter 14.

Descriptive Research

Descriptive research is concerned with status. The most prevalent descriptive research technique is the survey, most notably the questionnaire. Other forms of surveys include the personal interview, online polling, and the normative survey. Chapter 15 provides detailed coverage of these techniques. The following sections briefly describe three types of survey research techniques.

> **descriptive research**—A type of research that attempts to describe the status of the study's focus. Common techniques are questionnaires, interviews, normative surveys, case studies, job analyses, observational research, developmental studies, and correlational studies.

Questionnaire

The main justification for using a questionnaire is the need to obtain responses from people, often from a wide geographical area. The questionnaire usually strives to secure information about present practices, conditions, and demographic data. Occasionally, a questionnaire asks for opinions or knowledge. Online polling has become an increasingly common approach for questionnaire research in recent years.

Interview

The interview and the questionnaire are essentially the same technique insofar as planning and procedures are concerned. Obviously, the interview has certain advantages over the questionnaire. The researcher can rephrase questions and ask additional ones to clarify responses and secure results that are more valid. Becoming a skilled interviewer requires training and experience. Telephone interviewing has been used for decades. It costs half as much as face-to-face interviews and can cover a wide geographical area, which is generally a limitation in personal interviews. We discuss some other advantages of the telephone interview technique in chapter 15.

Normative Survey

A number of notable normative surveys have been conducted in the fields of physical activity and health. The normative survey generally seeks to gather performance or knowledge data on a large sample from a population and to present the results in the form of comparative standards, or norms. The *AAHPER Youth Fitness Test Manual* (American Association for Health, Physical Education and Recreation, 1958) is an outstanding example of a normative survey. Thousands of boys and girls ages 10 to 18 throughout the United States were tested on a battery of motor fitness items. Percentiles were then established for comparative performances to provide information for students, teachers, administrators, and parents. The AAHPER youth fitness test was developed in response to another survey, the Kraus-Weber test (Kraus & Hirschland, 1954), which revealed that American children scored dramatically lower on a test battery of minimum muscular fitness when compared with European children.

Other Descriptive Research Techniques

Among the other forms of descriptive research are the case study, the job analysis, observational research, developmental studies, and correlational studies. Chapter 16 provides detailed coverage of these descriptive research procedures.

Case Study

The case study is used to provide detailed information about an individual person, institution, or community, and so on. It aims to determine unique characteristics about the subject or condition. This descriptive research technique is used widely in such fields as medicine, psychology, counseling, and sociology. The case study is also a technique used in qualitative research.

Job Analysis

This type of research is a special form of case study. It is done to describe the nature of a particular job, including the duties, responsibilities, and preparation required for success in the job.

Observational Research

Observational research is a descriptive technique in which behaviors are observed in the participants' natural setting, such as the classroom or play environment. The observations are frequently coded, and then their frequency and duration are analyzed.

Developmental Studies

In developmental research, the investigator is usually concerned with the interaction of learning or performance with maturation. For example, a researcher may wish to assess the extent to which the ability to process information about movement can be attributed to maturation as opposed to strategy, or to determine the effects of growth on a physical parameter such as aerobic capacity.

Developmental studies can be undertaken by what is called the *longitudinal method*, whereby the same participants are studied over a period of years. Obvious logistical problems are associated with longitudinal studies, so an alternative is to select samples of participants from different age groups to assess the effects of maturation. This is called the *cross-sectional approach*.

Correlational Studies

The purpose of correlational research is to examine the relationship between performance variables, such as heart rate and ratings of perceived exertion; the relationship between traits, such as anxiety and pain tolerance; or the correlation between attitudes and behavior, as in the attitude toward fitness and the amount of participation in fitness activities. Sometimes correlation is employed to predict performance. For example, a researcher may wish to predict the percentage of body fat from skinfold measurements. Correlational research is descriptive in that a cause-and-effect relationship cannot be presumed. All that can be established is that an association is (or is not) present between two or more traits or performances.

Epidemiological Research

Another form of descriptive research that has become a viable approach to studying problems concerning health, fitness, and safety is the epidemiological research method. This type of research pertains to the frequencies and distributions of health and disease conditions among populations. Rate of occurrence is the basic concept in epidemiological studies. The size of the population being studied is an important consideration in examining the prevalence of such things as injuries, illnesses, or health conditions in a specified at-risk population.

Although cause and effect cannot be established by incidence and prevalence data, a strong inference of causation can often be made through association. Chapter 17 is devoted to epidemiological research.

Experimental Research

Experimental research has a major advantage over other types of research in that the researcher can manipulate treatments to cause things to happen (i.e., establish a cause-and-effect situation). As an example of an experimental study, assume that Virginia Reel, a dance teacher, hypothesizes that students would learn more effectively through the use of video. First, she randomly assigns students to two sections. One section is taught by the so-called traditional method (explanation, demonstration, practice, and critique). The other section is taught in a similar manner except that the students are filmed while practicing and can thus observe themselves as the teacher critiques their performances. After nine weeks, a panel of dance teachers evaluates both sections. In this study, the teaching method is the independent variable, and dance performance (skill) is the dependent variable. After the groups' scores are compared statistically, Virginia can conclude whether her hypothesis can be supported.

> **experimental research**—A type of research that involves the manipulation of treatments in an attempt to establish cause-and-effect relationships.

In experimental research, the researcher attempts to control all factors except the experimental (or treatment) variable. If the extraneous factors can be controlled, then the researcher can presume that the changes in the dependent variable are due to the independent variable. Chapter 18 is devoted to experimental and quasi-experimental research.

Qualitative Research

In the study of physical activity, qualitative research is the relatively new kid on the block, although it has been used for many years in other fields, such as anthropology and sociology. Researchers in education have been engaged in qualitative methods longer than researchers in our field have. As previously mentioned, several names are given to this type of research (*ethnographic*, *naturalistic*, *interpretive*, *grounded*, *phenomenological*, *subjective*, and *participant observational*). Some are simply name differences, whereas some have different approaches and points of focus. We have lumped them all together under *qualitative research* because that term seems to be the most commonly used in our field.

Qualitative research is different from other research methods. It is a systematic method of inquiry, and it follows the scientific method of problem-solving to a considerable degree, although it deviates in certain dimensions. Qualitative research rarely establishes hypotheses at the beginning of the study; instead, it uses more general questions to guide the study. It proceeds in an inductive process in developing hypotheses and theory as the data unfold. The researcher is the primary instrument in data collection and analysis. Qualitative research is characterized by intensive, firsthand presence. The tools of data collection are observation, interviews, and researcher-designed instruments (Creswell, 2009). Qualitative research is described in chapters 19 and 20.

Mixed Methods of Research

In mixed methods of research, both quantitative and qualitative approaches are included (or mixed) within a research effort. This approach, often viewed as a pragmatic one, suggests that both qualitative and quantitative techniques are useful when studying real-world phenomena. For capturing behavioral data, the notion is that a researcher should use the best approach; in this case, it is the mixed method in which quantitative techniques are integrated within a single study. In other words, the whole study comprises two smaller studies—one that is quantitative and the other qualitative.

Overview of the Research Process

A good overview of the research methods course, which serves well as an introduction to this book, is provided in figure 1.2. This flowchart provides a linear way to think about planning a research study. After the problem area is identified, reading and thinking about relevant theories and concepts, as well as a careful search of the literature for relevant findings, lead to the specification of hypotheses or questions. Operational definitions are needed

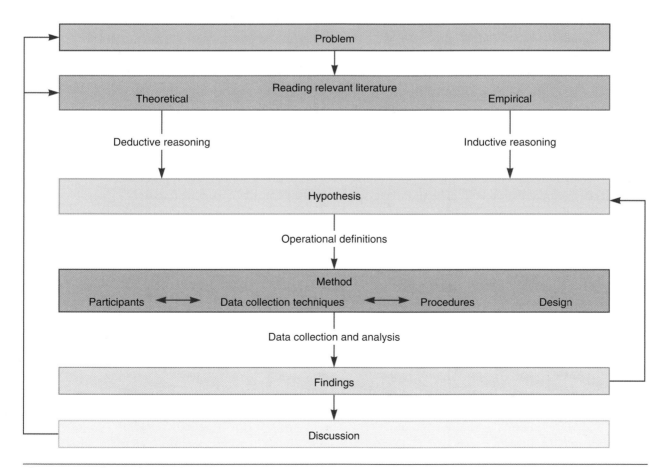

Figure 1.2 The total research setting.

in a research study so that the reader knows exactly what the researcher means by certain terms. Operational definitions describe observable phenomena that enable the researcher to examine empirically whether the predictions can be supported. The study is designed, and the methods are made operational. The data are then collected and analyzed, and the findings are identified. Finally, the results are related back to the original hypotheses or questions and discussed in relation to theories, concepts, and previous research findings.

Parts of a Thesis: A Reflection on the Steps in the Research Process

This chapter has introduced the research process. The theme has been the scientific method of problem-solving. Generally, a thesis or research article has a standard format. This feature is for expediency, so that the reader knows where to find the pieces of information, such as purpose, methods, and results. The format also reflects the steps in the scientific method of problem-solving. We now look at a typical thesis format and see how the parts correspond to the steps in the scientific method.

Sometimes, theses and dissertations are done in a chapter format in which each chapter represents a separate part of the research report (e.g., introduction). This model has been commonly used over the years. We believe that it is more appropriate for graduate students to prepare their theses or dissertations in a form suitable for journal publication because that is an important part of the research process. Writing the theses or dissertation in a journal format also prepares students for future writing. In chapter 22, we provide considerable detail about how to use a journal format for a thesis or dissertation and the value of doing so. Throughout this book, we refer to the typical parts of a research report. These can be considered either as parts of a journal paper or as chapters, depending on the format selected.

Introduction

In the introduction, the problem is defined and delimited. The researcher specifically identifies the problem and often states the research hypotheses. Certain terms critical to the study are operationally defined for the reader, and limitations and perhaps some basic assumptions are acknowledged.

The literature review may be in the introduction, or it may warrant a separate section. When it is part of the introduction, the literature review more closely adheres to the steps in the scientific method of problem-solving; that is, the literature review is instrumental in the formulation of hypotheses and the deductive reasoning leading to the problem statement.

Methods

The purpose of the methods section is to make the thesis format parallel to the data-gathering steps of the scientific method. First, the researcher explains how the data were gathered. The participants are identified, the measuring instruments are described, the measurement and treatment procedures are presented, the experimental design is explained, and the methods of analyzing the data are summarized. The major purpose of the methods section is to describe the study in such detail and with such clarity that a reader could replicate it.

The introduction and methods section often comprise the **research proposal** and are presented to the student's thesis committee before the research is undertaken. For the proposal, methods are often written in future tense and then changed to past tense when the final version of the thesis is completed. Of course, presenting methods in past tense in the proposal eliminates the need to make this conversion later in the research process. Discuss with your advisor which tense is preferred for the proposal. The research proposal also often contains some preliminary data demonstrating that the student has the required expertise to collect the data using the instrumentation needed.

research proposal—A formal preparation that includes the introduction, literature review, and proposed method for conducting a study.

Results

The results present the pertinent findings from the data analysis and represent the contribution to new knowledge. The results section corresponds to the step in the scientific method in which the meaningfulness and reliability of the results are scrutinized.

Discussion and Conclusions

In this last step in the scientific method, the researcher employs inductive reasoning to analyze the findings, interpret the findings relative to those of previous studies, and integrate them into a theoretical model. In this part of a research paper, the acceptability of the research hypotheses is judged. Then, based on the analysis and discussion, conclusions are usually made. The conclusions should address the purpose and secondary purposes that were specified in the introduction.

Qualitative and Mixed-Method Studies

When the qualitative or mixed-method approach is used, the format for the thesis or dissertation often varies from the preceding (i.e., introduction, method, results, and discussion). However, the general notion of explaining the problem, describing how data were collected, presenting the results, and providing a discussion remain the same. Chapters 19, 20, 21, and 22 provide a more in-depth discussion of this.

Summary

Research is simply a way of solving problems. Questions are raised, and methods are devised to try to answer them. There are various ways of approaching problems (research methods). Sometimes the nature of the problem dictates the method of research. For example, a researcher who wants to discover the origins of a sport would use the historical method of research. A researcher who wants to look at a problem from a particular angle may select a research method that can best answer the question from that angle.

Research on the topic of teaching effectiveness, for example, can be approached in several ways. An experimental study could be conducted in which the effectiveness of teaching methods in bringing about measurable achievement is compared. Or a study could be designed in which teachers' behaviors are coded and evaluated using some observational instrument. Or another form of descriptive research could be used that employs the questionnaire or the interview technique to examine teachers' responses to questions concerning their beliefs or practices. Or perhaps a qualitative study could be undertaken to observe and interview one teacher in one school systematically over an extended period to portray the teacher's experiences and perceptions in the natural setting. And of course, qualitative and quantitative approaches can be combined in a mixed-method format.

The point is that there is not just one way to do research. Some people do only one type of research. Some are critical of the methods used by others. However, anyone who believes that their type of research is the only "scientific" way to solve problems has a narrow-minded and downright foolish understanding of how to conduct research. Science is disciplined inquiry, not a set of specific procedures.

Basic research deals primarily with theoretical problems, and the results are not intended to have immediate application. Applied research, on the other hand, strives to answer questions that have direct value to the practitioner. There is a need to prepare proficient consumers of research as well as researchers. Thus, one purpose of a book on research methods is to help the reader understand the tools necessary both to consume and to produce research.

We presented here an overview of the nature of research. The scientific method of problem-solving was contrasted with "unscientific" methods by which people acquire information. Multiple research models were discussed to emphasize that there is not just one way to approach problems in our discipline and profession of kinesiology. We identified the five major types of research used in the study of physical activity: analytical, descriptive, experimental, qualitative, and mixed methods. These categories and the techniques they encompass are covered in detail in later chapters.

✓ Check Your Understanding

1. Look through some recent issues of *Research Quarterly for Exercise and Sport*. Find and read a research article of interest that is quantitative in nature and another that you believe is qualitative. Which of these did you find easier to understand? Why?

2. Find an article that describes a study that you would classify as an applied research study and another that you believe describes a basic research study. Defend your choices.

3. Think of two problems that need research in your field. From the descriptions of the types of research in this chapter, suggest how each problem might be researched.

Developing the Problem and Using the Literature

> **The library banned drinks after someone poured milk on the serials.**
>
> —Early Bird Books

Getting started is the hardest part of almost any new venture, and research is no exception. You cannot do any meaningful research until you have identified the area that you want to investigate, learned what has been published in that area, and figured out how you are going to conduct the investigation. In this chapter, we discuss ways to identify researchable problems, search for literature, and write the literature review.

Identifying the Research Problem

Of the many major issues facing the graduate student, a primary one is identifying a research problem. Research ideas may arise from a student's curiosity about some aspect of human performance, be stimulated by real-world settings, or be generated from theoretical frameworks. Regardless, a fundamental requirement for developing and appropriately limiting a good research problem is in-depth knowledge about the area of interest. But sometimes, as students become more knowledgeable about a content area, everything seems to be already known. Thus, although you want to become an expert, do not focus too narrowly, because doing so can limit topics. Relating your knowledge base to other areas often provides insight into significant areas for research.

Ironically, we ask students to start thinking about possible research topics in their research methods course, typically a course taken in the first semester (or quarter) of graduate school before students have had the opportunity to acquire in-depth knowledge. As a result, many of their research problems are ill-conceived, infeasible, unattainable, or superficial; lack a theoretical base, or are replications of earlier research. Although this shortcoming is considerable, the advantages of taking the research methods course early in the program are substantial in terms of success in other graduate courses. In this course, students learn the following:

- To approach and solve problems in a scientific way
- To search the literature

Problems That Have Not Been Resolved by Humankind

9. Is there ever a day when mattresses are not on sale?

8. If people evolved from apes, why are there still apes?

7. Why does someone believe you when you say there are four billion stars but checks when you say the paint is wet?

6. Why do you never hear father-in-law jokes?

5. If swimming is so good for your physique, how do you explain whales?

4. How do those dead bugs get into enclosed light fixtures?

3. Why does Superman stop bullets with his chest but duck when you throw a revolver at him?

2. Why do banks charge a fee on "insufficient funds" when they know there is not enough money in the account?

. . . drum roll . . .

1. Why doesn't Tarzan have a beard?

- To write in a clear, concise, scientific fashion
- To understand basic measurement and statistical issues
- To use an appropriate writing style
- To be intelligent consumers of research
- To appreciate the wide variety of research strategies and techniques used in an area of study

How, then, does a student without much background select a problem? As you devote ever-greater effort to thinking of a topic, you may become increasingly inclined to think that all the problems in the field have already been solved. Adding to this frustration is the pressure of time. To assure you that important questions have yet to be addressed, we have provided the Problems That Have Not Been Resolved by Humankind sidebar.

Guidelines for Finding a Topic

To help alleviate the problem of finding a topic, we offer the following suggestions. First, be aware of the research being done at your institution, because research spawns other research ideas. Often a researcher has a series of studies planned. Second, be alert for any controversial issues in some area of interest. Lively controversy prompts research in efforts to resolve the issue. In any case, be sure to talk to professors and advanced graduate students in your area of interest and use their suggestions to focus on a topic (the Look for Causes, Not Effects sidebar explains the importance of this advice when seeking a topic). Third, read a review paper (possibly in a review journal, research journal, or recent textbook). From there, read several research studies cited in the reference lists and locate other current research papers on the topic. Using all this information, make a list of research questions that appear unanswered or are logical extensions of the material you have read. Try to pick problems that are neither too complex nor too simple. The complex ones will take you forever and may not lead to clear outcomes, and you will never get your thesis done. No one cares about the simple ones.

Of course, no single problem necessarily meets all the criteria perfectly. For example, some theoretical problems may have limited direct application. Theoretical problems should be directed toward issues that may ultimately prove useful to practitioners. By honestly

Look for Causes, Not Effects

In an interesting paper, Salzinger (2001) pointed out that scientists should not just look at the effects they observe but also try to find the causes underlying them. Finding surprising results is always interesting, but the causes behind them are what we should seek. We often hear that the purpose of the scientific method is to disprove theories, or that we can never prove a theory. A single experiment, however, seldom disproves a theory. "Old theories, like old soldiers, just fade away when better theories come to take their place," wrote Salzinger (2001, p. B14). On the other hand, confirming a theory leads us toward underlying mechanisms; results explain why something happened.

We often read research results such as these:

- ESP is shown to work.
- Listening to Mozart improves spatial reasoning.
- People who go to places of worship live longer.

When we see headlines like these, we should ask ourselves why these things might happen. What are the explanations? Does religion have anything to do with living longer, or is it the fact that people who go to places of worship receive emotional reinforcement, network to learn about better physicians, are routinely social, are likely to walk to their designated place of worship, or any number of other explanations?

Salzinger (2001) suggested that we should spend more time thinking about what causes the results and the possible mechanisms and identify the critical ingredient that allows results to be applied in other contexts.

Reference

Salzinger, K. (2001, February). Scientists should look for basic causes, not just effects. *Chronicle of Higher Education, 157*(23), B14.

answering the questions set forth by McCloy (1930), a practical evaluation of the selected problem is possible (see the Criteria for Selecting a Research Problem sidebar).

Another approach to develop a research problem is to see how experts develop problems. In their book *The Creative Side of Experimentation* (1992), Snyder and Abernethy had well-established scholars in motor control, motor development, and sport psychology give first-person accounts of factors that influenced their career research programs. The editors also deduced themes that seemed to run through these scholars' research programs. Their analysis focused on such questions as these:

- What are common personal and professional characteristics of expert researchers?
- What types of experimentation do expert researchers perform?
- What strategies do expert experimenters use to enhance their ability to ask important questions?

In addition, Locke, Spirduso, and Silverman (2014) provided a comprehensive, 20-step approach to help graduate students identify topics and develop a proposal.

Using Inductive and Deductive Reasoning

The means for identifying specific research problems come from two methods of reasoning: inductive and deductive. Figure 2.1 provides a schema of the inductive reasoning process. Individual observations are tied together into specific hypotheses, which are grouped into more general explanations that are united into theory. To move from the level of observations to that of theory requires many individual studies that test specific hypotheses. But

Criteria for Selecting a Research Problem

In 1930, the first volume (issue no. 2) of *Research Quarterly* (now *Research Quarterly for Exercise and Sport*) had a paper by Dr. C.H. McCloy, one of our historically famous scholars. McCloy provided a list of questions by which a researcher can judge the quality and feasibility of a research problem. Those questions (listed here) are as valid and useful today as they were in 1930:

- Is the problem in the realm of research?
- Does it interest you?
- Does it possess unity?
- Is it worthwhile?
- Is it feasible?
- Is it timely?
- Can you attack the problem without prejudice?
- Are you prepared in the techniques to address the problem?

Our thanks to Jim Morrow, University of North Texas, for pointing out this information.

closed-loop theory—
A theory of motor skill learning advanced by Adams (1971) that proposes that information received as feedback from a movement is compared with some internal reference of correctness.

schema theory—A theory of motor skill learning advanced by R.A. Schmidt (1975) as an extension of Adams' (1971) closed-loop theory. The theory proposed to unify two more general explanations under one theoretical explanation.

even beyond the individual studies, someone must see how all the findings relate and then offer a theoretical explanation that encompasses all the individual findings.

An example of this process can be found in the motor learning and control area. Adams (1971) proposed a **closed-loop theory** of motor skill learning. Basically, the closed-loop theory suggests that information received as feedback from a movement is compared with some internal reference of correctness (assumed to be stored in memory). Then the discrepancies between the movement and the intended movement are noted. Finally, the next attempt at the movement is adjusted to approximate the movement goal more closely. Adams' theory ties together many previous observations about movement response. The theory was tightly reasoned but limited to slow-positioning responses. This limitation really makes it a more general explanation, according to figure 2.1.

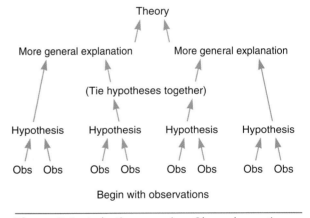

Figure 2.1 Inductive reasoning. Obs = observation.
Reprinted from R.L. Hoenes and B. Chissom, *A Student Guide for Educational Research* (Statesboro, GA: Vog Press, 1975), 22. By permission of Arlene Chissom.

Schmidt (1975) proposed a **schema theory** that extended Adams' reasoning to include more rapid types of movements, frequently called *ballistic tasks*. (Schema theory also deals with several other limitations of Adams' theory that are not important to this discussion.) The point is that schema theory proposed to unify two general explanations, one about slow movements and the other about ballistic (rapid) movements, under one theoretical explanation—clearly an example of inductive reasoning.

Reasoning must be careful, logical, and causal; otherwise, one of our examples of inappropriate induction may result:

A researcher spent several weeks training a cockroach to jump. The bug became well trained and would leap high in the air on the command "Jump." The researcher then began to manipulate his independent variable, which was to remove the bug's legs one at a time. After removing the first leg, the researcher said, "Jump," and the bug did. He then removed in turn the second, third, fourth, and fifth legs and said, "Jump." After each leg was removed, the bug jumped every time. After the researcher removed the sixth leg and gave the "Jump" command, the bug just lay there. The researcher's conclusion from this research was this: "When all the legs are removed from a cockroach, the bug becomes deaf." (Thomas, 1980, p. 267)

Figure 2.2 presents a model of deductive reasoning. Deductive reasoning moves from a theoretical explanation of events to specific hypotheses that are tested against (or compared with) reality to evaluate whether the hypotheses are correct. Using the previously presented notions from his schema theory (to avoid explaining another theory), Schmidt advanced a hypothesis frequently called **variability of practice**. Essentially, this hypothesis (reasoned or deduced from the theory) states that practice of a

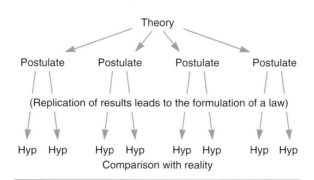

Figure 2.2   Deductive reasoning. Hyp = hypothesis.

Reprinted from R.L. Hoenes and B. Chissom, *A Student Guide for Educational Research* (Statesboro, GA: Vog Press, 1975), 22. By permission of Arlene Chissom.

variety of movement experiences (within a movement class), when compared with practicing a single movement, facilitates transfer to a new movement (still within the same class). Several researchers have tested this hypothesis, identified by deductive reasoning, and found it viable. In fact, within any given study, both inductive and deductive reasoning are useful. The total research setting was presented in chapter 1, figure 1.2. Review how the deductive and inductive processes operate within this scheme: That is, at the beginning of a study, the researcher deduces hypotheses from relevant theories and concepts and induces hypotheses from relevant findings from empirical research.

> **variability of practice**—A tenet of motor-skill learning advanced by R.A. Schmidt that states that the practice of a variety of movement experiences facilitates transfer to a new movement when compared with practicing a single movement.

Purpose of the Literature Review

A major part of developing and refining the research problem is reading what has already been published about the problem, which helps clarify what we know and don't know about a given topic. Much research may already have been done on the problem in which you are interested. In other words, the problem has been pretty much "fished out." We hope this will not be the case when you start this phase of the research process. (In many instances, your major professor can steer you away from a saturated topic. We urge you to work closely with them, because doing so will make your task more efficient.) Whatever the topic, past research is invaluable in planning new research.

Browsing in a library, either physically or electronically, confirms that information—and lots of it—exists. The dilemma lies in knowing how to locate and evaluate the information that you want and, ultimately, how to use the information after you have found it.

Reviews of literature serve several purposes. Frequently, they stimulate inductive reasoning. A scholar may seek to locate and synthesize all the relevant literature on a particular topic to develop a more general explanation or a theory to explain certain phenomena. For the dissertation or thesis, the review will be more focused, directly addressing the questions and methods for the study. An alternative way of analyzing the literature, mentioned in chapter 1, is meta-analysis (Schmidt & Hunter, 2014), which is discussed in detail in chapter 14.

The major problem of literature reviews is how all those studies can be related to one another in an effective way. Most frequently, authors attempt to relate studies by similarities and differences in theoretical frameworks, problem statements, methods (participants, instruments,

> In literature reviews, studies can be related to one another based on similarities and differences in theoretical frameworks, problem statements, methods, and findings.

treatments, designs, statistical analyses), and findings. Results are then determined by counting votes. For example, you would write, "From the eight studies with similar characteristics, five found no significant difference between the treatments; thus, this treatment has no consistent effect."

This procedure is most easily accomplished through use of a summary sheet, such as the one in table 2.1, which relates the frequency and intensity of exercise to the percentage of change in body fat. The conclusion from looking at these studies might be the following:

Exercising 20 minutes per day for 3 days per week for 10 to 14 weeks at 70% of maximal heart rate produces moderate losses of body fat (4 to 5%). More-frequent exercise bouts produce minimal increases, but less-frequent or less-intense exercise is substantially less effective in eliminating body fat.

Techniques of this type lend themselves to the development of the literature review around central themes or topics. This approach not only allows synthesis of the relevant findings but also makes the literature review interesting to read.

Identifying the Problem

As we have already discussed, the literature review is essential in identifying the specific problem. Your goal is to establish a well-focused and well-justified research question that is feasible to address. Often, the final statement of the problem, questions, and hypotheses cannot be written until the literature is reviewed and the study is situated in the previous literature. Of course, after locating a series of studies, the first task is to decide which studies are related to the topic area. This objective can frequently be accomplished by reading the abstract and, if necessary, some specific parts of the paper. After a few key studies are identified, a careful reading usually produces several ideas and unresolved questions. You will find it useful to discuss these questions with a professor or advanced graduate student from your area of specialization. Doing so can eliminate unproductive approaches or dead ends. After the problem is specified, an intensive library search begins.

TABLE 2.1

Sample Form for Synthesizing Studies (Hypothetical Example)

Study	Problem statement	Participant description	Instrument	Procedure and design	Finding
		Characteristics of studies			
Smith (1993)	Effects of rowing on body fat in overweight individuals	15 males, 15 females, college-age, BMI > 25	Skinfolds	Ergometer rowing 3 d/wk at 70% of (220 − age) HR for 12 wk	4% reduction in body fat
Johnson (2002)	Effects of running intensity on body fat	40 college-age males	Underwater weighing	Run 3 d/wk at 50% or 70% of (220 − age) HR for 10 wk	5% for 70% gp 2% for 50% gp
Andrews (2012)	Effects of exercise frequency on body fat	45 college-age females	DEXA	Run 2, 4, 6 d/wk at 75% of (220 − age) HR for 12 wk	1% for 2 d 4% for 4 d 5% for 6 d
Mitchell (2018)	Effects of cycling cadence on body fat	15 males, 15 females, physically active, young adults (18-36 YO)	DEXA	Ergometer cycling at 40, 60, or 80% of (220 − age) HR at 80 rpm for 20 min, 3 d/wk for 14 wk	1% for 40% 3% for 60% 4% for 80%

Developing Hypotheses or Questions

Hypotheses are deduced from theory or induced from other empirical studies and real-world observations. These hypotheses are based on logical reasoning and, when predictive of the study's outcome, are labeled *research hypotheses*. For example, after spending a good deal of time at the grocery store, we are able to put forward this hypothesis for you to test: The shortest line is always the slowest. If you change lines, the one you left will speed up, and the one you entered will slow down. In qualitative studies, more general questions often serve the same purpose as hypotheses do for quantitative studies.

Developing the Methods

Although considerable work is involved in identifying the problem and specifying hypotheses and questions, one of the more creative parts of research is developing the methods to answer the hypotheses. If the methods are planned and pilot-tested appropriately, the outcome of the study allows the hypotheses and questions to be evaluated. We believe that the researcher fails when a study's equivocal results are blamed on methodological problems. Post hoc methodological blame results from lack of (or poor) planning and pilot work before undertaking the research.

The review of literature can be extremely helpful in identifying methods that have been successfully used to solve particular types of problems. Valuable elements from other studies may include the characteristics of the participants, data collection instruments and testing procedures, treatments, designs, and statistical analyses. All or parts or combinations of the previously used methods are helpful as the researcher plans the study, but these should not limit the researcher in designing the study. Creative methodology is a key to good hypothesis testing. But neither other scholars' research nor creativity ever replaces the need to conduct thorough pilot work.

> Elements from other studies can help identify methods for the study design. These may include characteristics of the participants, data collection instruments, and aspects of testing and statistical analysis.

Basic Literature Search Strategies

The prospect of beginning a literature search can sometimes be frightening or depressing. How and where do you begin? What kind of sequence or strategy should you use in finding relevant literature? What services does the library offer in your search?

Authors of research texts have advanced strategies for finding pertinent information on a topic. We know of no single right way of doing it. The search process depends considerably on your initial familiarity with the topic. In other words, if you have virtually no knowledge about a particular topic, your starting point and sequence will be different from that of someone already familiar with the literature.

In recent years, a computer search has become a fundamental part of every literature search. Some people are inclined to begin by doing a computerized search. This strategy can prove fruitful, to be sure, but it is not foolproof. Gaining some familiarity with the topic and search descriptors will aid in developing and refining the search. For students who are not well grounded in the topic, certain preliminary (general) sources may be helpful in locating **secondary sources** with which to become familiar with the topic at hand; both preliminary and secondary sources prepare students to acquire and understand **primary sources**.

secondary sources— Sources of data in research in which authors have evaluated and summarized previous research.

primary sources—First-hand sources of data in research; original studies.

Article indices or databases, such as the *Education Index*, Web of Science, and PubMed, are preliminary (not primary or secondary) sources. These invaluable tools can provide the researcher with books and articles that relate to the problem. Textbooks and review articles are valuable *secondary* sources that can give the reader an overview of the topic and what has been done in the way of research. In fact, secondary sources such as scholarly books and review articles may be the starting point for the search, by which the student becomes aware of the problem in the first place. *Primary* sources are ultimately the most valuable

Primary sources, most commonly in the form of journal articles, are the most valuable for researchers.

for the researcher because the information is firsthand. Most primary sources in a literature review are journal articles reporting outcomes of specific research studies. Theses and dissertations are also primary sources, and those completed since 1997 are available online through ProQuest Dissertations & Theses Global.

Steps in the Literature Search

Before you begin your search, we strongly recommend that you take a little time to plan it. You should follow six steps when reviewing the literature. Performing these steps ensures thoroughness and makes the search more productive.

Step 1: Write the Problem Statement

We discuss the formal writing of the problem statement in chapter 3. At this point, you are merely trying to specify what research questions you are asking. For example, a researcher wants to find out whether high-intensity interval training is effective in improving aerobic fitness. More specifically, the researcher wants to contrast the efficacy of high-intensity, short bouts of interval training versus longer bouts of moderate-intensity aerobic training for increasing maximal oxygen consumption. By carefully defining the research problem, the researcher can keep the literature search within reasonable limits. Remember, the problem statement, questions, and hypotheses may change based on the search. Write the statement as completely (but concisely) as you can at this time.

Step 2: Consult Secondary Sources

This step helps you gain an overview of the topic, but you can omit it if you are knowledgeable about the topic. Secondary sources such as textbooks and review articles are helpful when students have limited knowledge about a topic and can profit from background information and a summary of previous research. They also can be helpful in identifying important primary sources that should be examined. A current review paper on the topic of interest is especially valuable.

Reviews of research on a particular topic are an excellent source of information for three reasons:

1. Some knowledgeable person has spent a great deal of time and effort in compiling the latest literature on the topic.

2. The author has not only found the relevant literature but also has critically reviewed and synthesized it into an integrated summary of what is known about the area.

3. The reviewer often suggests areas of needed research, for which the graduate student may be profoundly grateful.

Some examples of review publications are *Kinesiology Review*, *Annual Review of Medicine*, *Annual Review of Psychology*, *Review of Educational Research*, *Physiological Reviews*, *Psychological Review*, *Sports Medicine*, and *Exercise and Sport Sciences Reviews*. In addition, many research journals occasionally publish review papers. *Medicine & Science in Sports & Exercise*, one of the more prominent journals in kinesiology, is predominantly devoted to publishing original research articles (primary sources), but it occasionally publishes review articles. For example, Samuel L. Buckner and colleagues' "The Basics of Training for Muscle Size and Strength: A Brief Review on the Theory" was published in the March 2020 issue (vol. 52, pp. 645-653) of *Medicine & Science in Sports & Exercise*. Occasionally, a journal publishes a special issue that consists entirely of review papers.

Step 3: Determine Descriptors or Keywords

Descriptors or keywords are terms that help locate sources pertaining to a topic. As you talk with your major professor and read preliminary and secondary sources, record the words that describe the topic and save them for the initial aspects of the search. This list might be long, but having these terms can help you adjust the scope of the search and venture into related areas.

For the topic of high-intensity interval versus moderate-intensity aerobic training, obvious descriptors are *training*, *intensity*, *interval*, and *aerobic*. The combination of descriptors helps the researcher pinpoint related literature. Obviously, these terms by themselves are extensive. A search using these terms will likely generate a very long and unmanageable list of references. As you become more familiar with a topic, the list of relevant keywords will grow, but they will also become more focused. We discuss this further when we describe computer searching later in this chapter.

Step 4: Conduct Computer-Aided Searches

Several automated online search tools greatly expedite the literature search by providing effective and efficient access to indexes and information. Becoming familiar with multiple computerized databases will assist in your search. Most students doing a literature review or wanting to understand a topic use three or more databases. For example, a student in pedagogy will want to use Web of Science, ERIC, and ProQuest Dissertations & Theses Global because they cover different journals and the search protocols are slightly different. Your major professor and advanced graduate students can assist you in identifying databases that will be most helpful. If the library has orientation sessions for using databases, we strongly suggest that you attend. If you used computerized searches at another library or more than a couple of years earlier, you should take the time to learn about the databases and the interface at your current university—systems change quickly, and each university system is slightly

I'm having trouble organizing my literature review of Mickey Mouse.

© globalmoments

different. As you become familiar with each database, keep notes about how the database works (e.g., Can it use wildcards to signify multiple, possibly related terms, or does it work better if you use words from the thesaurus?).

Most databases permit users to save, email, and print information. In addition, many show the history of changes during a session. Keeping notes on the files saved and changes made during a search is helpful. It may be particularly helpful to have a folder on your computer where you save the list of marked articles from the search. Before saving (or emailing), add a note to the comments section so that you can easily identify it when you go back to it later.

In addition to the library information system at your institution, online search tools provide access to abstract databases and online indexing services in the natural sciences, humanities, and social sciences, many of which are identified in the following sections.

Abstracts

poster session—A method of presenting research at a conference in which the author places summaries of their research on the wall or on a poster stand and answers questions from passersby.

Concise summaries of research studies are valuable sources of information in the early stages of a literature search. Abstracts of papers presented at research meetings are available at national, district, and most state conventions. Abstracts of symposia sponsored by the SHAPE America Research Consortium, free communications, and **poster sessions** presented at the national SHAPE America convention are published each year in *Research Quarterly for Exercise and Sport Supplement*. *Medicine & Science in Sports & Exercise* publishes a special supplement of abstracts each year for papers to be presented at the annual meeting of the American College of Sports Medicine.

Dissertation & Theses Global is another abstract source that contains abstracts of dissertations from colleges and universities around the world. Sources of abstracts in related fields include *PsycInfo* (American Psychological Association), proceedings abstracts of The Physiological Society, *Biological Abstracts*, *Sociological Abstracts*, and *ERIC* (Education Resource Information Center).

Library Information Systems

The traditional card catalog with little trays of cards containing bibliographic information by author and subject has all but disappeared. In fact, the main university library where one of the authors works now has the only card catalog on campus located in an exhibit that focuses on the history of the library. All university libraries have gone to a computerized catalog system.

Computerized catalogs abound. Usually, the searcher first selects the type of search from a menu, such as author, title, keyword, or call number. When the source is found, the full display for the reference includes author, title, publication information, all the index terms, and the call number. Libraries in colleges and universities can also be accessed on the Internet by faculty and students using personal computers. As you become familiar with the computer operations, the search process becomes much faster and more productive. Remember, if you have questions about any library operations, ask a librarian. They tend to be remarkably helpful and courteous.

Indexes

Many useful, computerized indexes are available to the kinesiology researcher in most institutional libraries. Several indexes provide references to magazine and journal articles concerning specific topics. Some general indexes relevant to kinesiology, physical education, and exercise and sport science include *Education Index Retrospective*, *Reader's Guide Retrospective*, *New York Times Index*, and *Physical Education Index* (also known as *Sports Medicine & Education Index*). This last source provides a comprehensive subject index to domestic and foreign periodicals in the fields of dance, health education, recreation, sports, physical therapy, and sports medicine. Other databases or indexes that provide broad coverage

Some Databases Available for Computer Searches

Web of Science

Web of Science is maintained by Clarivate Analytics and is available to libraries on a subscription basis. It provides indexing coverage of more than 12,000 journals covering 256 disciplines in science, social science, arts, and humanities. Web of Science uses six online databases, including Science Citation Index Expanded, Social Sciences Citation Index, and Arts & Humanities Citation Index. Coverage includes publications dating back as early as 1900. Some of the elements that can be specified for a search are topic (i.e., keywords), document titles, author names, publication names, and year of publication.

PubMed

PubMed is a search engine that is maintained by the U.S. National Library of Medicine at the National Institutes of Health and is available at no cost to the public. It primarily uses the MEDLINE database, which contains more than 30 million citations from the life sciences, behavioral sciences, chemical sciences, and biomedical literature. The search process is highly flexible. Terms that can be used in a search can include author names, author affiliations, journal and article titles and dates, and keywords.

Scopus

Scopus is an abstract and citation database created and maintained by the publisher Elsevier that indexes literature from more than 34,000 peer-reviewed journals in the life, social, physical, and health sciences. Like Web of Science and PubMed, Scopus is a flexible search tool that allows for a variety of search strategies.

PsycInfo

PsycInfo, a database of abstracts from psychology and related fields, is produced by the American Psychological Association. More than 1,400 sources, which date back to the 1800s, include professional journals, books and book chapters, reports, and dissertations and theses.

ERIC

ERIC, which stands for Educational Resources Information Center, is sponsored by the Institute of Education Sciences of the U.S. Department of Education. It is the world's largest database of education information, containing more than 1.5 million records. Its basic indexes are *Resources in Education (RIE)* and *Current Index to Journals in Education (CIJE)*. In addition, ERIC produces a thesaurus containing thousands of index terms that can be used in locating references and in conducting a computer search.

Current Contents Connect

In its original form, Current Contents was a weekly print publication that provided the table of contents of scholarly journals in biology and medicine. The current version of this tool, Current Contents Connect, continues with the same purpose, covering over 10,000 journals. It serves as a "rapid alert" or "current awareness" source of newly published research literature for researchers. It is now one of the many databases included as part of Web of Science.

SPORTDiscus

This index is a valuable resource for both practical and research literature on sport, physical fitness, and physical education topics. The database contains more than 2.5 million records from journals, monographs, dissertations, and theses on topics including sports medicine, sport and exercise sciences, physical therapy, occupational therapy, nutrition, coaching, and physical education and fitness.

of research journals relevant to kinesiology researchers include but are not limited to Web of Science, PubMed, Scopus, PsycInfo, ERIC, Current Contents Connect, and SPORTDiscus. Brief descriptions of these indexes are included in the Some Databases Available for Computer Searches sidebar. All these indexes are available electronically and allow for variable strategies using different categories for search terms such as author names, publication names, keywords or descriptors, and year of publication.

Adjusting the Scope of the Search

As you become more sophisticated in your literature search strategies and techniques, you will likely need to adjust the scope of your search by revising keywords. Take note of keywords authors have identified of particularly relevant primary sources as you revise your keyword list.

In addition, you can narrow or broaden the scope of your search by using special words called *Boolean operators*. The two most common operators, or connectors, are the words *and* and *or*. To narrow the search, add another term with the word *and*. For example, in a proposed study on investigating changes in attitude toward teaching after the student-teaching experience, 2,237 items were listed under the descriptor "attitude change." For "attitude + teaching," 186 items were listed, and for "student teacher attitudes," 100 items were listed. When all were connected with the word *and*, 44 references were listed, which is a manageable number to examine. Figure 2.3 illustrates the Boolean logic with this example.

The word *or* broadens the search. Additional related terms can be connected with *or* so that the computer searches for more than one descriptor (see figure 2.4). For example, a researcher who seeks information about "practice teaching" (133 items) could broaden the search by using "practice teaching" (133 items) *or* "internship" (495 items) to obtain 628 related items. Search tools (e.g., Web of Science, PubMed, Scopus) do not always have the same rules for using Boolean operators. Use the help sections available in search tools for details as to how to use Boolean operators and other shortcuts.

If the completed search produces one or more articles that are often cited or considered key to the topic, it is possible with some search tools to identify where each article has been cited since publication. Web of Science (and some others) is a search tool that offers this support. For example, let's assume that an article by Zoe Chan and colleagues entitled "Gait retraining for the reduction of injury occurrence in novice distance runners: 1-year follow-up of a randomized controlled trial" (American

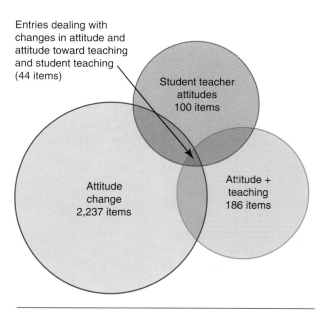

Entries dealing with changes in attitude and attitude toward teaching and student teaching (44 items)

Student teacher attitudes 100 items

Attitude change 2,237 items

Attitude + teaching 186 items

Figure 2.3 An illustration of the *and* connector to narrow a search.

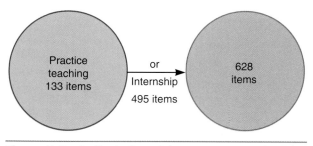

Practice teaching 133 items

or

Internship 495 items

628 items

Figure 2.4 An illustration of the *or* connector to broaden the search.

Journal of Sports Medicine, vol. 46, pp. 388-95, 2018) is especially relevant to your research topic. Web of Science data indicates this article has been cited 61 times (at the time of this writing, of course) and provides links to these publications, thereby providing quick access to additional relevant and more recently published literature.

Even though most search tools for leading indexes and databases are logical and easy to use, you may prefer to seek assistance with your first search. Most libraries have subject specialists available to assist you with finding and using one or more indexes or databases and selecting descriptors. Take advantage of this support, since it will likely help you become more efficient and productive.

Obtaining the Primary Sources

After you have a list of related references, you must obtain the actual studies and read them. Many electronic databases have abstracts in addition to the bibliographic information. The abstract is extremely helpful in deciding whether the article is worth retrieving. For most searches, reading the abstract will inform you whether the paper will be needed, and you can easily include this with the saved bibliographic reference with the citations you selected to look at further.

After determining which articles you want to retrieve, you need to obtain them. Virtually all journals are available in electronic form. Some journals have many years of electronic archives, whereas others have fewer years because they provided digital versions of papers. As you learn to use the databases available in your library, it is a good practice to learn how to retrieve journal articles electronically. Each library has a different way of doing this—everything from directly linking from the database to having to search for each journal separately—and we recommend becoming proficient in finding and retrieving papers at the library you will use.

Your library may not have all the journals that are on the list of references. You must consult your library's information system to see whether your library carries a particular journal. If the article is from an older volume, you may have to retrieve the physical copy and make a photocopy. If the library does not have the journal or volume that you need, you should use interlibrary loan or possibly obtain a fax or pdf file of the article through the service Ingenta Connect.

Table 2.2 lists journals that students studying physical activity often use. Each journal title is accompanied by its impact factor and 5-year impact factor from the *Journal Citation Reports* (Clarivate Analytics). The impact factor is a standardized measure of how often the journal's articles have been cited over the previous 2 years. The 5-year impact factor is a similar measure for the previous 5 years. The higher the impact factor is, the more citations there have been per article published by the journal over the period.

This simple and easily obtained metric may appear to be a valid way to evaluate journal quality. Many authorities, however, have criticized the use of impact factors to evaluate journals (e.g., see Joint Committee on Quantitative Assessment of Research, 2008; Kulinna, Scrabis-Fletcher, Kodish, Phillips, & Silverman, 2009; Lawrence, 2003; Sammarco, 2008; Silverman, Kulinna, & Phillips, 2014). Among the objections are that authors can cite themselves, that impact factors do not tell whether the citation was positive or negative, and that comparisons across fields are problematic because there are larger and smaller groups of researchers. In a recent paper, Larivière and Gingras (2009) did an extensive search and found that duplicate papers published in two journals had different citation rates. They suggested that some journals have a cumulative advantage in maintaining their impact factors. These and other bibliographic rating indexes are just one bit of information to consider in evaluating journals.

TABLE 2.2
2018 Impact Factors for Selected Journals

Journal title	Impact factor	5-year impact factor
ACSM's Health & Fitness Journal	1.000	0.957
Adapted Physical Activity Quarterly	1.109	2.381
American Educational Research Journal	3.170	4.861
American Journal of Physical Medicine & Rehabilitation	1.908	2.216
American Journal of Physiology—Heart and Circulatory Physiology	4.048	3.895
American Journal of Sports Medicine	6.093	7.006
Applied Physiology, Nutrition, and Metabolism (Physiologie Appliquée, Nutrition et Métabolisme)	3.455	3.018
Archives of Physical Medicine and Rehabilitation	2.697	3.618
Aviation, Space, and Environmental Medicine	0.933	1.004
Biology of Sport	2.202	2.004
British Journal of Sports Medicine	11.645	9.805
Child Development	5.024	6.151
Clinical Biomechanics	1.977	2.358
Clinical Journal of Sport Medicine	2.702	2.793
Clinics in Sports Medicine	2.178	1.942
Current Sports Medicine Reports	1.137	1.698
Epidemiology	4.719	6.118
Ergonomics	2.181	2.289
European Journal of Applied Physiology	3.055	3.060
European Journal of Sport Science	2.376	2.896
Exercise Immunology Review	6.455	6.250
Exercise and Sport Sciences Reviews	4.739	5.758
Experimental Brain Research	1.878	2.033
Gait & Posture	2.414	2.912
High Altitude Medicine & Biology	1.490	1.612
Human Biology	1.061	1.162
Human Factors	2.649	2.857
Human Movement Science	1.928	2.360
International Journal of Sport Nutrition and Exercise Metabolism	2.217	2.591
International Journal of Sport Physiology and Performance	3.979	4.133
International Journal of Sport Psychology	0.662	1.054
International Journal of Sports Medicine	2.132	2.773
Isokinetics and Exercise Science	0.452	0.546
Journal of Aging and Physical Activity	1.795	2.144
Journal of Applied Biomechanics	1.392	1.587
Journal of Applied Physiology	3.140	3.583
Journal of Applied Sport Psychology	2.203	2.346
Journal of Athletic Training	2.253	3.392
Journal of Back and Musculoskeletal Rehabilitation	0.814	1.080
Journal of Biomechanics	2.576	2.995
Journal of Curriculum Studies	1.420	2.102
Journal of Electromyography and Kinesiology	1.753	2.117
Journal of Experimental Psychology: Human Perception and Performance	2.939	2.976
Journal of Motor Behavior	1.313	1.549
Journal of Orthopaedic & Sports Physical Therapy	3.058	3.951

Journal title	Impact factor	5-year impact factor
Journal of Orthopaedic Trauma	1.826	2.376
Journal of the Philosophy of Sport	0.719	0.661
Journal of Physiotherapy (formerly Australian Journal of Physiotherapy)	5.551	6.380
Journal of Rehabilitation Medicine	1.907	2.183
Journal of Science and Medicine in Sport	3.623	4.198
Journal of Shoulder and Elbow Surgery	2.865	3.336
Journal of Sport & Exercise Psychology	2.434	3.508
Journal of Sport and Health Science	3.644	3.553
Journal of Sport Management	2.167	2.691
Journal of Sport Rehabilitation	1.500	1.819
Journal of Sports Medicine and Physical Fitness	1.302	1.395
Journal of Sports Science and Medicine	1.774	2.580
Journal of Sports Sciences	2.811	3.264
Journal of Strength and Conditioning Research	3.017	3.101
Journal of Teaching in Physical Education	1.775	2.062
Knee	1.762	2.081
Knee Surgery, Sports Traumatology, Arthroscopy	3.149	3.279
Medicina Dello Sport	0.393	0.350
Medicine & Science in Sports & Exercise	4.478	4.883
Motor Control	1.302	1.063
Movement Disorders	8.222	8.051
Pediatric Exercise Science	1.707	1.875
Perceptual and Motor Skills	1.049	1.023
Physical Education and Sport Pedagogy	2.035	2.721
Physical Therapy	3.043	3.599
Physical Therapy in Sport	2.000	2.355
Physician and Sportsmedicine	1.874	1.855
PLoS ONE	2.776	3.337
PM&R	1.902	2.062
Psychology of Sport and Exercise	2.710	3.662
Quest	1.819	2.176
Research Quarterly for Exercise and Sport	2.032	2.489
Scandinavian Journal of Medicine & Science in Sports	3.631	4.097
Science & Sports	0.684	0.779
Sociology of Sport Journal	1.418	1.541
Sports Biomechanics	1.714	1.989
Sport, Education and Society	1.962	2.418
Sport Psychologist	1.500	2.042
Sports Medicine	7.583	9.257
Sports Medicine and Arthroscopy Review	1.181	1.489
Strength and Conditioning Journal	0.986	1.021
Teaching and Teacher Education	2.411	3.218
Wilderness & Environmental Medicine	1.450	1.360

Adapted from Cardinal and Thomas (2005).

Using the Internet: The Good, the Bad, and the Ugly

A great deal of information may be found online when searching the Internet. The correctness and accuracy of information, however, are often difficult to verify. In fact, much information is wrong and misleading. You can search the Internet using keywords with one of the standard search engines. You will likely find many sites when you conduct the search. Separating the good and valid ones from the useless and biased ones is difficult. However, several sites consistently contain useful information on topics that students want to find.

- Internet Scout is a joint project from the National Science Foundation and the University of Wisconsin–Madison. It includes a considerable amount of useful scientific information on the biological sciences; physical sciences; and math, engineering, and technology.
- SOSIG, which stands for the Social Science Information Gateway, connects to more than 50,000 social science websites in such areas as anthropology, business, economics, education, environmental sciences, European studies, geography, government, law, philosophy, politics, psychology, and sociology.
- The WWW Virtual Library is organized using the Library of Congress subject headings, thus making it similar to your local college or university library. Headings include agriculture, business and economics, communications and media, computing and computer science, education, engineering, humanities and humanistic studies, international affairs, law, recreation, regional studies, social and behavioral sciences, and society.

Using your computer to search on a system such as Google Scholar can also be productive. For example, if you enter the term *schema theory*, you will find Richard Schmidt's original paper published in *Psychological Review* in 1975. This paper has been cited over 4,600 times. If you look down the list of papers in your search, you will see many papers that cite schema theory. If you have connections to your college or university library, you can download many of these papers at no cost.

Personal Computers

We have already mentioned that many faculty and students can access library sources and conduct literature searches from electronic devices in their own living quarters or offices. Your library should have information that explains how to log on and what information will be required to complete a search remotely.

You should also be aware of the ability to store bibliographic entries, abstracts, and even reprints of studies in the form of pdf files. Several commercial programs are available for reference management, including EndNote, Mendeley, and Reference Manager. After entering the bibliographic information and notes, you can retrieve the information using pertinent keywords, author names, or journal names. Additions, changes, and deletions are easily accomplished, and the number of items that can be stored is limited only by electronic storage space limits. Each new entry is automatically stored in alphabetical order, and the complete bibliography is always instantly accessible. Linking pdf files of articles directly to the notes is easy, and additional information can be found quickly. Many universities now make one or more of these programs available free or at a reduced cost. In addition, most libraries have classes to learn the program or consultants who can help if a problem is encountered. Storing information in the cloud is also an option (e.g., Amazon S3, Box, Dropbox, Google Drive, iCloud, Microsoft One Drive, pCloud).

Other Library Services

The library services available depend mostly on the size of the institution. Larger schools generally provide more financial support for the library, although some relatively small institutions have excellent library resources that provide outstanding support for the institutional and research aspects of the school.

Besides the usual services of the library information system (the computerized card catalog, reference and circulation departments, bibliographic collections, and stack areas), libraries also offer resources and services such as copy services, government document sections, digital repositories, special collections, and other special services. One such valuable service is the interlibrary loan, which enables you to get books, theses and dissertations, and journal articles that your library does not carry. An interlibrary loan usually takes a short time from the date it is requested, and most libraries are now delivering the requests as pdf files.

Many libraries offer guided tours, short orientation courses, self-guided tours, and tutorials in the use of electronic resources. Get acquainted with your library. The hours spent will be the wisest investment of your time as a graduate student.

Step 5: Read and Record the Literature

Collecting related literature is a major undertaking, but the next step is even more time consuming. You must read, understand, and record the relevant information from the literature, keeping in mind one of the many (anonymous) Murphy's laws: No matter how many years you save an item, you will never need it until after you have deleted it.

Whoever originated this quote was probably a researcher working on a literature review. You can count on the fact that if you discard one note from your literature search, that paper will be cited incorrectly in your text or reference list. If you delete your notes on an article because you do not believe that it is relevant to your research, your major professor, a committee member, or the journal to which you submit the paper will request inclusion of the article. Then, when you go back to locate the article in the library, either it will have been removed from the journal, the whole journal will be missing, or a professor will have checked it out and never returned it (and the librarian will not reveal the professor's name). Therefore, when you find a particular paper, take careful and complete notes, including exact citation information, and use a bibliographic program to link to the pdf file so that you are prepared if you need to see the article again.

To learn more about understanding the literature you read, try to decipher the scientific phrases in the Deciphering Scientific Phrases sidebar. After you understand what each phrase really means, you should have little difficulty understanding authors of research articles. Seriously, though, as a researcher, you should note the following information from research studies that you read:

- Statement of the problem (and maybe hypotheses)
- Characteristics of the participants
- Instruments and tests used (including reliability and validity information if provided)
- Testing procedures
- Independent and dependent variables
- Treatments applied to participants (if an experimental study)
- Design and statistical analyses
- Major findings
- Questions raised for further study
- Citations to other relevant studies that you have not located

When studies are particularly relevant to the proposed research, save a copy. If the article is not available electronically, photocopy it and write the complete citation on the title page if the journal does not provide this. Also, indicate in your electronic system which studies are saved electronically on your hard drive and which are photocopied.

When reading research articles, you should always be thinking critically about the justification for the research, the methods employed by the investigators, and the outcomes and conclusions noted. Use the series of questions in the Criteria for Critiquing a Research Paper sidebar when critiquing a study. You might also consider developing your own critique form that you can use during your review process. You will find that your critiquing skills will become stronger as you read more literature and become more familiar with your research topic.

Reviewers can occasionally go overboard in their critiques, much like the critique of Schubert's *Unfinished Symphony* presented in the sidebar A Critique of Schubert's Productivity.

To summarize, by using a computerized system for note-taking (or index cards if you prefer), you can record the important information about most studies and index it by topic. Always be sure to record the complete and correct citation with the appropriate citation style

Deciphering Scientific Phrases

What Was Said	What Was Meant
It has long been known that . . .	I haven't bothered to look up the original reference, but . . .
Of great theoretical and practical importance	Interesting to me
Although it has not been possible to provide definite answers to those questions . . .	The experiment didn't work out, but I figured I could at least get a publication out of it.
The W-PO system was chosen as especially suitable to show the predicted behavior.	The researcher in the next lab had some already made up.
Three of the samples were chosen for detailed study.	The results on the others didn't make sense.
Accidentally strained during mounting	Dropped on the floor
Handled with extreme care throughout the experiment	Not dropped on the floor
Typical results are shown.	The best results are shown.
Agreement with the predicted curve is *excellent* *good* *satisfactory* *fair*	Agreement with the predicted curve is *fair* *poor* *doubtful* *imaginary*
It is suggested that . . .	I think . . .
It is believed that . . .	I think . . .
It may be that . . .	I think . . .
It is clear that much additional work will be required before a complete understanding . . .	I don't understand it.
Let me make one thing perfectly clear.	A snow job is coming.
Unfortunately, a quantitative theory to account for these results has not been formulated.	Neither I nor anybody else understands it.
Correct with an order of magnitude . . .	Wrong
Thanks are due to Joe Glotz for assistance with the experiments and to John Doe for valuable discussion.	Glotz did the work, and Doe explained what it meant.

used by your institution (e.g., American Psychological Association [APA], Index Medicus). If using a computerized system, be sure to back up the data files and article pdf files on a USB flash drive or in the cloud.

Step 6: Write the Literature Review

After you have located and read the necessary information and have recorded the appropriate bibliographic data, you are ready to begin writing the literature review. The literature review has three basic parts:

1. Introduction
2. Body
3. Summary and conclusions

Criteria for Critiquing a Research Paper

I. Overall impression (most important): Is the paper a significant contribution to knowledge about the area?

II. Introduction and literature review
    a. Is the research plan developed within a reasonable theoretical framework?
    b. Is current and relevant research cited and properly interpreted?
    c. Is the statement of the problem clear, concise, testable, and derived from the theory and research reviewed?

III. Methods
    a. Are relevant participant characteristics described, and are the participants appropriate for the research?
    b. Is the instrumentation appropriate?
    c. Are testing or treatment procedures described in sufficient detail?
    d. Are the statistical analyses and research design sufficient?

IV. Results
    a. Do the results evaluate the stated problems?
    b. Is the presentation of the results clear and complete?
    c. Are the tables and figures appropriate?

V. Discussion
    a. Are the results discussed?
    b. Are the results related to the problem, theory, and previous findings?
    c. Is there excessive speculation?

VI. References
    a. Are all sources cited in the text included accurately in the reference list?
    b. Are all references cited in the text?
    c. Are all dates in the references correct, and do they match the text citation?

VII. Abstract
    a. Does it include a statement of the purpose; a description of participants, instrumentation, and procedures; and a report of meaningful findings?
    b. Is the abstract the proper length?

VIII. General
    a. Are keywords provided?
    b. Are running heads provided?
    c. Does the paper use nonsexist, bias-free language and provide for protection and appropriate labeling of human participants?

A Critique of Schubert's Productivity

A company CEO was given a ticket for a performance of Schubert's *Unfinished Symphony*. Unable to go, they passed the invitation to the company TQM (total quality management) coordinator. The next morning the CEO asked the TQM coordinator how they enjoyed the performance. Instead of voicing a few plausible observations, the TQM coordinator handed the CEO a memorandum that read as follows:

- For considerable periods, the oboe players had nothing to do. Their number should be reduced, and their work spread over the whole orchestra, thus avoiding peaks of inactivity.
- All 12 violins were playing identical notes. This seems an unnecessary duplication, and the staff of this section should be drastically cut. If a large volume of sound is required, this could be obtained through the use of an amplifier.
- Much effort was involved in playing the demi-semiquavers. This seems an excessive refinement, and it is recommended that all notes be rounded up to the nearest semiquaver. If this were done, it should be possible to use trainees instead of craftspersons.
- No useful purpose is served by repeating with horns the passage that has already been handled by the strings. If all such redundant passages were eliminated, the concert could be reduced from 2 hours to 20 minutes.

In light of the preceding, one can only conclude that had Schubert given attention to these matters, he would probably have had the time to finish his symphony.

The introduction should explain the purpose of the review and its organization. The body of the review should be organized around important topics. Finally, the review should summarize important implications and suggest directions for future research. The purpose of the review is to demonstrate that your problem requires investigation and that you have considered the value of relevant research in developing your hypotheses and methods—that is, you know and understand what other people have done and how that research relates to and supports what you plan to do.

The introduction to the review (or to topical areas within the review) is important. If these paragraphs are not well done and interesting, the reader may skip the entire section. Attempt to attract the reader's attention by identifying in a provocative way the important points to be covered.

The body of the literature review requires considerable attention. Relevant research must be organized; synthesized; and written in a clear, concise, and interesting way. No unwritten law dictates that literature reviews must be boring and poorly written, although we suspect that some graduate students work from that assumption. Part of the problem stems from graduate students' belief that they must find a way to make their scientific writing complex and circuitous as opposed to simple and straightforward. Apparently, the rule is never to use a short and simple word when a longer, more complex one can be substituted. In table 2.3, we provide a useful aid to potential research writers. We strongly recommend Gastel and Day's (2016) book, which can easily be read in a few hours. It provides many excellent and humorous examples that are valuable in writing for publication and for theses and dissertations.

Besides removing as much jargon as possible (using Gastel and Day's suggestions), you should be clear and to the point. We advocate the KISS principle (Keep it simple, stupid) as a basic tenet for writing. Many grammatical errors can be avoided by using simple declarative sentences.

TABLE 2.3

Examples of Avoiding Jargon When Writing

Jargon	Preferred usage
A considerable amount of	Much
A majority of	Most
A variety of	Many
Absolutely essential	Essential
Along the lines of	Like
An order of magnitude faster	10 times faster
Are of the same opinion	Agree
As a consequence of	Because
As a matter of fact	In fact (or leave out)
As a means of	To
As is the case	As happens
As to	About (or leave out)
At an earlier date	Previously
At the present time	Now
Based on the fact that	Because
Basic fundamentals	Basics
Benchmarking	Measuring
By means of	By, with
Is capable of	Can
Consensus of opinion	Consensus
Definitely proved	Proved
Despite the fact that	Although
Drive standards-based niches	Assess
Due to the fact that	Because
During the course of	During, while
End result	Result
Entirely eliminate	Eliminate
Exhibits a tendency to	Tends
Fabricate	Make
Fewer in number	Fewer
First of all	First
For the duration of	During
For the purpose of	For
For the reason that	Since, because
From the point of view that	For
Give consideration to	Consider
Give rise to	Cause
Has a requirement for	Needs
Has the capability of	Can
If at all possible	If possible
If conditions are such that	If, when
In a number of cases	Some
In a satisfactory manner	Satisfactorily

(continued)

Table 2.3 *(continued)*

Jargon	Preferred usage
In a timely manner	Quickly
In a very real sense	In a sense (or leave out)
In all cases	Always, invariably
In case	If
In connection with	About, concerning
In many cases	Often
In my opinion it is not an unjustified assumption that	I think (or leave out)
In order to	To
In respect to	About
In some cases	Sometimes
In terms of	About
In the course of	During, while
In the event that	If
In view of the fact that	Because, since
Inasmuch as	For, as
Initiate	Begin, start
Is defined as	Is
It has been reported by Smith	Smith reported
It is apparent that	Apparently
It is believed that	I think
It is clear that	Clearly
It is clear that much additional work will be required before a complete understanding	I do not understand it
It is evident that *a* produced *b*	*a* produced *b*
It is of interest to note that	(leave out)
It is often the case that	Often
It may be that	I think
It may, however, be noted that	But
It should be noted that	Note that (or leave out)
It was observed in the course of the experiments that	We observed
It would thus appear that	Apparently
Makes an attempt	Attempts
Many different	Many
Necessitates the inclusion of	Needs, requires
Needless to say	(Leave out, and consider leaving out whatever follows it)
Notwithstanding	Even though
Of great theoretical and practical importance	Useful
On a daily basis	Daily
On the basis of	By
On the grounds that	Since, because
On the part of	By, among, for
Our attention has been called to the fact that	We belatedly discovered that
Past history	Past
Perform	Do

Jargon	Preferred usage
Pertaining to	About
Postpone until later	Postpone
Potentialities	Chances
Prior to	Before
Protein determinations were performed	Proteins were determined
Provides guidance for	Guides
Quite unique	Unique
Rather interesting	Interesting
Red in color	Red
Referred to as	Called
Resultant effect	Result
Scaled back	Cut
Sufficient	Enough
Take into consideration	Consider
Terminate	End
The opinion is advanced that	I think
The question as to whether	Whether
The reason is because	Because
There can be little doubt that this is	This probably is
There is reason to believe	I think
This result would seem to indicate	This result indicates
Throughout the whole of the experiment	Throughout the experiment
Through the use of	By, with
Upon	On
Utilize	Use
Was of the opinion that	Believed
We have insufficient knowledge	We do not know
We wish to thank	We thank
Without further delay	Now
With reference to	About (or leave out)
With regard to	Concerning, about (or leave out)
With the possible exception of	Except
With the result that	So that

Based on Day and Gastel (2016).

Proper syntax (the way words and phrases are put together) is the secret to successful writing. Some examples of improper syntax and other unclear writing are highlighted in the Examples of Unclear Writing sidebar.

As mentioned previously, the literature review should be organized around important topics. These topics serve as subheadings in the paper to direct the reader's attention. The best way to organize the topics and the information within topics is to develop an outline. The more carefully the outline is planned, the easier the writing will be. A good task is to select a review paper from a journal or from a thesis or dissertation review of literature and reconstruct the outline that the author must have used. In looking at older theses and dissertations, we find that the literature review tends to be an historical account, often presented in chronological order. We suggest that you not select one of these older studies, because this

---

Examples of Unclear Writing

Unclear Questions to Children and Their Answers

- Name the four seasons. Salt, pepper, mustard, and vinegar.
- How can you delay milk turning sour? Keep it in the cow.
- What is the fibula? A small lie.
- What does *varicose* mean? Nearby.
- How is dew formed? The sun shines and makes the leaves perspire.
- What do fish do in school? They take debate.
- What do you see in Los Angeles when the fog lifts? UCLA.

---

style is cumbersome and usually poorly synthesized. Locke and colleagues (2014) provided detailed advice for preparing and writing the literature review.

To write a good literature review, you should write as you like to read. No one wants to read abstracts of study after study presented in chronological order. A more interesting and readable approach is to present a concept and then discuss the findings about that concept, documenting them by references to the research reports related to them. In this way, consensus and controversy can be identified and discussed. More relevant and important studies can be presented in detail, and several studies with the same outcome can be covered in one sentence.

In a thesis or dissertation, the two most important aspects of the literature review are criticism and completeness. The studies should not simply be presented relative to a topic. The theoretical, methodological, and interpretative aspects of the research should be criticized not study by study, but rather across studies. This criticism demonstrates your grasp of the issues and identifies problems that should be overcome in the study you are planning. Frequently, the problems identified by criticism of the literature provide justification for your research.

Completeness—not in the sense of the review's length, but rather of reference completeness—is the other important aspect of the literature review. You should demonstrate to your committee that you have located, read, and understood all the related literature. Many studies may be redundant and need only appropriate citing, but they must be cited. The thesis or dissertation is your passport to graduation because it demonstrates your competence; therefore, never fail to be thorough. This advice, however, applies only to the thesis or dissertation. Writing for publication or using alternate thesis or dissertation formats does not require emphasizing the completeness of the literature cited (note the use of the word *cited*, not *read*). Journals have limited space and usually want the introduction and literature review to be integrated and relatively short.

> Insightful criticism and completeness of references are the most important aspects of the literature review.

Summary

Identifying and formulating a researchable problem are often difficult tasks, especially for the novice researcher. This chapter offered suggestions to help you find suitable topics. Inductive and deductive reasoning for formulating research hypotheses were discussed. Inductive reasoning moves from observations to specific hypotheses to a more general theoretical model. Deductive reasoning moves from a theoretical explanation to specific hypotheses to be tested.

The steps in a literature search include writing the problem statement, consulting secondary sources, determining descriptors, searching preliminary sources to locate primary sources, reading and recording the literature, and writing the review.

There are no shortcuts to locating, reading, and indexing the literature and then writing the literature review. If you follow our suggestions, you can do it more effectively, but much hard work is still required. A good scholar is careful and thorough. Do not depend on what others report, because they are often incorrect. Look it up yourself.

No one can just sit down and write a good literature review. A careful plan is necessary. First, outline what you propose to write, write it, and then write it again. When you are convinced that the review represents your best effort, have a knowledgeable graduate student or faculty member read it. Then welcome their suggestions. Next, have a friend who is not as knowledgeable read it. If your friend can understand it, then your review is probably in good shape. Of course, your research methods professor will find something wrong or at least something they think should be different. Just remember that professors feel obligated to find errors in graduate students' work.

✓ Check Your Understanding

These exercises will help you locate, synthesize, organize, critique, and write a literature review. Suggestions from this chapter will help you complete these exercises.

1. Critique a research study in your area of interest using the questions in the Criteria for Critiquing a Research Paper sidebar.

2. Select a research paper from a journal. Construct the outline that the author probably used to write the introduction and literature review (whether presented together or separately).

3

Presenting the Problem

> You are entitled to your own opinions—but not your own facts.
>
> —Daniel Patrick Moynihan

In a thesis or dissertation, the first section or chapter introduces the problem. Indeed, it is often titled "Introduction." The introduction serves to convey the significance of the problem and set forth the dimensions of the study. This chapter discusses each of the following components, which are frequently required in the first part of a thesis or dissertation:

- Title
- Background and justification
- Purpose statement
- Hypothesis
- Definitions (more common when using the traditional format)
- Assumptions and limitations (also more common in the traditional format)
- Significance

Not all thesis advisors subscribe to the same thesis format, because there is no universally accepted one. Moreover, the nature of the research problem causes the format to vary. For example, an historical study would not adhere to the same format used in an experimental study, and section titles might differ between descriptive studies and qualitative studies. We merely present sections, each with a purpose and with specific characteristics, typically found in the introduction.

Choosing the Title

The importance of the title of a journal article or thesis should not be underestimated. When conducting a literature search, the title of a research study often provides the first opportunity to catch the attention of the reader. A title that is superficial, too long, or does not clearly and accurately reflect the focus of the study will probably not entice the reader to learn more about the research. Thus, careful consideration must be given to the title. Although discussing the title first may seem logical, it might surprise you to learn that titles are often not determined until after the study has been written. But at the proposal meeting, you must have a title (although it may be provisional), so we discuss it first.

Some writers, such as Gastel and Day (2016), claim that there is a trend toward shortening titles. But an analysis of more than 10,000 dissertations in seven areas of education failed to demonstrate such a trend (Coorough & Nelson, 1997). Many titles are, in essence, the statement of the problem (some even include the methods section). Following is an example of a too-lengthy title. (*Note:* The examples we use to represent poor practices are fictional. Frequently, they have been suggested by actual studies, but any similarity to a real study is purely coincidental.)

> **An Investigation of a Survey and Analysis of the Influence of PL 94-142 on the Attitudes, Teaching Methodology, and Evaluative Techniques of Randomly Selected Male and Female Physical Education Teachers in Public High Schools in Cornfield County, State of Confusion**

> A good title succinctly informs the reader of the study's content and stimulates the reader's interest in the research.

Such a title includes simply too much information. Day (1983, p. 10) humorously addressed this problem by reporting a conversation between two students. When one asked whether the other had read a certain paper, the reply was, "Yes, I read the paper, but I haven't finished the title yet." A better title for the previous example is "PL 94-142's Influence on Physical Education Teachers' Attitudes, Methodology, and Evaluations."

The purpose of the title is to convey the content, but this purpose should be accomplished as succinctly as possible. For example, "State Physical Activity Policies and Racial/Ethnic Disparities in Adult Physical Activity: 2006-13" (Merritt, 2015) is a good title because it tells the reader exactly what the study is about. It defines the purpose, which was to examine the effect of public policies on racial and ethnic disparities in physical activity, while also delimiting the research. However, beware of going to the other extreme in striving for a short title. A title such as "Professional Preparation" is not particularly helpful. It does not include the field or the aspects of professional preparation studied. The key to the effectiveness of a short title is whether it reflects the contents of the study. A title that is specific is more easily indexed and retrieved through electronic databases and is more meaningful for a reader who is searching for literature on a certain topic.

Avoid superfluous words and phrases such as *An Investigation of, An Analysis of,* and *A Study of.* They simply increase the length of the title and contribute nothing to the description of the content. Consider this title: "A Study of Three Teaching Methods." Half the title consists of superfluous words: *A Study of.* The rest of the title is not specific enough to be indexed effectively: Teaching what? Which methods?

Furthermore, always be aware of your audience. You can assume that your audience is reasonably familiar with the field, the accepted terminology, and the relevant problem areas. An outsider can question the significance of studies in any field. Some titles of supposedly scholarly works are downright humorous. Consider "The Nature of Navel Fluff" from the journal *Medical Hypotheses* (Steinhauser, 2009), which suggested that lint might provide a cleaning function for the navel.

For a rousing good time, peruse the titles of theses and dissertations completed at a college or university in any given year. For example, we found "The Phospholipid Distribution in the Testes of the House Cricket." How weird can you get? We are joking, of course. The point is that people tend to criticize studies done in other disciplines simply because they are ignorant about those disciplines. We'll give more attention to this topic in chapter 21.

Developing the Introduction: Background and Justification

The introduction section of a thesis or journal article justifies the need for the research. If an effective justification cannot be developed, there is little need for the research. You use the introduction to stimulate interest in the problem, justify the need for your study, and establish the significance of the research by documenting what is known about your research topic and highlighting where there are controversies, gaps, or equivocal outcomes in the research literature that need to be addressed. Your goal is to convince the reader that your research is important and makes a significant contribution to our knowledge about a topic. A well-written introduction then logically leads to a concluding paragraph in the journal article's introduction section or a brief standalone section in a thesis that states the specific purpose of the study. Hypotheses, when appropriate, are typically included immediately after the purpose statement.

Whether a research project reflects basic or applied research often influences how the significance of the study is judged. This should be kept in mind when developing the introduction section. In chapter 1, we explained that basic research does not have immediate social significance; it usually deals with theoretical problems and is conducted in a controlled laboratory setting. Applied research addresses immediate problems for the purpose of improving practice. It offers less control but ideally more real-world application. Consequently, basic and applied research cannot be evaluated by the same criteria. The significance of a basic research study obviously depends on the purpose of the study, but usually the criteria focus on the extent to which the study contributes to the formulation or validation of some theory. The worth of applied research must be evaluated on the basis of its contribution to the solution of some immediate problem.

How to Write a Good Introduction

A good introduction requires literary skill because it should flow smoothly yet be reasonably brief. Be careful not to overwhelm the reader with technical jargon, because the reader must be able to understand the problem to gain an interest in the solution. Therefore, an important rule is this: Do not be too technical. A forceful, simple, and direct vocabulary is more effective for purposes of communication than scientific jargon and the worship of polysyllables. Day (1983, pp. 147-148) related a classic story of the pitfalls of scientific jargon:

> **This reminds me of the plumber who wrote the Bureau of Standards saying that he had found hydrochloric acid good for cleaning out clogged drains. The Bureau wrote back, "The efficacy of hydrochloric acid is indisputable, but the corrosive residue is incompatible with metallic permanence." The plumber replied that he was glad the Bureau agreed. The Bureau tried again, writing, "We cannot assume responsibility for the production of toxic and noxious residues with hydrochloric acid and suggest that you use an alternative procedure." The plumber again said that he was glad the Bureau agreed with him. Finally, the Bureau wrote to the plumber, "Don't use hydrochloric acid. It eats the hell out of pipes."**

Audience awareness is important. Again, you can assume that readers are reasonably informed about the topic (or they probably would not be reading it in the first place). But even an informed reader needs some refresher background information to understand the nature of the problem, to be sufficiently interested, and to appreciate your rationale for studying the problem. You must remember that your audience has not been as completely and recently immersed in this area of research as you have been.

> Write the introduction so that readers know the study's purpose before reading the problem statement. Introductions should also create interest, provide background information, and explain the study's rationale.

The introductory paragraphs must create interest in the study; thus, your writing skills and knowledge of the topic are especially valuable in the introduction. The narrative should introduce the necessary background information quickly and explain the rationale behind the study. A smooth, unified, well-written introduction should lead to the problem statement with such clarity that the reader could predict the purpose of the study before reading it.

The following introductions were selected from research journals for their brevity of presentation and effectiveness. This is not to say that brevity in itself is a criterion, because some topics require more comprehensive introductions than do others. For example, studies developing or validating a theoretical model usually need longer introductions than do applied research topics. Furthermore, theses and dissertations (in the traditional format) almost always have longer introductions than journal articles simply because of the page-cost considerations in the latter.

Examples of Good Introductions

The following examples demonstrate some desirable features in an introduction; they include a general introduction, more specific background information, a mention of gaps in the literature and areas of needed research, and a logical progression leading to the problem statement. After you have read them, see whether you can write the purpose for each study.

Example 3.1 (from Teramoto & Golding, 2009)

From *Research Quarterly for Exercise and Sport* Vol. 80, pgs. 138-145, Copyright 2009 by the American Alliance for Health, Physical Education, Recreation and Dance, 1900 Association Drive, Reston, VA 20191

[General introduction] Research shows that participating in regular, vigorous physical activity is a key factor in reducing the risk of coronary heart disease (CHD) (Mazzeo et al., 1998). CHD is the leading cause of death in the United States (Rosamond et al., 2007). Physical inactivity is one of the CHD risk factors that can be modified (National Cholesterol Education Program [NCEP] Expert Panel, 2002).

[Background information] Elevated plasma levels of cholesterol and triglycerides are associated with the development of CHD. . . . Regular physical activity can improve certain lipid levels associated with the risk of CHD. . . .

[Lead-in] Although the research on positive effects of regular physical activity on plasma lipids is extensive, few studies have monitored the changes in plasma lipid levels in individuals who exercise regularly for more than 10 years.

Example 3.2 (from Ellingson, Meyer, Shook, et al., 2018)

From *Preventive Medicine Reports,* Vol. 11, pgs. 274-281, https://doi.org/10.1016/j.pmedr.2018.07.013, 2211-3355/Copyright 2018 by the authors. Published by Elsevier Inc. This is an open access article under the CC BY-NC-ND license (http://creativecommons.org/licenses/BY-NC-ND/4.0).

[General introduction] Much of the chronic disease burden in the United States is attributed to modifiable behavioral risk factors (e.g. diet, exercise) (Bauer et al., 2014). One such factor, excessive sedentary behavior, has recently received significant attention with evidence demonstrating deleterious effects for cardiometabolic health and all-cause mortality that may be independent of physical inactivity (Koster et al., 2012; Matthews et al., 2012) . . . .

[Background information and lead-in] Sedentary time also negatively influences mental health including increased risk for anxiety (Teychenne et al., 2015), depression (Teychenne et al., 2010), and lower levels of emotional wellbeing (Atkin et al., 2012a; Endrighi et al., 2015) in diverse populations including younger adults . . . . However, this research has largely relied on cross-sectional data. While longitudinal studies examining sedentary behavior and outcomes related to mental wellbeing exist (Hamer and Stamatakis, 2013;

Lucas et al., 2011; Sanchez-Villegas et al., 2008; Teychenne et al., 2014), these have been conducted primarily in older populations and have relied on self-report and surrogate measures for sedentary time (e.g., hours of TV viewing), which typically have poor validity (Atkin et al., 2012b). Additionally, these measures do not usually consider the potential health consequences of different accumulation patterns of sedentary time . . . . [G]iven the limitations of the sedentary assessments used to date, it is unknown whether the total amount of sedentary time or the amount accumulated in longer bouts is most problematic for mental wellbeing . . . . Lastly, research conducted in this area has largely focused on diagnosed mental health conditions (e.g., major depressive disorder) and less is known about the influence of sedentary time on sub-clinical mental health symptoms that affect a much larger segment of the population.

Stating the Research Purpose

A statement of the research purpose follows the introductory paragraphs, either as the concluding paragraph in a journal article or a standalone section in a thesis. The purpose statement should be clear and succinct, although this goal is more challenging if the study has several secondary purposes. To achieve clarity in the purpose statement, an important aspect to consider is sentence structure, or syntax. For example, suppose a researcher conducted a study "to compare sprinters and distance runners on anaerobic power, as measured by velocity in running up a flight of stairs." Observe the difference in meaning if the researcher had worded the purpose as "to compare the anaerobic power of sprinters and distance runners while running up a flight of stairs." The researcher would have to be in good shape to make those comparisons while running up stairs. Another example of faulty syntax was the case in which the purpose of the study was "to assess gains in quadriceps strength in albino mice using electrical stimulation." Those mice had to be awfully clever to use electrical instruments.

Returning to our two examples of good introductions, the purpose in example 3.1 from Teramoto and Golding (2009) was to examine well-documented longitudinal effects of regular exercise on reducing the risk of CHD especially for older adults' maintenance of an independent lifestyle. The purpose for the Ellingson et al. (2018) study (example 3.2) logically followed background information in the introduction that highlighted shortcomings of past research—to examine the longitudinal association of changes in sedentary time with changes in mood, stress, and sleep in healthy young adults. In addition, they contrasted the influence of sedentary time accumulated in longer (≥30 min) and short (< 30 min) bouts on these same mental health indicators.

We should note that a broader and more in-depth literature review is often included in the introductory section when using a traditional thesis or dissertation format and thus precedes the formal statement of the purpose. If this is the case, then a brief purpose statement should appear fairly soon in the introductory section before the literature review.

In summary, an effectively constructed introduction leads smoothly to the purpose of the study, which should be stated as clearly and concisely as the secondary purposes, or variables, allow it to be.

Identifying the Variables

Different sections of the thesis, dissertation, or journal article can be used to identify variables in the study. For an experimental or quasi-experimental study, the purpose statement may identify key independent, dependent, and categorical variables that shape the experimental design. Usually, some **control variables** (which could possibly influence the results and are kept out of the study) can also be identified here.

control variable—A factor that could possibly influence the results and that is kept out of the study.

categorical variable—A kind of independent variable, such as age, race, or sex, that cannot be manipulated; also called a *moderator variable*.

Independent and dependent variables were mentioned in chapter 1. The independent variable is the experimental, or treatment, variable; it is the cause. The dependent variable is what is measured to assess the effects of the independent variable; it is the effect. A **categorical variable**, sometimes called a *moderator variable* (Tuckman, 1978), is a kind of independent variable, such as age, race, or sex, that cannot be manipulated. A categorical variable is studied to determine whether its presence changes the cause-and-effect relationship of the independent and dependent variables.

Anshel and Marisi (1978) studied the effect of synchronous and asynchronous movement to music on endurance performance. One group performed an exercise in synchronization to background music, one group exercised with background music that was not synchronized to the pace of the exercise, and a third group exercised with no background music.

The independent variable was the background music condition. There were three levels of this variable: synchronous music, asynchronous music, and no music. The dependent variable was endurance performance, which was defined as the time the participant could exercise on a bicycle ergometer until exhaustion. In this study, the endurance performances of men and women under the synchronous, asynchronous, and no-music conditions were compared. The authors thus sought to determine whether men responded differently than women to the exercise conditions. Sex, then, represented a categorical variable. Not all studies have categorical variables.

The researcher decides which variables to manipulate and which variables to control and can control a variable's possible influence by keeping it out of the study. Thus, the researcher chooses not to assess a variable's possible effect on the relationship between the independent and dependent variables, so this variable is controlled. For example, suppose a researcher is comparing stress-reduction methods on the competitive state anxiety of gymnasts before dual meets. The gymnasts' years of competitive experience might have a bearing on their anxiety scores. The researchers have a choice: They can include this attribute as a categorical variable by requiring that half the participants have had a certain number of years of experience and that the other half have had less, or they can control the variable of experience by requiring that all participants have similar experience.

> Researchers control the influence of variables by deciding which variables should be manipulated and which should be controlled.

The decision to include or exclude some variable depends on several considerations, such as whether the variable is closely related to the theoretical model and how likely it is that an interaction will be present. Practical considerations include the difficulty of making a variable a categorical variable or controlling it (such as the availability of participants having a particular trait) and the amount of control the researcher has over the experimental situation.

In the study of the effects of synchronized music on endurance (Anshel & Marisi, 1978), the factor of fitness level was controlled by giving all the participants a physical working capacity test. Then, on the basis of this test, each person exercised at a workload that would cause a heart rate of 170 beats per minute (bpm). Thus, even though the ergometer resistance settings would be different from person to person, all participants would be exercising at approximately the same relative intensity; the differences in fitness were controlled in this manner. Another way of controlling fitness as a variable would be to assess participants on a fitness test and select just those with a certain level of fitness.

extraneous variable—A factor that could affect the relationship between the independent and dependent variables but that is not included or controlled.

Extraneous variables are factors that could affect the relationship between the independent and dependent variables but that are not included or controlled. The possible influence of an extraneous variable is usually brought out in the discussion section. Anshel and Marisi (1978) speculated that some differences in the performances of men and women might be due to the women's reluctance to exhibit maximum effort in the presence of a male experimenter. Consequently, this would be an extraneous variable. All the kinds of variables are discussed in more detail in chapter 18.

Presenting the Research Hypothesis

After you have stated the research problem, hypotheses (or a question for a qualitative study) are presented. The formulation of hypotheses was discussed in chapters 1 and 2. The discussion here is on the statement of the hypotheses and the distinction between research hypotheses and the null hypothesis. Remember that **research hypotheses** are the expected results. In the study by Anshel and Marisi (1978), a research hypothesis might be that endurance performance would be enhanced by exercising to synchronized music. The introduction produces a rationale for that hypothesis. Another hypothesis might be that exercising to asynchronous music would be more effective than exercising with no background music (because the pleasurable sensory stimuli blocks the unpleasant stimuli associated with the fatiguing exercise). As a further example, a researcher in cardiac rehabilitation might hypothesize that distance from the exercise center is more influential as a factor in patients' exercise adherence than the type of activities offered in the cardiac rehabilitation program. In the example given in chapter 1, a dance teacher hypothesized that the use of video in the instructional program would enhance the learning of dance skills. Here are a few humorous research hypotheses:

> **research hypothesis**—A hypothesis deduced from theory or induced from empirical studies that is based on logical reasoning and predicts the study's outcome.

- A plastic bag will never open from the end you try first.
- Pressing harder on the remote control will help if the batteries are weak.
- No matter the color of the bubble bath soap, the bubbles are always white.
- Glue will not stick to the inside of the bottle.

In contrast, the **null hypothesis** is used primarily in the statistical test for the reliability of the results; the null hypothesis says that there are no differences between treatments (or no relationships between variables). For example, any observed difference or relationship is due simply to chance (see chapter 7). The null hypothesis is usually not the research hypothesis, and the research hypothesis is what usually is presented. Generally, the researcher expects one method to be better than others or anticipates a relationship between two variables. One does not embark on a study if nothing is expected to happen. On the other hand, a researcher sometimes hypothesizes that one method is just as good as another. For example, in the multitude of studies done in the 1950s and 1960s on isometric versus isotonic exercises, it was often hypothesized that the upstart isometric exercise was just as effective at improving muscular strength as the traditional isotonic exercise. In a study on the choice of recreational activities for children with developmental disabilities, Matthews (1979) showed that most research in this area, which reported differences between children with and without a developmental disability, failed to consider socioeconomic status. Consequently, he hypothesized that there were no differences in the frequency of participation in recreational activities between children with developmental disabilities and those without when socioeconomic status was held constant.

> **null hypothesis**—A hypothesis used primarily in the statistical test for the reliability of the results that says that there are no differences between treatments (or no relationships between variables).

Furthermore, sometimes the researcher does not expect differences in some aspects of the study but does expect a difference in others. In a study of age differences in the strategy for recall of movement (Thomas, Thomas, Lee, Testerman, & Ashy, 1983), the authors hypothesized that because location is automatically encoded in memory, there would be no difference between younger and older children in remembering a location (where an event happened during a run). They hypothesized, however, that there would be a difference in remembering distance because the older child spontaneously uses a strategy for remembering whereas the younger child does not. As a second example, Muir, Haddad, van Emmerik, and Rietdyk (2019) studied how young (20-35 years old), middle-aged (50-64 years), and older adults (65-79 years) modified their walking patterns when stepping over obstacles of varying height as a way of exploring the effect of aging on locomotor control and mobility.

By increasing obstacle height, the investigators increased the challenge to the neuromuscular and cognitive systems for safely stepping over the obstacles. Under the less challenging obstacle conditions (i.e., no or low-height obstacles), Muir and colleagues predicted that the gait patterns (e.g., approach speed, obstacle clearance) of middle-aged participants would not differ from the patterns of the young participants. Under more challenging conditions (higher obstacle heights), however, middle-aged participants were expected to show gait adjustments different from young participants but similar to those exhibited by the older participants.

In summary, the formulation of hypotheses is an important aspect of defining and delimiting the research problem. Further, they must be carefully crafted such that they are testable using the outcomes of the research study.

Operationally Defining Terms

Operationally defining certain terms so that the researcher and the reader can adequately evaluate the results is common in both theses and journal articles, but the manner in which they are highlighted differs between the two formats. In a thesis or dissertation, operational definitions are often included in a standalone section in the introduction. In a journal article, operational definitions of key terms are sometimes integrated into the introduction if necessary when building the justification for the research, but more commonly are highlighted in the presentation of methods used to conduct the study. In particular, dependent variables often need to be operationally defined.

An **operational definition** describes an observable phenomenon, as opposed to a synonym definition or dictionary definition. To illustrate, a study such as Anshel and Marisi's (1978), which investigated the effects of music on forestalling fatigue, must operationally define fatigue. The author cannot use a synonym, such as *exhaustion*, because that idea is not concrete enough. We all might have our own ideas of what fatigue is, but if we are going to say that some independent variable affects fatigue, we must supply some observable evidence of changes in fatigue. Therefore, fatigue must be operationally defined. Anshel and Marisi did not use the term *fatigue*, but from their description of procedures, we can infer its operational definition as being when the participant was unable to maintain the pedaling rate of 50 revolutions per minute for 10 consecutive seconds.

Another researcher might define fatigue as the point at which maximal heart rate is achieved; still another might define fatigue as the point of maximal oxygen consumption. In all cases, though, the criterion must be observable.

A study dealing with dehydration must provide an operational definition, such as a loss of 5% of body weight. The term *obese* could be defined as having a body mass index of 30 kg/m^2 or greater. A study of teaching methods on learning must operationally define the term *learning*. The old definition, "a change in behavior," is meaningless in providing evidence of learning. Learning might be demonstrated by five successful throwing performances or some other observable performance criterion.

You may not always agree with the investigator's definitions, but at least you know how a particular term is being used. A common mistake of novice researchers is thinking that every term needs to be defined. (We have seen master's students define terms not even used in their studies!) Consider an example of an unnecessary definition in a study of the effects of strength training on changes in self-concept. *Self-concept* would need to be defined (probably as represented by some scale), but *strength* would not. The strength-training program used would be described in the methods section. Basically, operational definitions are directly related to the research hypotheses because, if you predict that some treatment will produce some effect, you must define how that effectiveness will be manifested.

operational definition—An observable phenomenon that enables the researcher to test empirically whether the predicted outcomes can be supported.

Basic Assumptions, Delimitations, and Limitations

Besides writing the introduction, stating the research problem and hypothesis, and operationally defining your terms, you may be required to summarize explicitly basic assumptions and limitations under which you performed your research.

Assumptions

Every study has certain fundamental premises that it could not proceed without. In other words, you must assume that certain conditions exist and that the behaviors in question can be observed and measured (along with other basic suppositions). A study in pedagogy that compares teaching methods must assume that the teachers involved are capable of promoting learning; if this assumption is not made, the study is worthless. Furthermore, in a learning study, the researcher must assume that the sample selection (e.g., random selection) results in a normal distribution with regard to learning capacity.

A study designed to assess an attitude toward exercise is based on the assumption that this attitude can be reliably demonstrated and measured. Furthermore, you can assume that the participants will respond truthfully, at least for the most part. If you cannot assume those things, you should not waste your time conducting the study.

Of course, the experimenter does everything possible to increase the credibility of the premises. The researcher takes great care in selecting measuring instruments, sampling, and gathering data concerning such things as standardized instructions and motivating techniques. Nevertheless, the researcher still must rely on certain basic assumptions.

Consider the following studies. Johnson (1979) investigated the effects of different levels of fatigue on visual recognition of previously learned material. Among his basic assumptions were that (a) the mental capacities of the participants were within the normal range for university students, (b) the participants understood the directions, (c) the mental task typified the types of mental tasks encountered in athletics, and (d) the physical demands of the task typified the exertion levels athletes commonly experienced.

Lane (1983) compared skinfold profiles of Black and white girls and boys and tried to determine which skinfold sites best indicated total body fat with regard to race, sex, and age. She assumed that (1) the skinfold caliper is a valid and reliable instrument for measuring subcutaneous fat, (2) skinfold measurements taken at the body sites indicate the subcutaneous fat stores in the limbs and trunk, and (3) the sum of all skinfolds represents a valid indication of body fat.

In some physiological studies, the participants are instructed (and agree) to fast or refrain from smoking or drinking liquids for a specified period before testing. Obviously, unless the study is conducted in some type of secured environment, the experimenter cannot physically monitor the participants' activities. Consequently, a basic assumption is that the participants will follow instructions.

Delimitations and Limitations

Every study has **limitations**. Limitations are shortcomings or influences that either cannot be controlled or are the results of the restrictions that the investigator has imposed. Some limitations refer to the scope of the study, which the researcher usually sets. These are often called **delimitations**. Kroll (1971) described delimitations as choices the experimenter makes to define a workable research problem, such as the use of one personality test in the assessment of personality characteristics. Moreover, in a study dealing with individual-sport athletes, the researcher may choose to restrict the selection of participants to athletes in just two or three sports, simply because all individual sports could not be included in one study.

limitation—A shortcoming or an influence that either cannot be controlled or is the result of the delimitations that the investigator has imposed.

delimitation—A limitation that the researcher imposes in the scope of the study; a choice that the researcher makes to define a workable research problem.

Thus, the researcher delimits the study. You probably notice that these delimitations resemble operational definitions. Although they are similar, they are not the same. For example, the size of the sample is a delimitation, but it is not included under operational definitions.

You can also see that basic assumptions are entwined with delimitations as well as with operational definitions. The researcher must proceed on the assumption that the restrictions imposed on the study will not be so confining as to destroy the external validity (generalizability) of the results.

Like operational definitions, the manner in which limitations and delimitations are presented differs between theses and journal articles. In a journal article, the scope of the study (delimitations) is usually highlighted by the statement of purpose and the description of the experimental design employed. Limitations most frequently are presented in the discussion of results. In theses and dissertations, limitations and delimitations are usually presented in the introduction. An examination of theses and dissertations studies shows numerous variations in organization. Some studies have delimitations and limitations described in separate sections. Some use a combination heading, some list only one heading but include both in the description, and still others embed these in other sections. As with all aspects of format, much depends on how the advisor was taught. Graduate colleges and schools often allow great latitude in format as long as the study is internally consistent. You will find considerable differences in format even within the same department. The Limitations and Explanations sidebar presents a few humorous explanations of limitations.

In Kroll's (1971) example of delimiting the scope of the study to just two sports to represent individual-sport athletes, an automatic limitation is how well these sports represent all individual sports. Moreover, if the researcher is studying the personality traits of these athletes and delimits the measurement of personality to just one test, a limitation is present. Furthermore, at least one inherent limitation in all self-report instruments in which participants respond to questions about their behavior, likes, or interests is the truthfulness of the responses.

Thus, you can see that limitations accompany the basic assumptions to the extent that the assumptions fail to be justified. As with assumptions, the investigator tries as much as pos-

OK, Thesis Committee, pay attention to my slides!

Limitations and Explanations

Limitations	Explanation
On a garbage truck	Yesterday's Meals on Wheels
On a plumber's truck	Repairing what your partner "fixed"
On a maternity room door	Push, push, push
On a fence door	Salespeople welcome, dog food is expensive
In a dog vet's waiting room	Be right back—sit, stay
In a funeral home	Drive carefully . . . we'll wait
In an eye doctor's office	If you can't read this, you're at the right place

sible to reduce limitations that might stem from faulty procedures. In Johnson's 1979 study of fatigue effects on the visual recognition of previously learned material, the participants first had to learn the material. He established criteria for learning (operational definition) and tried to control for overlearning (i.e., differences in the level of learning). Despite his efforts, however, Johnson recognized the limitation of possible differences in the degree of learning, which could certainly influence recognition.

In her study of skinfold profiles, Lane (1983) had to delimit the study to a certain number of participants in one part of Baton Rouge, Louisiana. Consequently, a limitation was that the children were from one geographic location. She also recognized that changes in body fat are associated with the onset of puberty, but she was unable to obtain data on puberty or other indices of maturation, and this therefore posed a limitation. Still another limitation was the inability to control possible influences on skinfold measurement, such as dehydration and other diurnal variations. Finally, because there are no internationally recognized standard body sites for skinfold measurement, generalizability may have been limited to the body sites used in this study.

You should not be overzealous in searching for limitations, lest you apologize away the worth of your study. For example, one of our advisees who was planning to meet with his proposal committee was overly apologetic with these anticipated limitations:

- The sample size may be too small.
- The tests may not represent the parameter in question.
- The training sessions may be too short.
- The investigator lacks adequate measurement experience.

As a result, the proposal was revised and the method was reassessed.

No study is perfect. You must carefully analyze the delimitations to determine whether the resulting limitations outweigh the delimitations. In addition, careful planning and painstaking methodology increase the validity of the results, thus greatly reducing deficiencies in a study.

Presenting the Significance of the Study in a Thesis or Dissertation

Early in this chapter, we mentioned that the introduction section of the journal article or thesis is used to persuade readers of the significance of the problem. In journal format, the significance is presented in the introduction. What we are discussing here is the traditional thesis format, where the significance is its own standalone section. Regardless, the perceived difficulty in writing the significance section may often be attributed to thinking only in terms of the practicality of the study—how the results can be immediately used to improve some aspect of the profession. Kroll (1971) emphasized the importance of maintaining continuity of the significance section with the introduction. Too often, the sections are written with different frames of reference instead of a continuous flow of thought. The significance section should focus on such things as contradictory findings of previous research, knowledge gaps in particular areas, and the study's potential contribution to practice. Difficulties in measuring aspects of the phenomenon in question are sometimes emphasized. The rationale for verifying existing theories may be the focus of the section in some studies, whereas in others, the practical application is the main concern. Generally, both theoretical and practical reasons are expressed, but the emphasis will vary according to the study.

Differences Between the Thesis and the Journal Article

In this chapter, we've described a number of differences between theses and journal articles in how certain introductory content is presented. Journal articles, including their introduction sections, are shorter in length than theses. At least two reasons account for this. The first is restrictions on length or word count for journal articles. Periodicals are concerned about publishing costs, so brevity is emphasized. Second, a kind of novice–expert ritual seems to be in operation for theses research. The novice is required to state the hypotheses explicitly, define terms, state assumptions, recognize limitations, and justify the study's worth in writing. Certainly, these steps are all part of defining and delimiting the research problem, and it is undoubtedly a worthwhile experience to address each step formally.

Research journal authors, on the other hand, need not explain the step-by-step procedure they used in developing the problem. Typically, a research journal has an introduction that includes a highly focused literature review. The length varies considerably, and some journals insist on brief introductions.

The purpose of the study is nearly always given, but it is usually not designated by a heading; rather, it is often the last sentence in the introduction. For example, in 30 articles in one volume of the *Research Quarterly for Exercise and Sport,* only one had a section titled "Purpose of the Study." The authors of 24 articles ended their introductions with sentences that began, "The purpose of this study was . . . ." Four indicated the purpose with sentences beginning, "This study was designed to . . ." or "The intent of the study was . . . ." One study did not state the purpose at all. In 29 cases, the authors and editors felt that the purpose was evident from the title and introduction.

Research hypotheses and questions are sometimes given but with little uniformity. Operational definitions, assumptions, and limitations of the study are rarely highlighted in the introduction. If it's a well-written article, you should be able to discern the operational definitions; the assumptions and limitations; and the independent, dependent, and categorical variables even though these steps are not specifically stated. Moreover, the significance of the study should be implicitly obvious if the author has written a good introduction.

There is a growing movement to replace the traditional thesis or dissertation with one or more published research papers (e.g., Patton, 2014). Often, these papers are collected into a bound volume labeled as a *thesis* or *dissertation*, but some institutions accept one or more of these publications in lieu of a thesis or dissertation. In later chapters, we provide a format for collecting published papers into a thesis or dissertation—a practice accepted by most university graduate schools.

Summary

This chapter discussed the information that is typically presented in the first section or chapter of a thesis or dissertation (excluding the review of literature) and the introduction of a journal article. First, we considered the length and substance of the study title. The importance of a good, short, descriptive title for indexing and searching the literature is sometimes overlooked.

The introduction of a research study often proves difficult to write. Conveying the significance of the study to the reader requires a great deal of thought, effort, and skill. If this section is poorly done, the reader may not bother to read the rest of the study.

The problem statement and the research hypotheses or questions commonly appear in most research studies, whether they are theses or dissertations, journal articles, or research grants. Operational definitions, assumptions, limitations and delimitations, and the significance of the study usually appear explicitly only when using the traditional format for the thesis or dissertation. Their purpose is to help (or force) the researcher to define and delimit the research problem succinctly. Operational definitions specifically describe how certain terms (especially dependent variables) are being used in the study. This is important for all studies, whether presented in a thesis or journal article format. Assumptions identify the basic conditions that must be assumed to exist for the results to have credibility. Delimitations relate to the scope of the study imposed by the researcher, such as the number and characteristics of the participants, the treatment conditions, the dependent variables used, and the way the dependent variables are measured. Limitations are possible influences on the results that are consequences of the delimitations or that cannot be completely controlled.

✓ Check Your Understanding

1. For each of the following brief descriptions of studies, write a title, the purpose or purposes, and three research hypotheses. The researchers assessed the following:
   a. Skill acquisition of three groups of fourth-grade boys and girls who had been taught by different teaching styles (A, B, and C).
   b. Self-concept of two groups of boys (a low-strength group and a high-strength group) before and after a strength-training program.
   c. Body composition (estimated percentage of fat), using the electrical impedance analysis method on participants at normal hydration and again after they had become dehydrated.
   d. Grade point averages of male and female athletes participating in major and minor (club) sports at large universities and small private colleges.
2. Locate five articles from research journals. For each, try to determine (a) the hypotheses, if not stated, (b) the independent variable or variables, (c) the dependent variable or variables, (d) an operational definition for the dependent variables, (e) at least two delimitations, (f) at least two limitations, and (g) a basic assumption.

4

Formulating the Method

> **The difference between failure and success is doing a thing nearly right and doing a thing exactly right.**
>
> —Edward Simmons

After completing the introduction, the researcher must describe the procedures used to address the research purpose. Typically, this section is labeled *Methods*. We give an overview of the four parts of the methods section here:

1. Participants
2. Instruments or apparatuses
3. Procedures
4. Design and analysis

For our purposes, let us assume that the journal format is used and that the literature review is included in the introduction of your thesis or dissertation, followed by the methods section. Much of the remainder of this book focuses on the methods.

- Important aspects of the study: the participants, instruments, procedures, design and analysis (described in this chapter)
- How to measure and analyze the results (part II)
- How to design the study (part III)
- How to write and present research (part IV)

The purpose of the methods section is to explain how the study was conducted. The standard rule is that the description should be thorough enough that a competent researcher could replicate the study.

Although qualitative studies are formatted somewhat differently from quantitative studies, they also present a methods section describing the participants and how the data were obtained. However, detailed explanations of the settings in which data were collected as well as descriptions of the participants are often required. We present more methodological details on qualitative studies in chapter 19.

How to Present Methods

Dissertations and theses differ considerably from published articles in terms of the methodological details provided. When using the journal format, however, you should place the additional methods details in an appendix. Journals try to conserve space, but space is not an issue in a dissertation or thesis. Thus, whereas standard techniques in a journal article are referenced only to another published study (in an easily obtainable journal), a thesis or dissertation should provide considerably greater detail in the appendix. Note that we indicated that a technique could be referenced to an easily obtainable journal. When writing for publication, use common sense in this regard. Consider, for example, this citation:

> Farke, F.R., Frankenstein, C., & Frickenfrack, F. (1921). Flexion of the feet by foot fetish feet feelers. *Research Abnormal: Perception of Feet, 22,* 1–26.

By most standards, this citation would not be considered easily obtainable. Therefore, if you are in doubt, present the details of the study or technique.

Furthermore, because theses and dissertations have appendixes, much of the detail that would clutter and extend the methods section should be placed there. Examples include exact instructions to **participants**, samples of tests and answer sheets, diagrams and pictures of equipment, sample data-recording sheets, and informed-consent agreements.

participants—People who are used as subjects in a study. In APA style, the term *participants* is used rather than *subjects*.

Why Planning the Methods Is Important

The purpose of carefully planning methods is to eliminate any alternative or rival hypotheses. This statement really means that when you design the study correctly and the results are as predicted, the only explanation is what you did in the research. Using a previous example to illustrate, our hypothesis is: Shoe size and mathematics performance are positively related during elementary school. To test this hypothesis, we go to an elementary school, measure shoe sizes, and obtain standardized mathematics performance scores of the children in grades 1 through 5. When we plot these scores, they appear as in figure 4.1, each dot representing a single child. Moving from the dot to the *x*-axis shows shoe size, whereas the *y*-axis shows math performance. "Look!" we say, "We are correct. As shoe sizes get larger, the children's mathematics performance increases. Eureka! All we need to do is buy the children bigger shoes and their mathematics performance will improve." But wait a minute! We have overlooked two things. First, a rival explanation is present: Both shoe size and mathematics performance are related to age. That connection really explains the relationship. As the kids get older, their feet grow larger and they perform better on mathematics tests. Second, just because two things are related does not mean that one causes the other. Correlation does not prove causation. Obviously, we cannot improve children's math performances by buying them bigger shoes.

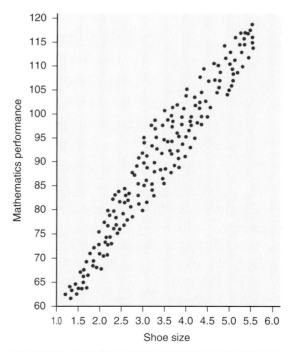

Figure 4.1 Relationship of mathematics performance and shoe size of children in grades 1 through 5.

In research, we want to use the **MAXICON principle**: Maximize true variance, or increase the odds that the real relationship or explanation will be discovered; minimize error variance, or reduce all the mistakes that could creep into the study to disguise the true relationship; and control extraneous variance, or make sure that rival hypotheses are not the real explanations of the relationship.

> **MAXICON principle**—A method of controlling any explanation for the results except the hypothesis that the researcher intends to evaluate. This is done by maximizing true variance, minimizing error variance, and controlling extraneous variance.

Two Principles for Planning Experiments

In an interesting paper, Cohen (1990) put forward two principles that make good sense when planning experiments. The first is *Less is more*. Of course, this principle seldom applies to the number of participants in a study, but it does apply to other aspects. For example, graduate students want to conduct meaningful studies that address and solve important problems. To do this, they often plan complex studies with many independent and dependent variables. From one perspective, this approach is good: The world of physical activity is truly complex. Students frequently start with useful ideas to study, but the ideas become so cumbersome that the studies often fail because of sheer complexity. Carefully evaluate the number of independent and dependent variables that are practically and theoretically important to your study. Do not let anyone convince you (except your major professor, of course, who can convince you of anything) to add additional variables just to see what happens. This action complicates your study and causes all types of measurement and statistical problems.

This idea leads to Cohen's second principle: *Simple is better*. This statement is true from the design to the treatments to the analysis to displaying data to interpreting results. Keep your study straightforward so that when you find something, you can understand and interpret what it means. Understanding your data is an important concept. Although fancy statistical procedures are impressive and informative, there is no substitute for plotting data graphically and evaluating it carefully. Summary statistics (e.g., mean, standard deviation) are helpful and informative, but they are no substitute for looking at the distribution of the original data. Summary statistics may not show us what we really need to know—things that become evident when we look at a graphic display of the data.

> Evaluating and graphically plotting the data can enhance your understanding and improve your interpretation of the data.

Describing Participants

This section of the methods describes how and why the participants were selected and which of their characteristics are pertinent to the study. These are questions to consider when selecting participants:

- Are participants with special characteristics necessary for your research?
  - Age (children, older adults)
  - Sex (females, males, or both)
  - Level of training (trained or untrained)
  - Level of performance (experts or novices)
  - Size (weight, adiposity)
  - Special types (athletes, cyclists, sedentary individuals)
- Can you obtain the necessary permission and cooperation from the participants?
- Can you find enough participants?

Of course, you want to select participants who will respond to the treatments and measures used in the study. For example, if you want to see the results of training a group of children in overhand throwing, selecting expert 12-year-old baseball pitchers as participants will not likely produce a change in measures of throwing outcome. An intense, long-term training program would be required to have any influence on these participants. Selecting children who are soccer players and have never played organized baseball would offer better odds for a training program to produce changes.

In experimental research, the interactions of participants, measures, and the nature of the treatment program are essential in allowing the treatment program to have a chance to work (Thomas, Lochbaum, Landers, & He, 1997). If you select participants who have high levels of physical fitness, subjecting them to a moderate-intensity aerobic training program will not produce changes in fitness. Also, participants high in physical fitness will have a small range of scores on a measure of cardiorespiratory endurance (e.g., $\dot{V}O_2max$). For example, you will not find a significant correlation between $\dot{V}O_2max$ and marathon performance in world-class marathon runners. Their range of scores in $\dot{V}O_2max$ is small, as is their range of scores in marathon performance times. Because the range of scores is small in both measures, no significant correlation will be found. This result does not mean that running performance and cardiorespiratory endurance are unrelated. It means that you have restricted the range of participants' performance so much that the correlation cannot be exhibited. If you had selected runners with more varied training backgrounds, a significant correlation would be found between running performance on a 5K run and $\dot{V}O_2max$. (We discuss procedures for selecting sample participants in chapter 6.)

> Participants, measures, and treatment programs are interrelated. Be sure to choose participants who will respond to the treatment program and have a broad enough range of results when measured with the chosen techniques.

What to Tell About the Participants

The exact number of participants should be given, as should any loss of participants during the time of study. In the proposal, some of this information may not be exact. For example, the following might describe the potential participants:

• *Participants:* For this study, 48 males, ranging in age from 21 to 34 years, will be randomly selected from a group ($N = 147$) of well-trained distance runners $\dot{V}O_2max = 60$ ml $\cdot$ kg^{-1} $\cdot$ min^{-1} or higher who have been competitive runners for at least 2 years. Participants will be randomly assigned to one of four groups ($n = 12$).

After the study is completed, details on the participants are available, so now this section might read as follows:

• *Participants:* In this study, 48 males, ranging in age from 21 to 34 years, were randomly selected from a group ($N = 147$) of well-trained runners $\dot{V}O_2max = 60$ ml $\cdot$ kg^{-1} $\cdot$ min^{-1} or higher who had been competitive runners for at least 2 years. The participants had the following characteristics (standard deviations in parentheses): age, $M = 26$ years (3.3); height, $M = 172.5$ cm (7.5); body mass, $M = 66.9$ kg (8.7); and $\dot{V}O_2max$, $M = 65$ ml $\cdot$ kg^{-1} $\cdot$ min^{-1} (4.2). Participants were randomly assigned to one of four groups ($n = 12$).

The participant characteristics listed are extremely pertinent in an exercise physiology study but not at all pertinent, for example, in a study of equipment used by children on the playground. The nature of the research dictates the participant characteristics of interest to the researcher. Carefully think through the inclusion and exclusion criteria that you will use in selecting participants and will report in your research. Look at related published research for ideas about important participant characteristics to report.

The characteristics of participants that you identify and report must be clearly specified. Note in the example that well-trained runners were exactly defined; that is, their $\dot{V}O_2max$ must be equal to 60 ml $\cdot$ kg^{-1} $\cdot$ min^{-1} or higher. When participants of different ages are to be used is another good example. Saying only that 7-, 9-, and 11-year-olds will be the participants is not sufficient. How wide is the age range for 7-year-olds? Is it ±1 month, ±6 months, or what? In the proposal, you may say that 7-, 9-, and 11-year-olds will be included in the study. At the time of testing, each age will be limited to a range of ±6 months. Then, when the thesis or dissertation is written, it may read as follows:

You may have to offer something to the participants to make being in the study worthwhile.

© Blend Images/fotolia

At each age level, 15 children were selected for this study. The mean ages are as follows (standard deviations in parentheses): the youngest group, 7.1 years (4.4 months); 9-year-olds, 9.2 years (3.9 months); and the oldest group, 11.2 years (4.1 months).

Protecting Participants

Most research in the study of physical activity deals with humans, although some research may use animals. Therefore, the researcher must be concerned about any circumstances in the research setting or activity that could harm humans or animals. In chapter 5, Ethical Issues in Research and Scholarship, we provide considerable details on what the researcher must do to protect both humans and animals used in research. Obtaining informed consent from humans and ensuring the protection and care of animals are particularly important.

Selecting and Describing Instruments

Careful planning is required when selecting the instruments, apparatuses, or tests used to collect data and generate the dependent variables in the study. Consider the following when selecting tests and instruments:

- What is the validity and reliability of the measures?
- How difficult is it to obtain the measures?

- Do you have access to the instruments, tests, or apparatuses needed?
- Do you know (or can you learn) how to administer the tests or use the equipment?
- Do you know how to evaluate participants' test performance?
- Will the tests, instruments, or apparatuses yield a reasonable range of scores for the participants you have selected?
- Will the participants be willing to spend whatever time is required for you to administer the tests or instruments?

For example, in a sport psychology study, you are interested in how a group of college athletes will be affected by a lecture on steroid use. In addition, you suspect that the players' attitudes might be modified by certain personality traits. So you select three tests—a steroid knowledge test, an attitude inventory about responsible drug use, and a trait personality measure—and administer them to all participants. The knowledge and attitude tests will probably be given before and after the lecture, while the trait personality test will be given only before the lecture (traits should not change, and this test is being used to stratify players in some way). In the instruments section, you should describe the three tests and probably put complete copies of each in the appendix (see chapter 5 about the ethical use of standardized tests). You also should describe the reliability (consistency) and validity (what the test measures) information that is available on each test with appropriate citations. Then you should explain the scoring sheets (place a sample in the appendix) and the scoring methods (but do not use inappropriate conversions for measurements like the ones in the Incorrect Metric Conversions sidebar).

Another example is a biomechanics study examining the effect of footwear heel height (e.g., flat shoes vs. high heels) on multiple characteristics of the walking pattern (e.g., step length, step rate, knee angle, ground-reaction force) in women. In the instruments section, you should describe the equipment used to capture the necessary data (e.g., motion capture system, force platform), including how the systems were calibrated and how data were sampled by computer. You should also describe how the dependent variables were computed.

It is worth noting at this point that a standalone instruments section within the methods section is optional. Many instrumentation systems and data capture and analysis procedures used in physical activity research are well established. When this is the case, detailed descriptions of such well-known equipment and well-established procedures are usually unnecessary. A brief description supported by references to relevant published literature should suffice. These descriptions can be integrated into the summary of procedures for data collection and analysis. A standalone instruments section is most appropriate when novel data collection apparatuses and testing procedures are employed. For example, if you developed specialized equipment to measure finger forces when grasping and lifting a glass, you must describe the design of the equipment, calibration procedures, and assessments of the validity and reliability of the force measurements. All the necessary information could be presented by use of both the instrument (or apparatus) part of the methods section and an appendix, thus allowing the methods section to flow smoothly.

Incorrect Metric Conversions

1 trillion microphones = 1 megaphone

1 million bicycles = 2 megacycles

2,000 mockingbirds = 2 kilomockingbirds

½ lavatory = 1 demijohn

1 millionth of a fish = 1 microfiche

454 graham crackers = 1 pound cake

10 rations = 1 decoration

10 millipedes = 1 centipede

10 monologues = 5 dialogues

Describing Procedures

In the procedures section, you should describe how the data were obtained, including all testing procedures and data analysis processes for obtaining measures of the variables of interest. This description is often organized chronologically as procedures are employed in the study. How tests were administered and who gave them are important features. You should detail the setup of the testing situation, participant preparation, and instructions given to the participants. If the study is experimental, then you should describe the treatments applied to the groups of participants.

Consider these points when planning procedures:

- Collecting the data
    - When? Where? How much time is required?
    - Do you have pilot data to demonstrate your skill and knowledge in using the tests and equipment and knowing how participants will respond?
    - Have you developed a scheme for data acquisition, recording, and analysis? (These are often computer controlled.)
- Planning the treatments (in quasi-experimental and experimental studies)
    - How long? How intense? How often?
    - How will participants' adherence to treatments be determined?
    - Do you have pilot data to show how participants will respond to the treatments and that you can administer these treatments?
    - Have you selected appropriate treatments for the type of participants to be used?

One of our favorite summaries of the problems encountered and solutions proposed is presented in the Errors in Experiments sidebar. These statements are extracted from an article by Martens (1973).

The procedures section contains most of the details that allow another researcher to replicate the study. (But these details must be useful, unlike the examples in the Statements in Procedures That Sometimes Get Twisted sidebar.) Tuckman (1978) outlined these details, which generally include the following:

- Specific order in which steps were undertaken
- Timing of the study (e.g., time taken for various procedures and time between procedures)
- Instructions given to participants
- Briefings, debriefings, and safeguards

Avoiding Methodological Faults: Importance of Pilot Work

No single item in this book is more important than our advice to pilot all your procedures. During our years as professors, editors, and researchers, we have seen abstracts of thousands of master's theses and doctoral dissertations. More than 75% of these research efforts are not publishable and do not contribute to theory or practice because of major methodological flaws that could have been easily corrected with pilot work. Sadly, this circumstance reflects negatively not only on the discipline and profession but also on the graduate students who conducted the research and the faculty who directed it. In addition, physical education, kinesiology, exercise science, and sport science have produced thousands of studies in which the discussions centered on methodological faults that caused the research to lack validity. Placing post hoc blame on the methodology for inadequate results is unacceptable.

Errors in Experiments

Martens' method derives from the basic premise that

> **In people experiments, people errors increase disproportionately to the contact people have with people.**

It is obvious that the most logical deduction from this premise is

> **To reduce people errors in people experiments, reduce the number of people.**

Although this solution might be preferred for its elegant simplicity, its feasibility can be questioned. Therefore, the following alternative formulation warrants consideration:

> **The contact between people testers and people subjects in people experiments should be minimized, standardized, and randomized.**

Reprinted by permission from R. Martens, "People Errors in People Experiments," *Quest* 20 (1973): 22.

Nearly all the problems could have been corrected by increased knowledge of the topic, better research design, and pilot work on the procedures.

Graduate students frequently seek information about appropriate procedures from related literature, and they should. Procedures for intensity, frequency, and duration of experimental treatments are often readily available, as is information about testing instruments and procedures. Remember, however, that procedures in one area do not necessarily work well in another, as the example in the Research Procedures May Not Generalize sidebar illustrates. Every thesis or dissertation proposal should present **pilot work** that verifies that all instruments and procedures will function as specified on the type of participants for which the research is intended. In addition, you must demonstrate that you can use these procedures and apparatuses accurately and reliably.

pilot work—Work undertaken to verify that you can correctly administer the tests and treatments for your study using appropriate participants.

Describing Design and Analysis

A study should be designed so that dependent variables change only when independent variables are changed.

Design is the key to controlling the outcomes from experimental and quasi-experimental research. The independent variables are manipulated to judge their effects on the dependent variable. A well-designed study is one in which the only explanation for change in the

Statements in Procedures That Sometimes Get Twisted

Clones are people, two.
Cole's law is thinly sliced cabbage.
Does the name Pavlov ring a bell?
Energizer bunny arrested and charged with battery.
Staph only in the microbiology lab.
A blood test of 20 pessimists was always b-negative.
Without geometry, life is pointless.
When two egotists fight, it's an I for an I.
Santa's helpers are subordinate clauses.

dependent variable is how the participants were treated (independent variable). The design and theory have enabled the researcher to eliminate all rival or alternative hypotheses. The design requires a section heading in the methods for experimental and quasi-experimental research. The plans for data analysis must also be reported. In most studies, some type of statistical analysis is used, but there are exceptions: for example, historical or qualitative research that uses other types of analysis.

Typically, the researcher explains the proposed application of the statistics. In nearly all cases, descriptive statistics are provided, such as means and standard deviations for each variable. If correlational techniques (relationships of variables) are used, then you name the variables to be correlated and the techniques to be used. For example, a researcher might write the following: "The degree of relationship between two estimates of percent body fat will be established by using Pearson r to correlate fat estimates from a BOD POD with

---

Research Procedures May Not Generalize

Dr. I.M. Funded was a good life scientist who studied the biochemistry of exercise in a private research laboratory. He had also done several studies with a colleague in sport psychology to determine whether some biochemical responses he had found were factors in psychological responses to exercise. Thus, he had a firm grasp of some of the social science techniques as well as those of life science.

Unfortunately, Dr. Funded's funding ran out, and he lost his job. A friend of his was the superintendent of a large school district. Dr. Funded went to his friend, Dr. Elected, and said, "I am a good scientist well trained in problem-solving techniques. Surely, you must need someone like me in your administrative structure. In addition, I have an undergraduate degree in physical education, so I am certified to teach, although I never have." Dr. Elected agreed to hire him as his teaching effectiveness supervisor because the school system was having difficulty identifying good teaching. Dr. Elected thought that perhaps a scientist with good problem-solving skills and the ability to make careful measurements could find a solution.

Dr. Funded decided that his first task was to identify some good teachers so that he could determine their characteristics. He would use some of the techniques he had acquired from his colleague in sport psychology to identify good teachers. He had learned that while questionnaires were effective in surveying large groups, interviews were more valid. Dr. Funded drew a random sample of six schools from the 40 in the district. Then, he randomly selected six teachers in each school and interviewed them. He used a direct interview question: "Are you a good teacher?" All 36 indicated that they were. So he went back to Dr. Elected, explained what he had done, and said, "You don't have a teaching problem. All of your teachers are good." (Of course, he noted that there could be some sampling error, but he was certain of his results.) Dr. Elected was not very happy with Dr. Funded's procedures and results and suggested that perhaps he needed more sophisticated techniques and strategies to identify good teachers.

Dr. Funded was slightly distraught but thought, "I have always questioned the techniques of those psychologists anyway. I will return to my life science techniques to determine the answer." He went back to the previously selected 36 teachers with a plan to draw blood, sample urine, and biopsy muscles (at four sites) once per week for four weeks. Immediately, 34 teachers said that they would not take part, but two who were triathletes agreed to participate. Dr. Funded noted that the participant mortality rate was about normal for biopsy studies, so the data should be generalizable. He collected all the data, did the correct chemical analyses, and reported back to Dr. Elected. He indicated that effective teachers had 84% slow-twitch fiber, higher-than-average amounts of hemoglobin per deciliter of blood, and a specific profile of catecholamines (epinephrine and norepinephrine) in the urine. In addition, good teachers trained for at least 100 miles (160 km) per week on the bicycle, 50 miles (80 km) running, and 7,500 meters swimming. Dr. Funded sat back smugly and said, "Techniques for the life sciences can be applied to solve many problems." Dr. Elected responded, "You are fired."

those from underwater weighing." In experimental and quasi-experimental studies, descriptive statistics are provided for the dependent measures, and the statistics for establishing differences between groups are reported. A researcher might write the following: "A *t* test was used to determine whether the energy cost of running differed when running barefoot vs. running with shoes."

The major problem that graduate students encounter in the description of statistical techniques is the tendency to inform everyone of their knowledge of statistics. Of course, that is not much of a problem for the new graduate student. But if your program of studies is a research-oriented one in which you take several statistics courses, your attitude may change rapidly.

> **Hiawatha, who at college majored in applied statistics, consequently felt entitled to instruct his fellow men on any subject whatsoever. (Kendall, 1959, p. 23)**

The point is for you to describe your statistical analyses but not to instruct in their theoretical underpinnings and proper use.

> **In terms of scientific progress, any statistical analysis whose purpose is not determined by theory, whose hypothesis and methods are not theoretically specified, or whose results are not related back to theory must be considered, like atheoretical fishing and model building, to be hobbies. (Serlin, 1987, p. 371)**

Establishing Cause and Effect

The establishment of cause and effect in an experimental study is much more than a statistical and design issue. The issue is one of logic. Remember, statistical tests are designed to retain or reject the null hypothesis. If the null hypothesis is rejected, then what hypothesis is true? Of course, the scholar hopes the research hypothesis is true, but this is difficult to establish. In science, the researcher seeks to explain that certain types of effects normally happen given specific circumstances or causes. For example, an effect may occur in the presence of something but not in its absence. Water boils in the presence of a high enough temperature but not in the absence of this specific temperature (given a specific air pressure).

People can (and do) differ in their opinions about what can be a cause-and-effect relationship or even whether such a relationship can exist. For example, whether you believe in universal laws, in destiny, in free will, or in an omnipotent deity is likely to influence your view of cause and effect.

Must causes be observable? If yes, then two criteria are needed for establishing cause and effect. First is the method of agreement. If an effect occurs when both A and B are present, and A and B have only C in common, then C is the likely cause (or at least part of it). Second is the method of disagreement. If the effect does not occur in E and F when C is the only common element absent, then C is the cause (or part of it). Thus, it is clear that the researcher's reasoning influences the establishment of cause and effect because the researcher's beliefs set the stage for what may be considered a cause or even whether one can exist. This suggests that the researcher's theoretical beliefs as well as the study's design and analysis are essential factors in establishing cause and effect (for a more detailed discussion of causation, see White, 1990).

Manipulation Effects

When using a treatment in experimental or quasi-experimental research, the participant is involved in the treatment on a regular basis. For example, if participants in a home exercise program are to exercise daily for at least 40 min, how can the researcher be sure they are doing so? Some sort of quantitative manipulation effect needs to be carried out to be certain. Maybe the researcher will ask each participant to wear a physical activity monitor every day during the training. Because this device records amount of movement, the researcher can determine whether a vigorous program of exercise was undertaken for the correct amount of time. Even a manipulation check such as the one described, however, has flaws: The participant could fail to exercise but ask someone else who was exercising to wear the device.

In any experimental study, researchers must have a plan as part of the methodology to check the manipulation of the independent variable in relation to the dependent variable. In a laboratory exercise study that involves one participant and one researcher, this is relatively easy. But as the research moves further into real-world settings, manipulation checks become increasingly important. In pedagogical, psychological, and sociological types of research related to physical activity, manipulation checks are nearly always essential to verify that participants perceived and responded to the treatments as the researcher intended.

Sometimes, manipulation checks can be done in a qualitative way by questioning participants either during or following the individual treatment sessions or after the treatment period is over. Although a quantitative manipulation check is preferred, a qualitative one is sometimes all that is possible or reasonable.

Fatal Flaws in Research

Every study proposal, particularly its methods, should be carefully evaluated for fatal flaws. That is, the researcher should ask, "Does this study lack a characteristic that will cause it to be rejected for publication regardless of the way it is conducted or what the outcomes are?" This is more easily said than done, but it again underscores the importance of careful planning and pilot testing of methods prior to any formal data collection and analysis. The following are some questions that can help to identify fatal flaws:

- Are all hypotheses logical when considering theory and study characteristics?
- Are all the assumptions made about the study good ones?
- Is there a sufficient number of the "right" participants selected for the study, especially in relation to the treatments and measurements used?
- Are the treatments intense enough and of sufficient length to produce the desired changes, especially considering the participants selected?
- Are all extraneous variables being controlled?
- Are all dependent variables appropriate for characterizing the response of participants to treatments?
- Are all measurements valid, reliable, and appropriate for the planned study, especially considering the characteristics of participants and treatment levels in experimental studies?
- Are data collection and storage procedures well planned and carefully done?

Even in famous studies, important errors have occurred. For example, Franklin Henry was one of the most famous researchers in kinesiology. His paper "Memory Drum Theory" (Henry & Rogers, 1960), which established the first real explanation for how movements are controlled, is the most highly cited paper ever to appear in *Research Quarterly for Exercise and Sport* (Cardinal & Thomas, 2005). Yet Fischman, Christina, and Anson (2008) reported a significant error in the method used. Specifically, they noted that one of the three movements used by Henry and Rogers to change the complexity of the motor task was not described accurately, making it impossible to replicate Henry and Rogers' experiment as described in their publication. Fischman and colleagues suggested that this discovery provides yet another lesson about the need for replication in science.

Interaction of Participants, Measurements, and Treatments

Correlational studies use selected participants to evaluate the relationship of two or more measures, whereas experimental studies use selected participants to evaluate whether a specific treatment causes a specific outcome. In both instances, selecting the appropriate participants, measurements, and treatments is extremely important. In correlational studies, the measurements must assess the characteristics of interest and produce an appropriate range of scores on the measures; if they do not, then the researcher has reduced the chances of discovering important relationships. For example, if the purpose of the study is to discover the relationship between anxiety and motor performance, then participants must exhibit a range of both anxiety and motor performance. If all participants have low levels of anxiety and have low motor skill, little opportunity exists to find a relationship.

The same idea is true in an experimental study to discover the influence of resistance training on jumping performance. If all the participants have been engaged in resistance training for several years and are excellent jumpers, a resistance-training program will not increase jumping performance. Said another way, the participants must show a range of performance levels on the measures of interest in correlational studies, and the measures must capture the characteristics that the researcher hopes to assess. In experimental studies, the participants must be at the right level to respond to the treatments, the treatments must be intense enough and frequent enough to produce the changes that the researcher hopes to see, and the measuring instruments must be responsive to the treatment changes in the participants.

Summary

This chapter provided an overview of the methods for the research study. We identified the major parts as participants, instruments or apparatuses, procedures, and design and analysis. The major purposes of the four parts of the methods section are to eliminate alternative or rival hypotheses and any explanation for the results (except the hypothesis that the researcher intends to evaluate). The MAXICON principle shows the way to accomplish this: (a) Maximize the true or planned sources of variation, (b) minimize any error or unplanned sources of variation, and (c) control any extraneous sources of variation. In later chapters, we detail how to do this from the viewpoints of statistics and measurement (part II) and of design (part III). We explain the final sections of the thesis, dissertation, or journal article in part IV.

✓ Check Your Understanding

1. Find an experimental study in a research journal consistent with your interests and critique the methods section. Comment on the degree to which the author provided sufficient information concerning the participants (including informed consent), instruments, procedures, and design and analysis.

2. Locate a survey study and compare and contrast its methods description with that in the study in problem 1.

5

Ethical Issues in Research and Scholarship

> When choosing between two evils, I always like to try the one I've never tried before.
>
> —Mae West

As a graduate student, you will encounter many ethical issues in research and scholarship. In this chapter, we draw your attention to many of these issues and provide a framework for discussion and decision making. But the choices are not always clear-cut. Most important to making good decisions is having good information and obtaining advice from trusted faculty members and, under some circumstances, your institution's office of research integrity. Due in part to increasing requirements for research ethics education by the National Institutes of Health and the National Science Foundation (Kalichman, Sweet, & Plemmons, 2014), many institutions offer and some require a course on research ethics. A single chapter in a research methods text cannot adequately cover the diversity of topics relevant to ethical practices in research, but we aim to introduce the new researcher to major topics including identifying misconduct in research, working with faculty and other graduate students, and using humans and animals as participants in research.

I just passed my ethics exam; of course, I cheated.

Seven Areas of Research Misconduct

The U.S. Department of Health and Human Services (HHS) Office of Research Integrity (ORI) has defined *research misconduct* for all U.S. agencies:

> **Research misconduct is fabrication, falsification, or plagiarism in proposing, performing, or reviewing research, or in reporting research results. (*Federal Register*, 42 CFR Part 93, May 17, 2005) It does not include honest error or differences of opinion.**

Shore (1991) identified seven areas in which research misconduct might occur; each is discussed in the following subsections. A 1993 issue of *Quest* (Thomas & Gill) included several thought-provoking articles on ethics in the study of physical activity. Steneck (2007) also is a valuable resource on the topic. For more resources on the responsible conduct of research (RCR), visit the website of the HHS ORI (https://ori.hhs.gov/).

Plagiarism

plagiarism—The act of using other people's ideas, concepts, writings, or drawings as your own.

Plagiarism is the act of using someone else's ideas, writings, or drawings. Of course, this is completely unacceptable in the research process (including writing). Citing a reference or source provides credit for ideas and information, but it does not give credit for words or phrasing. Even when a citation to information from a source is provided, it is not appropriate to use small, medium, or large sections of text from that source. Even changing a few words here or there while maintaining a source's basic sentence structure is insufficient. Directly quoting the words of someone else combined with a reference to the source is one approach to give appropriate credit, but it should be used sparingly. Thus, developing the skill to paraphrase content from other sources is important to avoid plagiarism. Plagiarism carries severe penalties at all institutions. A researcher who is publicly outed for plagiarism carries a lifelong stigma in their profession. No potential outcome is worth the risk involved.

On occasion, a graduate student or faculty member can inadvertently be involved in plagiarism. This situation generally occurs on work that is coauthored. If one author plagiarizes material, the other could be equally punished despite being unaware of the plagiarism. Although no means of protection is surefire (except not working with anyone else), never allow a paper with your name on it to be submitted (or revised) unless you have reviewed and approved the complete paper in its final form.

In scientific writing, originality is also important. Common practice is to circulate manuscripts and drafts of papers among scholars, including graduate students, who are known to be working in a specific area. If ideas, methods, findings, and so on are borrowed from them, proper credit should always be given.

Data Fabrication and Falsification

Based on information in "Publications Output: U.S. Trends and International Comparisons" reported by the National Science Foundation in 2019 (https://ncses.nsf.gov/pubs/nsb20206), articles published in peer-reviewed science and engineering journals increased from 1.8 million (2008) to 2.6 million (2018). Given this huge volume of research, it is not surprising to learn that scientists have occasionally been caught making up or altering research data. Of course, this action is completely unethical, and severe penalties are imposed on people who are caught. Pressure has been particularly intense in medical and health-related research because such research is often expensive, requires external funding, and involves risk. The temptation is great to make a little change here or there or to make up data because "I only need a few more participants, but I am running out of time." The odds of being detected in these types of actions are high, but even if you should get away with it, you will always know that you did it, and you will probably put other people at risk because of your actions.

Although graduate students and faculty may knowingly produce fraudulent research, established scholars are sometimes indirectly involved in scientific misconduct. This may occur when they work with other scientists who produce fraudulent data that follow the predicted outcomes (e.g., as in a grant-funded proposal that suggests which outcomes are probable). In these instances, the established scholar sees exactly what they expect to see in the data. Because this result verifies the hypotheses, the data are assumed to be acceptable. One of the most famous cases involved Nobel laureate David Baltimore, who coauthored a paper with principal authors Thereza Imanishi-Kari and David Weaver that was published in *Cell* in April 1986 (Weaver et al.). Baltimore checked the findings, but he saw in the data the expected outcomes and agreed to submit the paper. The fact that the data were not accurate subsequently led to Baltimore's resignation as president of Rockefeller University. Thus, although Baltimore was not the principal author of the paper, his career was seriously damaged.

Falsification can also occur with cited literature. Graduate students should be careful how they interpret what an author says. Work of other authors should not be "bent" to fit projected hypotheses. For that reason, graduate students should read original sources instead of relying on the interpretations of others, because those interpretations may not follow the original source closely.

Nonpublication of Data

Nonpublication of data refers to the exclusion of a selected subset of data from a study because they do not support the desired outcome. This tactic is sometimes known as "cooking" data. A thin line separates the action of eliminating "bad" data from "cooking" data. Bad data should be caught, if possible, at the time of data acquisition. For example, if a test value seems too large or small and the researcher checks the instrument and finds it out of calibration, eliminating this bad data is good research practice. But deciding that a value is inappropriate when data are being analyzed and then changing the value is cooking the data.

Another term applied to unusual data is **outlier**. Some people have called such data *outright liars*, suggesting that the data are bad. But extreme values now are sometimes trimmed. Just because a score is extreme does not mean that it is based on bad data. Although extreme scores can create problems in data analysis, trimming them automatically is a poor practice (see chapter 10 on nonparametric analysis for one solution to this problem). A researcher should always carefully explore potential causes of a participant's extreme response or outcome and have strong justification for declaring an outlier and eliminating data from further analysis.

outlier—An unrepresentative score; a score that lies outside of standard scores.

Perhaps the most drastic example of nonpublication of data is the failure to publish results that do not support projected hypotheses. Journals are often accused of a publication bias, meaning that only significant results are published. But authors should publish the outcomes of solid research no matter what findings support projected hypotheses. The results from well-planned studies based on theory and previous empirical data have important meaning regardless of whether the predicted outcomes are found.

Faulty Data-Gathering Procedures

A number of unethical activities can occur at the data collection stage of a research project. In particular, graduate students should be aware of issues such as these:

- Continuing with data collection from participants who are not meeting the requirement of the research (e.g., poor effort; failure to adhere to agreements about diet, exercise, rest)
- Using malfunctioning equipment
- Treating participants inappropriately (e.g., failing to follow the procedures and guidelines agreed to by the investigator and the human subjects committee)
- Recording data incorrectly

For example, a doctoral student whom we know was collecting data on running economy in a field setting. Participants returned several times to be video-recorded while repeating a run at varying stride lengths and rates. On the third day of testing, a male runner performed erratically during the run. When the experimenter questioned him, she learned that he was hung over from drinking with his buddies the previous evening. Of course, she wisely discarded the data and scheduled him for another run several days later. Had she not noted the unusual nature of his performance and questioned him carefully, she would have included data with skewed results because the participant was not adhering to previously agreed-upon conditions of the study.

Poor Data Storage and Retention

Based on a survey of authors for 516 biology-related journal articles, Vines and colleagues (2014) estimated that data availability declined 17% per year. The primary reasons for data unavailability were the data had either been lost or stored on media or in formats that were no longer accessible. Data must be stored and maintained as originally recorded and not altered. Lab notebooks documenting data collection and analysis should be stored in a secure and safe location. All original records should be maintained so that the original data are available for examination. Federal agencies and most journals suggest that original data be maintained for at least three years following publication of the results. We recommend that data be maintained indefinitely.

Data ownership is a related and equally important issue. A graduate student who expends countless hours collecting and analyzing data for a thesis or dissertation project might assume that they own those data. Ownership, however, typically belongs to the institution where the data were collected or, in the case of funded research through grants or contracts, to the funding agency. Government funding agencies usually allow the research institution to manage data, whereas private companies are more likely to retain the right to all data generated from contract funding (Steneck, 2007). A graduate student who wishes to take research data with them when leaving an institution should plan to leave the original data with their advisor or department and seek permission of their advisor to retain a copy of the data.

Misleading Authorship

A major ethical issue among researchers involves joint research projects or, more specifically, the publication and presentation of joint research efforts. Generally, the order of authorship for presentations and publications should be based on the researchers' contributions to the project. The first author is usually the researcher who developed the idea and the plan for the research. Other authors are normally listed in the order of their contributions (see Fine & Kurdek, 1993, for a detailed discussion and case studies). Some fields also recognize contributions by a "senior author," typically a well-established investigator closely involved in the research who appears last among the authors of a research article. It is not uncommon for the advisor or supervisor of a graduate student's research to serve as a senior author.

Although establishing whether a person warrants authorship credit and determining author order sounds easy enough, authorship decisions are difficult and can be contentious at times. Sometimes researchers make equal contributions and decide to flip a coin to determine who is listed first. Regardless, decide the order of authors at the beginning of a collaborative effort. This approach saves hard feelings later, when everyone may not agree on whose contribution was most important. If the contributions of various authors change over the course of the research project, discuss a change at that time.

A second issue is who should be an author (see Crase & Rosato, 1992; Grossman & DeVries, 2019; and Venkatraman, 2010, for good discussions of authorship). Studies occasionally have more authors than participants. In fact, sometimes the authors are also the participants. When you look at what participants must go through in some research studies, you can see why only a major professor's graduate students would allow such things to be done to them. Even then, they insist on being listed as authors as a reward. More seriously, the following two rules should help define authorship:

- *Authorship should involve only those who contribute directly to the research project.* According to Steneck (2007, p. 134),

> **authorship is generally limited to individuals who make significant contributions to the work that is reported. This includes anyone who: was intimately involved in the conception and design of the research, assumed responsibility for data collection and interpretation, participated in drafting the publication, and approved the final version of the publication.**
>
> **Many journals and professional organizations have established similar, but not identical, criteria for authorship credit. This listing does not necessarily include the laboratory director or a graduate student's major professor. The only thing that we advocate in the chain letter in the Chain Letter to Increase Publications sidebar is the humor.**

• *Technicians do not necessarily become joint authors.* Graduate students sometimes think that because they collect the data, they should be coauthors. Only when graduate students contribute to the planning, analysis, and writing of the research report are they entitled to be listed as coauthors. Even this rule does not apply to grants that pay graduate students for their work. Good major professors involve their graduate students in all aspects of their research programs; thus, these students frequently serve as coauthors and technicians.

Unacceptable Publication Practices

Another scientific dishonesty concern deals with coauthored or joint publications—specifically, those between the major professor and the graduate student. Major professors do (and should) immediately begin to involve graduate students in their research programs. When this happens, the general guidelines that we suggested earlier apply. But two conflicting forces are at work. A professor's job is to foster and develop students' scholarly ability. At the same time, pressure is increasing on faculty to publish so that they can obtain the benefits of promotions, tenure, outside funding, and merit pay. Being the first (senior) author is a benefit in these endeavors. As a result, faculty members want to be selfless and assist students, but they also feel the pressure to publish. This issue may not affect senior faculty, but it is certainly significant for untenured assistant professors. As mentioned previously, there are no hard-and-fast rules other than that everyone agree before the research is undertaken.

The thesis or dissertation is a special case. By definition, the thesis or dissertation is how a graduate student demonstrates competence to receive a degree. Frequently, for the master's thesis, the major professor supplies the idea, design, and much of the writing and editing. In spite of this, we believe that it should be regarded as the student's work. In other words, the student should be first author on any publication emanating from the thesis research.

Chain Letter to Increase Publications

Dear Colleague:

We are sure that you are aware of the importance of publications in establishing yourself and procuring grants, awards, and well-paying academic positions or chairpersonships. We have devised a way your curriculum vitae can be greatly enhanced with very little effort.

This letter contains a list of names and addresses. Include the top two names as coauthors on your next scholarly paper. Then remove the top name and place your own name at the bottom of the list. Send the revised letter to five colleagues.

If these instructions are followed, by the time your name reaches the top of the list, you will have claim to coauthorship of 15,625 refereed publications. If you break this chain, your next 10 papers will be rejected as lacking in relevance to real-world behavior. Thus, you will be labeled as ecologically invalid by your peers.

Sincerely,

Jerry R. Thomas, Professor

Jack K. Nelson, Professor

List as coauthors:

Jerry R. Thomas

Jack K. Nelson

I.M. Published

U.R. Tenured

C.D. Raise

The dissertation should always be regarded as the student's work, but second authorship for the major professor on either the thesis or the dissertation is acceptable under certain circumstances. The American Psychological Association's *Ethical Principles of Psychologists and Code of Conduct* (2017, section 8.12) has defined these circumstances adequately, and we recommend the use of their guidelines:

- Only second authorship is acceptable for the dissertation supervisor.
- Second authorship may be considered obligatory if the supervisor designates the primary variables, makes major interpretive contributions, or provides the database.
- Second authorship is a courtesy if the supervisor designates the general area of concern, is substantially involved in the development of the design and measurement procedures, or substantially contributes to the writing of the published report.
- Second authorship is not acceptable if the supervisor provides only encouragement, physical facilities, financial support, critiques, or editorial contributions.
- In all instances, agreement should be reviewed before the writing for publication is undertaken and at the time of the submission. If disagreements arise, they should be resolved by a third party using these guidelines.

dual publication— Having the same scientific paper published in more than one journal or other publication; this is generally unethical.

Authors must also be careful about **dual publication**. Sometimes, this circumstance is legitimate; for example, a scientific paper published by one journal may be reprinted by another journal or in a book of readings (this should always be noted). Authors may not, however, publish the same paper in more than one copyrighted original research journal. But what constitutes the "same paper"? Can more than one paper be written from the same set of data? The line is rather hazy. For example, Thomas (1986, pp. iv-v) indicated that

> **frequently, new insights may be gained by evaluating previously reported data from a different perspective. However, reports of this type are always classed as research notes whether the reanalysis is undertaken by the original author or someone else. This does not mean that reports which use data from a number of studies (e.g., meta-analyses, power analyses) are classed as research notes.**

Generally, good scientific practice is to publish all the appropriate data in a single primary publication. For example, if both psychological and physiological data were collected as a result of a specific experiment on training, publishing these separately may not be appropriate. Frequently, the main finding of interest is the interaction between psychological and physiological responses. But in other cases, the volume of data may be so large as to prohibit an inclusive paper. Sometimes the papers can be published as a series; at other times they may be completely separate. Another example is a large-scale study in which a tremendous amount of information is collected (e.g., exercise epidemiological or pedagogical studies). Usually, data are selected from computer records (or videos) to answer a set of related questions for a research report. Researchers may then use a different part (or even an overlapping part) of the data set to address another set of questions. More than one legitimate publication may then result from the same data set, but the authors should identify that more than one paper has been produced. Researchers who do not follow these general types of rules may be viewed as lacking scientific objectivity in their work and certainly as lacking in modesty (Gastel & Day, 2016).

A single primary publication should include either all appropriate data (e.g., psychological, physiological) or a notice that relevant data is published elsewhere.

Most research journals require that the author include a statement that the paper has not been previously published or submitted elsewhere while the journal is considering it. Papers published in one language may not be published as an original paper in a second language.

Ethical Issues Regarding Copyright

Graduate students should be aware of copyright regulations and the concept of fair use as it applies to educational materials. Copyrighted material is often used in theses and dissertations, and this use is acceptable if it is fair and reasonable. Often, graduate students want to use a figure or table in a thesis or dissertation. If you use a table or figure from another source, you must seek permission (see the Sample Copyright Permission Letter sidebar) from the copyright holder (for published papers, this is usually the author, but sometimes it's the research journal) and cite it appropriately (e.g., "used by permission of . . .").

The concept of fair use has four basic rules:

1. *Purpose.* Is the use to be commercial or educational? More leeway is given for educational use such as theses, dissertations, and published research papers.

2. *Nature.* Is copying expected? Copying a journal article for your personal use is expected and reasonable. Copying a complete book or standardized test, however, is not expected and is probably a violation of the fair-use concept.

3. *Amount.* How much is to be copied? The significant issue is the importance of the copied part, both in terms of the significance of the copied portion relative to the whole work and also the proportion of the source being copied.

4. *Effect.* How does copying affect the market for the document? Making a single copy of a journal article has little effect on the market for the journal, but copying a book (or maybe a book chapter) or a standardized test reduces royalties to the author and income for the publisher. That act is not fair use.

Few standard answers apply to fair use. Fair use is a flexible idea (or alternatively, a statement that cannot be interpreted rigidly). It's better to be safe than sorry when using material in your thesis or dissertation. If you have any doubt, seek permission.

Model for Considering Scientific Misconduct

Intention is often used as the basis for differentiating between scientific misconduct and mistakes. Drowatzky (1993, 1996) noted that the following model is often suggested:

Scientific misconduct → sanctions

Scientific mistakes → remedial activities

Sanctions for Scientific Misconduct

Sanctions are often imposed on those who are fraudulent in their scientific work. Internal sanctions imposed by the researcher's university have included the following:

- Restriction of academic duties
- Termination of work on the project
- Reduction in professorial rank
- Fines to cover costs
- Termination from the university
- Salary freeze
- Promotion freeze
- Supervision of future grant submissions
- Verbal reprimands
- Letter of reprimand (either included or not included in the permanent record)
- A monitoring of research with prior review of publications

Sample Copyright Permission Letter

Date

Permissions Editor (or name of an individual author)

Publisher (not needed if an individual)

Address

Dear _____:

I am preparing my thesis/dissertation, tentatively titled _____.

I would like permission to use the following material: _____

Title of article in journal, book, book chapter: _____

Author of article, book, or book chapter: _____

Title of journal or book (include volume and issue number of journal): _____

Editor if edited book: _____

Year of publication: _____

Place of publication and name of publisher of book or journal: _____

Copyright year and holder: _____

Page number(s) on which material appears: _____   □ to be reprinted   □ to be adapted

A COPY OF THE MATERIAL IS ATTACHED.

Table, figure, or page number in my thesis or dissertation: _____

I request permission for nonexclusive rights in all languages throughout the world. I will, of course, cite a standard source line, including complete bibliographic data. If you have specific credit line requirements, please make them known in the space provided below. _____

A duplicate copy of this form is for your files. Your prompt cooperation will be appreciated.

_____

If permission is granted, please sign the release below and return to:

YOUR SIGNATURE: _____

YOUR NAME AND ADDRESS: _____

• •

Permission granted, Signature: _____

Address: _____

Date: _____

Besides internal university sanctions for scientific fraud, sanctions may be imposed by the agency that funded the research, scholarly journals that published the work, and related scholarly or professional groups. In recent years, external sanctions have included the following:

- Revocation of prior publications
- Letters to offended parties

- Prohibition from obtaining outside grants
- Discontinuance of service to outside agencies
- Release of information to agencies and professional organizations
- Fines to cover costs
- Referral to legal system for further actions, including the potential for imprisonment

Responsibilities of Graduate Students

As a graduate student, you must become concerned with ethical issues. Of course, the issues are important for much more than just the biological science areas (such as exercise physiology, biomechanics, motor behavior, health promotion and exercise, and exercise psychology) or behavioral science areas (such as sport sociology, sport philosophy, sport history, sport psychology, and physical education pedagogy). Fraud, misrepresentation, inaccurate interpretation of data, plagiarism, unfair authorship issues, and unethical publications practices are problems that extend to any area of scholarship. Although these practices sometimes occur simply because the person is unethical, often pressures that exist in our system of higher education tempt researchers to behave unethically. Following are some examples of such pressures:

- The need to obtain external funding for research
- Pressure to publish scholarly findings
- The need to complete graduate degree work
- The desire to obtain rewards in higher education (e.g., promotion, merit raises)

Academic units should be encouraged to develop mechanisms for discussing ethical scholarship issues with graduate students. The approach might include seminars for graduate students focused on these issues. But some systematic means is needed to bring the issues to the forefront for discussion.

Philosophical Positions Underlying Ethical Issues

A person's basic philosophical position on ethical issues drives the decision-making process in research. Drowatzky (1993, 1996) summarized the ethical views that underpin the decision-making process:

- The individual is precious, and the individual's benefit takes precedence over society.
- Equality is of utmost importance, and everyone must be treated equally.
- Fairness is the overriding guide to ethics, and all decisions must be based on fairness.
- The welfare of society takes precedence over that of the individual, and all must be done for the benefit of society.
- Truth, defined as being true, genuine, and conforming to reality, is the basis for decision making.

Of course, several statements in this list are in direct conflict with each other and will lead to substantially different decisions depending on one's view. Discussing and evaluating these statements and what they imply about decisions in scholarship should enhance graduate students' understanding of important issues. The Philosophical Statements to Think About sidebar lists a few more philosophical statements for pondering.

Finally, reading some of the literature on fraud and misconduct in research is a sobering experience for anyone. A notable example is the special issue of *Ethics and Behavior* titled "Whistleblowing and Scientific Misconduct" (vol. 3, no. 1, 1993). This issue gives a fascinating account of the David Baltimore and Herbert Needleman cases, including overviews and responses by the whistleblowers and those accused of scientific misconduct. Drowatzky (1996) also offered numerous examples of misconduct and the problems associated with it. Finally, the website of the Office of Research Integrity provides a summary of numerous cases of research misconduct, one of which is from the physical activity field (https://ori.hhs.gov).

Working With Faculty

Ethical considerations among researchers and ethical factors in the graduate student–major professor relationship are two important topics (for more detailed discussions, see Löfström & Pyhältö, 2019; Roberts, 1993; and Roberts et al., 2001). Major professors should treat graduate students as colleagues. If we want our students to be scholars when they complete their graduate work, then we should treat them like scholars from the start, because graduate students do not become scholars on receipt of a degree. By the same token, graduate students must act like responsible scholars. This means producing careful, thorough, and high-quality work.

Selecting a Major Professor

Students should try to select major professors who share their views in their area of interest. Master's students frequently choose an institution based on location or the promise of financial aid. Doctoral students, on the other hand, should select an institution by evaluating the program's quality and the faculty in their area of specialization (see Hollingsworth & Fassinger, 2002, for a discussion). A listing and electronic link to doctoral programs in physical activity is available at https://www.americankinesiology.org. Do not choose your major professor hastily. If you are already at the institution, carefully evaluate the specializations available in your interest areas. Ask questions about faculty and learn whether they publish in those areas. Read some of their publications and determine your interest. What financial support, such as laboratories and equipment, is given to those areas? Also, talk to fellow graduate students. Finally, talk with the faculty members to determine how effectively you will be able to work with them.

We advocate a mentor model for preparing graduate students (particularly doctoral students) in kinesiology, physical education, exercise science, and sport science. To become a good researcher (or a good clinician), a student needs a one-on-one student–faculty relationship. This means several things about graduate students and graduate faculty.

First, graduate students need to be full-time students to develop the research and clinical skills needed for success in research and teaching. They need to work with a mentor in an

---

Philosophical Statements to Think About

1. The professor discovered that her theory of earthquakes was on shaky ground.
2. A backward poet writes inverse.
3. I planted some birdseed and a bird came up. What do I do with it?
4. I went to San Francisco and found someone's heart. Now what?
5. Do protons have mass? I didn't even know they were Catholic.
6. What is a free gift? Aren't all gifts free?
7. I used to be indecisive. Now I'm not sure.
8. How can there be self-help groups?
9. Is it my imagination, or do buffalo wings taste like chicken?

ongoing research program. This arrangement lends continuity to research efforts and pulls graduate students together into effective research teams. Theses and dissertation topics arise naturally from these types of settings. Additionally, more senior students become models and can offer assistance to novice graduate students. Expertise is acquired by watching experts, by working with them, and then by practicing the techniques acquired.

Conversely, for faculty members to be good mentors, they must have active research programs. Appropriate facilities and equipment must be available, and faculty members must have time to devote to research and graduate student mentoring. Potential graduate students should carefully investigate the situations into which they will place themselves, especially if they have a major interest in research (for a good description of mentors, see Bird, 2001; Newell, 1987).

If you are not yet enrolled at an institution, find out which institutions offer the specialties you are interested in. Request information from their graduate schools and departments and look at their websites. Read the appropriate journals (over the past 5 to 10 years) and see which faculty are publishing. After you narrow your list, explore financial support and plan a visit. Speak to the graduate coordinator for the department and the faculty in your area of interest. Sometimes you can meet faculty at conventions, such as those of SHAPE America (national or district meetings), the American College of Sports Medicine, the American Society of Biomechanics, and the North American Society for Psychology of Sport and Physical Activity.

After you select a major professor, you must select a committee. Normally, the master's or doctoral committee is selected in consultation with your major professor. The committee selected should be one that can contribute to the planning and evaluation of your work, not one that might be the easiest. It is preferable to wait a semester or quarter or two (if you can) before selecting a final committee. This interval gives you the opportunity to have several committee members as teachers and to evaluate whether you have common interests.

Changing Your Major Professor

What happens, however, if you have a major professor (or committee member) who is not ideal for you? First, evaluate the reason. You need not be best friends, but it is important that you and your major professor are striving for the same goals. Sometimes students' interests change. Sometimes people just cannot get along. If handled professionally, however, this situation should not be a problem. Go to your major professor and explain the situation as you perceive it and offer them an opportunity to respond. Of course, the conflict may be more personal. If so, use an objective and professional approach. If you cannot, or if this does not produce satisfactory results, the best recourse may be to seek the advice of the graduate coordinator or department chairperson.

Protecting Human Participants

Most research in the study of physical activity deals with humans, sometimes focusing on special populations (e.g., children, older adults, people with a disability). Therefore, the researcher must be concerned about any circumstances in the research setting or activity that could subject participants to risks or discomforts. Risks and discomforts do not refer only to physical injury or harm but also to psychological, emotional, or social risks or harm. Of course, researchers always run the risk of creating a problem. What must be balanced is the degree of risk, the participants' rights, and the potential value of the research in contributing to knowledge, to the development of technology, and to the improvement of people's lives.

There will always be a risk of harm. Researchers should balance the degree of risk, rights of participants, and value of the study.

What Should Research Participants Expect?

The HHS Office of Human Research Protections (OHRP) is responsible for ensuring the rights and welfare of human research participants are maintained. In 1991, a set of regulations governing the protection of research participants known as the "Common Rule" was adopted (Steneck, 2007). Revisions to these federal regulations that strengthened the protections for participants were made in 2017 and became effective in July 2018 (see section 45 CFR 46 of the Federal Registry). Prior to involving a person in research, an investigator must *inform* the potential participant of the details of the research and obtain the *consent* of the participant. Thus, this is referred to as *informed consent.*

Most institutions regulate this process of protecting participants in two ways. First, a researcher must complete training about protecting the rights and well-being of participants and obtaining informed consent prior to becoming an investigator on a research protocol. This training is usually done online. Second, researchers are required to submit a detailed application describing their project to their Institutional Review Board (IRB) for review and approval prior to initiating the project. Failure to follow these requirements will lead to loss of any information and data collected prior to IRB approval and may lead to formal reprimand or sanctions imposed on the researcher. The informed consent form to be presented to a potential participant must be included with the application. If participants are minors, then the researcher must obtain their parents' permission and the children's assent if they are old enough to understand.

Every university has slightly different guidelines for research with human participants and for obtaining consent. You should get these guidelines early and make sure your informed consent forms conform to them. Further, you should plan to submit your IRB application well in advance of the date you want to begin participant recruitment. IRB review processes typically take 4 to 6 wk, but they can be longer for poorly prepared applications and complicated protocols.

Federal law requires researchers to protect participants' rights and well-being.

Qualitative research (discussed in detail in chapter 19) lends itself to some potential ethical problems because of the close, personal interaction with participants. The researcher often spends a great deal of time with participants, getting to know them and asking them to share their thoughts and perceptions. Griffin and Templin (1989) raised ethical concerns about whether to share field notes, how to protect a participant's self-esteem without compromising accuracy in the research report when the two are in conflict, and what to do if you are told about (or observe) something illegal or immoral while collecting data. Locke, Spirduso, and Silverman (2014) discussed these issues further.

fieldwork—Methodology common in qualitative research in which data are gathered in natural settings.

There are no easy answers to situational ethics in **fieldwork**. Qualitative research sometimes includes people involved in illegal activities, such as people who have drug addictions or are members of unlawful groups or organizations, for example. Informed consent is impossible in some circumstances. Punch (1986, p. 36) made this point when he described his research with police. The patrol car that he was riding in was directed to a fight. As the police officers jumped out and started wrestling with the assailants, Punch wondered whether he should get involved by yelling "Freeze!" and then thrusting his head between the entangled limbs while chanting the Miranda rights to the best of his ability. Similarly, when Powermaker (cited by Punch) came in direct contact with a lynch mob, should she have flashed her academic identity card and explained to the crowd the nature of her presence? By these two examples, we are not implying that qualitative researchers are exempt from considerations such as informed consent and deception. We are simply pointing out that certain types of qualitative research situations present special ethical problems. We invite you to read the discussion by Punch and consult some of the sources he cited concerning this issue.

People with disabilities present a special issue as research participants. A participant with a disability is protected under the U.S. Right to Privacy Act. Thus, institutions providing care are prohibited from releasing the names of people with disabilities as potential research participants. The researcher must contact the institution about possible participants. The institution must then request permission from the participant or their guardians to release the participant's name and the nature of the disability to the researcher. If the participant or guardians approve, the institution allows the researcher to contact the participant or guardians to seek approval for the research to be undertaken. Although this procedure is cumbersome and varies from state to state, people clearly have the right not to be cited in studies and distinguished as having a disability unless they so choose.

Informed Consent

Obtaining informed consent from a participant is a critical step of the research process. Information about your project must be presented in language that is understandable to participants. The following are the basic elements of informed consent as outlined by the Common Rule (45 CFR 46) that must be shared with a potential participant:

- A description of the purposes of the research, the expected time commitment of the participant, and the procedures to be followed (i.e., what will the participant be asked to do and what will be done to the participant).
- A description of foreseeable discomforts and risks to the participant.
- A description of the benefits to the participant, if any, to be expected from participation.
- A disclosure of appropriate alternative procedures that would be advantageous for the participant. (This is often particularly relevant for student participants who may be offered course credit for their participation. Alternative ways to earn course credit must be provided.)
- A description of how confidentiality of records identifying a participant will be maintained.
- If the research involves more than minimal risk, descriptions of whether any compensation and medical treatments are available if injury occurs, what those consist of, and where additional information can be obtained.
- If participants will be compensated for their participation, a description of that compensation and how it will be provided.
- Contact information for people who can answer questions about the research (usually the lead investigator for the study) and a participant's rights (usually a contact within the IRB).
- A statement indicating participation is voluntary, that the participant may withdraw from the study at any time, and that there will be no penalties or loss of benefits brought against a participant who chooses to withdraw.
- For research that involves the collection of identifiable private information or biospecimens from a participant, a statement indicating how these may or may not be used in future research.

The researcher is required to comply with any institutional guidelines for the protection of human participants and for informed consent. A description of this compliance should be included under the participant heading in the methods section of the thesis or dissertation. Most journals also require a statement with regard to this issue. The form used for informed consent is normally placed in an appendix of the thesis or dissertation.

If I cover my ears, can I ignore this ethical issue?

© Jose Luis Pelaez Inc

Protecting Animal Subjects

Research involving animals is also carefully regulated. At the institutional level, this is usually coordinated by the Institutional Animal Care and Use Committee (IACUC), which has the responsibility to ensure the humane use of animals in research and teaching. Although animals may benefit from animal research, most research involving animals as participants is conducted for the benefit of humans. In addition, animals are obviously unable to provide consent (Steneck, 2007). Matt (1993), in her article "Ethical Issues in Animal Research," pointed out that this issue is not new, having been discussed in Europe for over 400 years and in the United States for over 100 years. According to Matt, the ground rules were established long ago when Descartes indicated that it was justifiable to use animals in research because they cannot reason and are therefore lower in the order of things than humans. Bentham (1970), however, said that the issue is not whether animals can think and reason but whether they perceive pain and suffer.

As Matt argued, far fewer animals are used in research than are slaughtered for food, held in zoos, and killed as unwanted pets in animal shelters. In fact, animal studies may have more stringent criteria for approval than studies with humans. Institutional review boards typically require that investigators demonstrate that animal studies add significant knowledge to the literature and are not replications, a requirement not placed on studies using humans as participants.

If animals are well treated, is their use in research justified? Matt (1993, pp. 46-47) says yes if the purposes of the research fall into one of five categories:

(1) drug testing, such as the development and testing of AIDS drugs; (2) animal models of disease, such as the development of animal models of arthritis, diabetes, iron deficiency, auto-immune dysfunction, and aging; (3) basic research, focused on examining and elucidating mechanisms at a level of definition not possible in human models; (4) education of undergraduate and graduate students in laboratories and lectures, with experience and information gained from the use of animal models; and (5) development of surgical

techniques, used extensively in the training of medical students and the testing of new surgical devices and procedures.

A careful consideration of these categories suggests few suitable alternatives (but see Zelaznik, 1993, for a discussion).

If animals are used as subjects for research studies in kinesiology, institutions require adherence to *Guide for the Care and Use of Laboratory Animals* (8th ed.) published by the National Research Council of the National Academies (NRC), as detailed in the Animal Welfare Act (PL 89-544), which was established in 1966. The Animal Welfare Act has been revised numerous times since its introduction and is overseen by the USDA. Also relevant is the U.S. Public Health Service policy Humane Care and Use of Laboratory Animals. Most institutions also support the rules and procedures for recommended care of laboratory animals outlined by the Association for Assessment and Accreditation of Laboratory Animal Care (AAALAC) International.

All these documents recognize that for advancements to be made in human and animal research, animals must be used. These animals must be well tended, and if their use results in the animals being incapacitated or sacrificed, this procedure must be done humanely.

> Although research advancement requires animal testing, animals must be well cared for and treated humanely.

Summary

This chapter addressed ethical issues that affect graduate students in their research and scholarly activities. We identified ethical issues and set the stage for you to think about and discuss these values as they influence your graduate and scholarly life.

Points that typically arise in the area of research misconduct include plagiarism, fabrication and falsification of data, nonpublication of data, data-gathering problems, data storage and retention issues, authorship controversies, and publication practices. We discussed copyright issues in research and publication. We also presented the model most frequently used to deal with research misconduct and some of the internal and external sanctions that have been imposed on people found guilty of scientific misconduct.

We discussed ethical and procedural issues in working relationships. How should graduate students select a major professor and committee members? Should students seek new major professors or committee members if they are unable to work effectively or amicably with their original choices?

Finally, we discussed ethics and procedures in the use of human and animal participants. These topics included policies and procedures for the protection of human and animal participants and obtaining informed consent from human participants.

✓ Check Your Understanding

1. Go online to your university's research office and find the instructions and forms for requesting the use of humans in research. Prepare a summary of the procedures for class discussion.

2. The following are examples or case studies for a variety of potential research misconduct scenarios. With the exception of the plagiarism example, names and institutions used should not be taken seriously; we are only trying to bring humor to our friends.

Plagiarism

- What is plagiarism? Can you recognize it when you see it? (We acknowledge Dr. Charlotte Bronson, Professor Emeritus and former Associate Vice President for Research at Iowa State University, who created the following case study on plagiarism.)

Case study. The following quote comes from a research publication entitled "Disease lesion mimics of maize: a model for cell death in plants" by Johal, Hulbert, & Briggs (1995): "Presently, the biological significance of this lesion mimicry is not clear, although suggestions have been made that they may represent defects in the plants' recognition of, or response to, pathogens" (p. 685).Consider whether each of the following examples of summarizing the content by Johal and colleagues represents fair use or plagiarism:

- The biological significance of lesion mimicry in plants is not currently known, although some researchers believe that they may represent defects in the ability of plants to recognize or respond to pathogens. *Is this plagiarism?*
- Currently, the biological significance of lesion mimicry in plants is not known, although suggestions have been made that they may represent defects in the plants' recognition of, or response to, pathogens (Johal et al., 1995). *Is this plagiarism?*
- Disease-like lesions in plants may be due to mutations in genes controlling the ability of plants to defend themselves against pathogens (Johal et al., 1995). *Is this plagiarism?*

Fabrication or Falsification of Data

Case study. Professor Wade has strength-training data on 20 older adult participants. As he was madly processing data to meet the ACSM abstract deadline, Professor Wade realized that his sample did not show a significant increase in strength. Being disappointed about this finding, he examined his data more closely and noticed that 15 participants appeared to increase strength substantially whereas five participants actually showed strength declines. Professor Wade concluded that these five participants must not have adhered to the training program, and he decided to drop them from the study. Using data on only the remaining 15 participants, he could now demonstrate statistically significant improvements in strength, and he wrote his abstract based on those 15 participants.

- Did Professor Wade act ethically?
- What is an outlier? How do you define an outlier?
- How long should you keep raw data available for others to review following publication?
- Are you obligated to provide your raw data for examination on request?

Authorship of Presentations and Publications

Case study. Professor Martin is well known for her research on the effects of exercise on bone density. In 2016, she was awarded a five-year research grant from NIH to study bone density. Graduate Assistant (GA) Jackson began work under Professor Martin's mentorship in 2018 and was immediately assigned the task of running one of the experiments on bone density that was outlined in the grant proposal. Professor Martin could never be found in the laboratory for data collection, but she regularly held lab meetings during which she discussed progress on data collection with GA Jackson. After completing data collection, GA Jackson organized and presented the data to Professor Martin, who was pleased with the data. Professor Martin then assigned GA Jackson the task of drafting a manuscript based on the data. After several iterations, both were pleased with the final product and concluded that the manuscript was ready for submission. Unfortunately, they had had no conversations about authorship.

- Should the authorship be Martin and Jackson or Jackson and Martin? Be prepared to present your justification for your preference.
- Professor Martin indicated to GA Jackson during the manuscript preparation phase that she would like to include GAs Powers, Cauraugh, Stelmach, and Thomas (the

other four GAs in Martin's lab) as coauthors because they had been involved in other aspects of the funded research (after all, they needed publications on their curricula vitae to secure prestigious postdoc positions).

- Is this request by Professor Martin reasonable? As a side note, an interesting exercise is to compute some oddball statistics for manuscripts with long lists of authors. Consider (a) words per author, and (b) the number of participants per author, and then consider just how significant each author's contributions might be.
- If the project for which GA Jackson collected the data was for his dissertation topic (an offshoot of Professor Martin's funded research line), should the authorship be Martin and Jackson, Jackson and Martin, or just Jackson?
- If Professor Martin hired Technician Magill to assist GA Jackson with data processing (for the original scenario), should the authorship be Martin and Jackson; Martin, Jackson, and Magill; Jackson and Martin; or Jackson, Martin, and Magill?
- How important is having some agreement about authorship at the beginning of the writing process?

Other Publication and Presentation Issues

Case study. Professor Sharp, an expert in forensic biomechanics, presented a research paper on the relationship between footprint spacing and body size at the 2018 annual meeting of American College of Sports Medicine (ACSM). Shortly after the meeting, during a moment of free time, he saw an announcement in one of his research magazines for a meeting titled "The Science of Forensic Biomechanics," sponsored by the Society of Police Detectives. He submitted the same abstract used at the ACSM annual meeting and subsequently presented the same research paper at the SFB meeting.

- Is this an ethical and acceptable practice?

Case study. Professor Sharp submitted a manuscript titled "The relationship between footprint spacing and body size" to the *Journal of Biomechanics*. He then used the same data but interpreted them from a somewhat different perspective. He prepared another manuscript titled "The relationship between footprint spacing and walking speed" and submitted it to the *Journal of Forensic Science* while his *JoB* manuscript was still under review.

- Did Professor Sharp do anything wrong?

Case study. Assistant Professor Roberts, a new faculty member at the University of Reallycold, has just completed a broad, multidisciplinary dissertation on the benefits of endurance training on psychological and physiological markers of health and well-being and biomechanical aspects of the running pattern. In each of three major areas (psychology, physiology, and biomechanics), she has four major dependent variables. Knowing that she needs a healthy publication record when she reaches the promotion and tenure review in five years, she decides to create 12 publications from her dissertation research ("The Benefits of Endurance Training on . . .").

- How do you react to her strategy for dealing with the dictum Publish or Perish?

Case study. Assistant Professor French, a new faculty member at the University of South Columbia, completed his doctorate at Big Time University under the direction of Dr. Samoht. His dissertation topic fit within the general scope of research activities being pursued by Dr. Samoht but reflected a unique focus, one that Dr. Samoht had not considered before French's work as a doctoral student. French submits a manuscript to *Medicine & Science in Sports & Exercise*, listing himself as the sole author of the paper and the University of South Columbia as his affiliation.

- Has French behaved ethically?
- Has he given appropriate credit to: (a) Professor Samoht, and (b) Big Time University?

Case study. Professor I.M. Kingman is an icon in the field of gerontology. He is the director of the Institute of Gerontological Research (IGR) at Jellystone University, a highly funded research lab in which several faculty, postdocs, and graduate students work. Professor Kingman requires that he be listed as an author on all manuscripts based on research completed at the IGR.

- Is Professor Kingman justified in his demand, or is this an example of ego gone wild?

Changing Your Major Professor

Case study. Graduate Assistant (GA) Lee has a strong interest in the mechanical behavior of muscle and was accepted into the exercise science doctoral program to study with Professor Silverman, an expert on muscle mechanics. After one year in the program, the chemistry between GA Lee and Professor Silverman is not great. In addition, GA Lee notices that one of his GA colleagues is getting some travel support from her mentor, Professor Moran, an expert on muscle energetics. GA Lee wants to continue to study muscle mechanics but thinks that he would like to do so under Professor Moran's direction rather than Professor Silverman's.

- Should GA Lee pursue a mentor change? If so, how should he go about doing so? What are GA Lee's obligations to Professor Silverman?

Juggling Multiple Job Offers

Case study. Postdoc Gallagher is pursuing several tenure-track job opportunities in kinesiology, following years of training as a graduate student and postdoctoral fellow. Her first choice is a faculty position at the University of Minnetonka. UM has an excellent reputation in kinesiology, the job description seems to match Gallagher's interests well, superb lab facilities are already in place, and her family lives within a few hours' drive. While Gallagher waits to hear from UM, she gets an invitation for an interview at Gator University in Florida. She accepts the invitation, and the interview goes well. She likes the faculty at Gator and thinks that there is good potential to build a solid research program at Gator despite the fact that the kinesiology program there and the university are not considered top tier. Gator calls and offers her a faculty position with a modest startup package. She is not excited about living in Florida (she is not fond of gators, mosquitoes, and humidity), but no other offers are pending, and time is getting short in the job search process. Thus, she accepts in writing the faculty offer from Gator. Two days after emailing her acceptance response to Gator, Gallagher receives an invitation to interview at the University of Minnetonka.

- How should Gallagher deal with this situation?

Assume for the moment that Gallagher accepted the invitation from UM to interview. While visiting UM, she is confronted by Professor Smith, who has a good friend and colleague at Gator. Professor Smith heard from his Gator colleague during a friendly telephone conversation that Gallagher had interviewed at Gator and had accepted their job offer.

- How would you expect the faculties at UM and Gator to react to this situation?
- If Gallagher had not yet received a written offer from Gator University but had only orally accepted an offer from Gator (i.e., she was waiting for the official offer to reach her by mail), would the situation be any different?

Statistical and Measurement Concepts in Research

Facts are stubborn, but statistics are more pliable.

—Mark Twain

In the following six chapters, we present some basic statistical and measurement techniques that are frequently used in physical activity research. We give more attention here to the basic statistical techniques than to other, more-complex methods. The underlying concepts of statistical techniques have been emphasized rather than any derivation of formulas or extensive computations. Because an understanding of how the basic statistical techniques work facilitates a grasp of the more advanced procedures, we have provided the computational procedures for most of the basic statistics and examples of their use. We use the procedures from Statistical Package for the Social Sciences (SPSS) to analyze data on 30 professional golf players from 2019's PGA Tour.

Chapter 6 discusses the need for statistics. We describe types of sampling procedures and summarize the basic statistics used in describing data, such as distributions, measures of central tendency, and measures of variation. Statistics can reveal two things about data: significance and magnitude of the effect.

Chapter 7 introduces the concepts of probability, effect size, and—most important—power. Using power analysis to evaluate studies you read and studies you plan is essential.

Chapter 8 pertains to interrelationships of variables. We review correlational techniques, such as the Pearson r for the relationship between only two variables. We explain partial correlation as a technique by which one can determine the correlation between two variables while holding the influence of additional variables constant. The use of correlation for prediction is discussed when using more than one variable to predict a criterion (multiple regression, logistic regression, and discriminant function analysis). Finally, we briefly present multivariate techniques of correlation: canonical correlation, factor analysis, and structural modeling.

Chapter 9 focuses on statistical techniques for comparing treatment effects on groups, such as different training methods or different samples. The simplest comparison of differences between two groups is the t test. Next, we describe analysis of variance as a means of testing the significance of differences between two or more groups. We also discuss the use of factorial analysis of variance, by which two or more independent variables can be compared. The use of repeated measures is common in our field, and we describe how they are analyzed. In addition, we provide an overview of multivariate techniques in this area, including discriminant analysis and multivariate analysis of variance.

Chapter 10 provides information on nonparametric techniques for data analysis. These are procedures in which the data fail to meet one or more of the basic assumptions of parametric techniques described in chapters 8 and 9. We present an approach to nonparametrics that parallels the procedures described in chapters 8 and 9. Thus, you do not need to learn a new set of statistics to use the techniques.

Finally, chapter 11 reviews many of the measurement issues that apply when conducting research. We focus on the validity and reliability of dependent variables. We present a short overview of measurement issues regarding data collected about physical performance, although complete coverage of this topic is not possible. In much research on physical activity, dependent variables may be affective and knowledge measures. Thus, we discuss these types of dependent variables.

After reading the six chapters in part II, you will not be a statistician or a measurement expert (unless you were one before you started). But if you read and study these chapters carefully and perhaps explore some of the references further, you should be able to comprehend the statistical analysis and measurement techniques of most research studies.

Becoming Acquainted With Statistical Concepts

The idea of learning statistics frightens many people. If you are intimidated, you need not be. Statistics is one of the few ways that data can be reported uniformly to allow relevant, accurate conclusions and comparisons to be made. Statistics are methodical, logical, and necessary, not random, inconsistent, or terrifying. Our approach to teaching statistics in this book is to acquaint you with the basic concepts and give you a working knowledge; it is not our purpose to make you a statistician, especially given this well-known saying that Mark Twain popularized: "There are three types of lies: lies, damned lies, and statistics."

> Don't be mean; be above average.

Why We Need Statistics

As a researcher and scientist, you have the responsibility of developing a research question, designing a study to answer that question, conducting analyses, and then interpreting your findings. In the stage where you are conducting analyses, you will need sufficient statistical expertise to identify the appropriate statistical tool, execute the analysis, and understand the findings. As Brownstein, Louis, O'Hagan, & Pendergast (2019, p. 59) explain, "The expertise of the statistician is needed both to understand the nuances of proper interpretation of the analytic results in context of the executed study, assumptions made, and modeling used and to guard against overinterpretation" (p. 59).

When we conduct a research study using quantitative techniques, it means we are collecting data from a sample (or samples) that can be expressed in numerical form. When we provide the mean (or average) for data from a sample, that is a statistic. Specifically, it is a descriptive statistic that *describes* the general nature of the data and so is helpful to us in terms of providing a quick understanding of that data. We also use statistics to make inferences from data we have collected from a sample to the larger population from which that sample was taken. Known as inferential statistics, in a very general sense, they can be used for the following: (a) comparing sample data to a standard (e.g., comparing the $\dot{V}O_2max$ for a sample of male college students to the average $\dot{V}O_2max$ of men in the United States),

(b) testing for relationships (e.g., assessing the relationship between aerobic fitness and cognitive performance in a sample), and (c) comparing between groups (e.g., comparing the means between a treatment group and a control group). To make such comparisons or to test these relationships, we use statistical tools to determine whether observed differences or relationships are likely due to chance or are indicative of a reliable relationship or a reliable difference that would then be expected to be evident in the larger population.

One intriguing aspect of statistical tests is that although each is unique in terms of the specific research design for which it is appropriate, all statistical tests are based upon the same general comparison, which is between the variance we can explain (true variance) and the variance we cannot explain (error variance). True variance is the variability among scores that can be logically attributed to the independent variable we are interested in. Error variance is the variability that cannot be explained by the independent variable. Instead, it is a consequence of things like measurement error or normal, individual differences between participants. Thus, for all quantitative statistics tests, the variance associated with statistical procedures is defined as follows:

Total (observed) variance = true variance + error variance

Thus, all statistical estimates are one of two arrangements of the variance formula:

Significance tests (t, F, and X^2) = true variance ÷ error variance

Correlation (r^2, R^2, % variance accounted for) = true variance ÷ total variance

Statistics is simply an objective means of interpreting a collection of observations. Various statistical techniques are necessary for describing the characteristics of data (descriptive statistics) and for testing relationships or differences between sets of data (inferential statistics). For example, if height and a standing long jump score were measured for each student in a seventh-grade class, you could add all the heights and then divide the sum by the number of students. The result (a statistic to represent the average height) is the **mean**, $M = \Sigma X / N$, where Σ means sum, X = each student's height, and N = number of students (considering this to be the entire population of interest); read this as "the sum of all Xs divided by N." The mean (M) describes the average height in the class. The mean is one measure of **central tendency**, meaning it is a single characteristic or statistic that represents the data.

An example of testing relationships of sets of data would be to measure the degree of association between height and the scores on the standing long jump. You might hypothesize that taller people can jump farther than shorter people. By plotting the scores (figure 6.1), you can see that taller people generally do jump farther than shorter people. But note that the relationship is not perfect. If it were, the scores would begin in the lower left-hand corner of the figure and proceed diagonally in a straight line toward the upper right-hand corner. One measure of the degree of association between two variables is called the **Pearson product moment coefficient of correlation** (also known as *Pearson r, interclass correlation*, or *simple correlation*). When two variables are unrelated, their correlation is approximately zero. In figure 6.1, the two variables (height and standing long jump scores) have a moderately positive correlation (r is probably between .40 and .60). We know it is positive because both variables change in the same direction. That is, as height increases, standing long jump distance also increases. We know it is moderate because the line is closer to a 45-degree angle than it is to a flat, horizontal line. We use the statistical tools of correlation and regression to determine if the observed relationship is due to chance or is a real relationship. Relationships and correlation are discussed in greater detail in chapter 8, but for now, you should know that researchers frequently want to investigate the relationships between variables.

mean—A statistical measure of central tendency that is the average score of a group of scores.

central tendency—A single value that provides an estimate of a group of values (data).

Pearson product moment coefficient of correlation—The most commonly used method of testing a relationship between two variables; also called *interclass correlation, simple correlation*, or *Pearson r*.

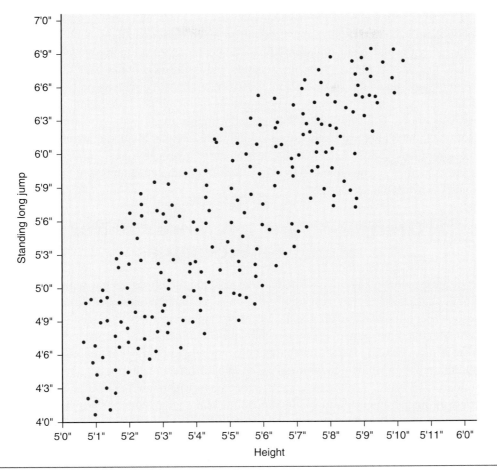

Figure 6.1   Relationship of height to standing long jump.

In addition to descriptive and correlational techniques, a third category of statistical techniques is used to measure differences between groups. Suppose you believe that weight training of the legs will increase the distance a person can jump. You divide a seventh-grade class into two groups and have one group participate for eight weeks in a weight-training program designed to develop leg strength (treatment group). The other students continue their regular activities (control group). You want to know whether the independent variable (group with two levels: weight training and regular activity, or treatment and control) produces a change in the dependent variable (standing long jump score). Therefore, you measure the two groups' scores in the standing long jump at the end of the eight weeks (treatment period) and compare their average performances. Here, a statistical technique to assess differences of two independent groups, a *t* test, would be used. By calculating *t*, you can judge whether the two groups were significantly different on their average long jump scores. Ways for assessing differences between groups are discussed in chapter 9.

t **test**—A statistical technique to assess differences between two groups.

Use of Computers in Statistical Analysis

Computers are extremely helpful in the calculation of statistics. Computers do not make the mistakes that can occur in hand calculations, and they are many times faster. Numerous software programs have been written to calculate statistics.

A number of statistical software packages are available, but the most popular ones for research are the Statistical Package for the Social Sciences (SPSS), the Statistical Analysis

System (SAS), and Stata. A free statistical software package, R, is also gaining in popularity. Most colleges and universities have one or more of these packages, which can be used on a personal computer. Your university or college has information about software options available to you as a student. Many universities and colleges provide user services or user consultant centers that offer advice about software availability and instructions for how to use it.

Most institutions and many kinesiology departments teach statistics courses in which these software packages are used. Statistics departments or other departments may also offer consulting services to help with the use of statistics for research projects. You should investigate the services your institution offers. Senior graduate students and faculty can also advise you on available services.

Description and Inference Are Not Statistical Techniques

inference—The generalization of results from a sample to some larger group.

sample—A group of participants, treatments, or situations selected from a larger population. The symbols used for sample statistics are Arabic letters (e.g., *n* for sample size).

At the beginning of this chapter, we stated that statistical techniques allow for the description of data characteristics, the testing of relationships, and the testing of differences. When we discuss *description* and **inference**, however, we are not discussing statistical techniques, although those two words are sometimes confused with statistical techniques. This confusion is the result of saying that correlations describe relationships and that cause-and-effect conclusions are inferred from techniques for testing differences of groups. These statements are not necessarily true. Any statistic describes the **sample** of participants for which it was calculated. If the sample of participants represents some larger group, then the findings can be inferred (or generalized) to the larger group. The statistic used, however, has nothing to do with inference. The method of selecting the sample, procedures, and context is what does or does not allow inference because these are the research design considerations that impact our ability to appropriately make inferences from our findings with our sample to statements about the population. We focus on the method of selecting the sample next.

The method used for selecting the sample, procedures, and context determines whether inference is allowed. The allowance of inference is not affected by the specific statistical tool being used.

Ways to Select a Sample

The sample is the group of participants, treatments, and situations on which the study is conducted. A key issue is how these samples are selected. In the following sections we discuss the types of sampling typically used in designing studies and using statistical analysis. The ideal ways to select a sample are through random sampling and stratified random sampling. Using these methods, each individual in the population has a known probability of being selected and, as a result, our confidence that the sample represents the population is high. Unfortunately, random sampling can be difficult, if not impossible, to use because of the challenges in getting contact information and access to an entire population. As a result, many researchers rely on voluntary samples, convenience samples, and snowball samples. When these methods are used, researchers will often use post hoc justifications to demonstrate that their sample is representative of the population.

Random Sampling

population—The entire group that is of interest relative to the research question(s) and from which a sample is taken. The symbols used for population statistics are Greek letters (e.g., μ, pronounced mu, for the mean) with the exception that *N* (note that this is capitalized to distinguish from sample size) is used for the number of people or non-human animals or items in a population.

The sample of participants might be randomly selected from some larger group, or a **population**. For example, if your college or university has 10,000 students, you could randomly select 200 for a study. You would assign each of the 10,000 students a numeric ID. The first numeric ID would be 00001, the second 00002, the third 00003, up through the last (the 10,000th), who would be 10000. Then, you can use a random number generator (search online for options) to create 200 random numbers between 1 and 10,000.

Of course, the purpose of random sampling is to select a sample so that it represents the larger population; that is important so that the findings from the sample can be inferred back to the population. The use of random sampling helps to ensure that the sample is representative of the population. From a statistical viewpoint, inference means that a characteristic, relationship, or difference found in the sample is likely also to be present in the population from which the sample was selected.

> The purpose of a random sample is to allow us to infer that the findings apply to the larger population.

Stratified Random Sampling

In **stratified random sampling**, the population is divided (stratified) on some characteristic before random selection of the sample. Returning to the previous example, the selection was 200 students from a population of 10,000; suppose your college is 30% freshmen, 30% sophomores, 20% juniors, and 20% seniors. You could stratify on class before random selection to make sure the sample is exact in terms of class representation. Here, you would randomly select 60 students from the 3,000 freshmen, 60 from the 3,000 sophomores, 40 from the 2,000 juniors, and 40 from the 2,000 seniors. This procedure still yields a total sample of 200.

> **stratified random sampling**—A method of stratifying a population on some characteristic before random selection of the sample.

Stratified random sampling might be particularly appropriate for survey or interview research. Suppose you suspect that attitudes toward exercise participation change over the college years. You might use a stratified random sampling technique to ensure you have good representation across class standings for interviewing 200 college students to test this hypothesis. Another example would be to develop normative data on a physical fitness test for grades 4 through 8 in a school district. Because you would expect performance to be related to age, you should stratify the population by age before randomly selecting the sample on which to collect normative data.

Voluntary Sample

Oftentimes, researchers rely on volunteers to participate in their studies. These volunteers are enlisted through various recruitment methods including flyer postings, local media and social media advertisements, mailed invitations, and face-to-face recruitment (within classes or community organizations). One challenge of recruiting is that volunteers are likely to be more interested in the study than people from a random sample. As a result, the researchers must acknowledge the limitation that the results from this sample may not generalize to the population. For example, if I am recruiting volunteers to participate in a study looking at the perception of the NFL's rule modifications to protect athletes from concussion, it is likely that I will end up with a sample that includes more NFL football fans than would be present in a random sample of the population. Hence, the findings might not generalize to the entire population of both NFL football fans and people who have no interest in NFL football.

Convenience Sampling

Sometimes researchers are interested in specific types of people who would not be present in large numbers if recruited from the general population. That is, perhaps a researcher is interested in comparing adults who are meeting the CDC's guidelines for aerobic and muscle-strengthening activities to those who are not meeting the guidelines. Since the percentage of adults meeting the guidelines is less than 20% in states like Tennessee, Kentucky, Indiana, and Arkansas (Blackwell & Clarke, 2018), researchers in those states might choose to purposefully recruit physically active participants from known sources such as fitness centers and community recreation centers. This might be necessary to ensure that sufficient numbers of physically active participants are enrolled in the study. However, the limitation is that findings for participants from fitness and community recreation centers may not generalize to others who are physically active and do not exercise at these types of facilities.

Snowball Sampling

When researchers are recruiting from even more select groups, for example, specific chronic diseases, they may start with convenience sampling but then also use the snowball sampling technique. This is when researchers ask their current participants to recruit other people they know who would be expected to be eligible for the study. So, for example, when researchers are recruiting participants with Parkinson's disease, they would ask their current participants to invite anyone they know through support groups or have met through the course of their treatments to also participate in the study. Like the other non-random techniques, the primary limitation of this recruitment method is that it hinders the extent to which findings would be expected to generalize to the population of people with Parkinson's disease.

Ways to Assign Participants to Groups

After recruiting the sample, the next step in experimental research is to identify a method for assigning participants to conditions (or treatments or groups). In experimental research, groups are formed from the sample. The issue here is not how the sample is selected but how the groups are formed within the sample. Chapter 18 discusses experimental research and true experimental designs.

Random Assignment

All true experimental designs require that the groups within the sample be randomly assigned or randomized. Although this requirement has nothing to do with selecting the sample, the procedures used for random assignment are the same. Each person in the sample is given a numeric ID. If the sample has 30 participants, the numeric IDs range from 01 to 30. In this case, suppose that three equal groups ($n = 10$) are to be formed. By using a random assignment generator (search online for options), you provide the numbers of participants and groups, and the program identifies which group each participant should be assigned to. This process allows you to assume that the groups are equivalent at the beginning of the experiment, which is one of several important features of good experimental design that is intended to establish cause and effect.

Random Matched Assignment

Another method of assigning participants to groups is to first match the participants for a characteristic that you want to make sure is represented equally across groups. For instance, if you believe that the participant's biological sex might influence the relationship between the independent variable and the dependent variable, you could randomly assign the men separately from the women to ensure equal numbers of men and women in each group.

Intact Groups

Although undesirable, sometimes researchers must assign participants based upon preexisting intact groups. In this situation, the study is not considered as using an experimental design. This would be necessary if researchers are unable to randomly assign participants to groups. For example, if comparing patients who have Parkinson's to people without Parkinson's, clearly this is an independent variable that cannot be randomly assigned. There are limitations inherent in the use of intact groups, but steps can be taken in the post hoc justification to help establish the extent to which the groups are similar.

Post Hoc Justifications

Frequently, the sample for research is not randomly selected; rather, the researcher attempts a **post hoc justification** that the sample represents some larger group. A typical example is showing that the sample does not differ in average age, racial balance, or socioeconomic status from some larger group. Of course, the purpose is to allow the findings on the sample to be generalized to the larger group. A post hoc attempt at generalization may be better than nothing, but it is not the equivalent of random selection, which allows the assumption that the sample does not differ from the population on the characteristics measured (as well as any other characteristics). In a post hoc justification, only the characteristics measured can be compared. Whether those are the ones that really matter is open to speculation.

Post hoc justification is also used to compare intact groups, or groups within the sample that are not randomly formed. However, in this case, the justification is that because the groups did not differ on certain measured characteristics before the study began, they can be judged equivalent. Of course, the same point applies: Are the groups different on some unmeasured characteristic that affects the results? This question cannot be answered satisfactorily. But, as before, a good post hoc justification of equivalence does add strength to comparisons of intact groups.

post hoc justification— An explanation that is provided after the event has occurred. As applied to a sampling method, this explanation is meant to demonstrate that the sample is representative of the population.

Difficulty of Random Sampling and Assignment: Good Enough?

As previously mentioned, in many studies of physical activity, random sampling procedures are just not possible. For example, when comparing experts with novice performers, typically, neither group is randomly sampled, nor are group memberships (expert and novice) randomly formed. The same is true when studying trained as opposed to untrained runners or cyclists. Often, we are interested in comparing the responses of different age groups, ethnic groups, and sexes to training programs designed to improve skill or increase physical activity levels. Obviously, these groups are seldom randomly selected and cannot be randomly formed. In many studies, sampling is not done at all; researchers are happy to have any participants who volunteer.

Sampling also applies to the treatments used for different groups of participants. How are the treatments selected? Do they represent some population of potential treatments? For example, if I am interested in dose-response relationships between exercise intensity and a cortisol response, which doses do I choose to test? I might select specific doses based upon common exercise intensity cut points (e.g., based upon ACSM definitions of light-, moderate-, and vigorous-intensity exercise). Alternatively, I might select doses based upon a desire to test a linear versus a curvilinear relationship (e.g., 20%, 40%, 60%, and 80% age-predicted max). What about the situational context under which participants are tested or receive the experimental treatment? Are these sampled in some way? Do they represent the potential situations?

The real answers to these questions about sampling are that we seldom randomly sample a population at any level: participants, treatments, or situations. Yet we hope to be able to use statistical techniques based on sampling assumptions and to infer that what we find applies to some larger group than the participants used in the study. What is really needed is a sample that is "good enough for our purpose" (Kruskal & Mosteller, 1979, p. 259). This concept is extremely important for research in physical activity. If we cannot disregard the strict requirements of random sampling that allow study outcomes to be generalized to some similar group, we will never be able to generalize beyond the characteristics of any particular study—location, age, race, sex, time, dose, and so on. Strictly speaking, there is no basis for the findings from a sample to be similar to a population that differs in any way from the sample (even an hour after the sampling). "A good-enough principle of sampling, however, would allow generalizations to any population for which the sample is representative enough" (Serlin, 1987, p. 366).

For the sample of participants, treatments, and situations to be good enough to generalize, it must be selected on some theoretical basis. For example, if the theory proposes that the cardiovascular system responds to training in a specific way in all untrained adults, using volunteer untrained undergraduates in a training study may be acceptable. "It is only on the basis of theory that one decides whether the experimental results can be generalized to the responder population, to the stimulus or ecological population, or both" (Serlin, 1987, p. 367). The best possible generalization statement is to say that the findings may be plausible in other participants, treatments, and situations, depending on their similarity to the study characteristics.

> A good sample leads to a generalization statement that asserts plausibility that the findings apply to a broader population.

Measures of Central Tendency and Variability

central tendency (measure of)—A single score that best represents all the scores.

variability—The degree of difference between each individual score and the central tendency score.

standard deviation—An estimate of the variability of the scores of a group around the mean.

variance—The square of the standard deviation.

Some of the more easily understood statistical and mathematical calculations are those that find the **central tendency** and **variability** of scores. When you have a group of scores, one number may be used to represent the group. That number is generally the mean, median, or mode. These terms are ways of expressing central tendency. Within the group of scores, each individual score differs to some degree from the central tendency score. The degree of difference is the variability of the score. Two terms that describe the variability of the scores are **standard deviation** and **variance**.

Central Tendency Scores

The statistic for the central tendency score with which you are probably familiar is the mean (*M*), or average:

$$\mu = \Sigma X\ /\ N \text{ for a population; } M = \Sigma X\ /\ n \text{ for a sample} \tag{6.1}$$

Thus, if you have the numbers 6, 5, 10, 2, 5, 8 , 5, 1, and 3, then

$$\mu = (6 + 5 + 10 + 2 + 5 + 8 + 5 + 1 + 3)\ /\ 9 = 45\ /\ 9 = 5$$

The number 5 is the mean and represents this population of numbers.

Sometimes, the mean may not be the most representative or characteristic score. Suppose we replace the number 10 in the previous set of scores with the number 46. The mean is now 9, a number larger than all but one of the scores. That number is not representative because one score (46) made the average high. In this case, another measure of central tendency is more useful. The **median** is defined as the value in the middle; the middle value is the value that occurs in the place (*N* + 1) / 2 when the values are put in order. In our example, arrange the numbers from lowest to highest—1, 2, 3, 5, 5, 5, 6, 8, 46—then count (9 + 1) / 2 = 5 places from the first score; the median is the value 5, which is a much more representative score than the value of 9. If *N* is an even number, the median value may be a decimal. For example, for scores 1, 2, 3, 4, the median is 2.5. The median is also useful if you have missing values for participants, which makes calculating the mean impossible.

median—A statistical measure of central tendency that is the middle score in a group.

mode—A statistical measure of central tendency that is the most frequently occurring score of the group.

Most often, you are interested in the mean of a group of scores. You may occasionally be interested in the median or perhaps in another measure of central tendency, the **mode**, which is defined as the most frequently occurring score. The previous example's mode is also 5 because it occurs three times. Some groups of scores may have more than one mode. The mode is particularly useful when your data is nonparametric. The mode is not commonly presented in empirical literature, but it would make sense to provide the mode if your data was not numeric; for example, to present data on career choices where the most frequently selected option was physical therapist.

Variability Scores

Another characteristic of a group of scores is the variability. Variability gives us an indication of how spread out (or clustered) the scores in a sample are around the mean. An estimate of the variability, or spread, of the scores for a sample can be calculated as the standard deviation (s):

$$s = \sqrt{\sum (X - M)^2/(n - 1)} \tag{6.2}$$

This formula translates as follows. Calculate the mean by equation 6.1, subtract the mean from each person's score ($X - M$), square the answer, sum (Σ) the squared scores, divide by the number of scores minus one ($n - 1$), and take the square root of the answer. Table 6.1 provides an example.

TABLE 6.1

Calculations of Mean and Standard Deviation

X	$X - M$	$(X - M)^2$
6	1	1
5	0	0
10	5	25
2	−3	9
5	0	0
8	3	9
5	0	0
1	−4	16
3	−2	4
$\Sigma X = 45$	$\Sigma (X - M) = 0$	$\Sigma (X - M)^2 = 64$

In table 6.1, the column labeled X contains the raw data, the column labeled $X - M$ contains the deviation scores, and the column labeled $(X - M)^2$ contains the squared deviations. The sum of the squared deviations is 64. So, the standard deviation for this sample would be:

$$M = \Sigma X/n = 45/9 = 5$$

$$s = \sqrt{\Sigma (X - M)^2/(n - 1)} = \sqrt{64/8} = \sqrt{8} = 2.83$$

The mean and standard deviation together are good descriptions of a set of scores. If the standard deviation is large, the mean may not be a good representation of the sample. But a small standard deviation indicates that most scores fall near the mean, and therefore the mean is a good representation of the sample. For a normal distribution (discussed later in the chapter), roughly 68% of a set of scores fall within ±1s, about 95% of the scores fall within ±2s, and about 99% of the scores fall within ±3s. Thus, if we know how far an individual score is from the mean in terms of standard deviation units (which is called a z *score*), we can use the unit normal table (see table 1 in the appendix) to identify the percentage of scores that are larger than the individual score (the number under the beyond column for a positive z score), which also tells us the likelihood of obtaining a score that is larger than the individual score.

$$z = \frac{X - \mu}{\sigma} \text{ for population; } z = \frac{x - M}{s} \text{ for sample} \qquad (6.3)$$

The purpose of equation 6.2 is to help you understand the meaning of the standard deviation. For use with a hand calculator and when sample sizes are larger, it is easier to first calculate the sum of squares (which is the sum of the squared deviations, SS) using a computational formula (see equation 6.4a) and then to calculate standard deviation for the sample (s) from that intermediate step (see equation 6.4b):

$$SS = \sum X^2 - \frac{(\sum X)^2}{n}$$

$$s = \sqrt{\frac{SS}{n-1}} \qquad \text{(6.4a and 6.4b)}$$

One final point for later consideration is that the square of the standard deviation is called the *variance*, or s^2.

Confidence Intervals

Confidence intervals are an effective technique that researchers use to interpret a variety of statistics such as means, medians, and correlations. Confidence intervals are also used in hypothesis testing. A confidence interval provides an expected upper and lower limit for a statistic (like the mean) at a specified probability level, usually either 95% or 99%. The sample's size and homogeneity of values and the level of confidence that the researcher selects affects the size, or length, of a confidence interval. Confidence intervals are based on the fact that any statistic (referring to, for example, the mean of a sample) possesses sampling error. This error relates to how well the statistic represents the target population. When we compute a mean for a sample, we are making an estimate of the mean of the target population because we believe the sample represents the population. But, remember that for a large population, there is a large number of possible random samples that could be selected from that population, so the same number of means could be computed. Creating a distribution of all possible sample means from a given population is referred to as the *sampling distribution of sample means*. A confidence interval provides a band around the individual sample mean you have obtained within which the estimate of the population mean is likely to fall.

> Confidence intervals should be used because statistics vary in how well they represent target populations.

A confidence interval (CI) of a statistic such as the mean employs the following information:

CI = observed statistic ± (standard error × specified confidence level value)

We will construct a confidence interval for a mean of a sample with the following characteristics: $n = 30$, $M = 40$, $s = 8$.

The observed statistic is the mean ($M = 40$). The **standard error** represents the variability of the sampling distribution of a statistic (in this case, the mean). Imagine that instead of one sample of $n = 30$, you had selected 100 samples of this size. The means for all the samples would not be the same; but, as you calculate means for an increasingly large number of samples, the histogram for the means would start to approximate a normal distribution. If you calculated the standard deviation of all of these means, you would have an estimate of the standard error. Fortunately, we do not have to draw hundreds of samples to calculate standard error; we simply divide our sample standard deviation by the square root of the sample size. Do not ask why; just accept it. (Have we ever lied to you before?) So, in our example, the standard error of the distribution of sampling means, or the standard error of the mean, is $S_M = \frac{8}{\sqrt{30}} = 1.46$.

> **standard error**—The variability of the sampling distribution.

OK, I've put in all the data, but this computer will not calculate the mean.

© Ivan Nakonechnyy/fotolia.

The last piece of information we need to construct a confidence interval is the value for the specified confidence level. This is easy: We just look it up in a table. For a 95% confidence interval, that would put 5% in the two outer tails of a normal distribution, which means we would have 2.5% (or 0.025) in each tail. If you look up 0.025 in the one tail beyond column or .05 in the two tail beyond column in Appendix table 1, you find the z score of 1.96. For a 99% confidence interval, that would put 1% in the two outer tails, which means we would have 0.5% (or 0.005) in each tail. Look up 0.005 the one tail beyond column and you find the z score of 2.58. Hence, if we knew that our sample had a normal distribution, we would use 1.96 for the 95% confidence level value, or 2.58 for 99%.

We can now construct the confidence interval (95% confidence) using equation 6.5:

CI = sample mean ± (standard error × confidence level table value)

$$= 40 \pm (1.46 \times 1.96) = 40 \pm 2.86 \tag{6.5}$$

We subtract and add 2.86 to 40 to obtain a confidence interval of 37.14 to 42.86. Thus, we are 95% confident that this interval includes the target population mean.

Confidence intervals are also used in hypothesis testing, such as applying the null hypothesis to the difference between two or more means. If we were comparing the means of two groups, the difference between the two means would be the observed statistic. The standard error would be the standard error of the difference (which combines the standard errors of the two means). The table value would be taken from the t table (Appendix table 4) using df of $(n_1 - 1) + (n_2 - 1)$ for the selected level of confidence (e.g., 95%). The calculated confidence interval would give an estimate of the range within which the true difference between these two population means would fall 95% of the time. Importantly, if the confidence interval does not include the value of 0, this tells us that the difference between the two samples is significant, meaning it's a reliable difference.

Frequency Distribution and the Stem-and-Leaf Display

In addition to using measures of central tendency and variability to help us understand your sample, it is also important to examine your data visually. This can be done using histograms or stem-and-leaf displays. The value of looking at your data visually is that it helps you to know if your data is normally distributed (a requirement for the use of parametric statistics) or not (in which case nonparametric techniques are required) and to identify any potential outliers. Parametric and nonparametric techniques will be described later in this chapter. Outliers are data points that appear to be extreme relative to the other data in your sample. In the previous example regarding calculation of the median, the score of 46 would certainly be considered an outlier when all of the other scores were in the range of 1-8.

A common technique for summarizing data is to produce a picture (called a *histogram*) of the distribution of scores by means of a **frequency distribution**. A simple frequency distribution just lists all the scores on the *x*-axis and the frequency of each score on the *y*-axis. If there is a wide range of values, a grouped frequency distribution is used in which scores are grouped into small ranges called **frequency intervals**. For example, figure 6.2*a* is a histogram of the scoring averages of 30 selected PGA players in 2019. The scoring averages of players are on the *x*-axis (bottom), and the frequency of each score is represented by the height of the column on the *y*-axis. You can see by looking at the figure that eight players averaged about 69.75 strokes per round. Note that a replication of the normal curve is placed over the actual distribution so that the two can be visually compared. One drawback to a grouped frequency distribution is that information is lost; that is, a reader does not know the exact score of each individual within a given interval.

Another effective way to provide information about the distribution of a set of data is called a **stem-and-leaf display**. This representation is similar to a grouped frequency distribution, but no information is lost.

Figure 6.2*b* provides a stem-and-leaf distribution of the same golf data. You can observe each player's actual scoring average—note that 11 players (identified from the frequency column) are listed by the first stem score of 70. Reading to the right, you then look at the leaf column to see that one player's scoring average is 70.5, four players average 70.6, two are at 70.7, three are at 70.8, and one is at 70.9. When the intervals are ordered from lowest to highest, you can turn the page on its side, look at the shape created by the leaf column, and see a type of graph similar to the histogram. This method also helps you visualize the normality of the distribution.

frequency distribution—A distribution of scores, including the frequency with which they occur.

frequency intervals—Small ranges of scores within a frequency distribution into which scores are grouped.

stem-and-leaf display—A method of organizing raw scores by which score intervals are shown on the left side of a vertical line and individual scores falling into each interval are shown on the right side.

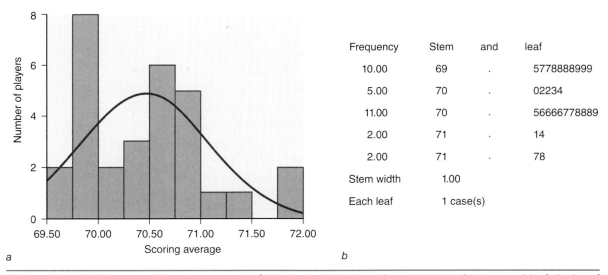

Frequency	Stem	and	leaf
10.00	69	.	5778888999
5.00	70	.	02234
11.00	70	.	56666778889
2.00	71	.	14
2.00	71	.	78
Stem width	1.00		
Each leaf	1 case(s)		

a   *b*

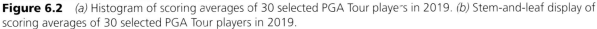

Figure 6.2 *(a)* Histogram of scoring averages of 30 selected PGA Tour players in 2019. *(b)* Stem-and-leaf display of scoring averages of 30 selected PGA Tour players in 2019.

Basic Concepts of Statistical Techniques

Besides measures of central tendency and variability, other slightly more complicated statistical techniques can be used. Before we explain each in detail, however, you must understand some general information about statistical techniques.

Two Categories of Statistical Tests

The two general categories of statistical tests are **parametric** and **nonparametric**. Using the various tests in each category requires meeting the assumptions for those tests. The first category, parametric statistical tests, has three assumptions about the distribution of the data:

1. The population from which the sample is drawn is normally distributed on the variable of interest.

2. The samples drawn from a population have the same variances on the variable of interest.

3. The observations are independent.

Certain parametric techniques have additional assumptions. The second category, nonparametric statistics, is called **distribution free** because the previous assumptions need not be met.

Whenever the assumptions are met, parametric statistics are often said to have more power, although there is some debate on this issue. Having more power increases the chances of rejecting a false null hypothesis—the outcome a researcher typically works toward. You frequently assume that the three criteria for use of parametric statistics are met. However, the assumption of normality can be tested by using estimates of **skewness** and **kurtosis**. (Only the meaning of these tests is explained here. Any basic statistics textbook provides considerably more detail. For a helpful discussion on skewness and kurtosis, see Kim, 2013.)

To understand skewness and kurtosis, first consider the normal distribution in figure 6.3. This is a **normal curve**, which is characterized by the mean, median, and mode being at the same point (center of the distribution). In addition, $\pm 1s$ from the mean includes 68% of the scores, $\pm 2s$ from the mean includes 95% of the scores, and $\pm 3s$ includes 99% of the scores. Thus, data distributed as in figure 6.3 would meet the first assumption for use of parametric techniques because it is normally distributed. Skewness of the distribution describes the direction of the hump of the curve (labeled A in figure 6.4) and the nature of the tails of the curve (labeled B and C). If the hump (A) is shifted to the left and the long tail (B) is shifted to the right (figure 6.4a), the skewness is positive. If the shift of the hump (A) is to the right and the long tail (C) is to the left (figure 6.4b), the skewness is negative. Kurtosis describes the vertical aspect of the curve, such as whether the curve is more or less peaked than the normal curve. Figure 6.5a shows a more peaked curve, and figure 6.5b shows a flatter curve. If a distribution appears to be abnormal due to skew or kurtosis, z tests can be applied to determine if the assumptions of normality can be assumed (see Kim, 2013).

Appendix table 1 shows a unit normal distribution (z) for a normal curve. The column z shows the location of the mean. When the mean is in the center of the distribution, its z is equal to .00; thus, .50 (50%) of the distribution is beyond (to the right of) the mean, leaving .50 (50%) of the distribution as a remainder (to the left of the mean). As the mean of the distribution moves to the right in a normal curve (say, to a z of $+1s$), .8413 (84%) of the distribution is to the left of the mean (remainder) and .1587 is to the right of (beyond) the mean. This table allows you to determine the percentage of the normal distribution included by the mean plus any fraction of a standard deviation. Suppose you want to know what percentage of the distribution would be included by the mean plus one-half (.50) of a standard deviation. Using Appendix table 1 and knowing that z = +0.5, you can see that it would be .6915 (remainder), or 69%.

parametric statistical test—A test based on data assumptions of normal distribution, equal variance, and independence of observations.

nonparametric statistical test—Any of a number of statistical techniques used when the data do not meet the assumptions required to perform parametric tests.

distribution free—A term used to describe nonparametric statistical tests, because the data distribution requirements for a parametric test do not have to be met.

skewness—A description of the direction of the hump of the curve of the data distribution and the nature of the tails of the curve.

kurtosis—A description of the vertical characteristic of the curve showing the data distribution, such as whether the curve is more peaked or flatter than the normal curve.

normal curve—A distribution of data in which the mean, median, and mode are at the same point (center of the distribution) and in which $\pm 1s$ from the mean includes 68% of the scores, $\pm 2s$ from the mean includes 95% of the scores, and $\pm 3s$ includes 99% of the scores.

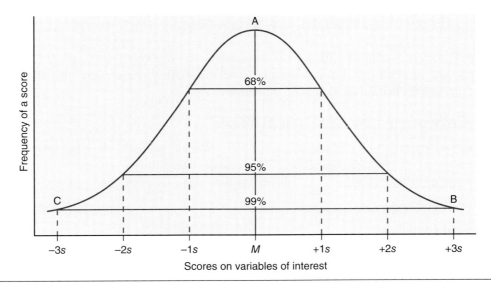

Figure 6.3   The normal curve.

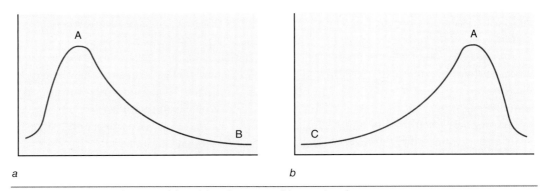

Figure 6.4   Skewed curves: *(a)* positive skewness; *(b)* negative skewness.

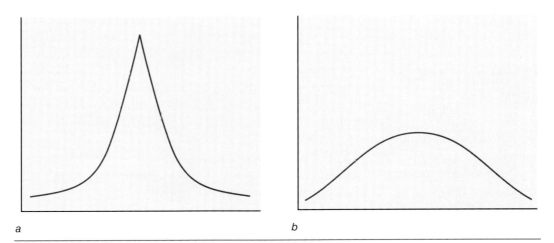

Figure 6.5   Curves with abnormal kurtosis: *(a)* more peaked; *(b)* flatter.

For chapters 8 and 9, consider that the three basic assumptions for parametric statistical tests have been met. This is done for two reasons. First, the assumptions are very robust to violations, meaning that the outcome of the statistical test is relatively accurate even with severe violations of the assumptions. Second, most of the research in physical activity uses parametric tests.

What Statistical Techniques Tell About Data

The statistical techniques presented in the next four chapters answer the following two questions about the data to which they are applied:

1. Is the effect or relationship of interest statistically significant? In other words, if the research is repeated, will the effect or relationship be there again—meaning, is it reliable?

2. How strong (or meaningful) is the effect or relationship of interest? This question refers to the magnitude, or size, of the effect or relationship.

Two important factors to consider about these questions are: (a) Question 1 is usually answered before question 2 because the strength of the relationship or effect may be of minimal interest until it is known to be reliable (significant); and (b) question 2 is always of interest when the answer to question 1 is yes (i.e., when the effect or relationship is significant) and question 2 is occasionally of interest without a significant effect because we may use the effect size to help provide direction for future research. Sometimes, in elating over the significance of effects and relationships, researchers lose sight of the need to look at the strength of these relationships. This point is particularly true in research where differences between groups are compared. The experimenter frequently forgets that relatively small differences can be significant. Finding a significant difference means only that the differences are reliable or that the same answer is likely to be obtained if the research is repeated. The experimenter then needs to look at the size of the differences to understand the magnitude of the effect and to interpret whether the findings are meaningful based upon a consideration of theory and cost-benefit considerations (this is discussed further in chapter 7). For each technique presented in the next three chapters, we first look at whether the relationship or effect is significant. Then we suggest ways to evaluate the strength of the relationship or effect as a first step to considering meaningfulness.

Distinction Between Correlation and Differences in Statistical Techniques

Statistical techniques can be usefully divided into two categories: (a) statistical techniques used to test relationships between variables in one group of participants (regression or correlation) and (b) techniques used to test differences between groups of participants (*t* tests and analysis of variance). This division is, strictly speaking, inaccurate because both sets of techniques are based on the general linear model and involve only different ways of entering data and manipulating variance components. An introduction to research methods, however, is neither the place to reform the world of statistics nor the place to confuse you. Thus, the techniques are considered as two distinct groups. Chapter 8 discusses the interrelationships of variables, and chapter 9 discusses the differences between groups. Learning the simple calculations underlying the easier techniques and then building on them helps you understand the more complex ones. Do not panic because statistics involves manipulating numbers. You can escape from this section with a reasonable grasp of how, why, and when the various statistical techniques are used in the study of physical activity.

Remember that correlation between two variables does not indicate causation (recall the discussion in chapter 4). Causation is not determined from any statistic or correlation. Cause and effect conclusions are established by theory, logic, and the total experimental situation, of which statistics is a part. As summarized by Pedhazur (1982, p. 579),

> **"Correlation is no proof of causation." Nor does any other index prove causation, regardless of whether the index was derived from data collected in experimental or in nonexperimental research. Covariations or correlations of variables may be suggestive**

of causal linkages. Nevertheless, an explanatory scheme is not arrived at on the basis of the data, but rather on the basis of knowledge, theoretical formulations and assumptions, and logical analyses. It is the explanatory scheme of the researcher that determines the type of analysis to be applied to data, and not the other way around.

There are some examples of actual correlational relationships that clearly illustrate that relationships cannot be used to infer causality. For example, look at the data in figure 6.6, which implies a relationship between suicides and U.S. spending on science, space, and technology from 1999-2009. Obviously, these two variables are not causally related, and this provides a good example of why correlation cannot be used to make determinations about causation.

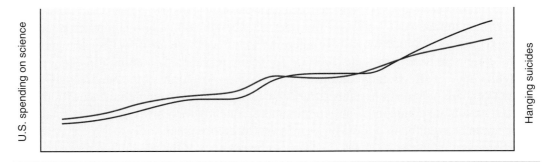

Figure 6.6 U.S. spending on science, space, and technology correlates with suicides by hanging, strangulations, and suffocation (correlation: 99.79% [r = 0.99789126]).

Adapted from T. Vigen, *Spurious Correlations* (Hachette Books, 2015).

Data for Use in the Remaining Statistical Chapters

As we indicated in the introduction to part II, we use a standard data set throughout the next several chapters. The data are from the PGA website, and we selected data from 2019 through the completion of the Masters Tournament. We then limited the data to that from the top 10 money winners, money winners 21 through 30, and money winners 41 through 50. The demographic data for the players are in table 6.2a: player's name, money won, world rank, player rank in three groups coded as follows: 1 = top 10, 2 = 21-30, and 3 = 41-50; and nationality coded as U.S. player (code = 1) and international player (code = 2). The performance data for the players are in table 6.2b: player's name, scoring average, driving distance, driving accuracy, greens in regulation, putts per round, scrambling, and sand saves.

TABLE 6.2A

Demographic Data for 30 of the Top PGA Money Winners in 2019

Player	Money to date	Rank	Grouping by money to date	U.S. (1) v international (2)
Rory McIlroy	$4,854,964	1	1	2
Matt Kuchar	$4,649,744	9	1	1
Xander Schauffele	$4,470,840	3	1	1
Dustin Johnson	$3,925,219	23	1	1
Brooks Koepka	$3,526,317	14	1	1

Player	Money to date	Rank	Grouping by money to date	U.S. (1) v international (2)
Paul Casey	$3,199,473	12	1	1
Francesco Molinari	$3,027,438	94	1	2
Gary Woodland	$3,020,845	7	1	1
Rickie Fowler	$2,853,167	4	1	1
Tiger Woods	$2,804,717	31	1	1
Jason Day	$2,058,571	5	2	2
Jon Rahm	$2,024,338	10	2	2
Keith Mitchell	$2,021,293	44	2	1
Patrick Cantlay	$1,961,090	30	2	1
J.B. Holmes	$1,937,035	79	2	1
Si Woo Kim	$1,830,373	32	2	2
Sungjae Im	$1,817,133	36	2	2
Adam Scott	$1,766,959	17	2	2
Lucas Glover	$1,629,133	24	2	1
Rafa Cabrera Bello	$1,579,919	61	2	2
Ian Poulter	$1,366,334	20	3	2
Adam Long	$1,307,493	149	3	1
Bubba Watson	$1,277,358	80	3	1
Danny Lee	$1,266,884	29	3	2
Scott Piercy	$1,249,254	28	3	1
Chez Reavie	$1,248,770	34	3	1
Jhonattan Vegas	$1,242,787	68	3	2
Ryan Palmer	$1,240,720	54	3	1
Branden Grace	$1,211,465	111	3	2
Abraham Ancer	$1,202,849	39	3	2

TABLE 6.2B

Performance Data for 30 of the Top PGA Money Winners in 2019

Player	Scoring average	Driving distance (yds)	Drive accuracy (%)	Greens in regulation (%)	Putts per round	Scrambling (%)	Sand saves (%)
Rory McIlroy	69.767	315.30	59.51	71.01	28.97	55.69	51.02
Matt Kuchar	69.910	290.80	71.73	75.76	29.61	61.98	58.70
Xander Schauffele	69.820	304.60	60.45	71.20	28.66	64.47	58.21
Dustin Johnson	69.536	310.80	54.59	70.83	28.66	66.67	71.79
Brooks Koepka	70.307	308.80	61.32	69.63	29.23	57.93	60.00
Paul Casey	70.554	299.40	70.81	70.47	29.37	58.42	43.66
Francesco Molinari	71.189	290.80	65.88	63.43	27.96	65.19	68.75
Gary Woodland	69.990	309.10	65.70	72.45	29.61	57.61	42.67
Rickie Fowler	69.871	304.10	61.79	69.17	28.45	59.91	48.57
Tiger Woods	69.729	297.80	64.64	75.56	29.65	60.23	40.74

(continued)

Table 6.2b *(continued)*

Player	Scoring average	Driving distance (yds)	Drive accuracy (%)	Greens in regulation (%)	Putts per round	Scrambling (%)	Sand saves (%)
Jason Day	70.021	304.30	63.90	70.93	28.37	62.42	51.06
Jon Rahm	69.911	306.70	63.75	69.03	28.70	62.78	50.00
Keith Mitchell	70.876	301.90	58.03	69.26	29.44	57.03	49.41
Patrick Cantlay	69.806	308.00	56.70	70.31	28.66	66.08	48.00
J.B. Holmes	71.473	305.50	55.28	68.06	29.40	60.43	55.56
Si Woo Kim	70.425	292.40	59.66	66.56	27.72	65.78	56.84
Sungjae Im	70.256	293.50	65.22	69.66	28.87	64.53	44.95
Adam Scott	70.209	299.60	61.20	68.10	29.19	62.36	58.00
Lucas Glover	69.813	294.30	65.47	72.22	29.23	70.45	55.36
Rafa Cabrera Bello	70.601	288.80	63.76	65.79	28.84	61.54	48.81
Ian Poulter	70.616	292.70	68.81	72.22	29.53	61.88	60.42
Adam Long	71.763	292.60	63.42	66.82	29.42	57.67	45.10
Bubba Watson	70.624	312.20	58.82	68.14	29.71	64.10	51.79
Danny Lee	70.877	300.70	56.70	70.20	29.07	59.32	58.82
Scott Piercy	70.813	290.90	69.19	71.30	28.75	62.90	44.74
Chez Reavie	70.914	287.00	73.79	70.77	29.20	57.85	46.05
Jhonattan Vegas	70.691	306.80	58.39	69.44	29.00	65.22	46.67
Ryan Palmer	70.703	297.40	58.11	72.88	29.06	56.63	46.27
Branden Grace	71.893	296.80	56.90	68.86	30.24	52.11	46.55
Abraham Ancer	70.798	293.70	68.79	68.48	28.61	64.75	43.24

Summary

Statistics are used to describe data, to determine relationships of variables, and to test for differences between groups. In this chapter, we have tried to make the point that the type of statistics used does not determine whether findings can be generalized; rather, it is sampling that permits (or limits) inference. Whenever possible, random sampling is the method of choice, but in behavioral research, the more important question may be whether the sample is good enough. In some types of research, such as surveys, stratified random sampling is desirable so that the study represents certain segments of a population. In experimental research, random assignment of participants to groups is definitely desirable so that the researcher can assume equivalence at the beginning of the experiment.

We began our coverage of statistical techniques with basic concepts such as measures of central tendency, variability, and normal distribution. Remember that statistics can do two things: establish significance and assess the magnitude of the effect. Significance means that a relationship or difference is reliable—that you could expect it to happen again if the study were repeated. The magnitude of the effect lets us know whether the effect is small, medium, or large. Meaningfulness refers to the importance of the results and is interpreted based upon the researchers' understanding of theory, context, and a cost-benefit consideration.

✓ Check Your Understanding

The following scores represent the number of pull-ups attained by a sample of 15 students:

8	4	13
11	6	2
6	7	5
5	0	4
8	6	5

1. Use equation 6.1 to calculate the mean.
2. Use equation 6.3 to calculate the standard deviation.
3. Use equation 6.4a and 6.4b to calculate the standard deviation, and see whether the answer is the same as that derived from equation 6.3.
4. Rearrange the scores from low to high. Use the formula $(n + 1)/2$ to find the median point. Count up from the bottom of that list of values to locate the median score.
5. Make a list of 50 names. Using an online random number generator, randomly select 24 subjects and randomly assign them to two groups of 12 each.
6. To construct a stem-and-leaf display, you first decide upon the intervals you want to use to organize the data. You identify the tens digits in your sample to identify all of the possible "stems" and write each stem once on the left of a vertical line. Then, you go through the data one score at a time and write the "leaf" on the right side of the vertical line directly next to the "stem" for that score. Construct a stem-and-leaf display using intervals of 10 scores (e.g., 10-19 and 20-29) for the following 30 scores:

50	42	64	18	41	30	48	68	21	48
43	27	51	42	62	53	45	31	13	58
60	35	28	46	36	56	39	46	25	49

7. Consult equation 6.5 and construct a confidence interval for the mean of a sample ($M = 50$, $s = 6$, $n = 100$). Use the 95% confidence level. Interpret the results.

7

Statistical Issues in Research Planning and Evaluation

> **Without data, we are just another person with an opinion.**
>
> —Andreas Schleicher

To plan your own study or evaluate someone else's study, you must first understand four concepts and their interrelationships: alpha, power, sample size, and effect size. In this chapter, we present these concepts and show how to use them in research planning and evaluation.

Probability

Before explaining these concepts, allow us to introduce another concept that is relevant to statistical techniques: **probability** (p). Probability asks what the odds are that certain things will happen. You encounter probability in everyday events. For example, what are the chances that it will rain? You hear from a weather report that the probability of rain is 90%. You wonder whether this means that it will rain in 90% of the places or, more likely, that the chances are 90% that it will rain where you are, especially if you are planning to play pickleball or tennis.

A concept of probability related to statistics is called **equally likely events**. For example, if you roll a die, the chances of the numbers from 1 to 6 occurring are equally likely (i.e., 1 in 6, unless you are playing craps in Las Vegas). Another pertinent approach to probability involves **relative frequency**. To illustrate, suppose you toss a coin 100 times. On average, you would expect heads 50 times and tails 50 times; the probability of either result is one-half, or .50. When you toss a coin 100 times, however, you may get heads 48 times, or .48. This is the relative frequency. You might perform 100 tosses 10 times and never get .50, but the relative frequencies you obtained for each of those "samples" of 100 tosses would be distributed closely around .50, and you would still assume the probability as .50.

In a statistical test, you sample from a population of participants and events. You use probability statements to describe the confidence that you place in the statistical findings. Frequently, you encounter a statistical test followed by a probability statement such as $p < .05$. This interpretation is that a difference or relationship of this size would be expected less than 5 times in 100 as a result of chance.

probability (p)—The odds that a certain event will occur.

equally likely events—A concept of probability in which the chances of one event occurring are the same as the chances of another event occurring.

relative frequency—A concept of probability concerning the comparative likelihood of two or more events occurring.

Hypothesis Testing

As you've learned in previous chapters, the scientific method is advanced by hypothesis testing. Once you have identified the relevant literature and your research question, the next step is to construct hypotheses that can be tested using data from your sample and the appropriate statistical test. With inferential statistics, we write a null hypothesis which says that there will be no relationship or no difference. We also generate an alternative hypothesis which says that there will be an effect, a relationship, or a difference. The alternative hypothesis is the one that we anticipate will be supported after conducting our experiment. After conducting the experiment and collecting the data, we use the appropriate statistical tool to test the null hypothesis. If the **test statistic** that we calculate from our data is larger than the **critical value** that we identified in advance, then our results show that the probability of obtaining a test statistic that large due to chance is less than alpha. In that case, we reject the null hypothesis, and we entertain the alternative hypothesis. We say that we entertain the alternative hypothesis because we can never know for sure that the alternative hypothesis is true after a single experiment.

Alpha

In research, the test statistic that you calculate is compared with a critical value for that statistic based upon the alpha level, the sample size, and whether the statistical test is being conducted as a one-tailed or a two-tailed test. The experimenter establishes an acceptable level of chance occurrence, **alpha (α)**, before the study. This level of chance occurrence can vary from low to high but can never be eliminated. For any given study, the probability of the findings being due to chance always exists. The value of alpha is typically set at .05, and this tells us that 5 times out of 100 we are willing to mistakenly conclude that our findings are statistically significant when, in fact, they are actually chance findings.

In behavioral research, the researcher commonly sets alpha (probability of chance occurrence) at .05 or .01 (the odds that the findings are due to chance are either 5 in 100 or 1 in 100). There is nothing magical about .05 or .01. These values are used to control for a **type I error**. In a study, the experimenter may make two types of error. A type I error is to reject the null hypothesis when the null hypothesis is true (and therefore should not have been rejected). For example, a researcher concludes that there is a difference between two methods of training, but there really is not. A **type II error** occurs when we do not reject the null hypothesis when the null hypothesis is false (and therefore should have been rejected). For example, a researcher concludes that there is no difference between the two training methods, but there really is a difference. Figure 7.1 shows a **truth table**, which displays type I and type II errors. As you can see, to accept a true null hypothesis or reject a false one is the correct decision. You control for type I errors by setting alpha. For example, if alpha is set at .05, then if 100 experiments are conducted, a true null hypothesis of no difference or no relationship would be incorrectly rejected on only 5 occasions. Although the chances for error still exist, the experimenter has specified them exactly by establishing alpha before the study.

To some extent the issue is this: If you had to make an error, which type of error would you be willing to make? The level of alpha reflects the type of error that you are willing to make. In other words, is it more important that you avoid concluding that one training method is better than another when it really is not (type I), or is it more important that you avoid the conclusion that one method is not any better than the other when it really is (type

test statistic (α)—The value that you calculate based upon the sample data you've collected and which is used to determine whether or not to reject the null hypothesis. Common test statistics are t, F, and r.

critical value (α)—When statistical tests are calculated in the absence of a computer, the test statistic calculated based upon sample data is compared to a critical value from a table. The critical value is identified based upon alpha, whether it is a one- or two-tailed test, and degrees of freedom (when relevant). If the test statistic is further from zero than the critical value, the null hypothesis is rejected.

alpha (α)—A level of probability (of chance occurrence) set by the experimenter before the study; sometimes referred to as *level of significance*.

type I error—Rejection of the null hypothesis when the null hypothesis is true.

type II error—Acceptance of the null hypothesis when the null hypothesis is false.

truth table—A graphic representation of correct and incorrect decisions regarding type I and type II errors.

	H_0 true	H_0 false
Accept	Correct decision	Type II error (β)
Reject	Type I error (α)	Correct decision

Figure 7.1 Truth table for the null hypothesis (H_0).

Adapted from Kirk (1995).

II)? For example, in a study of the effect of a cancer drug, the experimenter would not want to accept the null hypothesis of no effect if there were any chance that the drug worked. Thus, the experimenter might set alpha at .30 even though the odds of making a type I error would be inflated. The experimenter is making sure that the drug has every opportunity to show its effectiveness. On the other hand, setting alpha at a very low level, for example, .001, greatly decreases the odds of making a type I error and consequently increases the difficulty of detecting a real difference (a type II error). This might be important if the treatment has its own associated risks. So, in the previous example, if the cancer drug had the side effect of dramatically increasing the likelihood of having a heart attack, it might be important to use alpha of .001 to reduce the likelihood of mistakenly concluding that the drug has a positive effect for cancer because such a decision would have associated negative consequences through its increased risk of a heart attack.

We cannot tell you where to set alpha, but we can say that the levels .05 and .01 are widely accepted in the scientific community. If alpha is to be moved up or down, be sure to justify your decision, but also consider this: "Surely God loves the .06 nearly as much as the .05" (Rosnow & Rosenthal, 1989, p. 1277). This statement expresses the understanding that while .05 is a commonly used value for alpha, the choice of .05 is essentially arbitrary and, hence, a finding of $p < .06$ is not very different from $p < .05$. Therefore, it is important for scientists to not only test their hypotheses using an established value of alpha but also consider the magnitude of the observable effects relative to both the objective they are trying to achieve and any risks or costs of the treatment.

Even when experimenters set alpha at a specific level (e.g., .05) before the research, they often report the probability of a chance occurrence for the specific effects of the study at the level it occurred (e.g., $p = .012$). This procedure is appropriate (and recommended by organizations such as the American Psychological Association for their journal articles),

The probability of the ice cream falling off the cone is directly correlated to the newness of the floor covering.

© Thomas Perkins/fotolia

because the researchers are only demonstrating to what degree the level of probability exceeded the specified level.

It is debatable whether alpha should be specified before the research and the findings reported at the alpha that has been specified. Some argue that results are either significant at the specified alpha or they are not (not much is in between). Alpha is established as a criterion, and either the results meet the criterion or they do not. Although experimenters sometimes report borderline significance (if alpha is .05) by describing results for which p = .051 to .10 as borderline, this is typically only done when effect sizes are also reported and the implication for statistical power for future research is provided.

In fact, regardless of the specific value of p that is obtained, a sound approach is to report the exact level of probability (e.g., p = .024) associated with the test statistics (e.g., r, t) and then to estimate the magnitude of the difference or relationship by providing a measure of effect size. Using the statistical information (significance and effect size), the researcher should then interpret the findings within the theory that is guiding their work and the hypotheses that have been explicitly described in their study to determine meaningfulness. Rather than making the decision a statistical one, this approach places the responsibility of decision making where it belongs: on the researcher who has placed the study within a theoretical model and has considered related research. Much of the criticism of statistics has revolved around the blind application of statistical techniques to data without appropriate interpretation of the outcomes. For a comprehensive discussion of the pros and cons of binary decision making relative to alpha, see Harlow, Mulaik, and Steiger (2016).

Beta

beta (β)—The magnitude of a type II error.

Although the chance of making a type I error is specified by alpha, you may also make a type II error, the probability of which is determined by **beta (β)**. By looking at figure 7.2, you can see the overlap of the score distribution on the dependent variables for x (the sampling distribution if the null hypothesis is true) and y (the sampling distribution if the null hypothesis is false). In this figure, curve x represents the distribution of all possible sample means for a population supporting the null hypothesis while curve y represents the distribution of sample means for a population resulting in rejection of the null hypothesis. Specifying alpha allows for the identification of a critical value for the test statistic that equates to the rightmost 5% of the sampling distribution if H_0 is true. This critical value indicates that the

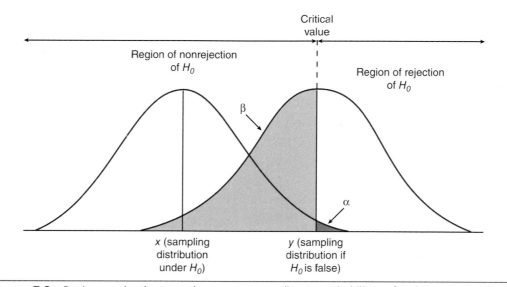

Figure 7.2 Regions under the normal curve corresponding to probabilities of making type I and type II errors.

mean of *y* (given a certain distribution) must be at a specified distance from the mean of *x* before the null hypothesis is rejected because the value of the calculated test statistic is bigger than the critical value. However, if the mean of *y* falls anywhere between the mean of *x* and the *y* that is specified by the critical value (identified by the lightly shaded region in the figure), you could be making a type II error (beta); that is, you do not reject the null hypothesis when, in fact, there is a true difference because some values from *y* fall in the lightly shaded region. As you can see, there is a relationship between alpha and beta; for example, as alpha is set increasingly smaller (i.e., by moving the critical value to the right), beta (the lightly shaded region) becomes larger. Similarly, by making alpha larger (i.e., by moving the critical value to the left), beta becomes smaller.

Meaningfulness (Effect Size)

Besides reporting the significance of the findings, scholars need to be concerned about the **meaningfulness** of the outcomes of their research. The meaningfulness of a difference between two means is in part determined by the magnitude of the difference, which can be estimated in many ways. But as Cohen (1988) suggests, **effect size** is the value that has gained the most attention. You may be familiar with using effect size (ES) in meta-analysis. (If not, you will be soon; see chapter 14 on research synthesis.) The general formula for ES is

$$ES = (M_1 - M_2)/s \tag{7.1}$$

This formula subtracts the mean of one group (M_1) from the mean of a second group (M_2) and divides the difference by the standard deviation (*s*). That places the difference between the means in the common metric called *standard deviation units*, which can be compared to guidelines for behavioral research suggested by Cohen (1988): 0.2 or less is a small ES, about 0.5 is a moderate ES, and 0.8 or more is a large ES. For a more detailed discussion of the use of effect size in physical activity research, see Thomas, Salazar, and Landers (1991).

Many authors, including Pek and Flora (2018) and Thomas (2014), have indicated the need to report some estimate of the magnitude of the effect (effect size) with all tests of significance.

> **meaningfulness**—The importance or practical significance of an effect or relationship.
>
> **effect size**—The standardized value that is the difference between the means divided by the standard deviation, also called *delta*. Effect size gives us an indication of the magnitude of the effect.

Power

Power is the probability of rejecting the null hypothesis when the null hypothesis is indeed false (e.g., detecting a real difference), or the probability of making a correct decision. Power ranges from 0 to 1. The greater the power, the greater the likelihood that you will detect a real difference or relationship. Thus, power increases the odds of rejecting a false null hypothesis. Of course, to a degree in behavioral research, *the null hypothesis is always false!*

What this statement reflects is that in behavioral research, the means of two groups are never exactly the same if you have enough precision in your measurement. Thus, if enough participants are obtained (one way of obtaining power), any two means can be declared significantly different even if the difference between the means is exceedingly small (e.g., 24.001 and 24.002). Remember what finding a significant difference means: If you do the study again, you will get about the same answer. The more interesting questions in behavioral research are the following:

- How large a difference is important in theory or practice?
- How many participants are needed to declare an important difference as significant?

Understanding the concept of power can answer these questions. If a researcher can identify the size of an important effect through previous research or even simply estimate an effect size (e.g., 0.5 is a moderate ES) and establish how much power is acceptable (e.g., a common estimate in the behavioral sciences is 0.8), then the size of the sample needed for a study can be estimated.

Figures 7.3 and 7.4 show the relationships of sample size for each group (*y*-axis), power (*x*-axis), and effect size (ES) curves (or lines, since most of these appear almost linear) when alpha is either .05 or .01, respectively.

Consider the following example. An investigator plans a study that will have two randomly formed groups, but she does not know how many participants are needed for each group to detect a meaningful difference between treatments. However, there are several related studies, and the investigator has calculated an average ES = 0.7 (using equation 7.1) favoring

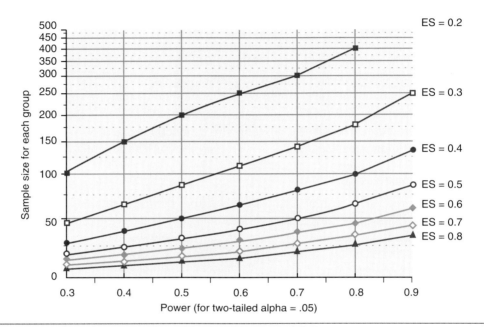

Figure 7.3   Effect size curves when alpha is .05 for a two-tailed test.

Republished with permission of Sage Publications, from *How Many Subjects? Statistical Power Analysis in Research,* C. Kraemer and S. Thiemann, 1987.

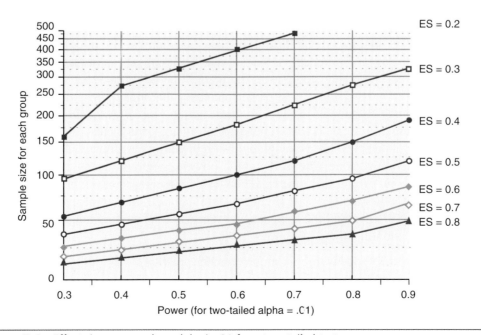

Figure 7.4   Effect size curves when alpha is .01 for a two-tailed test.

Republished with permission of Sage Publications, from *How Many Subjects? Statistical Power Analysis in Research,* C. Kraemer and S. Thiemann, 1987.

the experimental group from the outcomes in these studies. The investigator decides to set alpha = .05 and wants to keep beta at four times the level of alpha (thus, beta = .20) because Cohen (1988) suggested that in the behavioral sciences, the seriousness of type I to type II error should be in a 4-to-1 ratio. Because power is 1 − beta (1.0 − 0.2 = 0.8), power is set at 0.8 (often recommended as an appropriate power in behavioral research). When this information—alpha, ES, and power—is known, then the number of participants needed in each group can be estimated from figure 7.3 when alpha = .05. Follow along the 0.7 ES line to where it passes the x-axis (power) at 0.8. Then read across to the y-axis (sample size), and note that approximately 30 people would be needed in each group.

Note how this relationship works: As the number of people in each group (sample size, y-axis) is reduced, power is reduced (at the same ES). Thus, again looking at figure 7.3, if you reduce the sample size to 25 people (the dotted horizontal line), you would need an ES = 0.8 to maintain power at 0.8. Choose any of the ES lines and notice how as you move left to right along that line, sample size increases (the line is going up) as power increases (reading along the x-axis).

Going back to the initial example with power = 0.8 and ES = 0.7, now look at figure 7.4 to identify the necessary sample size when alpha = .01. Note that for the same level of power (0.8) and ES (0.7), the number of people per group increases from 30 (from figure 7.3, alpha = .05) to 50 (from figure 7.4, alpha = .01). So, if all else stays the same but a more stringent alpha is used (e.g., .05 to .01), a greater number of participants are required to ensure sufficient statistical power to detect a significant difference. Hence, with a more stringent (lower) alpha (e.g., .01), power is reduced, making it more difficult to detect a significant difference. A larger alpha such as .10 or .20 increases power; that is, it lessens the chance of making a type II error (these figures are not shown). Of course, it increases the chance of making a type I error (proclaiming that a difference is present when there is none). As stated before, the researcher has to decide which type of error is more important to avoid. A practical guideline to follow for applied research is that if the costs (with costs broadly defined to include considerations like financial cost, time required, and negative side effects) are the same for both treatments, go for power: Find out whether there is a difference. But if one treatment costs more than the other, avoid the type I error. You do not want to adopt a more costly treatment (cost, time, side effects) if it is not *significantly* better.

The relationships of alpha, sample size, and ES are important. For example, studies with a small sample size require a larger ES or alpha, or both.

The size of the sample (n) is extremely influential on power. Power increases with increased n. Table 7.1 illustrates this statement using an ES of 0.5 and alpha at .05. With a very small number of participants per group, for example, n = 10, the power would be .20 (off the chart in figure 7.3). This is only a 20% chance of detecting a real difference. Conversely, with a very large sample size, such as n = 100, the power is actually about .95, which is approaching certainty of finding a real difference.

TABLE 7.1

Relationship of Sample Size and Power (ES = 0.5, alpha = .05)

n	Power
10	.20*
20	.30
50	.70
75	.85
100	.95*

*These values are estimated from the chart in figure 7.3.

Keep in mind the relationships of alpha, sample size, and ES in planning a study. If you have access to only a small number of participants, then you need to have a really large ES or use a larger alpha, or both. Do not just blindly specify the .05 alpha if detecting a real difference is the main issue. Use a higher one, such as .20 or even .30. This approach is extremely pertinent in pilot studies.

In chapter 9 we discuss some of the things you can do to increase the ES. Sometimes, however, in planning a study, you may not be able to determine an expected ES from the literature. Here you can benefit from a pilot study to estimate the ES.

Our approach in chapter 9 is more **post hoc** because it involves reporting the effect size and variance accounted for in significant findings and interpreting whether this is a meaningful effect. The **a priori** procedures described previously are more desirable but not always applicable. (For a more detailed discussion of the relationship between significance, effect size, and meaningfulness, see Anderson, McCullagh, & Wilson, 2007; Kalinowski & Fidler, 2010; for more information about a priori and post-hoc calculations of power, see O'Keefe, 2007).

Finally, a summary of all of this is well presented in the short poem by Rosenthal (1991, p. 221) in the Achieving Power sidebar.

To summarize, knowledge of four concepts is needed for planning and evaluating any quantitative study: alpha, power, sample size, and effect size.

- *Alpha* establishes the acceptable magnitude of type I error (i.e., the chances of rejecting a true null hypothesis) and the level of significance selected. It is typically an arbitrary value—often .05 or .01 in the behavioral and biological areas of physical activity.

- *Power* represents the chances of rejecting a false null hypothesis. It is based on the magnitude of beta (i.e., the chances of making a type II error, accepting a false null hypothesis). Beta is typically set at $4 \times$ alpha, so if alpha is .05, then beta is .20. Power is typically established as $1 - $ beta or .80 in the behavioral and biological areas of physical activity.

- *Sample size* is the number of participants in the study being evaluated or planned.

- *Effect size* is the outcome of a study typically expressed in standard deviation units or as the percent of variance explained.

Most often in evaluating published studies, all the information to determine the four concepts is included, although everything might not be in final form (e.g., the effect size can be calculated from the means and standard deviations or converted from r).

Achieving Power

I. The Problem

Oh, F is large and p is small
That's why we are walking tall.
What it means we need not mull
Just so we reject the null.
Or chi-square large and p near nil
Results like that, they fill the bill.
What if meaning requires a poll?
Never mind, we're on a roll!
The message we have learned too well?
Significance! That rings the bell!

II. The Implications

The moral of our little tale?
That we mortals may be frail.
When we feel a p near zero
Makes us out to be a hero.
But tell us then is it too late?
Can we perhaps avoid our fate?
Replace that wish to null-reject
Report the *size* of the effect.
That may not insure our glory
But at least it tells a story
That is just the kind of yield
Needed to advance our field.

Republished by permission from Sage Publishing, *Psychological Science, Cumulating Psychology: An Appreciation of Donald T. Campbell*, R. Rosenthal, 1991.

Using Information in the Context of the Study

The item missing in these four concepts that is used to evaluate all research is one of greatest interest: **context**. How do the findings from this study fit into the context of theory and practice? In fact, the question of greatest interest to the reader of research (which should make it of considerable interest to the researcher) is one that often goes unanswered: Are effect sizes for significant findings large enough to be meaningful when interpreted within the context of the study, or for the application of findings to other related samples, or for planning a related study?

In planning research, the experimenter often has either previous findings or pilot research from which appropriate effect sizes of important variables can be estimated. If the experimenter sets alpha at .05 and power at .80, then the question becomes, *How many participants would be needed to detect an effect of a specific magnitude if the treatment works or the relationship is present?* Charts and tables in books (see figures 7.3 and 7.4) as well as computer programs such as G*Power (which is freely available) can be used to estimate the number of participants needed, given that the experimenter has established or estimated the other three components.

One of the major benefits of using effect sizes in informing your interpretation of the meaningfulness of data is that they are not directly sensitive to sample size. Certainly, effect sizes have some sensitivity to sample size because of its influence on the standard deviation size. Effect sizes are based on either the difference between the means (divided by the standard deviation) or the size of the correlation. Table 7.2 provides the magnitude of effect sizes as they reflect how two treatment groups overlap in their scores. Note that when an effect size is 0, the distribution of the scores of both groups overlaps completely and the midpoints are the same (50% of participants' scores on either side of the mean). As effect sizes increase, there is a net gain so that a greater percentage of the participants' scores in one group (typically the experimental) exceeds the midpoint (50%) of the distribution of the other group's scores (control). Effect sizes are often interpreted by their absolute size.

context—The interrelationships found in the real-world setting. (See also chapter 12, which addresses the use of context.)

TABLE 7.2

Interpretation of Effect Size Using the Distribution From the Normal Curve

Effect size	Percentage of experimental group difference above control group mean	Percentage – 50
0.0	50	0
0.1	54	4
0.2	**58**	**8**
0.3	62	12
0.4	65	15
0.5	**69**	**19**
0.6	73	23
0.8	**79**	**29**
1.0	84	34
1.2	88	38
1.5	93	43
2.0	98	48

Bolded rows indicate standard effect size constructs (0.2 = small, 0.5 = moderate, 0.8 = large).

Adapted from McNamara (1994).

In table 7.2, a small effect size (often labeled 0.2 or less) suggests that the experimental group has a net gain (as a result of the treatments) of 8% over the control group; that is, 58% of the distribution of the experimental group's scores is beyond the midpoint (50%) of the control group's scores. A moderate effect size (often labeled about 0.5) shows a net gain of 19% over the midpoint of the control group's scores, whereas a large effect size (often labeled 0.8 and above) has a net gain of 29%. This is an excellent way of describing the influence of changes in an experimental group compared with a control group. That is, if the experimental and control groups begin at similar points on the dependent variable, what percentage of experimental participants' scores are above the midpoint (50%) in the distribution of the control group's scores? Thus, the larger the effect size, the less the overlap between the distribution of scores in the two groups.

Finally, as we move into the next three chapters on statistics, remember that statistical analyses are of value only if the sample is neither too small nor too large. In very small samples, one unusual value can substantially influence the results. Moreover, variation within and between participants (the error variance) tends to be large, which reduces the size of the test statistic, resulting in few significant findings. On the other extreme of sample size, statistics has little value for very large samples because nearly any difference or relationship is significant with a big enough sample size. This is for two reasons: (a) The criterion for the test statistic decreases with increases in sample size (making it easier to find a test statistic that is bigger than the criterion); and (b) the error variance in significance tests is divided by the degrees of freedom. Degrees of freedom are calculated based on sample size, and this value is essentially equal to the number of subjects in your sample (e.g., for a single sample t-test, degrees of freedom = $n - 1$). Thus, you will notice that public health studies which often have extremely large samples sizes tend to discuss data analyses such as risk factors, percentage of the sample with a characteristic (prevalence, incidence, mortality rate), and odds ratios rather than significance because significance is not very informative with such large samples (see chapter 17).

In chapter 3, we discussed the research and null hypotheses. Remember, conceptually the testing of the null hypothesis is done by comparing a test statistic to a criterion value. The tables containing the criterion values are based on population estimates of how often randomly selected samples might differ if there really is no difference. Thus, $p = .05$ means that two randomly selected samples of a certain size (n) will differ by chance only 5 times in 100. Therefore, if we compare two samples and find that $p < .05$, that means the probability of obtaining this test statistic due to chance is less than 5 times out of 100, so the observed differences between the samples is unlikely to be a chance difference (well, it could be due to chance, but only 5 times out of 100). This then leads us to infer that the difference is a real difference between the samples that is reflective of the independent variable that distinguishes them. Two groups of humans or non-human animals, however, are hardly ever exactly alike on a measured variable, and if there are enough participants, p will be less than .05 because of the effects of n on the criterion and on the test statistic. Thus, in isolation, testing the null hypothesis is the wrong concept. The correct questions are *What is a meaningful difference (effect size) within the context of theory and application?* and *How many participants are required to reject a difference this large at* p < .05 *and power* > .80? Of course, $p < .05$ and power > .80 are arbitrary values.

Remember, statistical routines are just that: computer routines into which numbers are input and that provide standardized output based on those numbers. Just as the numbers we assign to variables do not know where they come from (i.e., someone has developed the procedures for assigning values to certain levels of characteristics), statistical computer programs do not know where the numbers come from, nor do they care. You as the researcher must verify that the numbers are good ones. You must verify that the measurements have been properly taken, that the data have been correctly recorded, and that the numbers (data) meet the assumptions used in the development of the tables to which they are being compared (e.g., normal distribution of data for the *F* table or the use of the chi-square table for nonnormal distributions).

Context is what matters with regard to meaningfulness. You must ask yourself, *Within the context of what I do, does an effect of this size matter?* The answer nearly always depends on who you are and what you are doing (and practically never on whether $p = .05$ or $.01$). Thus, having a significant (reliable) effect is a necessary, but insufficient, condition in statistics. To meet the criteria of being both necessary and sufficient, the effect must be significant and meaningful within the context of its use. Said another way,

- estimates of significance are driven by sample size;
- estimates of meaningfulness are driven by the size of the difference; and
- context is driven by how the findings will be used.

> An effect that is significant and meaningful is necessary and sufficient.

It is important that researchers provide context when interpreting the findings of their studies. Historically, quantitative researchers have been inconsistent in providing appropriate contexts for their studies, especially in terms of interpreting the meaningfulness of their findings. Some context may be provided in the introduction to papers when the rationale for the research is developed, but it is the researcher's responsibility to explain why their study is important and how it fits into the scheme of things. In the discussion section, the researcher should explain why the findings are important, how they support (or do not support) a theory, why the findings add to previous knowledge (or dispute it), what the logical subsequent research questions are, and why practitioners should care about the outcomes.

Qualitative researchers are typically much stronger than quantitative researchers at placing findings in context. See Biddle, Markland, Gilbourne, Chatzisarants, and Sparkes (2001) for a good discussion. Of course, this difference occurs because context drives qualitative researchers' research plans, data collection, data analyses, and interpretations of findings. We are not suggesting a switch to qualitative research by quantitative scholars; we have learned much, and much is yet to be learned, from quantitative research. But quantitative researchers can gain considerable benefit by learning to use context as the basis for planning and interpreting research, particularly important outcomes. Mixed-methods research takes advantage of the strength of both quantitative and qualitative methodologies. We discuss mixed methods in chapter 20 and recommend that you give them considerable thought as you plan your research.

Context: The Key to Meaning

Here is a simple example of context. Suppose we told you that we have developed (and advertised on TV) a program of physical training, the Runflex System, that will improve your speed in the 100 m dash by 100 ms. This system requires that you attach cables to your feet from flexible rods, lie on your back, and flex the rods by pulling your knees toward your chest. This system is guaranteed to work in six weeks or you get your money back. Testing by a major university laboratory in exercise physiology shows that the system does produce an average improvement of 100 ms in speed in the 100 m dash. Would you be willing to send us three easy payments of (US)$99.95 for this system? Right now you are thinking to yourself, *Are these people nuts? I don't even run 100 m dashes. Why would I pay that kind of money to improve my performance in an event I don't participate in?* Suppose, however, that you are the second fastest 100 m sprinter in the world, and the fastest sprinter runs 100 ms more quickly than you do. Would you buy our Runflex apparatus now? *Context matters!*

Summary

In this chapter, you learned the interrelationships of alpha, power, sample size, and effect size. Appropriate use of this information is the most important aspect of planning your own study or evaluating someone else's. Placing this information in the context in which you plan the research or plan to use the results allows others to interpret and use your research findings appropriately.

✓ Check Your Understanding

1. Locate a data-based paper from a research journal. Answer the following questions about this paper:
   a. What probability levels did the researchers use to test their hypotheses?
   b. What was the sample size for the group(s)?
   c. Did the authors report effect sizes in their results or discussion? If so, what were the effect sizes reported? If so, use figures 7.3 and 7.4 or G*Power to estimate power using your answers to a, b, and c above.
   d. Did the authors provide a discussion of the meaningfulness of their findings?
2. Using the information provided in table 7.2, draw a figure (such as figure 7.2) with two normal curves (one for an experimental group and one for a control group) that reflects an effect size of 0.2 (a small effect). Do this again for an effect size of 0.5 (a moderate effect). Repeat for an effect size of 0.8 (a large effect).

Relationships of Variables

> **When you are dissatisfied and would like to go back to youth, think of Algebra.**
>
> —Will Rogers

In chapter 6, we promised that after we had presented some basic information to help you understand statistical techniques, we would begin to explain in detail some specific techniques. We begin with correlation.

Correlation is a statistical technique used to determine the relationship between two or more variables. In this chapter, we discuss several types of correlation, the reliability and meaningfulness of correlation coefficients, and the use of correlation for predictions, including partial and semipartial correlations, multiple regression, logistic regression, and discriminant function analysis. Finally, we briefly overview multivariate forms of correlation: canonical correlation, factor analysis, and structural modeling.

correlation—A statistical technique used to determine the relationship between two or more variables.

step test—A test used to measure cardiorespiratory fitness involving the measurement of pulse rate after stepping up and down on a bench.

What Correlational Research Investigates

Often, a researcher is interested in the degree of relationship between, or correlation of, performances, such as the relationship between performances on a distance run and a **step test** as measures of cardiorespiratory fitness. Sometimes, an investigator wishes to establish the relationship between traits and behavior, such as how personality characteristics relate to participation in high-risk recreational activities. Still other correlational research problems might involve relationships between anthropometric measurements, such as skinfold thicknesses and fat percentages as determined by underwater weighing. Here, the researcher may wish to predict fat percentage from the skinfold measurements.

Correlation may involve two variables, such as the relationship between height and weight. It may involve three or more variables, such as when a researcher investigates the relationship between a criterion (dependent variable) such as cardiorespiratory fitness and two or more predictor variables (independent variables) such as body weight, fat percentage, speed, and muscular endurance. This technique is multiple correlation. Another technique, canonical correlation, establishes the relationships between two or more dependent variables and two or more independent variables. Factor analysis uses correlations of a number of variables to try to identify the underlying relationships or factors. Finally, structural modeling provides evidence for how variables may directly or indirectly influence other variables.

Understanding the Nature of Correlation

coefficient of correlation—A quantitative value of the relationship between two or more variables that can range from .00 to 1.00 in either a positive or negative direction.

The **coefficient of correlation** is a quantitative value of the relationship between two or more variables. The correlation coefficient can range from .00 to 1.00 in either a positive or negative direction. Thus, perfect correlation is 1.00 (either +1.00 or −1.00), and no relationship at all is .00 (see sidebar Correlation: Perfect and Not So Perfect for some lighthearted examples of perfect correlations and examples of no correlation).

Positive Correlation

positive correlation—A relationship between two variables in which a small value for one variable is associated with a small value for another variable, and a large value for one variable is associated with a large value for the other.

A **positive correlation** exists when a small value for one variable is associated with a small value for another variable and when a large value for one variable is associated with a large value for another. Strength and body weight are positively correlated: People with higher body weight are generally stronger (physically) than people who weigh less. However, while this is an example of a positive correlation, it is not perfect. In other words, with every generalization comes an exception. Therefore, some people with lower body weight will be stronger than some people with higher body weight and not as strong as some people who weigh even less.

Figure 8.1 is a graphic illustration of a perfect positive correlation involving boys. Notice that Bill's body weight is 70 lb (32 kg), and his strength measure is 150 lb (68 kg). Dick's data are weight = 80 lb (36 kg) and strength = 175 lb (79 kg); the increase continues through Tom's data, where weight = 110 lb (50 kg) and strength = 250 lb (113 kg). Thus, when the scores are plotted, they form a perfectly straight diagonal line. This is perfect positive correlation (*r* = 1.00). The relative positions of the boys' pairs of scores are identical in the two distributions. In other words, each boy is the same relative distance from the mean of each set of scores (the set of scores for weight and the set of scores for strength). Common sense tells us that perfect positive correlation does not exist in human traits, abilities, and performances because of individual variability and other influences.

Figure 8.2 illustrates a more realistic relationship between variables using the PGA data presented at the end of chapter 6. The data are from 30 PGA Tour players in 2019, and the correlation is between the percentage of greens hit in regulation (*x*-axis) and the number of

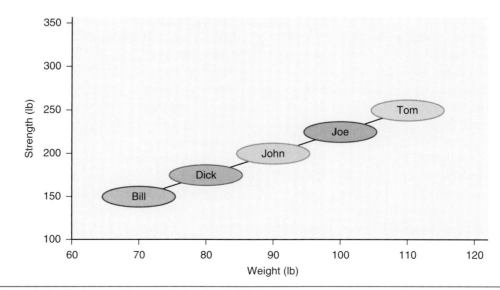

Figure 8.1 Perfect positive correlation (*r* = 1.00).

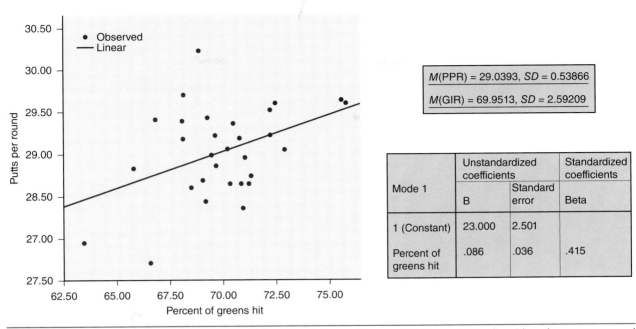

The table within the figure:

Mode 1	Unstandardized coefficients		Standardized coefficients
	B	Standard error	Beta
1 (Constant)	23.000	2.501	
Percent of greens hit	.086	.036	.415

M(PPR) = 29.0393, SD = 0.53866

M(GIR) = 69.9513, SD = 2.59209

Figure 8.2   Relationship between PGA Tour golfers' percentage of greens hit in regulation (x-axis) and putts per round (y-axis). Correlation is $r(28) = .415$, $p = .022$.

Correlation: Perfect and Not So Perfect

Quotes From Anonymous Famous People About Perfect Correlations ($r = 1.00$)

Smoking kills. If you are killed, you've lost a very important part of your life.

Traditionally, most of Australia's imports come from overseas.

It's bad luck to be superstitious.

Things are more like they are now than they ever were before.

The Internet is a great way to get on the net.

The president has kept all the promises he intends to keep.

China is a big country inhabited by many Chinese.

It's like déjà vu all over again.

The loss of life will be irreplaceable.

Students' Statements That Are Uncorrelated to Teaching Performance

Sometimes students write interesting things on the teaching evaluations for our research methods course. Here are a few of the statements that have an $r = .00$ to our teaching.

1. This class was a religious experience for me. I had to take it all on faith.
2. Textbook makes a satisfying "thud" when dropped on the floor.
3. Have you ever fallen asleep in one class and awakened in another?
4. Help! I've fallen asleep and can't wake up.
5. I'm learning by osmosis; I sleep with my head on the textbook.
6. Class is a great stress reliever. I was so confused I forgot who I was.
7. This class kept me out of trouble from 2:30 to 4:00 pm on Tuesday and Thursday.

putts per round (*y*-axis). The correlation is *r* = .415, which is significant at *p* = .02 with 28 degrees of freedom (number of participants, *N* – 2, called *df*). As can be observed, the data points (each point represents one person—read to the *x*-axis for the percentage of greens hit in regulation and to the *y*-axis for putts per round) generally progress from the lower left to the upper right, representing a relationship but not a perfect one. The line is called *the line of best fit* and is calculated to minimize the overall distance (or error) between the actual data point and the line in the *y*-dimension.

Negative Correlation

For an example of negative correlation, we plotted the PGA data for driving distance versus driving accuracy. The correlation is *r* = –.610, *p* < .001, *df* = 28. This means that the correlation is negative—driving the ball a greater distance is negatively related to driving accuracy percentage (i.e., driving the ball into the fairway). This relationship is shown in figure 8.3, in which the general pattern of the data points is from upper left to lower right. This is a **negative correlation**. A perfect negative correlation would be a straight diagonal line at a 45° angle (the upper left corner of the graph to the lower right corner). Figure 8.3 depicts a negative correlation

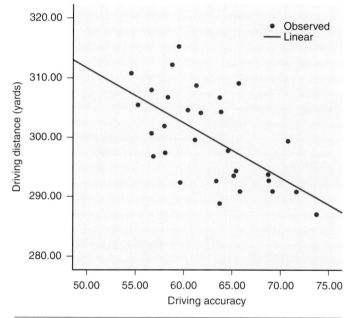

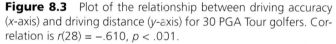
Figure 8.3 Plot of the relationship between driving accuracy (*x*-axis) and driving distance (*y*-axis) for 30 PGA Tour golfers. Correlation is *r*(28) = –.610, *p* < .001.

negative correlation— A relationship between two variables in which a small value for the first variable is associated with a large value for the second variable, and a large value for the first variable is associated with a small value for the second variable.

of a moderate degree (*r* = –.610), but an upper-left-to-lower-right pattern is still apparent.

Patterns of Relationships

Figure 8.4 is a hypothetical example of four patterns of relationships between two variables. Figure 8.4*a* is a positive relationship as previously described. Figure 8.4*b* is a negative relationship, also previously described. When virtually no relationship exists between variables, the correlation is .00, as shown in figure 8.4*c*. This example denotes independence between sets of scores meaning that changes in the *x* variable are not at all related to changes in the *y* variable. The plotted scores exhibit no discernible pattern at all. Finally, two variables may not have a linear relationship but can still be related, as in figure 8.4*d*, which shows a curvilinear relationship. The interpretation of the reliability and meaningfulness of correlations is explained later in this chapter.

Correlation and Causation

At this point, we must stress again that a correlation between two variables does not mean that one variable causes the other. In chapter 4, we used an example of a research hypothesis (albeit a dumb one) that achievement in math could be improved by buying a child larger shoes. This hypothesis resulted from a correlation between math achievement scores and

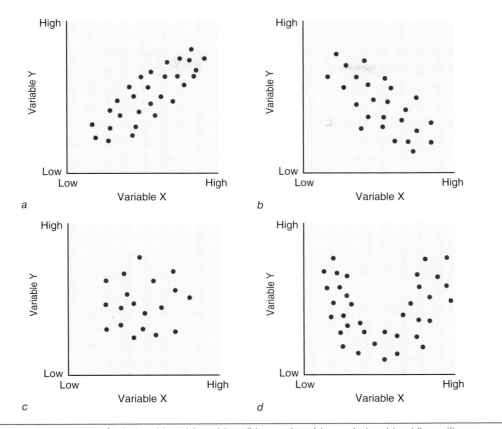

Figure 8.4   Patterns of relationships: *(a)* positive, *(b)* negative, *(c)* no relationship, *(d)* curvilinear.

shoe size in elementary school children (with age not controlled). This example helps to drive home the point that correlation does not mean causation. Just because two variables are related that does not mean that you can change one of the variables to *cause* a change in the other.

In Ziv's classic 1988 study on the effectiveness of teaching and learning with humor, the participants in the experimental group were told a story to illustrate the fact that correlations do not show a causal effect. In the story, aliens from another planet, who were invisible to earthlings, decided to study the differences between overweight people and healthy weight people. One alien observed the coffee-drinking behavior of overweight and healthy weight people at a cafeteria over an extended period. The alien then analyzed the observations statistically and decided that there was a correlation between coffee-drinking behavior and body weight. Healthy weight people tended to drink their coffee with sugar, whereas overweight people mostly drank theirs with an artificial sweetener. The alien concluded from this study that sugar makes humans thin, whereas artificial sweeteners increase body weight. Clearly, this is an incorrect causal conclusion drawn from variables that most likely represent a third untested variable, which is the person's effort (or lack thereof) toward weight loss.

We are not saying that one variable cannot be the cause of another; in fact, using our example of a negative correlation, it is likely that driving the golf ball farther does cause a decrease in driving accuracy (as determined by the percentage of fairways that the player hits), but establishing that relationship through a correlational analysis is not enough to establish cause and effect. For a cause-and-effect relationship to be identified, the two variables must be correlated. Correlation is a necessary but not sufficient condition for causation. The only way causation can be shown is with an experimental study in which an independent variable is manipulated to bring about an effect in the dependent variable.

Pearson Product Moment Coefficient of Correlation

> Although two variables must be correlated for a cause-and-effect relationship to exist, correlation alone does not guarantee such a relationship.

Pearson product moment coefficient of correlation—The most commonly used method of computing correlation between two variables; also called *interclass correlation, simple correlation, Pearson correlation, Pearson r,* or *r*.

Several times in the preceding discussion, we used the symbol r. This symbol denotes the **Pearson product moment coefficient of correlation** also more simply called the Pearson correlation. This type of correlation has one criterion (dependent) variable and one predictor (independent) variable. Thus, every participant has two scores that are included in the correlation, such as driving distance and driving accuracy. An important assumption for the use of r is that the relationship between the variables is expected to be linear, and therefore, a straight line is the best model of the relationship. When this is not true (e.g., figure 8.4*d*), r is an inappropriate way to analyze the data. If you were to apply the statistical formula to calculate r to the data in figure 8.4*d*, the outcome would be an r of about .00, indicating no relationship. Yet we can clearly see from the plot that a relationship does exist; it is just not a linear one.

The computation of the correlation coefficient involves the relative distances of the scores from the two means of the distributions. The computations can be accomplished with a number of formulas; we present just one. This formula is sometimes called *the computer method* because it involves operations similar to those performed by a computer. While the formula appears large and imposing, it actually consists of only three operations:

1. Sum each set of scores.
2. Square and sum each set of scores.
3. Multiply each pair of scores and obtain the cumulative sum of these products.

The formula is

$$r = \frac{n\Sigma XY - (\Sigma X)(\Sigma Y)}{\sqrt{n\Sigma X^2 - (\Sigma X)^2}\sqrt{n\Sigma Y^2 - (\Sigma Y)^2}} \tag{8.1}$$

To illustrate, we will again use the PGA Tour data but just those from the top 10 players based upon money won to date (example 8.1). Note that in figure 8.5, the correlation for the top 10 players is quite high, $r(8) = .843$ (see table 8.1 for SPSS output on this correlation). In the example, you can see the hand calculations of this r using equation 8.1. Note that N refers to the number of *paired* scores, not the total number of scores.

EXAMPLE 8.1

Known Values

2019 Data and Calculation of *r* for Greens Hit in Regulation and Putts per Round for Top 10 PGA Tour Players

Player	Greens hit in regulation (X)	X^2	Putts/round (Y)	Y^2	XY
Rory McIlroy	71.01	5,042.42	28.97	839.26	2,057.16
Matt Kuchar	75.76	5,739.58	29.61	876.75	2,243.25
Xander Schauffele	71.20	5,069.44	28.66	821.40	2,040.59
Dustin Johnson	70.83	5,016.89	28.66	821.40	2,029.99
Brooks Koepka	69.63	4,848.34	29.23	854.39	2,035.28
Paul Casey	70.47	4,966.02	29.37	862.60	2,069.70
Francesco Molinari	63.43	4,023.36	27.96	781.76	1,773.50
Gary Woodland	72.45	5,249.00	29.61	876.75	2,145.24
Rickie Fowler	69.17	4,784.49	28.45	809.40	1,967.89
Tiger Woods	75.56	5,709.31	29.65	879.12	2,240.35
SUM	**709.51**	**50,448.85**	**290.17**	**8,422.83**	**20,602.97**

Number of paired scores: $n = 10$

Sum of greens hit in regulation: $\Sigma X = 709.51$

Sum of putts per round: $\Sigma Y = 290.17$

Sum of greens hit in regulation squared: $\Sigma X^2 = 50{,}448.85$

Sum of putts per round squared: $\Sigma Y^2 = 8{,}422.83$

Sum of greens hit × putts: $\Sigma XY = 20{,}602.97$

WORKING IT OUT (EQUATION 8.1)

$$r = \frac{n \sum XY - (\sum X)(\sum Y)}{\sqrt{n \sum X^2 - (\sum X)^2} \sqrt{n \sum Y^2 - (\sum Y)^2}}$$

$$r = \frac{(10)20{,}602.97 - (709.51)(290.17)}{\sqrt{10(50{,}448.85) - 709.51^2}\sqrt{10(8422.83) - 290.17^2}}$$

$$r = \frac{206{,}029.7 - 205878.52}{\sqrt{1084.10}\sqrt{29.70}} = \frac{151.18}{179.43} = 0.84$$

TABLE 8.1
Correlation Between Greens Hit in Regulation and Putts per Round for Top 10 PGA Tour Players

			Model summary
R	**R^2**	**Adjusted R^2**	**Standard error of the estimate**
0.843	0.710	0.674	0.328

Note: The independent variable is greens hit in regulation.

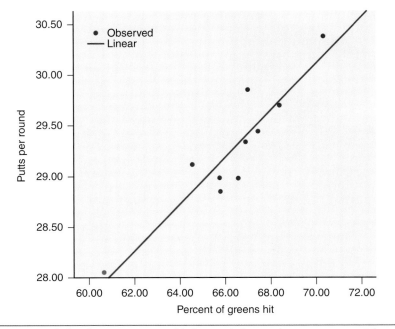

Figure 8.5   Correlation between greens hit in regulation and putts per round for top 10 PGA Tour players in 2019.

For the $(\Sigma X)^2$ and the $(\Sigma Y)^2$ values, the raw scores for X and Y are first summed and then that total is squared. Notice that these values are not the same as the ΣX^2 and ΣY^2 values, for which X and Y are squared first and then those values are summed. The ΣXY (the sum of the cross products of the X and Y scores) is calculated by multiplying each value of X (greens in regulation) by its corresponding value of Y (putts per round) and then adding those products. The sign of the ΣXY determines the direction of the correlation—that is, whether it is positive or negative.

In a correlation problem that simply determines the relationship between two variables, it does not matter which one is X and which is Y. If the investigator wants to predict one score from the other, then Y designates the criterion (dependent) variable (what is being predicted) and X the predictor (independent) variable. In this example, putts per round would be predicted from greens hit in regulation; it would not make much sense to predict in the other direction because hitting the green occurs before putting in golf. Prediction equations are discussed later in this section.

Let's turn our attention to another example from kinesiology. The relationship between a person's running distance in 12 minutes and their resting heart rate results in a negative correlation because a greater distance covered as well as a lower heart rate at rest are beneficial. In other words, both are indicative of better fitness. So, a person who has good cardiorespiratory endurance would have a high score on one test (the run) and a low score on the other (resting heart rate). So, whether the correlation value is negative or positive depends not only on the relationship but also on the scoring direction of the X and Y variables.

What the Coefficient of Correlation Means

So far, we have dealt with the direction of correlation (positive or negative) and the calculation of r. An obvious question that arises is: What does a coefficient of correlation mean in terms of being high or low, satisfactory or unsatisfactory? This seemingly simple question is not so simple to answer.

© Joe Raedle/Staff

Relationships are unique—and the future of some may be hard to predict.

Interpreting the Reliability of *r*

First, there are several ways of interpreting *r*. One criterion is its reliability, or **significance**. Does it represent a real relationship? That is, if the study were repeated, what is the probability of finding a similar relationship? For this statistical criterion of significance, simply consult a table. In using the table, select the desired level of significance, such as the .05 level. Then find the appropriate degrees of freedom *(df)*: This is based on the number of participants corrected for sample bias, which, for *r*, is equal to $n - 2$. Appendix table 2 contains the necessary correlation coefficients for significance at the .05 and .01 levels. Refer to the example of the correlation between greens hit in regulation and putts per round ($r = .843$). The degrees of freedom are:

> $n - 2 = 10 - 2 = 8$ (Remember, the variable *n* in correlation refers to the number of pairs of scores which is also the number of participants in the sample.)

When reading the table at 8 *df*, we see that a correlation of .6319 is necessary for significance of a two-tailed test at the .05 level (and .7646 at the .01 level). Therefore, we would conclude that our correlation of .843 is significant at $p < .01$ because 0.843 is greater than the critical value for .01, which is .7645. (We explain whether to find the correlation values under the column for a one-tailed or two-tailed test in the Interpreting *t* section in chapter 9.)

Another glance at Appendix table 2 reveals a couple of obvious facts. The correlation needed for significance decreases with increased numbers of participants *(df)*. In our example, we had only 10 participants (or pairs of scores). Very low correlation coefficients can be significant if you have a large sample of participants. At the .05 level, $r = .3809$ is significant with 25 *df*; $r = .2732$ is significant with 50 *df*; and $r = .1946$ is significant with 100 *df*. Although it's not shown in the table, with 1,000 *df*, a correlation as low as .08 is significant at the .01 level.

The second observation to note from the table is that a higher correlation is required for significance at the .01 level than at the .05 level. This statement should make sense. Remember, chapter 7 stated that the .05 level means that if 100 experiments were conducted, the null hypothesis (that there is no relationship) would be rejected incorrectly, just by chance, on 5 of the 100 occasions. At the .01 level, we would expect a relationship of this magnitude because of chance less than once in 100 experiments. Therefore, the test of significance at the .01 level is more stringent than at the .05 level, so a higher correlation is required for significance at the .01 level. When you use most computer programs to calculate statistics such as *r*, you are given the exact significance level in the printout, and then you compare this value to the alpha (α) level you have selected. So, if you selected α =.05 prior to conducting the analysis, any *p* values that were less than .05 would indicate a significant correlation.

significance—The reliability of or confidence in the likelihood of a statistic occurring again if the study were repeated.

Interpreting the Meaningfulness of *r*

The interpretation of a correlation for statistical significance is important, but because of the vast influence of sample size, this criterion is not always meaningful. As chapter 7 explained, statistics can answer two questions about data: Is the effect reliable? and What is the effect's magnitude? It is then up to the researcher to use this information and the context of each question to judge if the effect is meaningful.

The most commonly used criterion for interpreting the meaningfulness of the correlation coefficient is the **coefficient of determination** (r^2). In this method, the portion of common association of the factors that influences the two variables is determined. In other words, the r^2 indicates the portion of the total variance in one measure that can be explained, or accounted for, by the variance in the other measure.

coefficient of determination (r^2)—The squared correlation coefficient; used in interpreting the meaningfulness of correlations.

Figure 8.6 offers a visual depiction of this idea. Circle A represents the variance in one variable, and circle B represents the variance in a second variable. At the left is a relationship in which $r = .00$ (no overlap between the two); thus, $r^2 = .0$. At the right, $r = .71$; thus, $r^2 = .50$, a considerable overlap between the two.

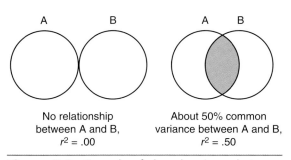

No relationship between A and B, $r^2 = .00$

About 50% common variance between A and B, $r^2 = .50$

Figure 8.6   Example of shared variance between variables.

For example, the standing long jump and the vertical jump are common tests of explosive power. We tend to think of them as interchangeable; that is, as measuring the same thing. Yet correlations between the two tests usually range between .70 and .80. Hence, the coefficients of determination range from .49 ($.70^2$) to .64 ($.80^2$). Usually, the coefficient of determination is expressed as the percentage of variation. Thus, $.70^2 = .49 = 49\%$ and $.80^2 = .64 = 64\%$.

For a correlation of .70 between the standing long jump and the vertical jump, only about half (49%) of the variance (or influence) in one test is associated with the other. Both tests involve common factors such as: (a) the explosive force of the legs with some flexion and extension of the trunk and swinging of the arms, (b) the influence a participant's body weight has on propelling the participant's body through space, (c) the ability to prepare psychologically and physiologically to generate explosive force, and (d) **relative strength**. For example, if $r = .80$, then one test's factors of performance inform, or explain, 64% of the other test's performance results.

But what about the unexplained variance ($1.0 - r^2$)? A correlation of .70 is made up of 49% common (explained) variance and 51% ($1.00 - .49$) error (unexplained) variance. What are some factors unique to each test? We cannot explain these differences fully, but some of the factors could be that (a) the standing long jump requires that the body be propelled forward and upward, whereas the vertical jump is only upward; (b) the scoring of the vertical jump neutralizes a person's height because standing reach is subtracted from jumping reach, but in the standing long jump, the taller person may have some advantage; and (c) perhaps more skill (coordination) is involved in the vertical jump because the person must jump, turn, and then touch the wall. This explanation is not intended to be any sort of mechanical analysis of the two tests. These are simply suggestions of some possible factors of common association, or explained variance, and some factors that might be unexplained or unique to each test of explosive power.

When we use the coefficient of determination to interpret correlation coefficients, it becomes apparent that a rather substantial relationship is needed to account for a great amount of common variance. A correlation of .71 is required to account for just half the variance in the other test, and a correlation of .90 accounts for only 81%. In some of the standardized tests used to predict academic success, correlations are generally quite low, often around .40. You can see by the coefficient of determination that a correlation of .40 accounts for only 16% of variance in academic success; therefore, unexplained variance is substantial. Still, these measures are often used as the criterion for admission to academic programs. Of course, the use of multiple predictors can greatly improve the estimate for success. The sizes of correlations can also be compared using the coefficient of determination. A correlation of .90 should not be interpreted as being three times larger than a correlation of .30; rather, it should be interpreted as being nine times larger because the comparison should be made in terms of the percent of variance explained. $.30^2 = .09$, or 9%, and $.90^2 = .81$, or 81%, so 81% is 9 times larger than 9%.

Interpreting the correlation coefficient is further complicated because whether a correlation is sufficient or inadequate again depends on the context. For example, if we are looking at the reliability (repeatability) of a test, a much higher correlation is needed than

relative strength—The measure of a person's ability to exert maximal force in relation to their size.

if we are determining simply whether a relationship exists between two variables. A correlation of .60 would not be acceptable for the relationship between two similar versions of an exercise knowledge test because we expect the two versions to be assessing the same content. However, a correlation of .60 between exercise knowledge and exercise behavior would be noteworthy because this suggests a relationship between two unique constructs. Similar to the conversation in chapter 7, context means everything when determining the meaningfulness of a significant relationship.

Z Transformation of *r*

Occasionally, a researcher wants to determine the average of two or more correlations. Trying to average the coefficients themselves is statistically unsound because the sampling distribution of coefficients of correlation is not normally distributed. In fact, the higher the correlation, in either a positive or a negative direction, the more skewed the distribution becomes. The most satisfactory method of approximating normality of a sampling distribution of linear relationships is by transforming coefficients of correlation to Z values. This is often called the **Fisher Z transformation**. *Note:* This Z should not be confused with the z used to refer to the height of the ordinate in the area of the normal curve.

Fisher *Z* transformation—A method of approximating normality of a sampling distribution of linear relationship by transforming coefficients of correlation to *Z* values.

The transformation procedure involves the use of natural logarithms, but we need not use Fisher's formula to calculate the transformations since they've already been done (see Appendix table 3). We simply consult the table and locate the corresponding Z value for any particular correlation coefficient.

Suppose, for example, that we obtained correlations between maximal oxygen consumption and a distance run (e.g., an eight-minute run–walk) on four groups of participants of different ages. We would like to combine these sample correlations to obtain a valid and reliable estimate of the relationship between these two measures of cardiorespiratory endurance. Table 8.2 shows the data for the following steps.

1. Convert each correlation to a Z value using Appendix table 3. For example, the correlation of .69 for the 13- and 14-year-olds has a corresponding Z value of 0.8480; the next correlation of .85 for the 15- and 16-year-olds has a Z value of 1.2562. We find the Z values for the other two age groups.

2. Then, weight the Z values by multiplying them by the degrees of freedom for each sample, which, in this process, is $N - 3$. So, for the 13- and 14-year-olds, the Z value of 0.8480 is multiplied by 27 for a weighted Z value of 22.90. We do the same for the other three samples.

3. Sum the weighted Z values and calculate the mean weighted Z value by dividing by the sum of degrees of freedom:

   $$137.41 / 135 = 1.02$$

4. Consult Appendix table 3 to convert the mean weighted Z value back to a mean correlation. We see that the corresponding correlation for a Z value of 1.02 is .77.

TABLE 8.2

Average of Correlation Coefficients by Use of the *Z* Transformation

Age group	N	*r*	Z	N – 3	Weighted Z
13-14	30	.69	0.8480	27	22.90
15-16	44	.85	1.2562	41	51.50
17-18	38	.70	0.8673	35	30.36
19-20	35	.77	1.0203	32	32.65
Total across age groups	—	—	—	135	137.41

Some authors declare that to average correlations by the Z-transformation technique, you must first establish that there are no significant differences between the four correlations. A comparison for differences could be made using a chi-square test of the weighted Z values (chi square is discussed in chapter 10). Other statisticians contend that averaging coefficients of correlation is permissible as long as the average correlation is not interpreted in terms of confidence intervals.

The Z transformation is also used for statistical tests (such as those for the significance of the correlation coefficient) and for determining the significance of the difference between two correlation coefficients. After reading chapter 9, you may wish to consult a statistics text, such as Salkind (2007), for further discussion on the use of the Z transformation for these procedures.

Using Correlation for Prediction

prediction equation—A formula to predict some criterion (e.g., some measure of performance) based on the relationship between the predictor variable(s) and the criterion; also called *regression equation*.

We have stated several times that one purpose of correlation is prediction. As an example of prediction, college entrance examinations are used to predict success in college. In other words, based upon how someone performs today on a standardized test, we believe we can predict their ability to perform well in their college courses. Sometimes we try to predict a criterion such as fat percentage by the use of skinfold measurements or maximal oxygen consumption by a distance run in a given amount of time. In studies of this type, the predictor variables (skinfold measurements or distance run) are less time-consuming, less expensive, and more feasible for mass testing than the criterion variables (underwater weighing or a $\dot{V}O_2$max test to assess maximal oxygen consumption, respectively); thus, a **prediction**, or regression, **equation** is developed.

Prediction is based on correlation. The higher the relationship is between two variables, the more accurately you can predict one from the other. If the correlation were perfect, which it never is with humans, you could predict with complete accuracy.

Working With Regression Equations

The greater the correlation between the variables, the more accurately one can be predicted from the other.

Of course, we do not encounter perfect relationships in the real world. However, in introducing the concept of prediction (regression) equations, it is often advantageous to begin with a hypothetical example of a perfect relationship.

Verducci (1980) provided one of the best examples in introducing the regression equation concerning monthly salary and annual income. If there are no other sources of income, we can predict with complete accuracy the annual income of, for example, teachers simply by multiplying their monthly salaries by 12. This is a mathematically true statement so clearly doesn't reflect relationships between variables that are actually distinct. The equation for prediction ($\hat{Y}$, predicted annual income) is thus $\hat{Y} = 12X$. The solid line in figure 8.7 illustrates this perfect relationship with data for monthly salary and annual salary from 5 teachers whose data are shown in table 8.3. By plotting the monthly salary (the X, or predictor variable), the predicted annual income (the $\hat{Y}$, or criterion variable) can be obtained using this predicted relationship between the variables. Thus, if we know that a teacher (e.g., Ms. Brooks) earns a monthly salary of $1,750, we can easily plot this on the graph by finding $1,750 on the horizontal (x) axis (the **abscissa**). Then we can use the solid line demonstrating the relationship to predict salary by observing that this point on the line is directly across from the vertical (y) axis (the **ordinate**) at $21,000. Thus, Ms. Brooks' annual salary can be predicted from the graph. In this example, her annual salary can also be predicted from monthly salary (X) using the formula:

abscissa—The horizontal (x) axis of a graph.

ordinate—The vertical (y) axis of a graph.

$$\hat{Y} = 12(1,750) = 21,000$$

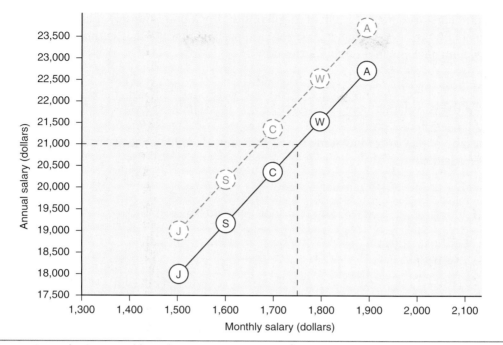

Figure 8.7 Plotting monthly and annual salaries with perfect *r*. *Note:* Letters refer to teachers' initial for last name.

TABLE 8.3

Data for Teachers

Teacher	Monthly salary (X)	Annual salary ($\hat{Y}$)
Mr. Jones (J)	$1,500	$18,000
Mrs. Sipe (S)	$1,600	$19,200
Mr. Campbell (C)	$1,700	$20,400
Ms. Wagner (W)	$1,800	$21,600
Mrs. Apple (A)	$1,900	$22,800

Thus, although we do not have access to Ms. Brooks' annual salary, we can predict it from the equation or the graph and based upon knowing her monthly salary.

Next, suppose that all teachers got an annual supplement of $1,000 for coaching or supervising cheerleaders or some other extracurricular activity. Now, the formula changes:

$$\hat{Y} = 1,000 + 12X$$

Ms. Brooks' annual income is now predicted as follows:

$$\hat{Y} = 1,000 + 12(1,750) = 22,000$$

All teachers' annual incomes could be predicted in the same manner and are displayed in figure 8.7 with the dashed line. This formula is the general formula for a straight line (remember back to middle school where you learned this) and is expressed as follows:

$$\hat{Y} = a + bX \tag{8.2}$$

where $\hat{Y}$ = the predicted score, or criterion; a = the intercept (the value of $\hat{Y}$ when $X = 0$); b = the slope of the regression line; and X = the predictor.

In this example, the b factor was ascertained by common sense because we know that there are 12 months in a year. The slope of the line (b) signifies the amount of change in $\hat{Y}$

that accompanies a change of 1 unit of X. Therefore, any X unit (monthly salary) is multiplied by 12 to obtain the $\hat{Y}$ value. In actual regression problems, we do not intuitively know what b is, so we must calculate it by this formula:

$$b = r(s_Y / s_X) \tag{8.3}$$

where r = the correlation between X and Y; s_Y = the standard deviation of $\hat{Y}$; and s_X = the standard deviation of X.

In our previous example including the $1,000 annual supplement, application of equation 8.3 uses these data:

X (monthly salary)   Y (annual income)

$M_X = 1{,}700$ $M_Y = 21{,}400$ (including $1,000 bonus)

$s_X = 141.42$ $s_Y = 1{,}697.06$

$r = 1.00$

therefore,

$$b = 1.00(1{,}697.06 / 141.42) = 12$$

The a in the regression formula indicates the intercept of the regression line on the y-axis. In other words, a is the value of Y when X = 0. On a graph, if you extend the regression line sufficiently, you can see where the regression line intercepts the y-axis. The a is a constant because it is added to each of the calculated bX values. Again, in our example, we know that this constant is 1,000. In other words, this is the value of Y even if there were no monthly salary (X). But to calculate the value of a, you must first calculate b. Then use the following formula:

$$a = M_Y - bM_X \tag{8.4}$$

where a = the constant (or intercept), M_Y = the mean of the Y scores, b = the slope of the regression line, and M_X = the mean of the X scores. In our example, this equates to the following:

$$a = 21{,}400 - 12(1{,}700) = 1{,}000.$$

Then the final regression equation becomes this:

$$\hat{Y} = a + bX, \text{ or } \hat{Y} = 1{,}000 + 12X$$

Next, let's use a more practical example in which the correlation is not 1.00. We can use the data from figure 8.2, in which the correlation between putts per round and greens in regulation was 0.42. The means and standard deviations are as follows:

X (greens in regulation) Y (putts per round)

$M_X = 69.95$ $M_Y = 29.04$

$s_X = 2.59$ $s_Y = 0.54$

$r = .415$

First, we calculate b as follows from equation 8.3 (see figure 8.2 for SPSS printout; data are within rounding error):

$$b = r(s_Y / s_X) = .415(0.54 / 2.59) = 0.09$$

Then, a is calculated as follows from equation 8.4:

$$a = M_Y - bM_X = 29.04 - 0.09(69.95) = 22.98$$

The regression formula (equation 8.2, $\hat{Y} = a + bX$) becomes the following:

$$\hat{Y} = 22.98 + 0.09X$$

For any PGA Tour player's greens in regulation (GIR), we can calculate the predicted putts per round (PPR). For example, a player who hits 60% of the GIR would have a predicted PPR of this:

$$\hat{Y} = 23.01 + 0.09(60) = 28.41$$

Conceptually, this relationship is counterintuitive because it suggests that the golfers who are hitting a higher percentage of GIR are also taking more PPR. The main difference between this example and that of monthly and annual salaries is that the latter had no error of prediction because the correlation was 1.00. When we predicted PPR from GIR, however, the correlation was less than 1.00, so an error of prediction is present.

Line of Best Fit

Figure 8.2 shows a straight line demonstrating the predicted relationship between the GIR and PPR scores. This **line of best fit** is used to predict $\hat{Y}$ from the X scores.

This line passes through the intersection of the X and Y means. You can confirm this by drawing a vertical line at $X = 69.95$ and a horizontal line at $Y = 29.04$. Note again our previous caution: This type of correlation is useful only if a straight line is the best fit to the data. This assumption that a straight line fits the data applies to all the statistical techniques that we discuss in chapters 8, 9, and 10, as does the assumption that the data are normally distributed.

Figure 8.2 shows this line of best fit. You can readily see that the scores do not fall on the straight line, as they did with the perfect correlation example. These differences between predicted ($\hat{Y}$) and actual (Y) scores represent the errors of prediction and are called **residual scores**. If we compute all the residual scores, the mean of these scores will always be zero, and the standard deviation of these residual scores is called the **standard error of prediction** ($S_{Y \cdot X}$).

With a small sample size (i.e., a small number of scores), it is easy to calculate the residuals by hand. But with a larger number of scores, a simpler way of obtaining the standard error of prediction is to use this formula:

$$S_{Y \cdot X} = S_Y \sqrt{1 - r^2} \tag{8.5}$$

The standard error of prediction is interpreted the same way as the standard deviation. In other words, the predicted value (PPR) of the player in our example, plus or minus the standard error of prediction, occurs approximately 68 times out of 100.

The larger the correlation is, the smaller the standard error of prediction is. Also, the smaller the standard deviation of the criterion is, the smaller the standard error of prediction is.

The line of best fit is sometimes called the *least-squares method*. This means that the calculated regression line is one for which the sum of squares of the vertical distances of every point from the prediction line is minimal. We will not develop this point here. Sum of squares is discussed in chapter 9.

Partial Correlation

The correlation between two variables is sometimes misleading. Interpretation may be difficult when there is little or no correlation between the variables other than that caused by their common dependence on a third variable.

line of best fit—The calculated regression line that results in the smallest sum of squares of the vertical distances of every point from the line.

residual scores—The difference between the predicted and actual scores that represents the error of prediction.

standard error of prediction—The computation of the standard deviation of all of the residual scores of a population; the amount of error expected in a prediction. Also called *standard error of estimate*.

partial correlation—A technique in which we assess the correlation between variables 1 and 2 after controlling for the effects of variable 3 on both of those variables.

spurious correlation—A relationship in which the correlation between two variables is due primarily to the common influence of another variable.

For example, many attributes increase regularly with age from 6 to 18 years, such as height, weight, strength, mental performance, vocabulary, and reading skills. Over a wide age range, the correlation between any two of these measures will almost certainly be positive and will probably be high because of the common maturity factor with which they are highly correlated. In fact, the correlation may drop to zero if the variability caused by age differences is eliminated. We can control this age factor in one of two ways. We can select only children of the same age, or we can remove the influence of the effects of age statistically.

The symbol for **partial correlation** is $r_{12 \cdot 3}$, which means the correlation between variables 1 and 2 with variable 3 controlled statistically (we could control for any number of variables, e.g., $r_{12 \cdot 345}$). The calculation of partial correlation of three variables is simple. Let us refer to the previous correlation of shoe size and achievement in mathematics. This is a good example of a **spurious correlation**, which means that the correlation between the two variables is due to the common influence of another variable (age or maturing). When the effect of the third variable (age) is removed, the correlation between shoe size and achievement in mathematics diminishes or vanishes completely. We label the three variables as follows: 1 = math achievement, 2 = shoe size, and 3 = age. Then, $r_{12 \cdot 3}$ is the partial correlation between variables 1 and 2 with 3 statistically controlled. Suppose the correlation coefficients between the three variables were $r_{12} = .80$; $r_{13} = .90$, and $r_{23} = .88$. The formula for $r_{12 \cdot 3}$ is the following:

$$r_{12 \cdot 3} = \frac{r_{12} - r_{13} r_{23}}{\sqrt{1 - r_{13}^2} \sqrt{1 - r_{23}^2}} \quad (8.6)$$

$$= \frac{.80 - (.90 \times .88)}{\sqrt{1 - .90^2} \sqrt{1 - .88^2}}$$

$$= .039$$

Thus, we see that correlation between math achievement and shoe size drops to about zero when age is statistically controlled.

The primary value of partial correlation is to develop a multiple-regression equation with two or more predictor variables. In the selection process, when a new variable is "stepped in," its correlation with the criterion is determined with the effects of the preceding variable statistically controlled. The size and the sign of a partial correlation may be different from the zero-order (two-variable) correlation between the same variables. The zero-order correlation is the original correlation between the two variables (without considering any other variables).

Semipartial Correlation

semipartial correlation—A technique in which we assess the correlation between variables 1 and 2 after controlling for the effects of variable 3 on one of those two variables.

In the previous section on partial correlation, the effects of a third variable on the relationship between two other variables were eliminated by using the formula for partial correlation. In other words, in $r_{12 \cdot 3}$, the relationship of variable 3 to the correlation of variables 1 and 2 is removed. In some situations, the investigator may wish to remove the effects of a variable from only one of the variables being correlated. This is called **semipartial correlation**. The symbol is $r_{1(2 \cdot 3)}$, which indicates that the relationship between variables 1 and 2 is determined after the influence of variable 3 on variable 2 has been eliminated.

Suppose, for example, that a researcher is studying the relationships between rating of perceived exertion (RPE, how hard a person feels they are working), heart rate (HR), and workload. Obviously, workload is going to be correlated with HR. The researcher wants to investigate the relationship between RPE and HR while controlling for workload. Regular partial correlation will show this relationship, but it will remove the effects of workload on the relationship between RPE and HR. But the researcher does not want to remove the effects of workload on the relationship of RPE and HR; they want to remove only the effects of workload on HR. In other words, the main interest is in the net effect of HR on RPE after the influence of workload has been removed. Thus, in semipartial correlation, the effect of workload is removed from HR but not from RPE.

Procedures for Multiple Regression

Multiple regression involves one dependent variable (usually a criterion of some sort) and two or more predictor variables (independent variables). The use of more than one predictor variable usually increases the accuracy of prediction. This should be self-evident. If you wished to predict scoring average on the PGA Tour, you would expect to get a more accurate prediction by using several skill results (e.g., driving distance, greens in regulation, putts per round) rather than by using only one.

The multiple correlation coefficient (R) indicates the relationship between the criterion and a weighted sum of the predictor variables. It follows then that R^2 represents the amount of the variance of the criterion that is explained or accounted for by the combined predictors. This is similar to the coefficient of determination (r^2), which was discussed earlier with regard to the common association between two variables. Now, however, we have the amount of association between one variable (the criterion) and a weighted combination of variables.

We wish to find the best combination of variables that gives the most accurate prediction of the criterion. Therefore, we are interested in knowing how much each of the predictors contributes to the total explained variance. Another way to say this is that we want to find the variables that most reduce the prediction errors. From a practical standpoint, in terms of the time and effort involved in obtaining measures of the predictor variables, it is desirable to find the fewest number of predictors that account for most of the variance of the criterion. Several selection procedures can be used for this purpose. (For additional information on multiple regression, see Cohen, Cohen, West, & Aiken, 2003.)

In deciding which predictors (or sets of predictors) to use, standard statistical packages offer several options. Table 8.4 includes the most common procedures. One option is simply to use all the predictor variables (i.e., to enter all predictor variables in the same step of the regression). But the fewer predictor variables that can be used while still obtaining a good prediction, the more economical and valuable the prediction equation is. To reduce the number of predictors to those that most efficiently account for the greatest percentage of the variance, sometimes researchers specify the order and combinations of predictor variables. In other words, they dictate which predictor variables should be entered into the regression

multiple regression—A model used for predicting a criterion from two or more independent, or predictor, variables.

R^2 (**R squared**)—A method of interpreting the strength of the relationship between the independent and dependent variables; the proportion of total variance that is due to the treatments.

R represents the multiple correlation coefficient, which indicates the relationship between the criterion (a dependent variable) and a weighted sum of the predictor variables (independent variables).

TABLE 8.4
Selection Procedures in Multiple Regression

Procedure	How it works
Full model	All prediction variables are included in the model.
Hierarchical	Prediction variables are entered as blocks (e.g., two together) and are determined in advance.
Forward	The predictor with the highest correlation is entered first; then the next highest is entered after accounting for the relationship with the first predictor.
Backward	All variables are entered; then the one with the least contribution is removed first, and then the one with the next least contribution.
Stepwise	A combination of forward and backward procedures in which checks are made at each step to see whether a previously entered predictor should be removed.
Maximum R^2	The program calculates the best variable combination of two, three, or more predictors.

equation on which step. This method, called hierarchical regression, should be based on theory and previous empirical evidence. The remaining four procedures listed in table 8.4 allow the statistical program to select the order of predictor variables and the number to be used based on statistical criteria. All of these procedures are designed to end up with fewer predictor variables than the full model (which includes all of the predictor variables).

Example 8.2 provides an example of multiple regression using several of the golf skill variables to predict scoring average (from the complete data set in table 6.2). The means and standard deviations of the criterion (scoring average) and the predictors (driving distance, driving accuracy, greens in regulation, putts per round, scrambling, and sand saves) are shown in the table. The procedure uses the backward regression model—all the predictors are included at the beginning (model 1 in the table), and a predictor that is unimportant (not a good predictor) is removed at each step (model 2 and model 3 in this example) until all the remaining variables are important (contribute to the prediction model). In this example, model 2 does not include sand saves, and model 3 (the final model) does not include driving accuracy. Thus, model 3 uses the combination of driving distance, greens in regulation, putts per round, and scrambling to predict scoring average. The four predictors in a linear composite (each weighted for its contribution) significantly predict scoring average, $F_{(4,25)} = 16.55, p = .000, R = .85$. This linear composite of four aspects of golf performance accounts for 72.6% of the variance (R^2) in scoring average. The prediction equation is in the table. But because the linear composite of predictors accounts for only 72.6% of the variance in scoring average, the prediction may not be a useful one.

EXAMPLE 8.2

Known Values

Multiple Regression Using Golf Skill Variables to Predict Scoring Average (Backward Multiple Regression)

Descriptive statistics			
Variable	Mean	Standard deviation	n
Scoring average	70.46	0.61	30
Driving distance	299.91	7.78	30
Driving accuracy	62.74	5.12	30
Greens hit in regulation	69.95	2.59	30
Putts/round	29.04	0.54	30
Scrambling	61.46	3.96	30
Sand saves	51.73	7.67	30

Model summary (d)		
Model	R	R^2
1	.867 (a)	.752
2	.866 (b)	.749
3	.852 (c)	.726

a. Predictors = driving accuracy, driving distance, greens in regulation, scrambling, putts per round, sand saves

b. Predictors = driving accuracy, driving distance, greens in regulation, scrambling, putts per round

c. Predictors = driving distance, greens in regulation, scrambling, putts per round

d. Criterion variable = scoring average

$F(4,25) = 16.55, p = .000$

Prediction Equation:

$\hat{Y} = 79.41 - (0.030)$ driving distance $- (0.159)$ greens in regulation $+ (0.481)$ putts per round $- (0.044)$ scrambling

Multiple-Regression Prediction Equations

The prediction equation resulting from multiple regression is basically that of the one-predictor regression model, $\hat{Y} = a + bX$. The only difference is that there is more than one X variable (or predictor); thus, the equation is as follows:

$$\hat{Y} = a + b_1X_1 + b_2X_2 \ldots + b_iX_i \qquad (8.7)$$

We will not delve into the formula for the calculation of the a and the bs for the selected variables. As we said before, researchers use computers for multiple-regression problems.

Some Problems Associated With Multiple Regression

The basic procedure to evaluate multiple regression is the same as that used in regression with only two variables: the size of the correlation. The higher the correlation, the more accurate the prediction. But some other factors should be mentioned.

One consideration with multiple regression is the problem of multicollinearity. This occurs when one or more of the independent variables (predictors) is highly correlated with another independent variable. The problem this causes is that the unique variance that can be predicted becomes small and the accuracy of the individual regression coefficients becomes questionable (as indicated by a large standard error).

A second limitation of prediction relates to generalizability. Regression equations developed with a particular sample often lose considerable accuracy when applied to others. This loss of accuracy in prediction is called **shrinkage**. The term **population specificity** also relates to this phenomenon. Shrinkage and the use of cross-validation to improve generalizability are further discussed in chapter 11. We need to recognize that as researchers seek more accuracy through selection procedures (forward, backward, stepwise, maximum R^2) that capitalize on specific characteristics of the sample, researchers are less able to generalize the findings from their sample to other populations. For example, a formula for predicting percentage of body fat from skinfold measurements developed with adult men would lose a great deal of accuracy if used for adolescents. Thus, the researcher should select the sample carefully with regard to the population for which the results are to be generalized.

In prediction studies, the number of participants in the sample should be sufficiently large. Usually, the larger the sample, the more likely it will be to represent the population from which it is drawn. Hence, one problem with small samples is that the regression results may not be representative of the population. Another related problem with small samples in multiple-regression studies is that the correlation may be spuriously high. A direct relationship exists between the correlation and the ratio of the number of participants to the number of variables. In fact, the degree to which the expected value of R^2 exceeds zero when it is zero in the population depends on two things: the size of the sample (n) and the number of variables (k). More precisely, it is the ratio $(k - 1) / (n - 1)$ that determines the expected value of R^2. To illustrate, suppose you read a study in which $R^2 = .90$. Impressive, right? The results would be meaningless, however, if the study had only 40 participants and 30 variables, because we could expect an R^2 of .74 just based on chance alone:

$$R^2 = (k - 1)/(n - 1) = (30 - 1)/(40 - 1) = .74$$

shrinkage—The tendency for the validity to decrease when the prediction formula is used with a new sample.

population specificity—A phenomenon whereby a regression equation that was developed with a particular sample loses considerable accuracy when applied to others.

You should be aware of this relationship between the number of participants and the number of variables when reading research that uses multiple regression. In the most extreme case, you can see that having the same number of variables as participants, $(k-1)/(n-1)$ yields an R^2 of 1.00! A participant-to-variable ratio of 10:1 or higher is often recommended.

Logistic Regression

Logistic regression is useful when the outcome (criterion) variable is expressed in only two categories (1 or 2, yes or no, female or male) and the predictor variable is continuous (e.g., running speed, strength, weight). In the logistic regression curve shown in figure 8.8, body weight in pounds is used to predict whether a middle school child is female or male in a sample of 100 (50 females and 50 males). The x-axis indicates body weight from 60 pounds to 120 pounds (27 kg to 54 kg), while the y-axis shows whether the sex of the child is female or male. Just as in regression, if the curved prediction line is read down to the x-axis, it represents the average weight at any point. When the line is read across to the

odds ratio—Expresses the relationship between an exposure (the continuous variable) and a binary outcome (typically a disease state such as "diagnosed with cancer" or "not diagnosed with cancer" or mortality such as "deceased" or "not deceased"). An odds ratio tells us the risk or probability of the outcome for someone with increased exposure relative to "normal" exposure.

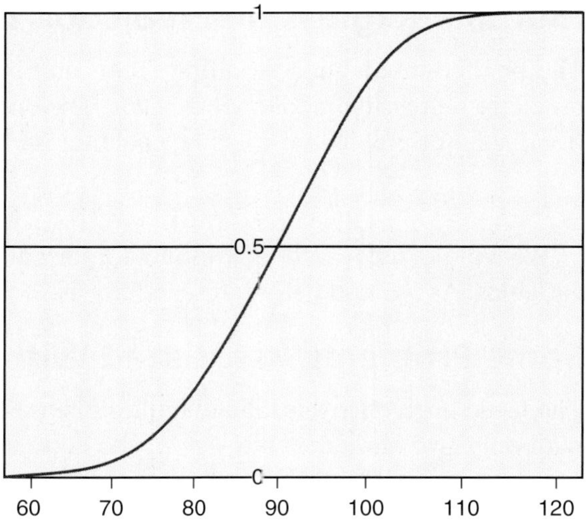

Figure 8.8 Logistic regression curve using body weight in pounds to predict the sex of middle school children.

y-axis, it represents the probability of the child being female (y value between 0 and 0.5) or male (y value between 0.5 and 1). However, the regression line in logistic regression is nonlinear. As you can see, the **odds ratio**, or probability, of the child being male increases with weight, but the outcome is simply male not some degree of being male.

Discriminant Function Analysis

Discriminant function analysis (DFA) is another type of prediction equation, but in this case, the variable being predicted is group membership. Each method listed in table 8.4 can be used to predict group membership when the groups are independent of one another. When employing DFA, the researcher uses data (a variety of predictor variables and the group each individual actually is in) to create an equation that can be used when group membership information is unavailable or nonexistent. The researcher can examine, based on the best predictor variables, which group an individual is predicted to be a member of and the overall success of predictions, or which predictions the equation got right.

To illustrate an example of using DFA, let's assume that a researcher wants to predict which undergraduate kinesiology majors will go to graduate school, which will go to professional school, and which will not continue formal education beyond the bachelor's degree. She wants to be able to predict the students' destinations so that she can incorporate more targeted career planning opportunities for the students. She has information on graduates for a five-year period and a variety of variables that she believes would help create a prediction equation (e.g., undergraduate GPA, GPA in the major, a motivation questionnaire from the junior year, an indication of future plans). She puts all the information into a computer and completes a stepwise DFA that will indicate the best predictors and how good the predictions are. If the best equation predicts around one-third of graduates correctly, it is not a good prediction because that could be obtained by chance. If, however, the prediction is much better, the equation can be considered valuable for predicting postgraduate education plans.

Moderators and Mediators

The term *moderator* is often used to describe a categorical variable thought to influence the relationship between two other variables (a predictor and a criterion or an independent variable and a dependent variable). When the term is used in an analysis of variance (ANOVA) model, the moderator is essentially an independent variable that is contributing to a significant interaction. When the term is used in correlational models, it describes a variable that affects the slope of the relationship between the predictor and the criterion. This term is used most commonly when relevant to a mediation analysis. A mediation analysis is conducted when researchers are interested in testing a variable as a potential pathway (or mechanism or mediator) between an independent variable and a dependent variable. An example is if an experimenter believed that physical activity participation resulted in reductions in body weight which explained the observed improvements in global self-esteem. In a full mediation model, changes in the mediator (body weight in this example) explain all of the change in the dependent variable (global self-esteem in this example). In a partial mediation model, the mediator only explains a portion of the change in the dependent variable. If we included age group (young adult, older adult) into this model as a moderator, we may find that the effects of physical activity participation on global self-esteem are different based upon age group such that there is a stronger relationship for the younger adults than for the older adults. In this example, age group would be a moderator. Baron and Kenny (1986) originally described the conceptual steps for conducting moderated-mediation analyses and statistically tested for mediation using regression analyses. More recently, the statistical analyses are typically conducted using Sobel's test or bootstrapping methods (see Hayes, 2009 for more information).

Multivariate Forms of Correlation

We have been discussing forms of correlation that are typically labeled as *univariate*, meaning that there are one or more predictor variables (independent variables) but only one criterion variable (dependent variable). In this short section, we introduce the multivariate forms of correlation. Our purpose is to provide enough information that you can read a paper using one of these procedures and understand what has been done.

Canonical Correlation

Canonical correlation is an extension of multiple correlation (several predictors and one criterion) to an analysis that has several predictors and several criteria, represented by the symbol R_c. In multiple correlation, a linear composite is formed of the predictor variables that maximally predict the single criterion variable. In canonical correlation, two linear composites are formed: one of the predictor variables and one of the criterion variables. These two composites are formed to maximize the relationship between them, and this relationship is represented by R_c. R_c^2 is interpreted just like R_c^2 from multiple correlation; it represents the shared variance between the two linear composites of variables.

For example, Atkinson, Short, and Martin (2018) collected survey data from college soccer players to identify the relationship between the players' perception of the efficacy of the coach and the players' perception of the efficacy of the team. The coach efficacy measure consisted of four subscales and the team efficacy measure was made up of five subscales. They used canonical correlation to identify the relationship between the subscales relevant to the coach and those relevant to the team. They found a significant canonical correlation indicating that the players' confidence in their coach's ability to motivate the team and to promote an effective game strategy was predictive of the perception that the team was more persistent, prepared, and united.

Factor Analysis

Many performance variables and characteristics are used to describe human behavior. Often, reducing a large set of performance and characteristic measures to a more manageable structure is useful. We have already discussed the likelihood that two performance measures might, to some extent, assess the same underlying characteristic. This represents the degree to which they are correlated. **Factor analysis** is an approach used to reduce a set of correlated measures to a smaller number of latent, or hidden, variables. Numerous procedures are grouped under the general topic of factor analysis. We do not discuss the techniques in detail, but we do provide a general explanation that will help you read and understand studies that use factor analysis. For informative explanations and examples of factor analysis, read Gunnell, Gareau, and Gaudreau (2016) and Pett, Lackey, and Sullivan (2003).

Factor analysis is performed on data from a group of individuals on whom a series of measurements have been taken. The researcher usually wants to describe a reduced number of underlying constructs and possibly select the one or two best measures of each construct. Factor analysis begins by calculating the intercorrelations of all the measures used—in other words, the correlation between all possible pairs of variables. Thus, if eight measures were used, the correlation would be determined between variables 1 and 2, variables 1 and 3, and so on, for a total of 28 correlations. The goal of factor analysis is to discover the factors (underlying or hidden constructs) that best explain a group of measurements and describe the relation of each measure to the factor or underlying construct. The two general types of factor analysis are (a) exploratory, in which many variables are reduced to an underlying set, and (b) confirmatory, which either supports or does not support a structure proposed from theory. Confirmatory factor analysis is the more useful technique.

For example, Marsh, Marco, and Abçý (2002) examined whether the Physical Self-Description Questionnaire (PSDQ) had similar factor structures when used with cross-cultural samples—high school students in Australia, Spain, and Turkey. Marsh et al. (2002) translated the PSDQ from English into Spanish and Turkish and used back-translation techniques to assess that the translations produced a similar questionnaire, or instrument, in the new languages. After children in each country completed the questionnaire, factor structures and fit statistics (such as **root mean square error of approximation [RMSEA]**) were compared and showed that the factor structure was satisfactorily similar when administered in all three languages. Thus, the research supported the use of the PSDQ in a variety of cultural environments. As we note in chapter 11, confirmatory factor analysis is often used to validate scores from attitude instruments or other questionnaires, and this study extended that into multiple countries.

Structural Equation Modeling

Path analysis and **linear structural relations (LISREL)** are structural, or causal, modeling techniques that are used to explain the way certain characteristics relate to one another and attempt to imply cause. You should remember from the discussion in chapter 4 and earlier in this chapter that cause and effect are not a statistical result but a logical one. That is, if the experimenter can argue theoretically that changing a certain characteristic should result in a specific change in behavior, and if the actual experiments (and statistical analysis) support this hypothesis, then a particular independent variable is often inferred to cause a dependent variable. Of course, this is true only if all other possible influences have been controlled.

The way variables influence one another is not always clear; for example, $X \rightarrow Y \rightarrow Z$, or $X \leftarrow Y \rightarrow Z$. In the first case, X influences Y, which in turn influences Z, but in the second case, Y influences both X and Z. This is a simple example. Path analysis and LISREL allow a more complex modeling of the way variables influence one another. But all you can say about these models is that they are either consistent or inconsistent with the data and hypotheses.

Whether they imply cause and effect depends on other factors, such as control of all other variables, careful treatments, logical hypotheses, and valid theories. For a useful overview and practical explanation of structural equation modeling, read Schumacker and Lomax (2004).

Consider two examples of the use of structural equation modeling in kinesiology. Sylvester, Curran, Standage, Sabiston, and Beauchamp (2018) analyzed data from 499 community adults who had completed questionnaires assessing exercise behavior, exercise motivation, and needs satisfaction. Using structural equation modeling, they tested a model in which psychological needs satisfaction was predicted to moderate the relationship between exercise variety, autonomous motivation, and exercise behavior. In this model, autonomous motivation is viewed as the mediator of the effects of variety on exercise behavior, and psychological needs satisfaction is predicted to moderate or change the nature of this relationship. This is called a *moderated mediated model*. It helps us to understand the complexity of how these variables interact to influence exercise behavior (figure 8.9).

Ste-Marie, Carter, Law, Vertes, and Smith (2016) manipulated the level of self-control among children who were learning a series of progressions on a mini-trampoline. Self-efficacy, interest and enjoyment, and performance were assessed at two acquisition times and then during retention. These authors used path analysis to understand the relationship between these psychological and behavioral variables at various points in time (figure 8.10).

The technique of LISREL was used because it allows related measures to be grouped (as in factor analysis) into components (e.g., exercise behavior), and it shows the interrelationships of the components in terms of magnitude and direction. The estimates of each component are grouped together in a linear equation (using their factor loadings). Then, a series of general linear equations (as in multiple regression) shows the components' relationships. Model-fitting indices are used to assess the nature of the fit. You can see, then, that structural modeling techniques provide a way to evaluate the complex relationships that exist in real-world data. By establishing the directions of certain relationships, researchers can make stronger inferences about which characteristics are likely to cause other characteristics.

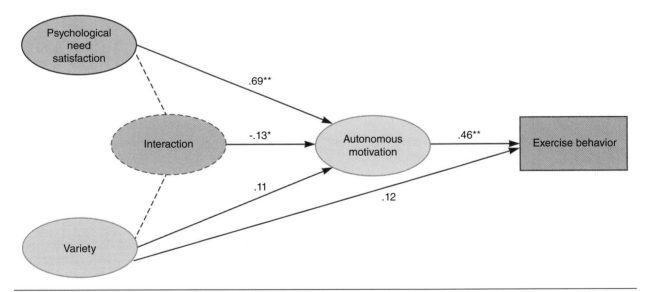

Figure 8.9   Example of a moderated mediation model to predict exercise behavior from variety.

Reprinted by permission from B.D. Sylvester et al., "Predicting Exercise Motivation And Exercise Behavior: A Moderated Mediation Model Testing the Interaction Between Perceived Exercise Variety and Basic Psychological Needs Satisfaction," *Psychology of Sport & Exercise* 36 (2018): 50-56.

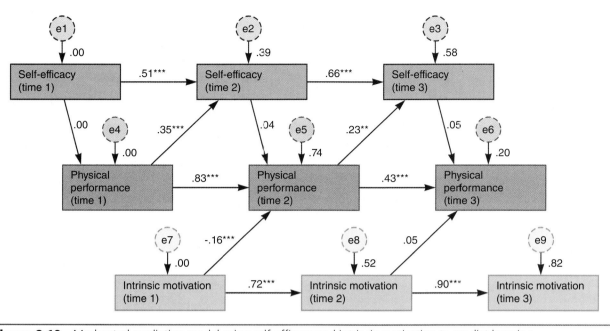

Figure 8.10 Moderated mediation model using self-efficacy and intrinsic motivation to predict learning.

Reprinted by permission from D.M. Ste-Marie, et al. "Self-Controlled Learning Benefits: Exploring Contributions of Self-Efficacy and Intrinsic Motivation Via Path Analysis," *Journal of Sports Sciences* 34, no. 17 (2016): 1650-1656.

Summary

We have explored some statistical techniques for determining interrelationships of variables. The simplest type of correlation is the Pearson r correlation, which describes the relationship between two variables. We introduced linear regression, which can be used to predict one variable from another. Correlation is interpreted for significance (reliability) and for variance explained (r^2), which indicates the portion of the total variance in one measure that can be explained or accounted for by the other measure.

Partial correlation is a procedure in which a correlation between two variables is obtained while the influence of one or more other variables is held constant. Semipartial correlation removes the influence of a third variable on only one of the two variables being correlated. Partial correlation (or semipartial correlation) is used in multiple correlation and in developing multiple-regression formulas. Logistic regression allows calculation of the odds ratio in a binary criterion variable as determined by a predictor variable. Discriminant analysis uses a group of variables to predict group membership.

In multiple regression, two or more predictor (independent) variables are used to predict the criterion variable. The most efficient weighted linear composite of predictor variables is determined through such techniques as forward selection, backward selection, stepwise selection, and maximum R^2.

Finally, we provided a brief overview of three multivariate correlational procedures: canonical correlation, factor analysis, and structural equation modeling. An example of the use of each technique was presented.

☑ Check Your Understanding

1. What is the correlation (*r*) between fat deposits on two body sites? Variable *X* is the triceps skinfold measure, and variable *Y* is the suprailiac skinfold measure. Try solving this problem by hand. Use the formulas provided for *r* and then check your work by using SPSS or another statistical package.

X	Y
16	9
17	12
17	10
15	9
14	8
11	6
11	5
12	5
13	6
14	5
4	1
7	4
12	7
7	1
10	3

2. Using Appendix table 2, determine whether the correlation obtained in problem 1 is significant at the .05 level. What size would the correlation need to be for significance at the .01 level if you had 30 participants? What is the percentage of common variance between the two skinfold measurements in problem 1?

3. Determine the regression formula (equation 8.2) to predict maximal oxygen consumption, $\dot{V}O_2$max, from scores on a 12-minute run. The information that you need follows:

X (12-minute run)   **Y ($\dot{V}O_2$max)**

$M_X = 2{,}853\ m$ $M_Y = 52.6\ \text{ml} \cdot \text{kg} \cdot \text{min}^{-1}$

$s_X = 305\ m$ $s_Y = 6.3\ \text{ml} \cdot \text{kg} \cdot \text{min}^{-1}$

$r = .79$

4. Using the prediction formula developed in problem 3, what is the predicted $\dot{V}O_2$max for a participant who ran 2,954 m in 12 minutes? For a participant who ran 2,688 m in 12 minutes?

5. What is the standard error of prediction (equation 8.5) for the prediction formula in problem 3? How would you interpret the predicted $\dot{V}O_2$max for the participants in problem 4?

Differences Between Groups

Statistical techniques are used for describing and finding interrelationships of variables, as we discussed in chapters 6 and 8. They are also used to detect differences between groups. The latter techniques are most frequently used for data analysis in experimental and quasi-experimental research. They enable us to evaluate the effects of an independent variable (cause or treatment) or categorical variable (sex, age, race, and so on) on a dependent variable (effect, outcome). Remember, however, that the techniques described in this chapter are not used in isolation to establish cause and effect but only to evaluate the influence of the independent variable. Cause and effect are not established by statistics but by theory, logic, and the total nature of the experimental situation.

How Statistics Test Differences

In experimental research, the experimenter may establish the levels of the independent variable. For example, the experiment might involve the investigation of the effects of exercise intensity on cardiorespiratory endurance. Thus, exercise intensity is the independent variable (or treatment factor), whereas a measure of cardiorespiratory endurance is the dependent variable. Intensity of training could have any number of levels. If it were evaluated as a percentage of $\dot{V}O_2max$, then it could be 30%, 40%, 50%, and so forth. The investigator would choose the number of levels and the intensity of the exercise at each level. For example, in a simple experiment, the independent variable comprises two levels of exercise intensity: 40% and 70% of $\dot{V}O_2max$. Each session is 30 minutes long, the frequency is three times per week, and the total number of weeks of training is 12, all of which are controlled for both groups. The dependent variable could be the distance that a person runs in 12 minutes.

> Statistics can only evaluate the influence of independent variables; they cannot establish cause and effect.

The purpose of the statistical test is to evaluate the null hypothesis at a specific level of probability (e.g., $p < .05$). In other words, do the two levels of treatment differ significantly ($p < .05$) so that these differences are not attributable to a chance occurrence more than 5 times in 100? The statistical test is always a test of the null hypothesis. All that statistics can do is reject or fail to reject the null hypothesis. Statistics cannot accept the research hypothesis. Only logical reasoning, good experimental design, and appropriate theorizing can do so. Statistics can determine only whether the groups are significantly different, not why they are different.

In using logical techniques to infer cause and effect after finding significant differences, you must be careful to consider all possibilities. For example, we might propose the theorem that all odd numbers are prime numbers (this example is from Ronen et al., cited in Scherr, 1983, p. 146). You know that prime numbers are those that can be divided only by 1 and by themselves. Thus, 1 is a prime number, 3 is a prime number, 5 is a prime number, 7 is a prime number, and so on. Using the induction technique of reasoning, every odd number is a prime number. However, in Ronen and colleagues' example, only a small number of levels of the independent variable (prime numbers) were sampled, and an error was made in inferring that all levels of the independent variable were the same. Had they included even one additional data point (e.g., 9), they would have recognized that their theorem was inaccurate.

When you use statistics that test differences between groups, you want to establish not only whether the groups are significantly different but also the strength of the association between the independent and dependent variables, or the size of the difference between two groups. The t and the F ratios are used throughout this chapter to determine whether groups are significantly different. The magnitude of the differences in standard deviation units is estimated by effect size (ES). Effect size (recall the discussion in chapter 7) is the standardized difference between two groups and is also used as an estimate of meaningfulness. In ANOVA techniques, R^2 is also used to establish meaningfulness. R^2 is the percentage of variance in the dependent variable accounted for by the independent variable.

> To establish whether groups are significantly different, remember two things: the significance of the association between the dependent and independent variables, and the size of the difference between the groups.

The uses of the t and the F distributions as presented in this chapter have four assumptions (in addition to the assumptions for parametric statistics presented in chapter 6; Kirk, 2013, is a good book to read for more information):

1. Observations are drawn from normally distributed populations.
2. Observations represent random samples from populations.
3. The numerator and denominator are estimates of the same population variance.
4. The numerator and denominator of F (or t) ratios are independent.

Although t and F tests are robust to (only slightly influenced by) violations of these assumptions, the assumptions still are important to consider. You should be sensitive to their presence and to the fact that violations affect the probability levels that may be obtained in connection with t and F ratios.

Types of *t* Tests

We discuss three types of t tests: a t test between a sample and a population mean (single sample t test), a t test for independent groups (independent samples t test), and a t test for dependent groups (dependent samples t test).

Single Sample *t* Test

First, we may want to know whether a sample of students differs from a larger population. For example, suppose that for a standardized knowledge test on physical fitness, the mean is 76 for a large population of first-year college students. When tested, a fitness class ($n = 30$) that you are teaching has a mean of 81 and a standard deviation of 9. Do your class members have significantly more knowledge about physical fitness than do the first-year college students?

The *t* test is a test of the null hypothesis, which states that there is no difference between the sample mean *(M)* and the population mean (μ), or *M* − μ = 0. Equation 9.1 is the *t* test between a sample and a population mean:

$$t = \frac{M - \mu}{S_M/\sqrt{n}}$$

(9.1)

where s_M = the standard deviation for the sample mean and *n* = the number of observations in the sample. Example 9.1 shows this formula applied to the mean and standard deviation of the example for the standardized fitness knowledge test.

EXAMPLE 9.1

Known Values

Population: *N* = 10,000

Fitness class: *n* = 30

Sample mean: *M* = 81

Population mean: μ = 76

Standard deviation: s_M = 9

Working It Out (Equation 9.1)

$$t = \frac{M - \mu}{S_M/\sqrt{n}}$$

$$t = \frac{81 - 76}{9/\sqrt{30}} = \frac{5}{1.64} = 3.05$$

Is the value of 3.05 significant? To find out, you will need to check Appendix table 4. To use the table, first work out the degrees of freedom *(df)* for the *t* test. Degrees of freedom are based on the number of participants with a correction for bias:

$$df = n - 1$$

(9.2)

In our example, the *df* are 30 − 1, or 29. Degrees of freedom are used to enter a *t* table to determine whether the calculated *t* is as large as or greater than the critical value of *t* found in the table. Note that across the top of Appendix table 4 are probability levels for one-tailed (read across top row) and two-tailed (read across second row) tests. Since we are predicting the direction of the difference (i.e., the question we asked is whether they have "more" knowledge, not simply whether they are different), we are using a one-tailed test. We want to know whether our *t* value is significant at α = .05, the level of significance that we set. Read across to the .05 level in the top row of the one-tailed test. Now read down the left side *(df)* to the row for the degrees of freedom that you calculated *(df* = 29). Read where the *df* = 29 row and the .05 column intersect to identify the critical value for this statistical test. Is the calculated value (*t* = 3.05) larger than this value (1.699)? Yes, it is. The value *t* is significant at *p* < .05 using a one-tailed test. Thus, our sample class is reliably (significantly) more knowledgeable than the population on the fitness test.

Independent *t* Test

The previous *t* test, applied to determine whether a sample differs from a population, is not used frequently because we don't often have access to population data. The most frequently used *t* test determines whether two sample means differ reliably from each other. This test is called *an independent samples* t *test* or simply an **independent *t* test**.

independent *t* test—A type of *t* test used to determine whether two sample means differ reliably from each other.

Suppose that we return to our example at the beginning of this chapter: Do two groups, exercising at different levels of intensity (40% and 70% of $\dot{V}O_2$max) for 30 minutes per day, three days per week for 12 weeks, differ from each other on a measure of cardiorespiratory endurance (12-minute run)? Let us further assume that 30 participants were randomly assigned to the exercise intensity levels to form two groups of 15 each.

Equation 9.3 is the t-test formula for two independent samples:

$$t = \frac{M_1 - M_2}{\sqrt{s_1^2/n_1 + s_2^2/n_2}} \tag{9.3}$$

The degrees of freedom for an independent t test are calculated as follows:

$$df = n_1 + n_2 - 2 \tag{9.4}$$

where n_1 = sample size for group 1 and n_2 = sample size for group 2. In our example, our equation appears as such:

$$df = 15 + 15 - 2 = 28 \ (\text{or } N - 2 = 30 - 2 = 28)$$

Example 9.2 shows the results.

Estimating the Meaningfulness of Treatments

How meaningful is the effect in our comparison of training intensities? Take a look at the data presented in example 9.2. You can see that after the training program, the group that trained at 70% intensity achieved an average 12-minunte run distance of 3,004 m as compared to 2,456 m for the group at 40% intensity. The question of meaningfulness, then, is this: Is the increase in cardiorespiratory endurance of running an additional 548 m (3,004 − 2,456) worth the additional work of exercising at 70% of $\dot{V}O_2$max as compared with 40% of $\dot{V}O_2$max? Given the total variation in running performance of the two groups, what we really want to know is (a) How much difference occurs between the groups relative to the observed variability (effect size)? and (b) What amount of the observed variation is accounted for by the difference in the two levels of the independent variable (70% vs. 40%)?

Effect Size

To estimate the degree to which the treatment influenced the outcome, use effect size (ES), the standardized difference between the means. Equation 9.5 (given previously as equation 7.1) is a way to estimate effect size (this concept was discussed in chapter 7 and is also used in meta-analysis, discussed in chapter 14):

$$ES = (M_1 - M_2)/s \tag{9.5}$$

where M_1 = the mean of one group or level of treatment, M_2 = the mean of a second group or level of treatment, and s = the standard deviation. The question is: Which standard deviation should be used? Considerable controversy exists over the answer to this question. Some statisticians believe that if there is a true control group (i.e., a group that gets no treatment), its standard deviation should be used because this value represents what would be expected in an untreated group. If there is no true control group (as in this example where both groups are exercising), then the pooled standard deviation (equation 9.6) should be used. Some advocate the use of the pooled standard deviation on all occasions. Either method can be defended, but when there is no true control group, we recommend that you use the pooled standard deviation

$$s_p = \sqrt{\frac{s_1^2(n_1 - 1) + s_2^2(n_2 - 1)}{n_1 + n_2 - 2}} \tag{9.6}$$

where

s_p = the pooled standard deviation, $s_1{}^2$ = the variance of group 1, $s_2{}^2$ = the variance of group 2.

Effect size can be interpreted as follows: An ES of 0.8 or greater is large, an ES around 0.5 is moderate, and an ES of 0.2 or less is small. Thus, the ES calculated in example 9.2 of 5.0 is a large value and would typically be judged as a meaningful treatment effect.

The other question we are interested in is: What percentage of the variance is accounted for by group assignment? The formula for this is the same for all t tests:

$$R^2 = \frac{t^2}{t^2 + df} \tag{9.7}$$

EXAMPLE 9.2

Known Values

	Group 1	Group 2
Mean distance run *(M)*	$M_1 = 3{,}004$ m	$M_2 = 2{,}456$ m
Standard deviation *(S)*	$S_1 = 114$ m	$S_2 = 103$ m
Number of subjects *(n)*	$n_1 = 15$	$n_2 = 15$

Working It Out (Equations 9.3, 9.5, 9.6, and 9.7)

$$S_p = \sqrt{\frac{s_1^2(n_1 - 1) + s_2^2(n_2 - 1)}{n_1 + n_2 - 2}}$$

$$S_p = \sqrt{\frac{(114)^2(15 - 1) + (103)^2(15 - 1)}{15 + 15 - 2}} = 108.64$$

$$ES = (M_1 - M_2)/s$$

$$ES = \frac{3{,}004 - 2{,}456}{108.64} = 5.0$$

$$t = \frac{3{,}004 - 2{,}456}{\sqrt{(114^2/15) + (103^2/15)}}$$

$$t(28) = 13.73$$

$$p < .005$$

$$R^2 = \frac{t^2}{t^2 + df} = \frac{13.2^2}{13.2^2 + 28} = \frac{188.51}{188.51 + 28} = 0.87$$

The independent t test is a common statistical technique in research studies. Sometimes it is the main technique used, and sometimes it is just one of several statistical tests in a study.

Cornett, Duski, Wagner, Wright, and Stager (2017) were interested in the relationship between maturational timing (i.e., age at menarche) and swimming performance (i.e., a standardized performance score, PPS) among collegiate athletes. Cornett et al. (2017) identified one group of swimmers as earlier maturing and another group as later maturing Their results showed that "later-maturing swimmers had significantly higher PPS (t_{158} = 2.10, P = 0.037, d = 0.33) . . . than the earlier-maturing group" (p. 14). The subscript by the t (158) is the *df*, which is $N - 2$; thus, we know the two groups comprised 160 people. The computed t was 2.10, and this was significant at p = .037. Cornett et al. (2017) used Cohen's d to present the ES of 0.33, considered as a small-to-moderate effect. They also described their results by saying that "later-maturing collegiate swimmers in our sample had standardized performance scores that were 5.2% higher than the earlier-maturing swimmers" (p. 15). They further interpreted this finding by converting their performance measure into an estimate that the later-maturing athletes were 1.68% faster and equating that time difference to approximately a body-length difference in the finish in a 100-yard race.

Checking for Homogeneity of Variance

All techniques for comparisons between groups assume that the variances (standard deviation squared) of the groups are equivalent. Although mild violations of this assumption do not present major problems, serious violations are more likely and more problematic if group sizes are not approximately equal. Formulas given here and used in most computer programs allow unequal group sizes. The homogeneity assumption should be checked, however, if group sizes are very different or even when variances are very different (these techniques are not presented here but are covered in basic statistical texts).

Dependent *t* Test

dependent *t* test—A test of the significance of differences between the means of two sets of scores that are related, such as when the same participants are measured on two occasions.

We have now considered use of the *t* test to evaluate whether a sample differs from a population and whether two independent samples differ from each other. A third application is called a *dependent samples* t *test* which is known simply as a **dependent *t* test** or a paired samples *t* test, which is used when the two groups of scores are related in some manner. Usually, the relationship takes one of two forms:

- Two groups of participants are matched on one or more characteristics and thus are no longer independent.

- One group of participants is tested twice on the same variable, and the experimenter is interested in the change between the two tests.

The formula for the dependent *t* test is:

$$t = \frac{M_D - \mu_D}{s_{M_D}}$$ (9.8)

Notice that the top part of this equation is comparing the mean of the difference scores for the sample to the mean of the difference scores for the population. But, to test the null hypothesis, we would be testing against an expectation of no difference in the population. So, we test against $\mu_D = 0$. To calculate t, we calculate the mean of the differences scores for the numerator and then calculate the standard error of the differences scores for the denominator. The standard error for the difference scores is calculated based upon the standard deviation of the difference scores, and this is calculated just as you would for any sample (i.e., using equation 6.2 with X replaced by D and M_D representing the mean for the difference scores). Thus, this formula for the standard deviation for difference scores can be written as follows:

$$s_D = \sqrt{\sum (D - M_D)^2 / (n - 1)}$$ (9.9)

The formula for the standard error for the differences scores then is written as follows:

$$s_{MD} = \frac{s_D}{\sqrt{n}}$$ (9.10)

The degrees of freedom for the dependent t test are as follows:

$$df = n - 1 \tag{9.11}$$

where n = the number of *paired* observations. To compute the t test for a dependent samples t test, we first calculate the difference scores for each participant. This serves to subtract out any differences between individuals and to focus instead on the effect of the treatment. The raw-score formula is then used with the difference scores (D):

$$t = \frac{\sum D}{\sqrt{[n \sum D^2 - (\sum D)^2]/(n - 1)}} \tag{9.12}$$

where D = the posttest minus the pretest for each participant (or the differences between the scores of two matched individuals) and n = the number of paired observations. Let us work out an example to illustrate the statistical technique. Ten dancers are given a jump-and-reach test (the difference between the height on a wall they can reach and touch while standing and the height they can jump and touch). Then they take part in 10 weeks of dance activity that involves leaps and jumps three days per week. The dancers are again given the jump-and-reach test after the 10 weeks. Our research hypothesis is that the 10 weeks of dance experience will change jumping skills as reflected by a change in jump-and-reach scores. The null hypothesis (H_0) is that the difference between the pretest and posttest of jumping is not significantly different from zero, $H_0 = M_{post} - M_{pre} = 0$. Example 9.3 shows how to calculate the dependent t value.

EXAMPLE 9.3

Subject	Pretest score (cm)	Posttest score (cm)	Posttest–pretest D	D^2
1	12	16	4	16
2	15	21	6	36
3	13	15	2	4
4	20	22	2	4
5	21	21	0	0
6	19	23	4	16
7	14	16	2	4
8	17	18	1	1
9	16	22	6	36
10	18	23	5	25

Known Values

Sum of pretest scores:	$\sum X_{pre} = 165$ cm
Sum of posttest scores:	$\sum X_{post} = 197$ cm
Sum of D:	$\sum D = 32$
Sum of D^2:	$\sum D^2 = 142$
Number of paired observations:	$n = 10$
$t_{crit}(9) = 2.262$	($\alpha = .05$, two-tailed)
Posttest mean:	$M_{post} = 19.7$
Pretest mean:	$M_{pre} = 16.5$

Working It Out (Equation 9.12)

$$t = \frac{\sum D}{\sqrt{[n \sum D^2 - (\sum D)^2]/(n-1)}}$$

$$t = \frac{32}{\dfrac{10(142) - (32)^2}{10 - 1}} = \frac{32}{\sqrt{\dfrac{1,420 - 1,024}{9}}} = \frac{32}{6.63} = 4.83$$

The results indicate that the posttest mean (19.7 cm) was significantly different from the pretest mean (16.5 cm), $t(9) = 4.83$, $p < .05$, two-tailed. We conducted this test as a two-tailed test because we predicted a change but did not predict whether this would be an increase or a decrease. Because $4.83 > 2.262$, the null hypothesis can be rejected, and if everything else has been properly controlled in the experiment, we can conclude that the dance training was associated with a reliable change in the height of jumping performance of +3.2 cm. Importantly, we would not draw a conclusion of cause and effect because our study did not use an experimental design.

We could estimate the magnitude of the effect by using equation 9.5 (replacing $M_1 - M_2$ with M_D and using s_D in place of s) and the percent of variance explained by using equation 9.7 as shown in example 9.4.

EXAMPLE 9.4

Known Values

$M_D = 3.2$

$s_D = 2.10$

$t = 4.83$

$df = 9$

Working It Out

$$Cohen's\ d = \frac{M_D}{S_D} = \frac{3.2}{2.10} = 1.52$$

$$R^2 = \frac{t^2}{t^2 + df} = \frac{4.83^2}{4.83^2 + 9} = 0.72$$

The results show that the ES is a large effect size, 1.52, and the treatment explained approximately 72% of the variance. The question of meaningfulness is then interpreted by the value of this magnitude of improvement relative to the "cost" of committing to the 10-week program.

An example of a dependent samples t test in the literature is provided by Hutchinson, Jones, Vitti, Moore, Dalton, and O'Neil (2018) who investigated the influence of self-selected music on the remembered pleasure of an exercise session. The same participants were tested while exercising with self-selected music and when no music was provided. One of the comparisons was reported as: "There was a significant difference for remembered pleasure, $t(16) = 4.181$, $p = .001$, $d = .72$, with higher scores reported for the music condition ($M = 53.53 \pm 29.30$) compared with the no-music condition ($M = 32.35 \pm 29.53$)" (p. 86). Because the df for a dependent samples t test is equal to the number of pairs of scores minus 1, this tells us that we had pairs of scores for 17 participants.

Interpreting *t*

You have now learned to perform the calculations that determine differences between groups. What do the results mean? Are they significant? Does it matter whether they are or are not? We answer these questions in this section by further explaining the difference between one-tailed and two-tailed *t* tests and discussing the aspects of the *t* test that influence power in research.

One-Tailed Versus Two-Tailed *t* Tests

We have already introduced the fact that *t* tests (and correlations) may be conducted as one-tailed tests (to test a directional hypothesis) or two-tailed tests (to test a nondirectional hypothesis). Allow us to provide further discussion about this distinction as it relates to statistical power. Refer again to Appendix table 4, the table of *t* values to which you compare the values calculated. Remember, you decide the alpha (probability) level (we have been using .05), calculate the degrees of freedom for *t*, and read the table *t* value at the intersection of the column (probability) and row (*df*). If your calculated *t* exceeds the table value (which is called the *critical value*), it is significant at the specified alpha and degrees of freedom. The critical values provided relative to the alpha levels for two-tailed tests (see second row of header) are used to conduct a **two-tailed *t* test** when we assume that the difference between the two means could favor either mean. Sometimes we might hypothesize that group 1 will be better than group 2 or, at worst, no poorer than group 2. In this case, the test is a **one-tailed *t* test**; that is, it can go in only one direction. Then, when looking at Appendix table 4 for a one-tailed *t* test, we use the alpha levels for one-tailed tests (see top row of header). You will notice that the .05 level for a one-tailed test is located in the .10 column for the two-tailed test, and the value for the .01 level for the one-tailed test is in the .02 column for the two-tailed test. Generally, in behavioral research, however, we are not so sure of the direction of our results that we can employ the one-tailed *t* table.

two-tailed *t* test—A test that assumes that the difference between the two means could favor either group.

one-tailed *t* test—A test that assumes that the difference between the two means lies in one direction only.

t Tests and Power in Research

In chapter 7, *power* was defined as "the probability of rejecting the null hypothesis when the null hypothesis is false." Obtaining power in research is desirable because the odds of rejecting a false null hypothesis are increased. The independent *t* test is used here to explain three ways to obtain power (besides setting the alpha level). These ways apply to all types of experimental research.

Consider the formula for the independent *t* test:

$$t = \frac{M_1 - M_2}{\sqrt{\dfrac{s_1^2}{n_1} + \dfrac{s_2^2}{n_1}}}$$ (9.13)

Note that we have placed numbers 1, 2, 3 beside the three horizontal levels of this formula. These three levels represent what can be manipulated to increase or decrease power.

The first level ($M_1 - M_2$) gives power if we can increase the difference between M_1 and M_2. You should see that if the second and third levels remain the same, a larger difference between the means increases the size of *t*, which increases the odds of rejecting the null hypothesis and thus increases power. How can the difference between the means be increased? The logical answer is by applying stronger, more concentrated treatments. One example would be to use a 12-week treatment instead of a six-week treatment to give the treatment a better chance to show its effect. The 12-week treatment should move the means of the experimental and control groups farther apart than the six-week treatment does. The result will be less overlap in their distributions.

The second level of the independent t formula is $s_1{}^2$ and $s_2{}^2$, or the variances (s^2) for each of the two groups. Recall that the standard deviation represents the spread of the scores about the mean. If this spread becomes smaller (scores distributed more closely about the mean), the variance is also smaller. If the variance term is smaller and the first and the third levels remain the same, the denominator of the t test becomes smaller and t becomes larger, thus increasing the odds of rejecting the null hypothesis and increasing power. How can the standard deviation and thereby the variance be made smaller? There are two answers. The first is to select groups that are homogeneous. You might do this by, for example, restricting your sample to females or to a single age group (i.e., 10-year-olds). The second answer is to apply the treatments consistently. The more consistently the treatments are applied to each participant, the more similar they become in response to the dependent variable. This groups each distribution more tightly around its mean, thus reducing the standard deviation (and thereby the variance).

Finally, the third level (n_1, n_2) is the number of participants in each group. If you increase n_1 and n_2 and the first and second levels remain the same, the denominator becomes smaller (note that n is divided into s^2) and t becomes larger, thus increasing the odds of rejecting the null hypothesis and obtaining power. Obviously, you can increase n_1 and n_2 by placing additional participants in each group.

Of course, power can also be influenced by varying the alpha (i.e., if you set alpha at .10 instead of .05, it increases power). But in doing this, we increase the risk of rejecting a true null hypothesis (i.e., we increase the chance of making a type I error).

> To increase power (and increase the odds of rejecting a false null hypothesis), consistently administer strong treatments, minimize variability within treatment groups, increase alpha, or use a larger sample size. Using appropriate research design and statistical analysis also increases power.

Keep in mind that your choice of design and the statistical techniques you use can influence power. We point this out as we discuss various statistical tests, but a simple example using tests that we have already addressed makes the point. We indicated that a dependent t test is used when you have either matched pairs or the same participants being measured twice. The dependent t test must be used because the independent t assumes that the groups comprise different participants and the error term is attributed to sampling error. The dependent t reduces this error. Another way of looking at this is that we increase power by using the dependent t when the situation warrants it. To illustrate, if you were to calculate an independent t test on the scores in example 9.3, you would find the t to be 2.32, whereas the dependent t was 4.83. Even with fewer degrees of freedom for the dependent t, we increased power by obtaining a higher t with that statistical test. We will see that we can also increase power by using techniques such as analysis of covariance and factorial analysis of variance.

> A significant t ratio means that true variance significantly exceeds error variance.

In summary, obtaining power is desirable because it increases the odds of rejecting a false null hypothesis. Power can be obtained by recruiting homogeneous samples, using strong treatments, administering those treatments consistently, using as many participants as feasible, varying alpha, and using an appropriate research design and statistical analysis. Remember, however, that a second question always arises, even in the most powerful experiments. After the null hypothesis is rejected, the strength of the effects must be evaluated and the meaningfulness interpreted.

true variance—The portion of the differences in scores that is (theoretically) real.

error variance—The portion of the scores that is attributed to participant variability.

The t ratio has a numerator and a denominator. From a theoretical point of view, the numerator is regarded as **true variance**, or the real difference between the means. The denominator is considered **error variance**, or variation about the mean. Thus, the t ratio is

$$t = \frac{true\ variance}{error\ variance}$$

where true variance = $M_1 - M_2$ and error variance = $\sqrt{s_1^2/n_1 + s_2^2/n_2}$ for an independent samples design. If no real differences exist between the groups, true variance = error variance, or the ratio between the two is true variance / error variance = 1.0. When a significant t ratio is found, we are really saying that true variance exceeds error variance to a significant degree. The amount by which the t ratio must exceed 1.0 for significance depends on the number of participants (df) and the alpha level established.

The estimate of the strength of the relationship (R^2) between the independent and dependent variables is represented by the ratio of true variance to total variance.

$$R^2 = \frac{true\ variance}{total\ variance}$$

Thus, R^2 represents the proportion of the total variance that is due to the treatments (true variance).

Effect size is also an estimate of the strength or meaningfulness of the group differences or treatments. Effect size places the difference between the means in standard deviation units, $(M_1 - M_2) / s$. Figure 9.1 shows how the normal distribution of two groups differs for two different effect sizes, 0.5 and 1.0. In figure 9.1a, the standardized difference between the group means is ES = 0.5; that is, the group 2 mean falls 0.5 standard deviation to the right of the group 1 mean. If the group 2 mean had fallen to the left, the ES would be −0.5. Figure 9.1b shows the distribution when ES = 1.0; that is, the group 2 mean is 1 standard deviation to the right of group 1. Notice that there is less overlap between the two groups' distribution of scores in figure 9.1b than in figure 9.1a. Said another way, when scores on the dependent variable were grouped by the independent variable in figure 9.1b, the means were farther apart, and there was less distribution overlap than in figure 9.1a.

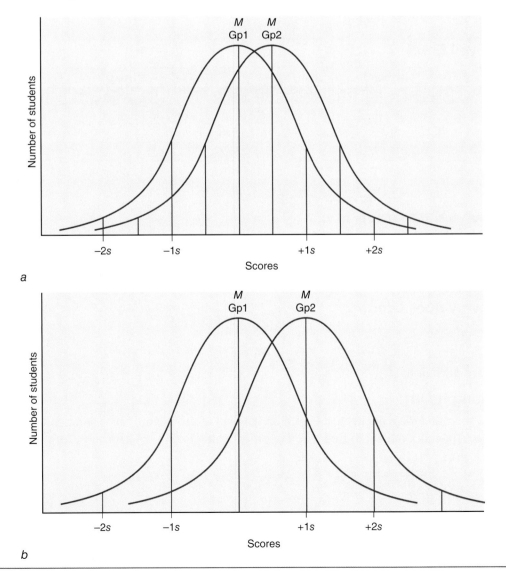

Figure 9.1 Distribution of scores for groups 1 and 2 for which ES = 0.5 (a) and 1.0 (b).

Effect size is sometimes interpreted as a percentile change attributed to the treatment. For example, in figure 9.1b, the treatment group (group 2) mean was 1.0s higher than the control group mean. If we consult Appendix table 2, we see that a z of 1.00 (a distance of 1s above the mean) shows that only .1587 (16%) of the scores are higher than 1.0. In other words, the percentile rank for such a score is 84 (100 −16). The z score for the group 1 (the control group) mean is at 0s, therefore $z = 0$. Consulting Appendix table 2, we see that 50% of scores are greater than the control group mean. Consequently, in interpreting an effect size of 1.0, we can infer that the treatment improved average performance by 34 percentile points (i.e., treatment group = 84th percentile, control group = 50th percentile; 84 − 50 = 34 percentile points).

Relationship of *t* and *r*

As we mentioned previously, our separation of statistical techniques into the two categories of interrelationships of variables (chapter 8) and differences between groups (this chapter) is artificial because both sets of techniques are based on the general linear model. A brief demonstration with *t* and *r* should make the point. But the idea can be extended into the more sophisticated techniques discussed in chapter 8 and later in this chapter. Example 9.5 shows the relationship between *t* and *r*. Group 1 has a set of scores (dependent variable) for five people, as does group 2.

EXAMPLE 9.5

Group 1		Group 2	
Participant	Dependent variable	Participant	Dependent variable
Patty	1	Sally	6
Lucy	2	Schroeder	7
Charlie	3	Snoopy	8
Marcie	4	Linus	9
Spike	5	Woodstock	10

Known Values Group 1

Sum:	$\Sigma X_1 = 15$
Mean:	$M_1 = 3$
Standard deviation:	$s_1 = 1.58$

Known Values Group 2

Sum:	$\Sigma X_2 = 40$
Mean:	$M_2 = 8$
Standard deviation:	$s_2 = 1.58$

Working It Out

1. Conduct an independent *t* test (equation 9.3) and test for significance as a two-tailed test (use Appendix table 4 to find that the critical value for $t(8) = 2.306$).

$$t = \frac{3 - 8}{\sqrt{\frac{1.58^2}{5} + \frac{1.58^2}{5}}} = 5.0^*$$

$$df = (n_1 + n_2) - 2 = 8$$

$$t(8) = 5.0, p < .05$$

*The sign is not important for this example.

2. Assign each participant a dummy code that stands for their group. A dummy code is one where you simply use a 0 to represent group membership for one group (Group 1 in this example) and a 1 to represent group membership for the other group (Group 2 in this example).

Group 1			Group 2		
Participant	Dummy code	Dependent variable	Participant	Dummy code	Dependent variable
A	0	1	F	1	6
B	0	2	G	1	7
C	0	3	H	1	8
D	0	4	I	1	9
E	0	5	J	1	10

3. Apply the correlation formula (equation 8.1). If we treat the dummy-coded variable as X and the dependent variable as Y and ignore group membership (10 participants with two variables), then the correlation formula used earlier can be applied to the data.

$$r = \frac{n \sum XY - (\sum X)(\sum Y)}{\sqrt{n \sum X^2 - (\sum X)^2} \sqrt{n \sum Y^2 - (\sum Y)^2}}$$

*The sign is not important for this example.

4. Convert r to t. In this example $t = 5.0$, the same as the t test done on the group means (within rounding error).

$$t = \sqrt{\frac{r^2}{(1 - r^2)/(n - 2)}} = \sqrt{\frac{.87^2}{(1 - .87^2)/(10 - 2)}} = \sqrt{\frac{.757}{.030}} = 5.0$$

The point is that r represents the relationship between the independent and dependent variables, and a t test can be applied to r (see step 4 in example 9.5) to evaluate the reliability (significance) of the relationship.

There are two sources of variance: true variance and error variance (true variance + error variance = total variance). The t test is the ratio of true variance to error variance, whereas r is the square root of the proportion of total variance accounted for by true variance. To get t from r only means manipulating the variance components in a slightly different way. This is because all parametric correlational and differences-between-groups techniques are based on the general linear model. This result can easily be shown to exist in the more advanced statistical techniques. (For a thorough treatment of this topic, see Cohen et al., 2003.)

You must understand this basic concept because increasingly, researchers are using regression techniques to analyze experimental data. These data have traditionally been analyzed using the techniques discussed in this chapter. We have, however, demonstrated that what is traditional is not required. What is important is that the data be appropriately analyzed to answer the following questions:

- Are the groups significantly different?
- Does the independent variable account for a meaningful proportion of the variance in the dependent variable?

Significance is always evaluated as the ratio of true variance to error variance, whereas the percentage of variance accounted for is always the ratio of true variance to total variance.

Analysis of Variance

Using *t* tests is an accurate way to determine differences between two groups, but often experimenters work with more than two groups. This section explains how to use analysis of variance to detect differences between two or more groups.

Simple Analysis of Variance

analysis of variance (ANOVA)—A test that allows the evaluation of the null hypothesis between two or more group means.

The concept of simple (sometimes called *one-way* but seldom considered simple by graduate students) **analysis of variance (ANOVA)** is an extension of the independent *t* test. In fact, *t* is just a special case of simple ANOVA in which there are two groups. Simple ANOVA allows the evaluation of the null hypothesis for two or more group means with the restriction that the groups represent levels of the same independent variable. When conducting ANOVAs, we now work with formulas that partition the total variance observed in the data into two sources. The first source is what can be explained by the treatment (i.e., by differences between levels of the independent variable) and the second is that which is due to unexplained variance.

We are going to use an example from the PGA Tour golf data (chapter 6, tables 6.2a and 6.2b) for ANOVA. We can compare three groupings of player performance (based on money won to date) with the dependent variable of scoring average. Table 9.1 shows the means, standard deviations, and 95% confidence intervals (CIs) for each group. Also displayed is the ANOVA, showing that the three groups are significantly different, $F(2, 27) = 8.717$, $p < .001$. The variance accounted for can be estimated by using $SS_{Between}/SS_{Total} = 4.281/10.911 = .39$; in other words, 39% of the variance in scoring average is accounted for by the grouping of players based on money won.

Why not do a *t* test between groups 1 and 2, a second *t* test between groups 1 and 3, and a third *t* test between groups 2 and 3? The reason is that this approach would violate an assumption concerning the established alpha level (let it be $p = .05$). The .05 level means that there is 1 chance in 20 of a difference because of sampling error, assuming that the groups of participants on which the statistical tests are done are from independent random samples. In our case, this assumption is not true. Each group has been used in two comparisons (e.g., 1 vs. 2 and 1 vs. 3) rather than only one. Thus, we have increased the chances of making a type I error (i.e., alpha is no longer .05). Making this type of comparison, in which the same group's mean is used more than once, is an example of increasing the experiment-wise error rate (EER), discussed later in this chapter. Simple ANOVA allows all three group means to be compared simultaneously, thus keeping alpha at the designated level of .05.

TABLE 9.1

ANOVA of Three Performance Groups (Top 10, 21-30, and 41-50) With Scoring Average as the Dependent Variable

Simple ANOVA								
	Descriptors				95% CI for mean			
Top 10	*N*	*M*	*s*	Standard error	Lower bound	Upper bound	Min.	Max.
21-30	10	70.0637	.49101	.15527	69.7160	70.4186	69.54	71.19
2.00	10	70.3391	.52907	.16731	69.9606	70.7176	69.81	71.47
3.00	10	70.9692	.46439	.14685	70.6370	71.3014	70.62	71.89
Total	**30**	**70.4585**	**.61339**	**.11199**	**70.2295**	**70.6876**	**69.54**	**71.89**

ANOVA					
Scoring ave.	SS	*df*	*MS*	*F*	Significance
Between groups	4.281	2	2.141	8.717	.001
Within groups	6.630	27	.246		
Total	**10.911**	**29**			

Calculating Simple ANOVA

Table 9.2 provides the formulas for calculating simple ANOVA and the F ratio. This method, the so-called ABC method, is simple:

$A = \Sigma X^2$: Square each participant's score, sum these squared scores (regardless of which group the participant is in), and set the total equal to A.

$B = (\Sigma X)^2/N$: Sum all participants' scores (regardless of group), square the sum, divide by the total number of participants (N), and set the answer equal to B.

$C = (\Sigma X_1)^2/n_1 + (\Sigma X_2)^2/n_2 + ... + (\Sigma X_k)^2/n_k$. Sum all scores in group 1, square the sum, and divide by the number of participants in group 1 (n_1); do the same for the scores in group 2 (n_2), and continue for however many groups (k) there are. Then add all the group sums and set the answer to C.

NOTE: In this example, you can see that a lowercase n is used to denote the number of participants within a group and an uppercase N is used to denote the total number of participants in the study.

Next, fill in the summary table for ANOVA using A, B, and C. Thus, the between-groups (true variance) **sum of squares** *(SS)* is equal to C − B; the between-groups degrees of freedom *(df)* is the number of groups minus one ($k - 1$); the between-groups variance, or mean square (MS_B), is the between-groups sum of squares divided by the between-groups *df*. The same is done within groups (error variance) and then for the total. The F ratio is MS_B/MS_W (i.e., the ratio of true variance to error variance).

sum of squares—A measure of variability of scores; the sum of the squared deviations from the mean of the scores.

Example 9.6 shows how the formulas in table 9.2 are used. The scores for groups 1, 2, and 3 are the sums of two judges' skill ratings for a particular series of movements. The groups are randomly formed from 15 junior high school students. Group 1 was taught with video, and the teacher made individual corrections while the student viewed the video. Group 2 was taught with video, but the teacher made only general group corrections while students viewed the video. The teacher taught group 3 without the benefit of video. From looking at the formulas in table 9.2, you should be able to see how each number in this example was calculated.

TABLE 9.2
Formulas for Calculating Simple ANOVA

$$A = \Sigma X^2$$

$$B = \frac{(\Sigma X)^2}{N}$$

$$C = \frac{(\Sigma X_1)^2}{n_1} + \frac{(\Sigma X_2)^2}{n_2} + ... + \frac{(\Sigma X_k)^2}{n_k}$$

Summary table for Simple ANOVA

Source	SS	df	MS	F
Between (true)	$C - B$	$k - 1$	$(C - B)/(k - 1)$	MS_B/MS_W
Within (error)	$A - C$	$N - k$	$(A - C)/(N - k)$	
Total	$A - B$	$N - 1$		

Note: X = a participant's score, N = total number of participants, n = number of participants in a group, k = number of groups, SS = sum of squares, df = degrees of freedom, MS = mean square.

EXAMPLE 9.6

Known Values

	Group 1		Group 2		Group 3	
	X	X^2	X	X^2	X	X^2
	12	144	9	81	6	36
	10	100	7	49	7	49
	11	121	6	36	2	4
	7	49	9	81	3	9
	10	100	4	16	2	4
Σ	50	514	35	263	20	102
M	10	—	7	—	4	—

Working It Out

$$A = \Sigma X^2 = 514 + 263 + 102 = 879$$

$$B = (\Sigma X)^2/N = (50 + 35 + 20)^2/15 = (105)^2/15 = 11{,}025/15 = 735$$

$$C = (\Sigma X_1)^2/n_1 + (\Sigma X_2)^2/n_2 + (\Sigma X_3)^2/n_3 = (50)^2/5 + (35)^2/5 + (20)^2/5$$

$$= 2{,}500/5 + 1{,}225/5 + 400/5 = 825$$

	Summary table for ANOVA			
Source	SS	df	MS	F
Between	90	2	45	10.00*
Within	54	12	4.5	
Total	144	14		

*$p < .05$

Your main interest in example 9.6, after you make sure that you understand how the numbers were obtained, is the F ratio of 10.00. Appendix table 5 contains F values for the .05 and .01 levels of significance. Although the numbers in the table are obtained the same way as in the t table, you use the table in a slightly different way. The F ratio is obtained by dividing MS_B by MS_W. Note in example 9.6 that the term MS_B (the numerator of the F ratio) has 2 df ($k - 1 = 2$), and the term MS_W (the denominator of the F ratio) has 12 df ($N - k = 12$). Notice also that Appendix table 5 has degrees of freedom across the top (numerator) and down the left-hand column (denominator). For our F of 10.00, read down the 2 df column to the 12 df row; there are two numbers where the row and the column intersect. The top number (3.88) in dark print is the table F value for the .05 level, whereas the bottom number (6.93) in light print is the table F value for the .01 level. If our alpha had been established as .05, you can see that our F value of 10.00 is larger than the table value of 3.88 at the .05 level (it is also larger than the value of 6.93 at the .01 level). So our F is significant and could be written in the text of an article as $F(2,12) = 10.00$, $p < .01$ (read this as: F with 2 and 12 degrees of freedom equals 10.00 and is significant at less than the .01 level).

Labban and Etnier (2011) used a simple ANOVA to compare long-term memory performance between those who exercise prior to performing a memory task, those who exercised after performing a memory task, and a control group. They reported that "A one-way ANOVA revealed significant between-groups differences in mean delayed recall score, F(2, 45) = 4.37, $p < .05$, $\eta_p^2 = 0.16$." The degrees of freedom for the F-test tell us that there were 3 groups with 48 total participants, because (2, 45) are first for the between-groups variation $(k - 1)$ and then for the within-groups variation $(N - k)$. The calculated F was 4.37 and was significant at the $p < .05$ level. The effect size is provided as $\eta_p{}^2 = 0.16$.

Follow-Up Testing

In our example of ANOVA with three groups, we now know that the three group means have significant differences $(M_1 = 10, M_2 = 7, \text{and } M_3 = 4)$. But we do not know which groups are different from which (e.g., whether groups 1 and 2 differ from group 3 but not from each other). So we next perform a follow-up test. One way to do this is to use t tests between groups 1 and 2, groups 1 and 3, and groups 2 and 3. But the same problem that we discussed earlier exists with alpha (type I error is increased). Fortunately, there are several follow-up tests that protect the EER (type I error, discussed later in this chapter). These methods include those of Scheffé, Tukey, Newman-Keuls, Duncan, and several others—see Keppel and Wickens (2004) for both conceptual explanations and calculations of the various techniques. Each test is calculated in a slightly different way, but they all are conceptually similar to the t test in that they identify which pairs of groups differ from each other. The Scheffé method is the most conservative in that it identifies fewer significant differences; following Scheffé is Tukey's method. Duncan is the most liberal method, identifying more significant differences. The Newman-Keuls method falls between the extremes but has some other problems. In the aforementioned study by Labban and Etnier (2011), they followed up their significant finding as follows: "Post-hoc testing using Tukey's HSD revealed that only the exercise-prior group's long-term recall performance was significantly better than the control group, $p < .05$. Performance differences between exercise-prior and exercise-after groups did not reach statistical significance $(p = .09)$" (p. 717).

For our purpose, one example should suffice, so we explain the use of the Scheffé method of making multiple comparisons of means. Scheffé (1953) is a widely recognized multiple-comparison technique. Because we believe that researchers should be conservative when using behavioral data, we generally recommend the Scheffé method for follow-up testing. Other multiple-comparison techniques, however, are appropriate for various situations.

The Scheffé technique has a constant critical value for the follow-up comparison of all means when the F ratio from a simple ANOVA (or a main effect in a factorial ANOVA, discussed next) is significant. Scheffé controls type I error (alpha inflation) for any number of appropriate comparisons. Equation 9.14 is the formula for the critical value (CV, required size of the difference) for significance using the Scheffé technique:

$$CV = \sqrt{(k-1)F_{a;k-1,N-k}}\ \sqrt{2(MS_w/n)} \qquad (9.14)$$

where k = the number of means to be compared, F_α = the table F ratio (from Appendix table 5) for the selected alpha (e.g., .05) given dfs of $k - 1$ (for between groups) and $N - k$ (for within groups). For the second half of the equation, MS_w is obtained from the ANOVA summary table (SS_W/df_W), and n = the number in a single group.

Example 9.7 shows the Scheffé technique applied to the previous simple ANOVA shown in example 9.6. You find the critical value for F, or the F ratio, in Appendix table 5 at the intersection of the column $df = 2$ $(k - 1 = 3 - 1 = 2)$ and the $df = 12$ row $(N - k = 15 - 3 = 12)$ from the dfs in the summary ANOVA in table 9.1; the table F ratio is 3.88. Solving the left part of formula 9.14 in example 9.7, we find that the Scheffé value for significance at the .05 level is 2.79. For the second part of formula 9.14, we obtain $MS_W = 4.5$ (from table 9.1) and the n for each group is 5. Thus, $\sqrt{2(4.5/5)} = 1.3$. Hence, formula 9.14 becomes the following:

CV = (2.79)(1.34) = 3.74

This value is the size of the difference between means needed in order to be significant at $p < .05$.

Example 9.7 arranges the means from highest (10) to lowest (4). The observed difference for each of the comparisons is computed; then each observed difference is compared with the needed difference (CV) for significance. By following these steps, you can see that only the comparison between groups 1 and 3 was significant at $p < .05$. We could conclude that the techniques used in group 1 were significantly better than those used in group 3 but not significantly better than those used in group 2. Also, the techniques used in group 2 were not significantly better than those used in group 3.

EXAMPLE 9.7

Working It Out

1. Order the means from highest (10) to lowest (4).
2. Compute the differences between all the means.

Group 1 ($M = 10$)	Group 2 ($M = 7$)	Group 3 ($M = 4$)	Observed difference
10	7	—	3
10	—	4	6
—	7	4	3

3. Compute the critical value (difference needed) for significance at alpha = .05 (equation 9.14):

$$CV = \sqrt{(k-1)F_{a;k-1,N-k}}\ \sqrt{2(MS_w/n)}$$

$$= \sqrt{(3-1)3.88}\sqrt{2(4.5/5)}$$

$$= (2.79)(1.34) = 3.74$$

4. Compare the CV of 3.74 with each observed difference to determine which difference or differences are significant. In this case, only the difference between groups 1 and 3 (observed difference = 6) is as large as or larger than the CV of 3.74.

---

planned comparison— A planned comparison of groups prior to the experiment, rather than as a follow-up of a test such as ANOVA.

The researcher may also use **planned comparisons** to test for differences. Planned comparisons are planned (i.e., testable hypotheses are developed) before the experiment. Thus, an experimenter might postulate a test between two groups before the experiment because, in theory, this comparison is important and should be significant. But the number of planned comparisons in an experiment should be small ($k - 1$, where k = the number of groups or treatment levels of an independent variable) relative to the total number of possible comparisons.

Determining the Meaningfulness of Results

Now that we know that F is significant for our example study and have followed it up to see which groups differ, we should answer our second question: What percentage of variance is accounted for by our treatments, and, given the context, what then is the meaningfulness of our results? One way to get a quick idea is to refer to table 9.1 and divide the variance attributed to the treatments by the total variance:

$(SS_{Between}/SS_{Total}) = 90/144 = .625$, meaning 62.5% of the variance is accounted for by the treatments.

This measure of effect size is called eta-squared (η^2). Although this method is sufficient for a quick estimate, it is only interpretable within a study (i.e., to compare the percent of variance explained by multiple independent variables tested with the same sample). A more popular measure of effect size is partial eta-squared ($\eta^2_{partial}$), which is recommended for use with ANOVA (Lakens, 2013; Nesselroade Jr & Grimm, 2019).

$$n_p^2 = \frac{SS_{effect}}{SS_{effect} + SS_{error}} \qquad (9.15)$$

When you are analyzing a study with only one independent variable, partial eta-squared and eta-squared are the same. Thus, if we use this formula with the data from example 9.6, we have this:

$$n_p^2 = \frac{90}{90 + 54} = 0.625$$

Thus, $\eta^2_{partial}$ indicates that 62.5% of the variance is accounted for by the treatments which is the same answer we found when using the formula for eta-squared.

Putting our statistics together, we could say that the treatment was significant, $F(2, 12) = 10.00$, $p < .01$, and accounted for a large proportion of the variance, $\eta^2 = 62.5\%$. In addition, a follow-up Scheffé test indicates that group 1 had the best performance and was significantly different ($p < .05$) from group 3. However, group 1 was not significantly different from group 2, nor was group 2 significantly different from group 3.

Summarizing Simple ANOVA

One final point to recall before leaving our discussion of simple ANOVA is that t is a special case of F when there are only two levels of the independent variable (two groups). In fact, this relationship ($t^2 = F$) is exact, within rounding error (see example 9.8).

EXAMPLE 9.8

Known Values

	Control group			Experimental group		
	X	$(X - M)^2$	X^2	X	$(X - M)^2$	X^2
	2	.09	4	8	4	64
	4	2.89	16	7	1	49
	3	.49	9	5	1	25
	3	.49	9	4	4	16
	2	.09	4	7	1	49
	1	1.69	1	5	1	25
	1	1.69	1	6	0	36
Σ	16	7.43	44	42	12	264
M	2.3	—	—	6.0	—	—
S	1.11	—	—	1.41	—	—

Working It Out

1. Find t (equation 9.3).

$$t = \frac{M_1 - M_2}{\sqrt{(s_1^2/n_1) + (s_2^2/n_2)}} = \frac{2.3 - 6}{\sqrt{(1.11^2/7) + (1.41^2/7)}} = \frac{3.7}{0.68} = 5.44^*$$

*The sign is not important to report.

2. Compute the simple ANOVA and complete the summary table.

$$A = \Sigma X^2 = 44 + 264 = 308$$

$$B = (\Sigma X)^2/N = (58)^2/14 = 240.29$$

$$C = (\Sigma X_1)^2/n_1 + (\Sigma X_2)^2/n_2 = (16)^2/7 + (42)^2/7 = 36.57 + 252 = 288.57$$

	Summary table for ANOVA			
Source	**SS**	**df**	**MS**	**F**
Between	48.28	1	48.28	29.80
Within	19.43	12	1.62	
Total	**67.71**	**13**		

Note: $t^2 = F$; $(5.44)^2 = 29.6 \cong 29.8$ (within rounding error).

In other words, there is little need for t because F will handle two or more groups. But t remains in use because it was developed first, being the simplest case of F. Also, t is convenient and can be computed by hand because M and s are already calculated for each group to serve as descriptive statistics.

Factorial ANOVA

factorial ANOVA—An analysis of variance in which there is more than one independent variable.

Up to this point, examples of two (t) or more levels (simple ANOVA) of one independent variable have been discussed. In fact, in our examples so far, all other independent variables have been controlled except the single independent variable to be manipulated and its effect on a dependent variable. But, in fact, we can manipulate more than one independent variable and statistically evaluate the effects on a dependent variable by using **factorial ANOVA**. As you read the literature, you will see that many more studies use factorial ANOVA than use one-way, or simple, ANOVA. Theoretically, a factorial ANOVA may have any number of factors (two or more) and any number of levels within a factor (two or more), but we seldom encounter ANOVA with more than three factors. This is another good place to apply the KISS principle (Keep it simple, stupid) because it is extremely difficult to interpret findings when the data are influenced by four or more variables simultaneously.

Components of Factorial ANOVA

main effects—Tests of each independent variable when all other independent variables are held constant.

For the purpose of explaining, we consider the simplest form of the factorial ANOVA, which uses only two independent variables and only two levels of each variable. The two-way ANOVA is specifically designated as a 2×2 ANOVA, indicating that each of the two independent variables has two levels. All two-way ANOVAs have two main effects and one interaction. **Main effects** are tests of each independent variable when the other is disregarded (and controlled). Look at table 9.3 and note that the first independent variable (IV_1, intensity) has two levels, labeled A_1 (high) and A_2 (low). In our example, this IV represents the level of intensity for training. The second independent variable (IV_2, fitness) represents participants with low (B_1) or high (B_2) fitness levels. The numbers in the cells represent mean attitude scores of each of the four groups toward the type of training. Thus, we have one group comprising high-fitness people training at a high-intensity level and one group of high-fitness people training at a low-intensity level. The low-fitness participants were also split into the two intensity programs.

We can test IV_1's main effect by comparing the row means (M_{A1} and M_{A2}), because IV_2 (B_1 and B_2), or participants' fitness level, is equally represented at each A level. Actually, an SS with $A - 1$ df and MS for the between-group and within-group sources of variance are computed. Thus, we can compute an F for main effect A (intensity).

TABLE 9.3
Factorial (2 × 2) ANOVA Model

		IV$_2$ (fitness)		
		B_1 (low)	B_2 (high)	
IV$_1$ (intensity)	A_1 (high)	$M = 10$	$M = 40$	$M_{A1} = 25$
	A_2 (low)	$M = 30$	$M = 20$	$M_{A2} = 25$
		$M_{B1} = 20$	$M_{B2} = 30$	

Summary table				
Source	**SS***	**df**	**MS**	**F**
A (intensity)	—	$A - 1 = 2 - 1 = 1$	SS_A / df_A	MS_A / MS_{error}
B (fitness)	—	$B - 1 = 2 - 1 = 1$	SS_B / df_B	MS_B / MS_{error}
AB (interaction)	—	$(A - 1)(B - 1) = 1$	SS_{AB} / df_{AB}	MS_{AB} / MS_{error}
Error	—	$(N - 1) - [(A - 1) + (B - 1) + (A - 1)(B - 1)]$	SS_{error} / df_{error}	
Total	**—**	**N – 1**		

For simplicity, formulas for computing sums of squares are not given here.

The same holds true for IV$_2$, fitness level, by comparing the column means (M_{B1} and M_{B2}). You can see that the two levels of A (IV$_1$), intensity of training, are equally represented in the two levels of B. Therefore, the main effect of high- and low-fitness participants' attitudes toward exercise can be tested by the F ratio for B.

In a study of this type, the main interest usually lies in the interaction. We want to know whether attitude toward high- or low-intensity exercise programs (factor A) depends on (is influenced by) the participants' fitness level (factor B). This effect is tested by an F ratio for the interaction ($A \times B$) of the two IVs, which evaluates the four cell means: M_{A1B1}, M_{A1B2}, M_{A2B1}, and M_{A2B2}. Unless some special circumstance exists, interest in testing the main effects is usually nullified by a significant interaction. The significant interaction tells us that what happens in one independent variable depends on the level of the other. Thus, normally it makes little sense to evaluate main effects when the interaction is significant.

This particular factorial ANOVA is labeled a 2 (intensity of training) $\times$ 2 (level of fitness) ANOVA (read "2-by-2 ANOVA"). The true variance can be divided into three parts:

- True variance because of A (training intensity)
- True variance because of B (fitness level)
- True variance because of the interaction of A and B

Each of these true variance components is tested against (divided by) error variance to form the three F ratios for this ANOVA. Each F ratio has its own set of degrees of freedom so that it can be checked for significance in the F table (Appendix table 5).

Main effects (of more than two levels) can be followed up with a Scheffé test, but a follow-up test is not required for an independent variable with only two levels. In this case, if the F is significant, you only have to see which has the higher M. If the F for the interaction is significant, meaning that attitude toward high- or low-intensity training depends on the participants' fitness level, it will be reflected in a plot, as shown in figure 9.2. You can see in table 9.3 that there is no mean difference for the main effect of training intensity (both A_1 and A_2 are the same, $M = 25$). The plot in figure 9.2, however, reflects a significant interaction. The low-fitness group's scores for mean attitude toward low-intensity training were higher than their scores for mean attitude toward high-intensity training. The opposite was shown for the high-fitness participants. They preferred the high-intensity program.

This example shows how using a particular type of statistical test increases power. If we had just used a t test or simple ANOVA, we would not have found any difference in attitude toward the two levels of intensity (both means were identical, $M = 25$). When we added another factor (fitness level), however, we were able to discern attitude differences that depended on participants' fitness levels.

As you can see, all we have done to follow up the significant interaction is to verbally describe the plot in figure 9.2. Considerable disagreement exists among researchers about how to follow up significant interactions. Some researchers use a multiple-comparison-of-means test, such as Scheffé, to contrast the interaction cell means. These multiple-comparison tests, however, were developed for contrasting levels within an independent variable and not for cell means across two or more independent variables; thus, using these tests may be inappropriate. Other researchers use a simple main effects follow-up test. If an interaction is to be followed up with a statistical test, using simple main effects via the Scheffé procedure is the preferred technique.

Our preference for evaluating an interaction is to do as we have done previously—plot the interaction and describe it. This takes into account the true nature of an interaction; that is, what happens in one independent variable depends on the other. But you are likely to encounter all these ways of testing interactions (and probably some others) as you read the research literature. Just remember that the researcher is trying to show you how the two or more independent variables interact. Also remember that follow-ups on main effects are usually unnecessary (or at least the interpretation must be qualified) when the interaction is significant.

Figure 9.3 shows a nonsignificant interaction. In this case, both groups preferred the same type of program over the other; hence, the lines are parallel. Significant interactions show deviations from parallel (as shown in figure 9.2). The lines do not have to cross to reflect a significant interaction. Figure 9.4 shows a

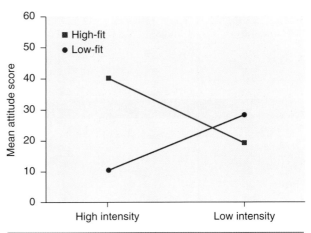

Figure 9.2 Plot of the interaction for a two-way factorial ANOVA.

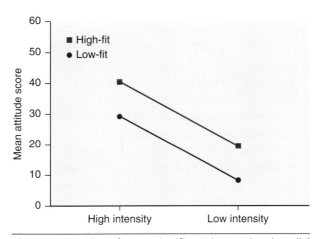

Figure 9.3 Plot of a nonsignificant interaction (parallel lines) in which both fitness groups prefer low-intensity programs over high-intensity programs.

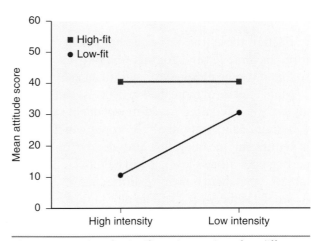

Figure 9.4 Plot of a significant interaction of no difference in high-fit people's preferences but a significant difference in intensity preference among low-fit people.

significant interaction in which those in the high-fitness group liked both forms of exercise equally, but there was a decided difference in preference in the members of the low-fitness group, who preferred the low-intensity program over the high-intensity program.

Determining the Meaningfulness of Factorial Analysis Results

Having answered the first statistical question about a factorial ANOVA (Are the effects significant?), we now turn to the magnitude of the effects. What percentage do the independent variables and their interaction account for in the dependent variable's variance? The three formulas that follow (from Lakens, 2013) provide the test for each component of ANOVA:

$$Main\ effect\ for\ A{:}\ n_p^2 = \frac{SS_A}{SS_A + SS_{error}} \tag{9.16}$$

$$Main\ effect\ for\ B{:}\ n_p^2 = \frac{SS_B}{SS_B + SS_{error}}$$

$$Main\ effect\ of\ A \times B{:}\ n_p^2 = \frac{SS_{A \times B}}{SS_{A \times B} + SS_{error}}$$

These formulas involve the proportion of true variance to all sources of variance other than the effect of interest.

Each n_p^2 represents the percentage of variance accounted for by that component of the ANOVA model. Of course, effect sizes can be calculated according to the previous equations (9.5) for any two means for which you want to compare main effects.

Summarizing Factorial ANOVA

Frequently, an independent variable's levels are categorical, or classified, rather than random. For example, we could do a simple ANOVA in which the independent variable's levels (or groups) involve novice and expert tennis players. Or we could compare high-fitness and low-fitness participants' attitudes toward training intensity (see our example in table 9.3). That is, we are interested in whether preference toward type of training programs affected high-fitness and low-fitness participants differently. Participants are randomly assigned to the first independent variable's intensity level of program; they cannot be randomly assigned to the categorical variable of fitness level. Common categorical variables are sex and age. A study might look at the effects of the levels of a treatment (independent variable) to which male and female (independent variable) or 6-, 9-, and 12-year-olds (independent variable) participants are randomly assigned. Clearly, the interest in this type of factorial ANOVA is the following interaction: Does the effectiveness of treatment differ according to the participant's sex (or age group)? If all three variables were included in one study, then we would be able to test for main effects for treatment, sex, and age group; for two-way interactions of treatment × sex, treatment × age group, and sex × age group; and for a three-way interaction of treatment × sex × age group.

Examples of factorial ANOVA are plentiful in the research literature. Gromeier, Koester, and Schack (2017) compared overarm throwing performance between boys and girls from three age groups. The data were analyzed using a 3 (age group) × 2 (sex) factorial ANOVA. The following is a sample description of the results of the factorial ANOVA: "The analysis showed no significant effect of gender on quantitative performance; $F(1, 90) = 1.207$, $p = 0.275$. The analysis yielded a significant effect of age on quantitative performance; $F(2, 90) = 23.848$, $p < 0.001$. The interaction effect was not significant, $F(1, 90) = 0.582$, $p = 0.561$" (p. 5). The significant main effect for age was followed up using t tests with a Bonferroni correction (explained later in this chapter).

Repeated-Measures ANOVA

Much of the research in the study of physical activity involves studies that measure the same dependent variable more than once. For example, a study in sport psychology might investigate whether athletes' **state anxiety** (how nervous they feel at the time) differs before and after a game. Thus, a state-anxiety inventory would be given just before and immediately after the game. The question of interest is, *Does state anxiety change significantly after the game?* A dependent *t* test could be used to see whether significant changes occurred, but this is also the simplest case of **repeated-measures ANOVA**.

Another study might measure participants on the dependent variable on several occasions. Suppose we want to know whether the distance that children (between ages 6 and 8) throw a ball using an overhand pattern increases over time. We decide to measure the children's distance throw every three months during the two years of the study. Thus, each child's throwing distance was assessed nine times. We now have nine repeated measures (at the beginning and in eight three-month intervals) on the same children. We would use a simple ANOVA with repeated measures on a single factor. Basically, the repeated measures are used as nine levels of the independent variable, which is time (24 months).

The most frequent use of repeated measures involves a factorial ANOVA in which one or more of the factors (independent variables) are repeated measures. An example is an investigation of the effects of knowledge of results (KR) on skilled motor performance. Three groups of participants (three levels of the independent variable, KR) receive different types of KR (no KR, whether they were short or long of the target, and the number of centimeters they were short or long of the target). The task is to position a handle that slides back and forth on a track (called a linear slide) as close to the target as possible. But the participants are blindfolded, so they cannot see the target. They have only verbal KR to correct their estimates of where the target is on the track.

In this type of study, participants are usually given multiple trials (assume 30 trials in this example) so that the effects of the type of KR can be judged. The score on each trial is distance from the target in centimeters. This type of study is frequently analyzed as a two-way factorial ANOVA with repeated measures on the second factor. Thus, a 3 (levels of KR)

Significant differences may be in the eye of the beholder.

Alkir/iStock/Getty Images

× 30 (trials) ANOVA with distance error as the dependent variable is used to analyze the data. The first independent variable (level of KR) is a true one (three groups are randomly formed). The second independent variable (30 trials) is repeated measures. Sometimes this ANOVA is called a two-way factorial ANOVA or a mixed ANOVA with one between-subjects factor (levels of KR) and one within-subjects factor (30 repeated trials). Although an *F* ratio is calculated for each independent variable, the major focus is usually on the interaction. For example, do the groups change at different rates across the trials?

> **Repeated-measures designs identify and separate individual differences from the error term, increase power, require fewer participants, and study a phenomenon across time.**

Advantages of Repeated-Measures ANOVA

Repeated-measures designs have three advantages (Gravetter & Wallnau, 2017). First, they provide the experimenter the opportunity to control for individual differences among participants, probably the largest source of variation in most studies. In between-subjects designs, which are completely randomized, the variation among people goes into the error term. Of course, this tends to reduce the *F* ratio unless it is offset by a large *N*. Remember that to calculate the error term of the *F* ratio, divide the variation among participants by the degrees of freedom, which are based on the number of participants. In repeated-measures designs, variation from individual differences can be identified and separated from the error term, thereby reducing it and increasing statistical power. As you can deduce from the first advantage, repeated-measures designs are more economical because they require fewer participants. Finally, repeated-measures designs allow the study of a phenomenon across time. This feature is particularly important in studies of changes in, for example, learning, fatigue, forgetting, performance, and aging.

Problems of Repeated-Measures ANOVA

Several problems adversely affect repeated-measures designs, including the following:

- *Carryover effects.* Treatments given earlier influence treatments given later.
- *Practice effects.* Repeated trials, in addition to the treatment, or testing effect, cause participants to improve at the task (dependent variables).
- *Fatigue.* Boredom or fatigue adversely influences participants' performance.
- *Sensitization.* Repeated exposure heightens participants' awareness of the treatment.

Note that some of these problems may be the variables of interest in repeated-measures designs. Carryover effects may be of interest to a researcher of learning, whereas increased fatigue over trials may intrigue an exercise physiologist.

The tricky part of repeated-measures designs involves how to analyze the data statistically. We have already mentioned ANOVA models with repeated measures. Unfortunately, repeated-measures ANOVA has an assumption—**sphericity**—beyond the ones we have given for all previous techniques. Sphericity "can be thought of as an extension of the homogeneity of variance assumption across the repeated variable" (Maurissen & Vidmar, 2016, p. 79). If the design has a between-subjects factor, the pooled data (across all participants) must exhibit sphericity. How well the data meet these assumptions is best estimated by the statistic epsilon (ε). Epsilon ranges from 1.0 (perfect sphericity, the assumption is met) to 0.0 (complete violation). An epsilon above .75 is desirable in repeated-measures experiments. Most of the widely used statistical packages (e.g., SPSS and SAS) have repeated-measures functions that provide both estimates of epsilon and tests that should be used to evaluate the *F* ratio for the repeated-measures factors in the designs. The failure to meet the sphericity assumption results in an increase in type I error; that is, the alpha level may be considerably larger than the researcher intended. Several statisticians (Armstrong, 2017; Haverkamp & Beauducel, 2017) have suggested that multivariate techniques may be more appropriate methods of analysis. Two additional points are important (Pedhazur, 1982):

> **sphericity**—An assumption that repeated measures are uncorrelated and have equal variance.

- When the assumptions are met, ANOVA with repeated measures is more powerful than the multivariate tests.
- If the number of participants is small, only the ANOVA repeated-measures test can be used.

If you are contemplating conducting a study that uses a repeated-measures design, an additional source of reading should be helpful to you. Verma (2016) provides a tutorial on the use of repeated measures for univariate and multivariate data that gives specific examples of how the data may be analyzed and evaluated.

Components of Repeated-Measures ANOVA

Table 9.4 provides the sources of variation in a one-way repeated-measures ANOVA, which is an analysis also known as subject × trials ANOVA or within-subjects ANOVA.

The values we used to arrive at the statistics in table 9.4 are the same as in example 9.6. Recall that in example 9.6, 15 participants were randomly assigned to treatment group 1, 2, or 3. Suppose that, in a different study, five participants were each measured on three trials of a task (or were given different treatments), and we coincidentally got the same 15 measurements in example 9.6. In effect, rather than having 15 participants in groups 1, 2, and 3, we now have five participants who were measured on trials 1, 2, and 3 (or treatments 1, 2, and 3). The known values for table 9.4 would look identical to those in example 9.6, except that the columns could be labeled Trial 1, Trial 2, and Trial 3 (or Treatments 1, 2, and 3) instead of Group 1, Group 2, and Group 3. We are using the same 15 values so that when table 9.4 and example 9.6 are compared, you can see why the repeated-measures design results in increased economy. Notice that the sum of squares and degrees of freedom are the same. Also, the between-groups effect in example 9.6 and trials effect in table 9.4 are the same. But the sum of squares for the within-group effect (error) in example 9.6 is divided into two components in the repeated-measures analysis in table 9.4. The residual effect (estimating error) has a sum of squares of 28 with 8 df, and the subjects effect has a sum of squares of 26 with 4 df. This results in an F ratio for the repeated-measures ANOVA (in table 9.4) that is larger than the F ratio for the simple ANOVA (in example 9.6), despite the repeated-measures ANOVA having only one-third as many participants. The sum of squares for subjects in table 9.4 is not tested and simply represents the normal variation among subjects. Thus, across the three trials, the participants' mean performance decreased significantly (trial 1 = 10, trial 2 = 7, and trial 3 = 4). This analysis has all the strengths and weaknesses of repeated-measures designs previously discussed.

The epsilon estimate from this analysis, shown at the bottom of table 9.4, was obtained by running a repeated-measures computer program from SPSS. This epsilon is what we examine relative to the criterion of 0.75 to determine if we meet the assumption of sphericity or fail to meet the assumption. If epsilon is less than 0.75, the degrees of freedom is adjusted according to the following formula:

$$df \times \varepsilon = \text{adjusted } df.$$

This is done for the degrees of freedom in both the numerator and the denominator of any F ratio that includes the repeated-measures factor.

TABLE 9.4

Summary of Repeated-Measures ANOVA

Source	SS	df	MS	F
Trials (T)	90	(T − 1) = 2	45.0	12.86*
Subjects (S)	26	(S − 1) = 4	6.5	
Residual (error)	28	(S − 1)(T − 1) = 8	3.5	
Total	**144**	**N − 1 = 14**		

*p < .05.

Note: $\varepsilon = 1.00$; no adjustment of dfs needed.

The **Geisser-Greenhouse correction** is a conservative approach to this adjustment and can be done as follows (Maurissen & Vidmar, 2016):

$$\theta = 1/(k - 1) \tag{9.17}$$

Geisser-Greenhouse correction—A conservative approach to the adjustment of the epsilon estimate in repeated-measures ANOVA that calculates adjusted degrees of freedom to find an *F* ratio to determine significance.

where k = the number of repeated measures. Then we multiply the degrees of freedom for trials by this value, $\theta(k - 1)$, as well as the degrees of freedom for error, $\theta(n - 1)(k - 1)$. These degrees of freedom are adjusted to the nearest whole *df* used to look up the *F* ratio in the *F* table. If the *F* ratio is significant, then the effect likely is a real one. SPSS provides output relative to Mauchly's Test of Sphericity that identified whether this test is significant ($p < .05$) indicating that there is a violation of the sphericity assumption necessitating an adjustment to degrees of freedom or not significant ($p > .05$) indicating that no adjustments are necessary. When the test of sphericity is significant, SPSS also provides various adjustments to degrees of freedom to account for this violation.

It is probably safe to say that most ANOVA models in experimental research reported in national journals in our field use factorial ANOVA with repeated measures on one or more of the factors with earlier studies using multivariate techniques when violations of sphericity were identified and more recent studies using degree of freedom corrections. An example of a study using multivariate techniques is by Weiss, McCullagh, Smith, and Berlant (1998), who examined the role of peer mastery and coping models on children's swimming skills, fear, and self-efficacy. If the sphericity assumption was violated ($\varepsilon < .75$), the authors used a multivariate repeated-measures method. If sphericity was not violated, they used a 3 (peer model type) $\times$ 3 (assessment period) factorial ANOVA with repeated measures on the last factor. An example of a study using a degree of freedom correction is by Hutchinson, Zenko, Santich, and Dalton (2020), who were interested in differences in how the intensity of a resistance training protocol implemented during a session (either increasing intensity or decreasing intensity within the session) influences participant's affective responses to the exercise. They used a fully within-subjects design so participants performed both treatment conditions and affect was measured at three time points. They used a Greenhouse-Geisser correction in cases where the sphericity assumption was violated.

Analysis of Covariance

Analysis of covariance (ANCOVA) is a combination of regression and ANOVA. The technique is used to adjust the dependent variable for some distractor variable—the **covariate**—which is a variable that could affect the treatments.

analysis of covariance (ANCOVA)—A combination of regression and ANOVA that statistically adjusts the dependent variable for some distractor variable called the *covariate*.

covariate—A distractor variable that is statistically controlled in ANCOVA and MANCOVA.

Using ANCOVA

Suppose we want to evaluate a training program's effects to develop leg power on the time required to run 50 m. We know that reaction time (RT) influences the 50 m dash time because runners who begin more quickly after the start signal have an advantage. We form two groups, measure RT in each group, train one with our power development program, use the other as a control group, and measure each participant's time for running 50 m. ANCOVA might be used in this study to analyze the data. There is an independent variable with two levels (power training and control), a dependent variable (50 m dash time), and an important distractor variable, or covariate (RT).

Analysis of a covariance is a two-step process in which an adjustment is first made for the 50 m dash scores of runners according to their RT. A correlation (r) is calculated between RT and the time in the 50 m dash. The resulting prediction equation, 50 m dash time = $a +$ (b)RT (the familiar formula $a + bX$), is used to calculate each runner's predicted 50 m dash time ($\hat{Y}$). The difference between the actual 50 m dash time (Y) and the predicted time ($\hat{Y}$)

is the residual $(Y - \hat{Y})$. A simple ANOVA is then calculated using each participant's residual score as the dependent variable (1 *df* for within-group sums of squares is lost because of the correlation). This allows an evaluation of 50 m dash speed with RT controlled.

Analysis of covariance can be used in factorial situations and with more than one covariate. The results are evaluated as in ANOVA except that one or more distractor variables are controlled. Also, ANCOVA is frequently used in situations in which a pretest is given, some treatment is applied, and then a posttest is given. The pretest is used as the covariate in this type of analysis. Note that in the preceding section on repeated-measures ANOVA, we indicated that this same situation could be analyzed by repeated measures. In addition, ANCOVA is used when comparing intact groups because the groups' performances (dependent variable) can be adjusted for covariates on which they differ. If ANCOVA is warranted, it can increase the power of the *F* test (making it easier to detect a difference) because of the ability to control for a variable that could affect the results.

Limitations of ANCOVA

Although ANCOVA may seem to be the answer to many problems, it does have limitations. In particular, its use to adjust final performance for initial differences can result in misleading interpretations (Lord, 1969). In addition, if the correlations between the covariate and the dependent variable are unequal across the treatment groups, then standard ANCOVA (there are nonstandard ANCOVA techniques) is inappropriate.

Diekfuss, Rhea, Schmitz, Grooms, Wilkins, Slutsky, and Raisbeck (2019) explored the effects of attentional focus on balance control by assessing balance at a pretest and then after a seven-day period. Participants were randomly assigned to a control group that was inactive during the seven-day period or to an internal or external group that practiced the task daily using an internal or external attentional focus, respectively. For balance at the retention test, ANCOVA was used to assess differences as a function of condition (between-subjects factor) with pretest performance used as a covariate.

Experiment-Wise Error Rate

Sometimes researchers make several comparisons of dependent variables using the same participants. Usually, a multivariate technique (discussed in the next section) is the appropriate solution, but when dependent variables are combinations of other dependent variables (e.g., cardiac output is heart rate $\times$ stroke volume), a multivariate model using all three dependent variables is inappropriate. (This book is not the appropriate place to explain why. For more detail, see Thomas, 1977.) Thus, an ANOVA among three groups might be calculated separately for each dependent measure (i.e., three ANOVAs). The problem is that this procedure results in increasing the alpha that has been established for the experiment. One of two solutions is appropriate for adjusting alpha. The first, the Bonferroni technique, simply divides the alpha level by the number of comparisons to be made:

$$\alpha_{EW} = \alpha / c \tag{9.18}$$

where $\alpha_{EW} = alpha$ corrected for the EER, α = alpha, and c = the number of comparisons. If, for example, $\alpha = .05$ and $c = 3$, then the alpha for each comparison is $.05 / 3 = .017$. This means that the *F* ratio would have to reach an alpha of .017 to be declared significant.

An example of this approach can be seen in a manuscript by Cinque et al. (2017). Using *t* tests, Cinque et al. made comparisons of various career statistics for NFL players relative to having or not having an ACL injury. The authors used the Bonferroni procedure to adjust the alpha level for their six comparisons. With alpha at .05, the adjusted alpha was $.05 / 8 = .008$. Consequently, to be significant, the *t* tests had to reach $\alpha < .008$.

The second option is to leave the overall alpha at .05 but to calculate the upper limit that the alpha might be:

$$\alpha_{UL} = 1 - (1 - \alpha)^k \qquad\qquad\qquad\qquad (9.19)$$

where α_{UL} = alpha (upper limit) and k = the number of groups. Again, using the example with three comparisons and alpha at .05, $\alpha_{UL} = 1 - (1 - .05)^3 = .14$. Thus, the hypotheses are really being tested somewhere between an alpha of .05 if the dependent variables are perfectly correlated and an alpha of .14 if they are independent. When researchers make multiple comparisons using the same participants, they should either adjust alpha to the EER or at least report the upper limit on their alpha.

Understanding Multivariate Techniques

Up to this point, we have discussed experimental research examples testing for differences in a single dependent variable between groups based upon one or more independent variables. Multivariate cases have one or more independent variables and two or more dependent variables. For example, it seems likely that when independent variables are manipulated, they influence more than one thing. The multivariate case allows more than one dependent variable. Using univariate techniques, which allow only one dependent variable repeatedly when several dependent variables are present, increases the EER (sometimes called *probability pyramiding*) in the same way as doing multiple t tests instead of simple ANOVA when there are more than two groups. But in some instances, using univariate techniques in a study that has multiple dependent variables is acceptable or the only choice. As with all research, theory should drive decision making. You might not want to include a theoretically important dependent variable in a multivariate analysis when less theoretically important dependent variables could disguise its importance. Also, a multivariate technique is sometimes just not feasible because the study has a small number of participants. Following are multivariate techniques often used in experimental studies:

- Discriminant analysis
- Multivariate ANOVA (MANOVA) and special cases with repeated-measures designs
- Multivariate ANCOVA (MANCOVA)

Remember that the general linear model still underlies all the techniques, and we are still attempting to learn two things: Are we evaluating something significant (reliable)? What is the magnitude of the effect so we can infer meaningfulness from significant findings?

Discriminant Analysis

We use discriminant analysis when we have one independent variable (two or more levels) and two or more dependent variables. The technique combines multiple regression and simple ANOVA. In effect, discriminant analysis uses a combination of the dependent variables to predict or discriminate between the levels of the independent variable, which, in this case, is group membership. In chapter 8's discussion of multiple regression, we used several predictor variables in a linear combination to predict a criterion variable. In essence, discriminant analysis does the same thing except that several dependent variables are used in a linear combination to predict the group to which a participant belongs. This prediction of group membership is the equivalent of discriminating between the groups (recall how t could be calculated by r). The same methods used in multiple regression to identify the important predictors are used in discriminant analysis. These include forward, backward, and stepwise selection techniques. For greater detail, as well as a useful and practical description of discriminant analysis, read Barton, Yeatts, Henson, and Martin (2016).

Forward, Backward, and Stepwise Selection

As mentioned in chapter 8, the forward selection technique enters the dependent variables in the order of their importance; that is, the dependent variable with the greatest contribution to separation of the groups (discriminating between or predicting group membership the best) is entered first. By correlation techniques, the effect of the first dependent variable on all others is removed, and the dependent variable that contributes the next greatest amount to the separation of the groups is entered at step 2. This procedure continues until all dependent variables have been entered or until some criterion that the researcher has established for stopping the process is met.

The backward selection procedure is similar except that all the dependent variables are entered, and the one that contributes least to group separation is removed. This continues until the only variables remaining are those that contribute significantly to the separation of the groups.

The stepwise technique is similar to forward selection except that at each step, all the dependent variables are evaluated to see whether each still contributes to group separation. A dependent variable that does not contribute is stepped out (removed) from the linear combination, just as in multiple regression.

Example of Discriminant Analysis

O'Connor, Larkin, and Williams (2016) conducted a study in which 127 elite youth male soccer players provided information relative to their sport history and performed tasks to assess perceptual–cognitive skills. They then participated in a tournament during which 22 players were selected to receive scholarships at a selective residential program for soccer.

Discriminant analysis was applied to determine which variables predicted selection. A stepwise discriminant analysis resulted in four predictive variables that accounted for 57.6% of the variance. The significant predictors were recent match-play performance, the region where they played soccer, the number of other sports they engaged in, and performance on the decision-making portion of the perceptual–cognitive task. Using these variables, researchers were able to predict who would be selected for the camp with 97% accuracy.

Discriminant analysis may be followed up with univariate techniques to determine which groups differed from one another on each of the dependent variables selected. There are several ways to approach this follow-up, but for simplicity, you could perform the Scheffé test for the three groups on the first dependent variable. Then you could use ANCOVA for the three groups on the second dependent variable using the first dependent variable as a covariate. This would provide adjusted means of the second dependent variable (means of the second dependent variable corrected for the first). A Scheffé test could be done for the adjusted means using the adjusted mean square for error. This procedure, called a *stepdown F technique*, is continued through each dependent variable using the previously stepped-in dependent variables as covariates.

Summarizing Discriminant Analysis

O'Connor et al. (2016) conducted a discriminant analysis to predict group assignment based upon predictor variables. This is a common application of discriminant analysis. However, it should be obvious that this does not determine cause and effect but rather reflects correlations or relationships. If determination of cause and effect is the purpose of the research, different experimental designs would be needed.

Multivariate Analysis of Variance

From an intuitive point of view, multivariate analysis of variance (MANOVA) is a straightforward extension of ANOVA. The only difference is that the *F* tests for the independent variables and interactions are based on an optimal linear composite of several dependent variables. There is no need to consider simple MANOVA here, because this is discriminant analysis (one independent variable with two or more levels and two or more dependent variables). For a practical discussion, definition, and example of MANOVA, see Haase and Ellis (1987).

Using MANOVA

The mathematics is complex for factorial MANOVA, but the idea is simple. An optimal combination (linear composite) of the dependent variables is made that maximally accounts for (predicts) the variance associated with the independent variables. The variance associated with each independent variable is then separated out (as in ANOVA), and each independent variable and interaction on the optimal linear composite is tested. The associated F and degrees of freedom for each test are interpreted in the same way as for ANOVA. There are several ways to obtain F in MANOVA: Wilks' lambda, Pillai's trace, Hotelling's trace, and Roy's greatest characteristic root. We point this out only because authors sometimes identify how the MANOVA Fs were obtained. For your purposes, these distinctions are not important: Just remember that the MANOVA F ratios are similar to ANOVA F ratios.

After MANOVA has been used, a significant linear composite of dependent variables is identified, which separates the levels of the independent variable(s). Then, the important question usually is *Which of the dependent variables contribute significantly to this separation?* Using the discriminant analysis and the stepdown F procedures discussed in the previous section as follow-up techniques is one of many ways to answer this question. This method works well for the main effects but not so well for interactions. Interactions in MANOVA are most frequently handled by calculating factorial ANOVA for each dependent variable, although this procedure fails to consider the interrelationships of the dependent variables.

Example of MANOVA

As an example of the use and follow-up of MANOVA, consider French and Thomas' (1987) experiment. One aspect of this study was to evaluate the influence of two groupings of age level (8- to 10-year-olds and 11- to 12-year-olds) and expertise level (expert and novice players within each age level) on basketball knowledge and performance. A basketball knowledge test and two basketball skill tests (shooting and dribbling) were given to all the children. Following is French and Thomas's description of these results (p. 22):

> A 2 × 2 (age league × expert/novice) MANOVA was conducted on the scores of the knowledge test and both skill tests. The results of the MANOVA indicated significant main effects for age league, $F(3, 50) = 5.81, p < .01$, and expert/novice, $F(3, 50) = 28.01, p < .01$, but no significant interaction. These main effects were followed up by a stepdown procedure using a forward selection discriminant analysis. The alpha level used as a basis for stepping in variables was set at .05. The discriminant analysis for age league revealed that knowledge was stepped in first, $F(1, 54) = 8.31, p < .01$. Neither skill test was entered. Older children ($M = 79.5$) possessed more knowledge than younger children ($M = 64.9$). The discriminant analysis for expert/novice revealed that shooting was stepped in first, $F(1, 54) = 61.40, p < .01$; knowledge was stepped in second, $F(1, 53) = 5.51, p < .05$; and dribbling was not entered. Child experts ($M = 47.2$) performed significantly better than novices ($M = 25.7$) in shooting skills. The adjusted means for knowledge showed that child experts ($M = 77.1$) possessed more basketball knowledge than novices ($M = 64.2$).

If more than two levels had existed within an independent variable in the MANOVA (i.e., suppose three age-league levels had been used), then following the discriminant analysis, there would have been three means (one for each age-league level). To test among them for significance, it would be appropriate to use a Scheffé test for this follow-up, just as we did for ANOVA.

Multivariate Analysis of Covariance

Conceptually, multivariate analysis of covariance (MANCOVA) represents the same extension of ANCOVA that MANOVA does for ANOVA. In MANCOVA, there are one or more independent variables, two or more dependent variables, and one or more covariates. Recall the earlier explanation of ANCOVA. A variable was used to adjust the dependent variable by correlation, and then ANOVA was applied to the adjusted dependent variable. In MANCOVA, each dependent variable is adjusted for one or more covariates. Then an optimal linear composite that best discriminates between the independent variables is formed from the adjusted dependent variables, just as in MANOVA. Follow-up procedures are the same as in MANOVA except that the linear composite of adjusted dependent variables is used. This technique is seldom used and typically used incorrectly. Normally, there is no reason to adjust a dependent variable for a linear composite of covariates. We have purposely not provided an example of MANCOVA use from the physical activity literature. MANCOVA is a fine technique if you understand what it does and you specifically want to test for that. But many researchers misunderstand MANCOVA and do not use it properly. Of course, the same is true of ANCOVA; for a useful discussion of the issues, see Porter and Raudenbush (1987).

Repeated Measures With Multiple Dependent Variables

Rather frequently, experiments have multiple dependent variables that are measured more than once (over time). For example, in an exercise adherence study, both physiological and psychological variables might be measured once a week during a 15-week training program. If there were two training groups (different levels of training) and a control group, each with 15 participants, all of whom were measured every week (i.e., 15 times) on two psychological and three physiological measures, we would have a design that is three levels of exercise $\times$ 15 trials (3×15) for five dependent variables. This design offers several options for analysis.

We could do five 3×15 ANOVAs with repeated measures on the second factor. Here, we would follow the repeated-measures procedures described earlier in this chapter. But we would be inflating the alpha by doing multiple analyses on the same subjects. Of course, the alpha could be adjusted by the Bonferroni technique ($\alpha = .05/5 = .01$), but this fails to consider the potentially substantial and possibly very interesting interrelationships of the dependent variables. Schutz and Gessaroli (1987) provide an excellent tutorial on how to handle this problem. The following is a brief discussion of their study, but if you are using this design and analysis, you should read their complete study and example.

Two options are available for this analysis: multivariate mixed-model (MMM) analysis and doubly multivariate (DM) analysis. Which is used depends on the assumptions that your data meet. Using the study previously described, three levels of exercise $\times$ 15 trials for five dependent variables, the MMM analysis treats the independent variable (levels of exercise) as a true multivariate case by forming a linear composite of the five dependent variables to discriminate between the levels of the independent variable. If this composite is significant, it can be followed up by the stepdown F procedure (for an alternative procedure, see Schutz & Gessaroli, 1987). For the repeated-measures factor (and interaction), a linear composite is formed and treated as a regular repeated-measures analysis for each trial. This means that the sphericity assumption must be met as previously described, and the epsilon can be used to test this assumption with the same standards previously described. The interpretation of the resulting F ratio for main effects for groups, trials, and the group $\times$ trial interaction are the same as in other designs. The question usually posed here is *Do the groups change at different rates across the trials on the linear composite of dependent variables?* If the sphericity assumption can be met, MMM analysis is preferred because most authors believe that it offers more power. This assumption is difficult to meet, however, especially if there are more than two dependent variables measured on more than three trials.

The DM analysis does not require that the sphericity assumption be met. The analysis is the same for the independent variable of exercise. In the repeated-measures part of the analysis, however, a linear composite is formed not only of the dependent variables at each trial but also of the 15 trials, which themselves become a linear composite; thus the name *doubly multivariate*. The interpretations of the *F*s for the two main effects and interaction remain essentially the same, but follow-ups become more complex.

McCullagh and Meyer (1997) compared four methods of providing information (physical practice with feedback, learning model with model feedback, correct model with model feedback, and learning model without model feedback) on learning correct form in the free-weight squat lift. There were two dependent variables (outcome and form) and five trials. The authors analyzed the data using MANOVA with repeated measures. Univariate ANOVAs and post hoc comparisons were done as follow-ups for significant *F*s.

Summary

This chapter presented techniques used in situations where differences between groups are the focus of attention. These techniques range from the simplest situation of two levels of an independent variable with one dependent variable to complex multivariate techniques involving multiple independent variables, multiple dependent variables, and multiple trials. These techniques can be categorized as follows:

• A *t* test is used to determine how a group differs from a population, how two groups differ, how one group changes from one occasion to the next, and how several means differ (the Scheffé or some other multiple comparison test).

• ANOVA shows differences between the levels of the following: (1) one independent variable (simple ANOVA), (2) two or more independent variables (factorial ANOVA), and (3) independent variables when there is a distractor variable, or covariate (ANCOVA).

• MANOVA is used when there is more than one dependent variable. Discriminant analysis is the simplest form of MANOVA (when there is only one independent variable and two or more dependent variables). More complex MANOVAs involve two or more independent variables and two or more dependent variables.

• MANCOVA is an extension of MANOVA in which there is one or more covariates.

• Repeated-measures MANOVA is used when one or more of the independent variables are repeated measures.

Table 9.5 provides an overall look at the techniques and their appropriate use presented here and in chapter 8. Techniques for interrelationships of variables parallel each technique

TABLE 9.5

Comparison of Statistical Techniques From Chapters 8 and 9

Description	Relationships of variables	Differences of groups
1 IV (2 levels) → 1 dv		Independent *t* test
1 predictor → 1 criterion	Pearson *r*	
2 or more IVs → 1 dv		Factorial ANOVA
2 or more predictors → 1 criterion	Multiple regression	
1 IV (2 or more levels) → 2 or more dvs		Discriminant analysis
2 or more IVs → 2 or more dvs		MANOVA
2 or more predictors → 2 or more criteria	Canonical correlation	

Note: IV = independent variable; dv = dependent variable.

for differences between groups. In fact, the relationship between *t* and *r* that we demonstrated earlier in this chapter holds because ANOVA techniques are the equivalent of multiple regression. Each technique evaluates our two basic questions: *Is the effect or relationship significant? What is the magnitude of the effect?* Once we answer these two questions, knowledge of the context can allow us to answer the question: *Is the effect or relationship meaningful?*

Some ideas we've presented here are complex and may not be easily understood in one reading. We provided some suggested readings and practice problems that should be helpful. If you do not feel confident in your understanding of this material, reread the chapter, do the problems, and consult some of the suggested readings. This review is important because parts III and IV assume that you understand part II.

✓ Check Your Understanding

1. Report the statistical part of a study that uses an independent *t* test. Calculate ω^2 for *t* in this study.

2. Find a study that uses a multiple-comparison test (e.g., Newman-Keuls, Duncan, or Scheffé) as a follow-up to ANOVA. Explain what prompted the researchers to use this test and their findings.

3. A researcher wishes to compare the effectiveness of three desk arrangements (rows, clusters, and circles) on classroom behavior (frequencies of observed, on-task behaviors). Elementary school children (*N* = 18; NOTE: again, in this instance, the capital *N* denotes that this is the number of participants in the entire study and the lowercase *n* denotes the number of participants in one of the conditions or groups) were randomly assigned to three groups of desk arrangements for instruction. The *M*, *s*, and *n* for each group are as follows:

Group	*M*	*s*	*n*
Rows	6.0	1.4	6
Clusters	6.8	1.9	6
Circles	9.5	1.0	6

Here is a partial summary of the ANOVA:

Source	SS	df	MS	F
Between	40.1			
Within	34.3			
Total	**74.4**			

a. Complete the table. Use table 9.1 and example 9.6 for help.
b. Test the *F* for significance at the .05 level using Appendix table 5.
c. Compute ($\eta^2_{partial}$) for meaningfulness.
d. Use the Scheffé method to determine where the differences occur in the three groups (consult example 9.7).
e. Write a short paragraph interpreting the results of your data analysis.

4. Find a study that used a two-way factorial ANOVA. Explain what the authors found for each main effect and for the interaction. Did they report the effect sizes for their effects? If so, provide those. If not, did they provide the necessary data to calculate an effect size? If so, provide that calculation. If not, explain what is needed.

5. Locate a study that used MANOVA in its data analysis. Identify the independent and dependent variables.

10

Nonparametric Techniques

If you have always done it that way, it is probably wrong.

—Charles Kettering

In the preceding chapters, various parametric statistics were described. Recall that parametric statistics make assumptions about the normality and homogeneity of the variance of the distribution. Another category of statistics, nonparametric statistics, is also referred to as *distribution-free statistics* because no assumptions are made about the distribution of scores. Recall the assumptions from chapter 6 for parametric statistics: the population from which the sample is drawn is normally distributed on the variable of interest, the samples drawn from a population have the same variances on the variable of interest and the observations are independent. When the assumptions are not met, a solution is to use nonparametric techniques. Annual U.S. household income is an example where we have data on the population and the variable is of interest to most of us! U.S. Census Bureau reported the median household income at $68,703 for 2019, while the mean was $98,088 (Semega, Koolar, Shrider, & Creamer, 2021). The difference between the mean and the median suggests the data is not normally distributed in the population. For most households $30,000 is a meaningful amount of money! Thus, a study examining physical activity and household income must consider income as nonparametric because the population from which the sample is drawn does not meet the first assumption.

Nonparametric statistics are versatile because they can deal with ranked scores and categories. This feature can be a definite advantage when dealing with variables that do not lend themselves to precise interval-type or ratio-type data (which are more likely to meet parametric assumptions), such as categories of responses on questionnaires and affective behavior-rating instruments. You might think of these categories like a multiple-choice question. The answers are limited to specific responses, for example always, frequently, sometimes, and never. Data from quantitative research are frequently numerical counts of events that can be analyzed with nonparametric statistics. This might include physical activity data counting how many times a participant moved from one quadrant to another quadrant on a field or court, a variable estimating the amount of activity in a sport setting. The goalie might not change quadrants while an offensive player with the ball would move through several quadrants.

Because nonparametric statistics can deal with ranked scores and categories, they can be useful when working with data that do not meet parametric assumptions for normal distribution. Questionnaire responses and behavior-rating techniques are two examples of data that can be analyzed using nonparametric statistics.

The main drawback to nonparametric statistics that has often been voiced is that they are less powerful than parametric statistics. As you recall, power refers to the ability of a statistical test to reject a null hypothesis that is false. It should be pointed out that there is no agreement regarding the supposed power advantage of parametric tests (see Brunnstrom & Barkowsky [2018] for a more thorough discussion). Another drawback in the use of nonparametric tests has been the lack of statistical software for the more complex statistical tests, such as multivariate cases.

In chapter 6, we provided you with a set of procedures to judge the normality of data so that you can determine whether to use parametric or nonparametric approaches. This included looking at the distribution of the data (e.g., using a stem-and-leaf or histogram plot to evaluate whether the data fit a normal curve) and evaluating the skewness and kurtosis of the distribution. Deciding whether data meet the normality assumption is difficult (Thomas, Nelson, & Thomas, 1999). For example, how normal do data have to be to use parametric techniques? On numerous occasions a researcher's data will not meet the assumptions of normality. Micceri (1989), for example, maintained that much of the data in education and psychology are moderately or largely nonnormal, and thus nonparametric tests should be considered. Moreover, sometimes the only scores available are frequencies of occurrence or ranks (which often are not normally distributed), in which case the researcher should use nonparametric tests.

In this chapter, we present two categories of nonparametric analyses. The first approach is used to analyze the frequency of responses that are in categories, as in these examples:

- How many highly fit children are girls and how many are boys?
- How many former athletes participate in recreational sports after age 30?
- How many nonathletes participate in recreational sports after age 30?
- How many highly fit women regularly participate in swimming, running, or cycling?

© aerogondo/fotolia

What's normal may depend on your job.

Second, we present a standard approach to analyzing ranked data when the ranks are not normally distributed. To use any of the general linear model (GLM) techniques described in chapters 8 and 9, the data must be normally distributed, and the data points must fit a straight line. Following are examples of data that might not meet the normality assumption:

- Questionnaire responses of *Strongly agree, Agree, Neutral, Disagree,* and *Strongly disagree* can be considered ranked data; the responses are ranked from 1, *Strongly agree,* to 5, *Strongly disagree.*

- Data such as time, velocity, acceleration, and counts (e.g., how many push-ups you can do) may not be normally distributed.

In these instances, the use of parametric correlational and ANOVA procedures (including *t* tests) would be inappropriate because the assumptions for using the *r, t,* and *F* tables are not met.

In this chapter, we provide a set of procedures for rank-order data that are parallel to the parametric procedures of correlation and ANOVA (including *t* tests). The ideas underlying these procedures are identical to the ones presented in chapters 8 and 9. The procedures use the same computer programs (e.g., SPSS, SAS) to analyze the ranked data, but instead of using the *r, t,* and *F* tables, the statistic *L* is calculated and then compared to a chi-square table, which does not require a normality assumption.

Chi Square: Testing the Observed Versus the Expected

Data are often sorted into categories, such as sex, age, grade level, treatment group, or some other **nominal** (categorical) **measure**. A researcher is sometimes interested in evaluating whether the number of cases in each category is different from what would be expected on the basis of chance, some known source of information (such as census data), or some other rational hypothesis about the distribution of cases in the population. **Chi square** provides a statistical test of the significance of the discrepancy between the observed and the expected results.

nominal measure—A method of classifying data into categories, such as sex, age, grade level, or treatment group.

chi square—A statistical test of the significance of the discrepancy between the observed and the expected results.

The formula for chi square is the following:

$$\chi^2 = \Sigma[(O - E)^2/E] \qquad (10.1)$$

where O = the observed frequency and E = the expected frequency. Thus, the expected frequency in each category—often labeled *cells*, as in a table divided into four equal boxes—is subtracted from the observed (or obtained) frequency. This difference is then squared (which means that all differences will be positive), and these values are divided by the expected frequency for their respective categories and then summed.

For example, tennis coach Roger Roundball believes in jinxes. He believes that one court at his university is definitely unlucky for his team. He has kept records of matches on four courts over the years and is convinced that his team has lost significantly more matches on court 4 than on any other. Roger has set out to prove his point by comparing the number of losses on each of the four courts. The same number of matches has been played on each court, and his team has lost 120 individual matches on these courts during this period. Theoretically, it would be expected that each court would have one fourth, or 30, of the losses. The observed frequencies and the expected frequencies are shown in example 10.1.

EXAMPLE 10.1
Known Values

		Court number				
		1	**2**	**3**	**4**	**Total**
Observed number of losses	$O =$	24	34	22	40	120
Expected number of losses	$E =$	30	30	30	30	120

Working It Out (Equation 10.1)

	Court number			
	1	**2**	**3**	**4**
$(O - E)$	−6	+4	−8	+10
$(O - E)^2$	36	16	64	100
$(O - E)^2/E$	1.20	0.53	2.13	3.33

$$X^2 = \Sigma(O - E)^2/E] = 7.19$$

The resulting chi square is then interpreted for significance by consulting Appendix table 6. Because there are four courts, or cells (c), the number of degrees of freedom is $c - 1 = 3$. The researcher finds the critical value for 3 df, which is 7.82 for the .05 probability level. Roger's calculated value of 7.19 is less than this, so the null hypothesis that there were no differences between the four courts in the number of losses is not rejected. The observed differences could be attributed to chance. Roger probably still believes in his heart that he is right (called *cardiac research*).

In some cases, the expected frequencies for classifications can be obtained from existing sources of information, as in the following example. A new assistant professor, Dr. Niceperson, is assigned to teach the large sections of an introductory kinesiology course. After a few semesters, the department chairperson begins hearing rumors that Dr. Niceperson is too lenient in her grading practices. The department chairperson uses chi square to compare the grades of Dr. Niceperson's 240 students with the department's prescribed normal-curve grade distribution: 3.5% As and Fs, 24% Bs and Ds, and 45% Cs. If Dr. Niceperson were adhering to this normal distribution, she would be expected to have given 8 As and 8 Fs, (3.5% × 240), 58 Bs and 58 Ds (24% × 240), and 108 Cs (45% × 240). The observed frequencies (Dr. Niceperson's grades) and the expected frequencies (the department's mandated distribution) are compared using chi square in example 10.2.

EXAMPLE 10.2
Known Values

		Grade					
		A (3.5%)	**B** (24%)	**C** (45%)	**D** (24%)	**F** (3.5%)	**Total**
Observed number of grades given	$O =$	21	75	114	28	2	240
Expected number of grades given	$E =$	8	58	108	58	8	240

Working It Out (Equation 10.1)

	Grade				
	A	**B**	**C**	**D**	**F**
$(O - E)$	13	17	6	−30	−6
$(O - E)^2$	169	289	36	900	36
$(O - E)^2/E$	21.13	4.98	.33	15.52	4.50

$$X^2 = \Sigma(O - E)^2/E] = 46.46$$

The chairperson, who is always fair, does not want to make a bad decision, so she decides to use the .01 level of probability. Appendix table 6 shows that for 4 *df* (there are five grades, or cells), a chi square of 13.28 is needed for significance at the .01 level. The obtained chi square of 46.46 exceeds the value, indicating a significant deviation from the expected grade distribution. Clearly, Dr. Niceperson's grades show too many As and Bs and too few Ds and Fs. The chairperson does the only fair thing by firing Dr. Niceperson on the spot.

Contingency Table

Often, a problem involves two or more categories of occurrences and two or more groups (a two-way classification). A common example is in the analysis of the results of questionnaires or attitude inventories in which there are several categories of responses (e.g., *Agree, No opinion, Disagree*) and two or more groups of respondents (e.g., exercise adherents and nonadherents). This type of two-way classification is called a **contingency table**.

To illustrate, suppose a group of athletes and a group of nonathletes respond to the following statement on a fair play inventory: "A baseball player who traps a fly ball between the ground and his glove should tell the umpire that he did not catch it." In the previous examples of one-way classification, the expected frequencies were determined by some type of rational hypothesis or source of information. In a contingency table, the expected values are computed from the **marginal totals**.

Example 10.3 shows the participants' responses.

contingency table—A two-way classification of occurrences and groups that is used for computing the significance of the differences between observed and expected scores.

marginal totals—A method of calculating rows and columns by summing the row or column and dividing by the overall total; these represent one variable independent of the other variables.

EXAMPLE 10.3

Known Values

Observed responses	Agree	No opinion	Disagree	Total
Athletes	30	46	124	200
Nonathletes	114	80	56	250
Total	**144**	**126**	**180**	**450**

Working It Out

1. Find the expected values (column total × row total) / *N*.

Expected responses	Agree	No opinion	Disagree	Total
Athletes	144 × 200 / 450 = 64	126 × 200 / 450 = 56	180 × 200 / 450 = 80	200
Nonathletes	144 × 250 / 450 = 80	126 × 250 / 450 = 70	180 × 250 / 450 = 100	250
Total	**144**	**126**	**180**	**450**

2. Compute X^2 (equation 10.1).

Responses	$O - E$	$(O - E)^2$	$(O - E)^2 / E$
Athletes agree	−34	1,156	18.06
Nonathletes agree	34	1,156	14.45
Athletes no opinion	−10	100	1.79
Nonathletes no opinion	10	100	1.43
Athletes disagree	44	1,936	24.20
Nonathletes disagree	−44	1,936	19.36
			$X^2 = 79.29$

A total of 144 respondents agreed with the statement. Because there are 450 people in all, 144 / 450, or 32% of the total group, agreed with the statement. Thus, if there were no difference between athletes and nonathletes on fair play (the null hypothesis) as reflected by this statement, then 32% of the athletes ($0.32 \times 200 = 64$) and 32% of the nonathletes ($0.32 \times 250 = 80$) would be the expected frequencies for these two cells.

A much faster method of calculating these expected frequencies is simply to multiply the column total by the row total for each cell and divide by the total number (N), as we worked out in step 1 of example 10.3. Chi square is computed in the same manner as in the examples of one-way classification.

The degrees of freedom for a contingency table are $(r-1)(c-1)$, where r stands for rows and c for columns. Here, we have two rows and three columns, so $df = (2-1)(3-1) = 2$. As before, the investigator then looks up the table value for significance (in this study, the investigator had decided on the .01 level) and sees that a chi square of 9.21 is needed. The obtained chi square of 79.29 is clearly significant. This result tells us that the null hypothesis is rejected: Athletes and nonathletes respond to the statement in a significantly different way. After inspecting the table, we could conclude that a significantly greater proportion of nonathletes agreed that the player should tell the umpire about trapping the ball and, conversely, that a higher proportion of athletes thought that the player should not.

Restrictions in Using Chi Square

Although we indicated that nonparametric statistics do not require the same assumptions concerning the population as do parametric statistics, some restrictions apply in using this technique. The observations must be independent, and the categories must be mutually exclusive. By this we mean that an observation in any category should not be related to or dependent on other observations in other categories. You could, for example, ask 50 people about their activity preferences. If each gave three preferences, you would not be justified in using a total (N) of 150 because the preferences of any participant would likely be related and chi square thus inflated. Moreover, an observation can be placed in only one category. The observed frequencies are exactly that: numbers of occurrences. Ratios and percentages are not appropriate. Another related point is that the totals of the expected and observed frequencies for any classification must be the same. In example 10.3, notice that these frequency totals for athletes are the same. The same is seen for nonathletes and for the column totals.

Chi square is usually not applicable for small samples. The expected frequency for any cell should not be less than 1.0. Furthermore, some statisticians claim that no more than 20% of the cells can have expected values of less than 5, although opinions vary on this point. Some say that no cell should have a frequency of less than 5, whereas others allow as many as 40% of the cells to have a frequency of less than 5. A common tactic in cases in which several cells have expected frequencies less than 5 is to combine adjacent categories, thereby increasing the expected values.

Researchers generally agree that a 2×2 contingency table should have a **Yates' correction for continuity**, which subtracts 0.5 from the difference between the observed and the expected frequencies for each cell before it is squared:

$$\text{Corrected } X^2 = \Sigma[(O - E - 0.5)^2/E] \tag{10.2}$$

Another limitation imposed on the 2×2 contingency table is that N should be at least 20.

Finally, the expected distribution should be logical and established before the data are collected. In other words, the hypothesis (probability, equal occurrence, census data, and so on) precedes the analysis. Researchers are not allowed to look over the distribution and then conjure up an expected distribution that fits their hypothesis.

Yates' correction for continuity—A method of correcting a 2 × 2 contingency table by subtracting 0.5 from the difference between the observed and expected frequencies for each cell before it is squared.

Contingency Coefficient

Several correlational techniques can be used in situations in which data are discrete, or not continuous. You can compute the relationship between dichotomous variables such as sex and race by using a **contingency coefficient**. The test of significance is chi square. Recall the contingency table for use in detecting differences between groups or sets of data, described in the previous section. The contingency table can also be used to determine relationships. You can have any number of rows and columns in a contingency table. After chi square is computed, the contingency coefficient C can be computed:

$$C = \sqrt{X^2/(N + X^2)} \qquad (10.3)$$

If X^2 is significant, C is also significant. Examining the data establishes the direction of the relationship. Several limitations affect the ability of the contingency coefficient to estimate correlation. In general, a researcher needs several categories and many observations to obtain a reasonable estimate.

> **contingency coefficient**—A method of computing the relationship between dichotomous variables such as sex and race.

Multivariate Contingency Tables: The Loglinear Model

Categorical data can be analyzed in combination with other variables. In other words, contingency tables can be studied in more than two dimensions. The researcher can therefore identify associations of many variables, such as the interrelationships of age, sex, skill level, and teaching method. This approach is similar to parametric multivariate analysis. With continuous quantitative data, however, variables are expressed as linear composites, whereas with categorical variables, the researcher deals with contributions to the expected frequencies within each cell of the multivariate contingency table. Any given cell represents the intersection of several marginal proportions.

Loglinear models are used to analyze multivariate contingency tables. Relative frequencies are transformed to logarithms, which are additive and similar to the sum of squares in ANOVA. Main effects and interactions can be tested for significance. The probability of membership in a particular category can be predicted as a function of membership in other categories using a logistic regression equation based on the logarithmic odds of membership, **logit**.

> **loglinear models**—A system that analyzes multivariate contingency tables by transforming relative frequencies into logarithms.
>
> **logit**—The probability of membership in a particular category as a function of membership in other categories in multivariate contingency tables.

The loglinear model has considerable potential application in research with categorical data (Schutz, 1989). This type of analysis is attracting much attention from researchers and theoreticians. The loglinear analysis provides a means for sophisticated study of the interrelationships of categorical variables. Readers who are interested in pursuing this topic are directed to the text by Eye and Mun (2012).

Procedures for Rank-Order Data

Most books that report nonparametric procedures (including the first four editions of this book) propose a series of rank-order techniques:

- Mann-Whitney U test—analogous to the parametric independent t test
- Wilcoxon matched-pairs signed-ranks test—analogous to the parametric dependent t test
- Kruskal-Wallis ANOVA by ranks—analogous to the parametric one-way ANOVA
- Friedman two-way ANOVA by ranks—analogous to the parametric repeated-measures ANOVA
- Spearman rank-difference correlation—analogous to the parametric Pearson r

You will still see these techniques used in the research literature, so you need to understand what each one does; our list shows you how they compare with parametric procedures. Eventually, as better methods become available, most of these techniques will pass from use, but that change will take time. Researchers who have been trained to use these procedures often do not keep up with current statistical methods; we acknowledge that staying current is difficult because researchers have their own research areas to keep up with. One jokester was heard to say that it takes 30 to 40 years for a new statistical approach to come into widespread use. This slow pace of adoption occurs because researchers rarely change the statistical procedures they use, so we have to wait for them to retire and for a more recently trained group to begin using the new procedures.

We can replace all the rank-order techniques previously listed with a standardized approach to rank-order data in which the distribution of data is not normal (for greater detail, see Thomas et al., 1999). The techniques parallel the parametric methods described in chapters 8 and 9. The procedures for all techniques involve these steps:

1. Change all the data values to ranks (most statistical software can do this).

2. Run the standard parametric statistical software (e.g., correlation, multiple-regression ANOVA) on the ranked data.

3. Calculate the L statistic as the significance test, and compare it to the chi-square table (Appendix table 6).

The nonparametric test statistic, L, used for evaluating significance for all these procedures was developed by Puri and Sen (1969, 1985), and represents the test of the null hypothesis between X and Y where X can be groups or variables and Y is variables. This method can be used in place of all the techniques previously discussed and assumes that the data fit a straight line (as all the techniques in chapters 8 and 9 did) *but does not assume normal distribution of data*. L is easy to calculate using the following formula:

$$L = (N - 1)r^2 \tag{10.4}$$

> The nonparametric test statistic, L, can replace all previously discussed techniques, assumes that the data fit a straight line, and is easy to calculate.

where N = the number of participants and r^2 = the proportion of true variance = $SS_{regression}$ / SS_{total}. The proportion of true variance in correlation is r^2 and in multiple regression is R^2. In a t test, $r^2 = t^2/(t^2 + df)$, *where* $df = (n^1 + n^2)$. In ANOVA, it is the sum of squares (SS) for the effect of interest (e.g., between groups in one-way ANOVA or for each factor in factorial ANOVA; each of these is the true variance) divided by the total sum of squares (these values are shown in the ANOVA printout from standard statistics programs).

When the L statistic is calculated for each significance test (either for correlation or ANOVA), it is compared to the chi-square table: Appendix table 6 with $df = pq$, where p = the number of df for the independent (or predictor) variable(s) and q = the number of df for the dependent (or criterion) variable(s). Thus, for a correlation, r, between one predictor (e.g., abdomen skinfold) and one criterion (e.g., body weight), the pq df is $1 \times 1 = 1$. In a t test, $df = p$ (number of groups − 1) $\times$ q (number of dependent variables); thus, again, the pq $df = 1 \times 1 = 1$. Additional examples of how to calculate L and df are provided throughout the rest of this chapter.

Correlation

In this section, we provide sample ranked data and show how to calculate correlation and multiple regression using standard correlational techniques. We then calculate the L and df for each example. In the simple correlation example, we also show how to decide whether to use parametric or nonparametric analysis, but we do not repeat this process for subsequent procedures (you can obtain this logic for each procedure by reading Thomas et al., 1999).

Simple Correlation

We are interested in calculating the correlation between two skinfold measurements, biceps and forearm, that were collected on 157 participants. Descriptive data are given in figure 10.1, along with a histogram showing the data distribution with the normal curve overlaid. Recall from chapter 6 that skewness and kurtosis provide information about the nature of the distribution. Here they are given in the form of a z score, where 0.0 indicates a normal distribution, and positive or negative values indicate specific variations from normality in the distribution. For the biceps skinfold, the skewness is above +1.0, indicating that the hump of the curve is shifted to the left. The kurtosis is nearly +2.0, indicating that the distribution is very peaked. For the forearm skinfold, the hump of the curve is also shifted left, and the curve is extremely peaked. Neither data set seems to be normal in its distribution. Although no set rule defines how normal or nonnormal data must be, we do know that parametric techniques are not as resistant to nonnormality as once thought. In our view, these data are nonnormal

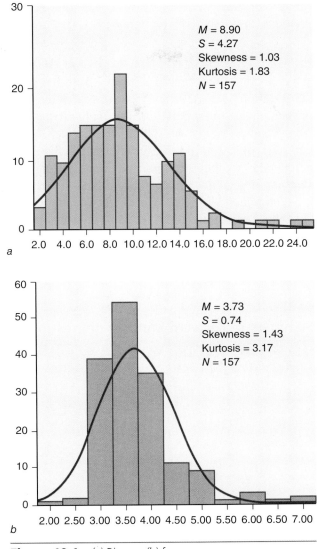

Figure 10.1 *(a)* Biceps, *(b)* forearm.

enough to suggest the use of rank-order procedures rather than parametric ones.

If we use a computer program for parametric correlation to determine the relationship between these two variables (using the original data), $r = .26$, and the test of significance is $F(1, 155) = 11.46, p < .001$. But we are likely violating the assumption that data are normally distributed. If we rank subjects from the smallest biceps skinfold measurement (number 1) to the largest (number 157) and do the same for the forearm skinfold, we can then use the same computer program to correlate the ranked data; the correlation is actually larger, $r = .28$. If we square this value, we get .0784. If we then use equation 10.4 to calculate the significance test to replace F, we find that

$$L = (N - 1)r^2 = (157 - 1)(.28)^2 = (156)(.0784) = 12.23$$

This L has $df = 1$ because there is one X variable (biceps skinfold) and one Y variable (forearm skinfold), so $pq = 1 \times 1 = 1$. If we look up this L (Appendix table 6) as chi square with $df = 1$, we find a value of 10.83 for $p = .001$. Our value exceeds that, so it is significant at $p < .001$. Thus, we have now used a rank-order procedure that does not violate the assumption for normality, and our r and test of significance for the r, the L statistic, were both larger than the ones we would have obtained if we had used the parametric statistics. Although this circumstance will not always occur, this approach to nonparametric analysis does have good power compared with parametric procedures when data are not normal.

Multiple Regression

We can extend the same procedures from the previous section to multiple regression. In this example, we use four skinfolds—abdomen, calf, subscapular, and thigh—to predict the percentage of fat determined from hydrostatic weighting. (The data in this example are from Thomas, Keller, & Holbert, 1997. We thank these authors for allowing us to use their data.) All four skinfold measurements are changed to ranks, as are the measurements of fat percentage. We then run the regular multiple-regression program from SPSS using forward-stepping procedures on the ranked data. The correlations for all ranked pairs of variables (e.g., abdomen with calf skinfold, abdomen with subscapular skinfold) range from .45 to .74. These values are about the same as for the original (unranked) data, for which the correlations range from .41 to .78. Table 10.1 summarizes the multiple-regression results for the ranked data (note that all test statistics have been changed to L). The overall multiple regression, R, was higher for the ranked data (.82) than for the original unranked data (.80).

TABLE 10.1

Summary of Forward Multiple Regression for the Ranked Data

Step	Variable	R	R^2	β	df	L
1	Subscapular skinfold	.68	.46	.332	1	35.83*
2	Calf skinfold	.77	.60	.602	2	20.29*
3	Abdomen skinfold	.80	.64	.321	3	7.85*
4	Thigh skinfold	.82	.68	−.327	4	9.56*

*$p < .05$ for all

$L(4) = 53.01$, $p < .001$, for the linear composite of predictors

Reprinted by permission from J.R. Thomas, J.K. Nelson, and K.T. Thomas, "A Generalized Rank-Order Method for Nonparametric Analysis of Data From Exercise Science: A Tutorial," *Research Quarterly for Exercise and Sport* 70, no. 1 (1999): 11-23.

Differences Between Groups

We can apply the same logic and procedures from the previous two sections when we want to test differences between groups. Here we provide examples of a t test, one-way ANOVA, and factorial ANOVA. If you want to extend these procedures to repeated-measures ANOVA and multivariate ANOVA, see Thomas et al. (1999). In these examples, we use data from Nelson, Yoon, and Nelson (1991); we appreciate their permission to use their data.

t Test

Data are modified push-up scores for 90 boys and 90 girls in the upper grades of elementary school. The distribution of data is positively skewed and somewhat peaked. Following are the means and standard deviations for the girls and boys and the t test comparing them:

Boys	$M = 18.6$	$s = 9.9$
Girls	$M = 12.7$	$s = 9.8$

$t(178) = 4.00$, $p < .001$

We changed the data to ranks from 1 to 180, disregarding whether the scores were for boys or girls. We then calculated t on the ranked data; it was 3.77. But we need to change the t to L by using equation 10.4. First we have to calculate r^2. Following are those calculations:

$$r^2 = t^2/(t^2 + df) = 3.77^2/(3.77^2 + 178) = 14.21/(14.21 + 178) = .0739$$

$$L = (N - 1)r^2 = (179)(.0739) = 13.23$$

L is tested as a X^2 with $df = pq$, where $p =$ the number of groups $- 1$ ($2 - 1 = 1$) and $q =$ the number of dependent variables (1), so $1 \times 1 = 1$ df. Consulting the chi-square table (Appendix table 6), we see that our calculated value of 13.23 exceeds the table value of 10.83 at the .001 level; thus, our L is significant at $p < .001$.

You can practice this procedure by applying it to the data in chapter 9's example 9.8. Use the data for groups 1 and 2, with five participants in each group ($N = 10$). Rank the scores from lowest to highest, disregarding whether the participant is in group 1 or 2. Then calculate the mean and standard deviation for each group and apply equation 9.3 to calculate t. The t value that you find will be 2.79. Then obtain r from t:

$$r^2 = t^2/(t^2 + df) = 2.79^2/(2.79^2 + 8) = .49$$

We then calculate L using formula 10.4:

$$L = (N - 1)r^2 = (9)(.49) = 4.41$$

This L has 1 df [(groups $- 1$) $\times$ number of dependent variables]. With 1 df, the critical X^2 at the .05 level from Appendix table 6 is 3.84. Because our value is larger than that, the two groups are significantly different.

One-Way ANOVA

Nelson et al. (1991) also used grade level—grades 4, 5, 6—to evaluate the data from modified push-ups. We use this as the example for one-way ANOVA. Grade level is the independent variable with three levels (this is really a categorical variable, but for ANOVA, we treat it like an independent variable), and the modified push-up score is the dependent variable. Here are the descriptive data for each group:

Grade 4	$M = 12.1$	$s = 10.0$
Grade 5	$M = 16.3$	$s = 9.6$
Grade 6	$M = 18.5$	$s = 10.2$

We then change the data to ranks, disregarding the grade level of the children, so we have 180 scores (number of children in all three grades) ranked from the highest modified push-up score (ranked number 1) to the lowest (ranked number 180). We then run the one-way ANOVA program on the ranked data. Table 10.2 shows these results. But note the printout shows F where L appears in table 10.2. Therefore, to calculate the L, we first calculate R^2 (the proportion of true variance in ANOVA) using this formula:

$$R^2 = SS_{between}/SS_{total} = 31,646.61/485,412.00 = .065$$

Then, we calculate L as the following:

$$L = (N - 1)R^2 = (179)(.065) = 11.64$$

TABLE 10.2
Summary of ANOVA Data for the Ranked Scores

Source	df	Sum of squares	Mean squares	L	p
Between groups	2	31,646.61	15,823.30	11.64	.01
Within groups (error)	177	453,765.39	2,563.65	—	—
Total	179	485,412.00	—	—	—

Reprinted by permission from J.R. Thomas, J.K. Nelson, and K.T. Thomas, "A Generalized Rank-Order Method for Nonparametric Analysis of Data From Exercise Science: A Tutorial," *Research Quarterly for Exercise and Sport* 70, no. 1 (1999): 11-23.

The L has 2 df [(number of groups − 1) × number of dependent variables]. The value needed for significance at the .01 level with 2 df is 9.21 in Appendix table 6; our value exceeds that, so it is significant.

You can also calculate an example by hand if you return to chapter 9's example 9.7 and use all three groups. As before, rank the scores from 1 to 15, disregarding which group the participants are in. Then work the one-way ANOVA as was done in example 9.8, except replace the original scores with the ranks. Calculate R^2 and L as previously described. You will find $R^2 = .65$ and $L(2) = 9.20$. The value for significance at the .02 level is 7.82 in Appendix table 6, and the calculated value exceeds that, so it is significant.

Factorial ANOVA

Continuing with the same data set, we have a modified push-up score for girls and boys at each of three grade levels. Thus, we can rank the modified push-up scores from 1 to 180 as before and calculate a factorial ANOVA where one factor is sex (girls and boys) and the other is grade level (4, 5, and 6). This becomes a 2 (sex) × 3 (grade level) factorial ANOVA with the modified push-up score as the dependent variable.

These ranked data were analyzed using regular parametric statistical software. We found the outcome shown in table 10.3. We calculated R^2 using the same formula for each factor and for the interaction:

$$R^2{}_{grade} = SS_{grade}/SS_{total} = 31{,}646.61/485{,}412.00 = .065$$

$$R^2{}_{sex} = SS_{sex}/SS_{total} = 35{,}814.01/485{,}412.00 = .074$$

$$R^2{}_{grade \times sex} = SS_{grade \times sex}/SS_{total} = 10{,}865.20/485{,}412.00 = .022$$

We then used the R^2 to calculate L for each factor and for the interaction. In Appendix table 6, the grade effect with 2 df is significant at the .005 level, the sex effect with 1 df is significant at the .001 level, but the interaction with 2 df is not significant.

TABLE 10.3
Summary of Factorial ANOVA Data for the Ranked Scores

Source	df	Sum of squares	Mean squares	L	p
Grade	2	31,646.61	15,823.30	11.64	.001
Sex	1	35,814.01	35,814.01	13.21	.001
Grade × sex	2	10,865.20	5,432.60	4.01	—
Residual	174	407,086.18	2,339.58	—	—
Total	**179**	**485,412.00**	—	—	—

Reprinted by permission from J.R. Thomas, J.K. Nelson, and K.T. Thomas, "A Generalized Rank-Order Method for Nonparametric Analysis of Data From Exercise Science: A Tutorial," *Research Quarterly for Exercise and Sport* 70, no. 1 (1999): 11-23.

Extension to Repeated-Measures ANOVA and Multivariate ANOVA

Although we do not provide examples here, the use of ranked data and the L statistic can be extended to ANOVA models with repeated measures and multivariate ANOVA. Conceptually, the ideas are identical, and you follow procedures similar to the ones used here, including ranking the data, using standard computer software, and changing the test statistic to L instead of F. If you need one of these techniques, see Thomas et al. (1999) for the steps to follow and ways to interpret the results.

Summary

In this chapter, we provided statistical tests for data that do not meet the assumptions of parametric data (chapters 8 and 9). We included chi-square tests for frequency data and a set of procedures for ranked data that are parallel to parametric procedures. If distributions are not normal, the data are changed to ranks. The ranked data can be analyzed by standard SPSS and SAS statistical packages. One test, the L statistic, is calculated and compared with a chi-square table for significance. These procedures can be applied to all parametric linear models.

✓ Check Your Understanding

1. Compute chi square for the following contingency table of activity preferences of men ($n = 110$) and women ($n = 90$). Determine the significance of chi square at the .01 level using Appendix table 6. When calculating the expected frequencies, round off to whole numbers.

Group	Racquetball	Weight training	Aerobic dance
Men	35	45	30
Women	28	13	49

   a. Write a brief interpretation of your results.

   b. What would be the critical value for significance at the .05 level if you had five rows and four columns?

2. Locate a published study that uses chi square in the analysis. Identify the dependent variable.

11

Measuring Research Variables

The average lifespan of a major league baseball is seven pitches.

—Lyrics Quote #5494509

A basic step in the scientific method of problem-solving is the collection of data; therefore, an understanding of basic measurement theory is necessary. (We should point out that although measurement is discussed here as a research tool, measurement itself is an area of research.) In this chapter, we discuss fundamental criteria for judging the quality of measures used in collecting research data: validity and reliability. We explain types of validity and ways by which evidence of validity and reliability may be established. (The trustworthiness of qualitative research is discussed in chapter 19.) We conclude with some issues concerning the measurement of movement, of written responses used in paper-and-pencil instruments, of affective behavior, of knowledge, and of item response theory.

Validity

In gathering the data on which results are based, we are also greatly concerned with the validity of the measurements we are using. If, for example, a study seeks to compare training methods for producing strength gains, the researcher must have scores that produce a valid measure of strength to evaluate the effects of the training methods. **Validity** of measurement indicates the degree to which the test or instrument measures what it is supposed to measure. Thus, validity refers to the soundness of the interpretation of scores from a test, the most important consideration in measurement.

Different measures have different purposes. Consequently, there are different kinds of validity. We consider four basic types of validity as **logical, content, criterion,** and **construct**.

Although we list logical validity as a separate type of validity, the American Psychological Association and the American Educational Research Association consider logical validity a special case of content validity.

validity—The degree to which a test or instrument measures what it purports to measure; can be categorized as *logical, content, criterion,* or *construct validity*.

logical validity—The degree to which a measure obviously involves the performance being measured; also known as *face validity*.

content validity—The degree to which a test—usually in educational settings—adequately samples what was covered in the course.

criterion validity—The degree to which scores on a test are related to some recognized standard or criterion.

construct validity—The degree to which a test measures a hypothetical construct; usually established by relating the test results to some behavior.

Logical Validity

Logical validity is sometimes referred to as *face validity*, although measurement experts dislike that term. Logical validity is claimed when the measure obviously involves the performance being measured. In other words, the test is valid by definition. A static balance test that consists of balancing on one foot has logical validity. A speed-of-movement test, in which the person is timed while running a specified distance, must be considered to have logical validity. Occasionally, logical validity is used in research studies, but researchers prefer to have more objective evidence of the validity of measurements.

Content Validity

Content validity pertains largely to learning in educational settings. A test has content validity if it adequately samples what was covered in the course. As with logical validity, no statistical evidence can be supplied for content validity. The test maker should prepare a table of specifications, or test blueprint, before making up the test. The topics and course objectives, as well as the relative degree of emphasis accorded to each, can then be keyed to a corresponding number of questions in each area.

A second form of content validity occurs with attitude instruments. Often, a researcher wants evidence of independent verification that the items represent the categories for which they were written. When this happens, experts (in many cases 20 or more) are asked to assign each statement to one of the instrument categories. These categorizations are tallied across all experts, and the percentage of experts who agreed with the original categorization is reported. Typically, an agreement of 80 to 85% would indicate that the statement represents the content category.

Concurrent and Predictive Validities

concurrent validity—A type of criterion validity in which a measuring instrument is correlated concurrently, or at about the same time with some criterion.

Measurements used in research studies are frequently validated against some criterion. The two main types of criterion validity are concurrent validity and predictive validity. **Concurrent validity** involves correlating an instrument concurrently with some criterion. The scores from many physical performance measures are validated in this manner. Popular criterion measures include already validated or accepted measures, such as judges' ratings and laboratory techniques. Concurrent validity is usually employed when the researcher wishes to substitute a shorter, more easily administered test for a criterion that is difficult to measure.

As an illustration, maximal oxygen consumption is regarded as the most valid measure of cardiorespiratory fitness. Measurement, however, requires a laboratory, expensive equipment, and considerable time for testing; furthermore, only one person can be tested at a time. Let us assume that a researcher, Douglas Bag, wishes to screen people by their fitness levels before assigning them to experimental treatments. Rather than using such an elaborate test as maximal oxygen consumption, Douglas determines that it would be advantageous to use a stair-walking test that he has devised. To determine whether it is a valid measure of cardiorespiratory fitness, he could administer both the maximal oxygen consumption and stair-walking tests to a group of participants (from the same population that will be used in the study) and correlate both tests' results. If a satisfactory relationship exists, Doug can conclude that his stair-walking test is valid.

Written tests can also be validated in this way. For example, a researcher might wish to use a 10-item test of anxiety that can be given in a short period rather than a lengthier version. To establish the validity of the shorter version, the researcher would give both versions to the same sample to compare the scores. Another example is that judges' ratings serve as criterion measures for some tests; for instance, sport skills and behavioral inventories are sometimes validated this way. Significant amounts of time and effort are required to secure competent judges, teach them to use the rating scale, test for agreement among judges, arrange for a sufficient number of trials, and so on. Consequently, judges' ratings cannot be used routinely to evaluate performances. The use of some skills tests would be more economical. Furthermore, the skills tests usually provide knowledge of results and measures of progress for the students. The skills tests could be initially validated, however, by giving the test and having judges rate the individual participants on those skills. A validity coefficient can be obtained by correlating the scores on the skills tests with the judges' ratings.

Choosing the criterion is critical in the concurrent validity method. All the correlation can tell you is the degree of relationship between a measure and the criterion. If the criterion is inadequate, then the concurrent validity coefficient is of little consequence.

Predictive Validity

When the criterion is some later behavior, such as using entrance examinations to predict later success, **predictive validity** is the major concern. Suppose a physical education instructor wished to develop a test that could be given in beginning gymnastics classes to predict success in advanced classes. Students would take this test while they were in the beginner course. At the end of the advanced course, those test results would be correlated with the criterion of success (grades, ratings, and so on). In trying to predict a certain behavior, a researcher should try to ascertain whether there is a known base rate for that behavior. For example, someone might attempt to construct a test that would predict which female students might develop bulimia at a university. Suppose the incidence of bulimia is 10% of the female population at that school. Knowing this, the base rate at which one could be correct in predicting that females were not bulimic is 90%. If the base rate is very low or very high, a predictive measure may have little practical value because the increase in predictability is negligible.

Chapters 8 and 16 discuss aspects of prediction in correlational research. Multiple regression is often used because several predictors are likely to have a greater validity coefficient than the correlation between any single test and the criterion. An example of this is the prediction of fat percentage from **skinfold measurements**. The researcher could use the **underwater, or hydrostatic, weighing** technique to measure the criterion—fat percentage; however, this method requires special equipment and expertise. Thus, there is value in identifying how to use skinfold measures to predict fat percentage. Several skinfold measures are taken, and multiple regression is used to determine the best prediction equation. The researcher hopes to use the skinfold measures in the future if the prediction formula demonstrates an acceptable validity coefficient.

One limitation of such studies is that the validity tends to decrease when using the prediction formula with a new sample. This tendency is called **shrinkage**. Common sense suggests that shrinkage is more likely to occur when the original study used a small sample and especially when it included numerous predictors. In fact, if the number of predictor variables is the same as the sample size, you can achieve perfect prediction. The problem is that the correlations are unique to the sample, and when the prediction formula is applied to another sample (even similar to the first one), the relationship does not hold. Consequently, the validity coefficient decreases substantially—in other words, shrinkage occurs.

In **cross-validation**, a technique recommended to help estimate shrinkage, the same tests are given to a new sample from the same population to check the formula's accuracy. For example, a researcher might administer the criterion measure and predictor tests to a sample of 200 people. Then, she would use multiple regression on the results from 100 of the participants to develop a prediction formula. The researcher then applies the formula to the other group of 100 participants to see how accurately it predicts the criterion for them. Because the researcher has the actual criterion measures for these people, she can ascertain the amount of shrinkage by correlating (Pearson r) the predicted scores with the actual scores. A comparison of the R^2 from the multiple prediction with the r^2 between the actual and the predicted criteria yields an estimate of shrinkage.

Expectancy Tables

A problem of interpretation often arises when using a formula for prediction. We frequently encounter the question of how large a predictive validity coefficient must be before it provides useful information. For example, correlation between the attainment of advanced degrees and scores on the Graduate Record Examination (GRE) is typically quite small ($r < .40$). When we examine r^2 for meaningfulness, we find only 16% or less common variance, which is discouraging. But when we look at the relationship from a different perspective, such as what percentage of students who score highly on the GRE attain a PhD and what percentage of students who score poorly attain the degree, we see that the GRE does have good validity. We can get this type of perspective by using an **expectancy table**.

An expectancy table can be used for predicting the probability of types of performance such as academic and job-related performance. An expectancy table is easily constructed, comprising a two-way grid that contains the probability of a person with a particular assessment score attaining some criterion score. As an example, assume that we have developed a test that purportedly measures good sport behavior, called the Jolly Good Show Inventory. We ask 60 students to complete the test and categorize their scores based on ranges of 10 (e.g., 0-9, 10-19). We wish to see how well our inventory relates to ratings of fair play by judges who have both observed a group of 60 students at play (for hours and hours under various forms of competitive situations) and rated the group on sport behavior using categories from "lousy" to "excellent."

Our first step is to simply tally the number of students who fall in each unique cell based on the inventory score and the judges' ratings. For example, three students scored between 50 and 59 on the inventory. One of these students was rated average, one good, and one excellent. Table 11.1*a* shows the results of this step. We do this for all of the students. Next, we convert the cell frequencies to percentages of the total number of students in each row. For example, one of the three students scoring between 50 and 59 was rated average; thus the percentage is $1 / 3 = 33\%$. The percentages are shown in table 11.1*b*. That is all there is to

TABLE 11.1

Expectancy Table for Jolly Good Show Inventory Scores to Ratings of Good Sport Behavior

11.1*a* Frequency of Ratings for Each Predictor Score Level

Test score	Lousy	Poor	Average	Good	Excellent	Totals
50-59	—	—	1	1	1	3
40-49	—	1	2	4	2	9
30-39	—	4	8	2	—	14
20-29	2	5	7	5	—	19
10-19	3	6	2	—	—	11
0-9	3	1	—	—	—	4
						60

11.1*b* Expectancy Table After Converting Frequencies to Percentages

Test score	Lousy	Poor	Average	Good	Excellent
50-59	—	—	33	33	33
40-49	—	11	22	44	22
30-39	—	29	57	14	—
20-29	11	26	37	26	—
10-19	27	55	18	—	—
0-9	75	25	—	—	—

it. We can see, for example, that of the nine students who scored in the 40-to-49 interval on the inventory, 22% were rated as excellent in sport behavior. Further, none of the top-scoring students (50 to 59) was rated below average, and conversely, none of the lowest-scoring students (0 to 9) was given ratings higher than poor.

In this example, we are using an expectancy table to provide evidence of the concurrent validity of our inventory. That is, the inventory results compare favorably with the judges' results, suggesting the inventory is an acceptable measure of sport behaviors. Sometimes, we can use information from an expectancy table to predict success, for example, a college student's GPA based on an aptitude test they took in high school. As with all situations involving criterion validity, the availability and relevance of the criterion and its susceptibility to reliable measurement are key issues.

Construct Validity

Many human characteristics are not directly observable. Rather, they are hypothetical constructs that carry a number of associated meanings concerning how a person who possesses them to a high degree would behave differently from someone who possesses them to a low degree. Anxiety, intelligence, sporting behavior, creativity, and attitude are a few such hypothetical constructs. Because these traits are not directly observable, measurement poses a problem. Construct validity is the degree to which scores from a test measure a hypothetical construct and is usually established by relating the test results to some behavior. For example, certain behaviors are expected of someone with a high degree of positive sporting behavior. Such a person might be expected to compliment the opponent on shots made during a tennis match. For an indication of construct validity, a test maker could compare the number of times a person scoring high on a test of sporting behavior complimented the opponent with the number of times a person scoring lower on the test did so.

The **known group difference method** is sometimes used in establishing construct validity. For example, construct validity of a test of anaerobic power could be demonstrated by comparing the test scores of sprinters and jumpers with those of distance runners. Sprinting and jumping require greater anaerobic power than distance running does. Therefore, the tester could determine whether the test differentiates between the two kinds of track performers. If the sprinters and jumpers score significantly better than the distance runners do, this finding would provide some evidence that the test measures anaerobic power.

known group difference method—A method used for establishing construct validity in which the test scores of groups that should differ on a trait or ability are compared.

An experimental approach is occasionally used in demonstrating construct validity. For example, a test of cardiorespiratory fitness might be assumed to have construct validity if it reflected gains in fitness following a conditioning program. Similarly, the originator of a motor skills test could demonstrate construct validity through its sensitivity in differentiating between groups of children with and without instruction.

Correlation can also be used in establishing construct validity. Hypothesized structures or dimensions of the trait being tested are sometimes formulated and verified with factor analysis. The tester also uses correlation to examine relationships between constructs; for example, when it is hypothesized that someone with high scores on the test being developed (e.g., cardiorespiratory fitness) should also do well on some total physical fitness scale. Conversely, those with low scores on the cardiorespiratory test would be expected to do poorly on the fitness test.

Another example of using correlational techniques to assess construct validity is in the development of an attitude instrument. A discussion of the procedures for attitude research can be found in the paper by Silverman (2017). Confirmatory factor analysis is used to see whether the data from a pilot study fit the proposed model of constructs. If the data do not fit, additional pilot testing occurs until the model fits established standards. Examples of this technique with attitude measurement can be found in physical education and other areas (e.g., Keating & Silverman, 2004; Kulinna & Silverman, 1999; Subramaniam & Silverman, 2000).

A valid test is reliable: It yields the same results on successive trials.

Actually, all the other forms of validity we have discussed are used for evidence of construct-related validity. Indeed, using evidence from all the other forms is usually necessary to provide strong support for the validity of scores for a particular instrument and the use of its results.

Reliability

observed score—In classical test theory, an obtained score that comprises a person's true score and error score.

true score—In classical test theory, the part of the observed score that represents the person's real score and does not contain measurement error.

error score—In classical test theory, the part of an observed score that is attributed to measurement error.

An integral part of validity is reliability, which pertains to the consistency, or repeatability, of a measure. A test cannot be considered valid if it is unreliable. In other words, if the test is inconsistent—if you cannot depend on successive trials to yield the same results—then the test cannot be trusted. Of course, test scores can be reliable yet not valid, but an unreliable test can never be valid. For example, weighing yourself repeatedly on a broken scale would give reliable results but not valid ones. Test reliability is sometimes discussed in terms of **observed score, true score,** and **error score**.

A test score obtained by a person is the observed score. It is not known whether this is a true assessment of this person's ability or performance. Measurement error may occur because of the test directions, the instrumentation, the scoring, or the person's emotional or physical state. Thus, an observed score theoretically consists of the person's true score and error score. Expressed in terms of score variance, the observed score variance consists of true score variance plus error variance. The tester's goal is removing error to yield the true score. The coefficient of reliability is the ratio of true score variance to observed score variance. Because true score variance is never known, it is estimated by subtracting error variance from observed score variance. Thus, the reliability coefficient reflects the degree to which the measurement is free of error variance.

$$\text{Observed score (O)} = \text{True score (T)} + \text{Error score (E)} \qquad (11.1)$$

$$\text{Reliability} = T/O = (O - E)/O \qquad (11.2)$$

Sources of Measurement Error

Measurement error can come from four sources: the testing, the instrumentation, the scoring, and the participant.

Errors in testing are related to how clear and complete the directions are, how rigidly the instructions are followed, whether supplementary directions or motivation are applied, and so forth. Measurement error due to instrumentation includes such obvious examples as inaccuracy and the lack of properly calibrated mechanical and electronic equipment. It also refers to a test's inadequacy to discriminate between abilities and to the difficulty of scoring some tests. Errors in scoring relate to the competence, experience, and dedication of the scorers and to the nature of the scoring itself. When the experimenter performs scoring, the extent to which the experimenter is familiar with the behavior being tested and the test items can greatly affect scoring accuracy. Carelessness and inattention to detail produce measurement error. Measurement error associated with the participant includes many factors, including mood, motivation, fatigue, health, fluctuations in memory and performance, previous practice, specific knowledge, and familiarity with the test items.

Expressing Reliability Through Correlation

The degree of reliability is expressed by a correlation coefficient, ranging from 0.00 to 1.00. The closer the coefficient is to 1.00, the less error variance it reflects and the more the true score is assessed. Reliability is established in several ways, which are summarized later in this chapter. The correlation technique used for computing the reliability coefficient differs from that used for establishing validity. Pearson r is used to establish validity and is often called **interclass correlation**. This coefficient is a bivariate statistic, meaning that it is used to correlate two variables, as when determining validity by correlating judges' ratings with skill test scores. But interclass correlation is not appropriate for establishing reliability because reliability requires the correlation of two values for the same variable. When a test is given twice, the scores on the first test are correlated with the scores on the second test to determine their degree of consistency. Here, the two test scores are for the same variable, so interclass correlation should not be used. Rather, **intraclass correlation** is the appropriate statistical technique. This method uses ANOVA to obtain the reliability coefficient.

Interclass Correlation

There are three main weaknesses of Pearson r (interclass correlation) for reliability determination. The first is that, as mentioned previously, the Pearson r is a bivariate statistic, whereas reliability involves univariate measures. Second, the computations of Pearson r are limited to only two scores, X and Y. Often, however, more than two trials are given, and the tester is concerned with the reliability of multiple trials. For example, if a test specifies three trials, the researcher must either (a) give three more trials and use the average or best score of each set of trials for the correlations, or (b) correlate the first trial with the second, the first with the third, and the second with the third. In the first case, extra trials must be given solely for reliability purposes; in the second case, meaningfulness suffers when several correlations of trials are shown. The third weakness is due to the interclass correlation's inability to provide a thorough examination of different sources of variability on multiple trials. For example, changes in means and standard deviations from trial to trial cannot be assessed with the Pearson r method but can be analyzed with intraclass correlation.

Intraclass Correlation

Intraclass correlation provides estimates of systematic and error variance. For instance, by having participants perform repeated trials of the same task, testers can examine the trials' systematic differences. The last trials may differ significantly from the first trials because of

interclass correlation— The most commonly used method of computing correlation between two variables; also called *Pearson r* or *Pearson product moment coefficient of correlation.*

intraclass correlation— An ANOVA technique used for estimating the reliability of a measure. This technique must be used for reliability because of the focus on the relationship between two measures of the same variable.

Measuring anything is possible when you don't know what you're talking about.

a learning phenomenon or fatigue effect (or both). If the tester recognizes this, then perhaps they can exclude initial (or final) trials or use the point at which performance diminishes as the score. In other words, through ANOVA, the tester can truly examine test performance from trial to trial and then select the most reliable testing schedule.

The procedures leading to the calculation of intraclass correlation *(R)* are the same as those of simple ANOVA with repeated measures, discussed in chapter 9. The components are subjects, trials, and residual sums of squares *(SS)*. Table 11.2 shows a summary of an ANOVA example for five participants who performed three trials of the same task. We use these data to compute *R*.

TABLE 11.2

Summary of ANOVA for Reliability Estimation (Three Trials)

Source	SS	df	MS	F
Subjects	14.9	4	3.73	—
Trials	7.6	2	3.80	5.94*
Residual	5.1	8	0.64	—

*$p < .05$

M (trial 1) = 2.4; M (trial 2) = 3.8; M (trial 3) = 4.0

An *F* for trials determines whether the three trials have any significant differences. Refer to Appendix table 5 and read down the 2 *df* column to the 8 *df* row. Our *F* of 5.94 is greater than the table *F* of 4.46 at the .05 level of probability. Thus, significant differences exist. At this point, we should mention that opinions vary about what should be done with trial differences (Baumgartner, Jackson, Mahar, & Rowe, 2016). Some test authorities argue that the test performance should be consistent from one trial to the next and that any trial-to-trial variance should be attributed simply to measurement error. If we decide to do this, the formula for *R* is the following:

$$R = (MS_S - MS_E)/MS_S \tag{11.3}$$

in which MS_S is the mean squares for subjects (from table 11.2) and MS_E is the mean squares for error, which is computed as follows:

$$MS_E = \frac{SS \text{ for trials } + SS \text{ for residual}}{df \text{ for trials } + df \text{ for residual}} \tag{11.4}$$

$$= \frac{7.6 + 5.1}{2 + 8}$$

$$= 1.27$$

Thus, $R = (3.73 - 1.27) / 3.73 = .66$.

Another way of dealing with significant trial differences is to discard any trials that are noticeably different from the others (Baumgartner & Jackson, 1991). Then a second ANOVA is calculated on the remaining trials, and another *F* test is computed. If *F* is not significant, *R* is calculated using equation 11.3, in which trial variance is considered a measurement error. If *F* is still significant, additional trials are discarded, and another ANOVA is conducted. The purpose of this method is to find a measurement schedule that is free of trial differences (i.e., to find a nonsignificant *F* that yields the largest possible criterion score) and that is most reliable. This method is especially appealing when there is an apparent trend in trial differences, as when a learning phenomenon (or release of inhibitions) is evident from increased scores on a later trial or trials. For example, if five trials on a performance test yield mean scores of 15, 18, 23, 25, and 24, you might discard the first two trials and compute another analysis on the last three trials. Similarly, a fatigue effect may be evidenced by a decrease in scores on the final trials in some types of tests.

At the bottom of table 11.2, note that the mean of trial 1 is considerably lower than the means of trials 2 and 3. This may suggest a large learning effect on this task with performance becoming more reliable by trials 2 and 3. Therefore, we discard the first trial and compute another ANOVA on trials 2 and 3 only (table 11.3 shows the results). The F for trials in the table is nonsignificant, so first, we compute MS_E by combining the sums of squares for trials and residual and dividing by their respective degrees of freedom:

$$MS_E = (0.1 + 3.4)/(1 + 4) = 0.7$$

Then we compute R with equation 11.3:

$$R = (2.85 - 0.7) / 2.85 = 0.75$$

We see that R is considerably higher when we discard the first trial.

TABLE 11.3
Summary of ANOVA for Reliability With Trial 1 Discarded

Source	SS	df	MS	F
Subjects	11.4	4	2.85	—
Trials	0.1	1	0.10	0.11*
Residual	3.4	4	0.85	—

*$p > .05$

The same procedures we just outlined also determine **intertester (interrater) reliability**. Thus, the objectivity of judges or different testers is analyzed by intraclass R, and judge-to-judge variance is calculated in the same way as trial-to-trial variance. Of course, ANOVA designs that are more complex can be used in which trial-to-trial, day-to-day, and judge-to-judge sources of variance all can be identified. Baumgartner and colleagues (2016) discussed some models that can be used for establishing reliability.

intertester (interrater) reliability—The degree to which different testers can obtain the same scores on the same participants; also called *objectivity*.

Methods of Establishing Reliability

Establishing reliability is easier than establishing validity. We first look at three types of coefficients of reliability: stability, alternate forms, and internal consistency.

Stability

The coefficient of **stability** is determined by administering the **test–retest method** on one day and repeating it a day or so later. The interval may be governed to some extent by how strenuous the test is and whether more than a day's rest is needed. Of course, the interval cannot be so long that actual changes in ability, maturation, or learning occur between the two test administrations. The test–retest method is used frequently with fitness and motor performance measures but less often with pencil-and-paper tests. This method is one of the most severe tests of consistency because measurement errors are likely to be more pronounced when the two test administrations are separated by a day or more.

Intraclass correlation should be used to compute the coefficient of stability of the scores on the two tests. Through ANOVA procedures, the tester can determine the amount of variance accounted for by the separate days of testing, test trial differences, participant differences, and error variance.

stability—A coefficient of reliability measured by the test–retest method for assessments taken on different days.

test–retest method—A method of determining stability in which a test is given one day and then administered exactly as before, usually with an interval of one day or so.

alternate-forms method—A method of establishing reliability involving the construction of two tests that both supposedly sample the same material; also called *parallel-form method* or *equivalence method*.

internal consistency—An estimate of the reliability that represents the consistency of scores within a test.

same-day test–retest method—A method of establishing reliability in which a test is given twice to the same participants on the same day.

split-half technique—A method of testing reliability in which the test is first divided in two; usually one half comprises odd-numbered items, and the other half comprises even-numbered items. The two halves are then correlated by the totals of correctly answered even- and odd-numbered questions.

Kuder-Richardson (KR) method of rational equivalence—Formulas KR-20 and KR-21 used to estimate test reliability from a single-test administration.

coefficient alpha technique—A method used to estimate the reliability of multiple-trial tests; also called *Cronbach alpha coefficient*.

Spearman-Brown prophecy formula—An equation one step up from the split-half technique used to estimate the entire test's reliability rather than just one half or the other.

Flanagan method—A process used to estimate reliability of two halves of a test, whereby the variances of the test halves are analyzed in relation to the test's overall variance.

Alternate Forms

The **alternate-forms method** of establishing reliability involves the construction of two tests that supposedly sample the same material. This method is sometimes referred to as the *parallel-form method* or the *equivalence method*. The two tests are given to the same people. Ordinarily, a period of time elapses between the two administrations. The scores on the two tests are then correlated to obtain a reliability coefficient. The alternate-forms method is widely used with standardized tests, such as those of achievement and scholastic aptitude. It is rarely used with physical performance tests, likely because constructing two sets of sound physical test items is more difficult than writing two sets of questions.

Some test experts maintain that theoretically, the alternate-forms method is the preferred method for determining reliability. Any test is only a sample from a universe of possible test items. Thus, the degree of relationship between two such samples should yield the best estimate of reliability.

Internal Consistency

Reliability coefficients can be obtained by several methods that are classified as **internal consistency** techniques. Some common methods are the **same-day test–retest, split-half technique, Kuder-Richardson (KR) method of rational equivalence,** and **coefficient alpha technique**.

The same-day test–retest method is used almost exclusively with physical performance tests. In written tests, this technique has the tendency to influence practice effects and recall and to result in spuriously high correlations. Testing on the same day results in a higher reliability coefficient than performing a test–retest on separate days. We would certainly expect more consistency of performance within the same day than on different days. Intraclass correlation is used to analyze trial-to-trial (internal) consistency.

The split-half technique has been widely used for written tests and occasionally in performance tests that require numerous trials. The test is divided in two, and the two halves are then correlated. A test could be divided into first and second halves, but this approach is usually not deemed satisfactory. People may tire near the end of the test, or easier questions may be placed in the first half. Usually, the odd-numbered questions are compared with the even-numbered ones; in other words, the total of odd-numbered questions a person answered correctly is correlated with the total of even-numbered questions the person answered correctly.

Because the correlation is between the two halves of the test, the reliability coefficient represents only half of the total test; that is, behavior is sampled only half as thoroughly. Thus, the **Spearman-Brown prophecy formula**, a step-up procedure, estimates the reliability for the entire test because the total test is based on twice the sample of behavior (twice the number of items). The formula is as follows:

$$\text{Corrected reliability coefficient} = \frac{2 \times \text{reliability for } 1/2 \text{ test}}{1.00 + \text{reliability for } 1/2 \text{ test}} \tag{11.5}$$

If, for example, the correlation between the even- and the odd-numbered items was .85, the corrected reliability coefficient would be this:

$$\frac{2 \times .85}{1.00 + .85} = \frac{1.70}{1.85} \cong .92$$

Another split-half method is the **Flanagan method**, which analyzes the variances of test halves in relation to the test's overall variance. No correlation or Spearman-Brown step-up procedure is involved.

In the Kuder-Richardson (KR) method of rational equivalence, one of two formulas, known as *KR-20* and *KR-21*, is used for items scored dichotomously (i.e., correct or incorrect). Only one test administration is required, and no correlation is calculated. The resulting coefficient represents an average of all possible split-half reliability coefficients.

Highly regarded by many test experts, the KR-20 involves the proportions of students who answered each item correctly and incorrectly in relation to the total score variance. The KR-21 is a simplified, less-accurate version of the KR-20.

The coefficient alpha technique, sometimes referred to as the *Cronbach alpha coefficient* (see Cronbach, 1951), is a generalized reliability coefficient that is more versatile than other methods. One particularly desirable feature of coefficient alpha is its usage with items that have various point values, such as essay tests and attitude scales that have as possible answers *Strongly agree* and *Agree*. The method involves calculating variances of the parts of a test. The parts can be items, test halves, trials, or a series of tests, such as quizzes. With dichotomous items, the coefficient alpha results in the same reliability estimate as KR-20 (in fact, KR-20 is just a special case of coefficient alpha). The results of test halves are equivalent to the Flanagan split-halves method. With trials or a series of tests, the results are the same as intraclass correlation. Coefficient alpha is arguably the most commonly used method of estimating reliability for standardized tests.

Intertester Reliability (Objectivity)

Objectivity, sometimes called *intertester* (or *interrater*) *reliability*, is a form of reliability that pertains to the testers. This facet of reliability is the degree to which different testers can achieve the same scores on the same subjects.

To establish the degree of objectivity, more than one tester is needed to gather data. Then researchers analyze the scores with intraclass correlation techniques to obtain an intertester reliability coefficient. It is possible to assess a number of sources of variance in one analysis, such as variance caused by testers, trials, days, participants, and error (for a discussion of the calculations involved, see Safrit, 1976).

Some forms of research in physical activity involve the observation of certain behaviors in real-world settings, such as during physical education classes or sport participation. This approach involves the use or development of some sort of coding instrument. Most frequently, the instrument has a series of categories into which motor behaviors may be coded. Researchers observe the behaviors using techniques such as event recording, time sampling, and duration recording (see chapter 16 for descriptions).

In these types of scales, as with any measure, validity and reliability are important. But it is usually much more difficult to obtain consistency in recording activities in a physical education class or sport than in recording error measurements from a laboratory task such as a vertical jump. Consequently, observational researchers are concerned about coder consistency. Typically, coders are trained to a criterion level of reliability, and reliability is checked regularly throughout the project. **Interobserver agreement (IOA)** is a common way to estimate reliability among coders and is often used in event recording and time sampling. Agreements are behaviors that are coded identically and disagreements are behaviors that are coded differently. IOA, which is typically reported as the percentage of agreement, uses the following formula:

$$IOA = \frac{\text{agreements}}{(\text{agreements} + \text{disagreements})} \tag{11.6}$$

objectivity—The degree to which different testers can achieve the same scores on the same subjects; also known as *intertester* (or interrater) reliability.

interobserver agreement (IOA)—A common way of estimating reliability among coders by using a formula that divides the number of agreements in behavior coding by the sum of the agreements and disagreements.

Standard Error of Measurement

Previous chapters touched on the idea of standard error several times regarding confidence intervals, the t and F tests, and interpretation of significance levels. Chapter 8 also discussed standard error of prediction in correlational research. The standard error of measurement is an important concept in interpreting the results of measurement. Sometimes we get too carried away with the aura of scientific data collection and fail to realize that the possibility of measurement error always exists. For example, maximal oxygen consumption ($\dot{V}O_2$max) has been mentioned several times in this book as the most valid measure of cardiorespiratory fitness. Field tests of fitness are frequently validated through correlation with $\dot{V}O_2$max. We must be careful, however, not to consider $\dot{V}O_2$max as a perfect test that is error free. Every test yields only observed scores; we can obtain only estimates of a person's true score. It is much better to think of test scores as falling within a range that contains the true score. The formula for the standard error of measurement ($S_{Y \cdot X}$) follows:

> **Although a test never yields a person's true score, test scores should be considered as within the range containing the true score.**

$$S_{Y \cdot X} = S\sqrt{1.00 - r} \tag{11.7}$$

where s is the standard deviation of the scores and r is the reliability coefficient for the test. In the measurement of percentage of body fat for female adults, assume that the standard deviation is 5.6% and the test–retest correlation is .83. The standard error of measurement would be the following:

$$S_{Y \cdot X} = 5.6\sqrt{1.00 - .83} = 2.3\%$$

Assume further that a particular woman's measured fat percentage is 22.4%. We can use the standard error of measurement to estimate a range within which her true fat percentage likely falls.

We expect standard errors to be normally distributed and therefore interpret them in the same way as standard deviations. About two-thirds (68.26%) of all test scores fall within ±1 standard error of measurement of their true scores. In other words, a person's true score has about a 68% chance of being found within a range of the obtained score plus or minus the standard error of measurement. In the example of the woman who had an obtained score of 22.4% fat, chances are 2 in 3 that her true fat percentage is 22.4% ± 2.3% (standard error of measurement), or somewhere between 20.1% and 24.7%.

We can be more confident if we multiply the standard error of measurement by 2 because about 95% (95.44%) of the time, the true score can be found within a range of the observed score plus or minus two times the standard error of measurement. Thus, in the present example, we can be about 95% certain that the woman's true fat percentage is 22.4% ± 4.6% (±2.3% × 2); that is, between 17.8 and 27.0%.

From the formula, we see that the standard error of measurement is governed by the variability of the test scores and the reliability of the test. If we had a higher reliability coefficient, the error of measurement would obviously decrease (see table 11.4). In the present example, if the reliability coefficient was .95 (r), the standard error of measurement would be only 1.3%. With reliability of .83 but with a smaller standard deviation, such as 4.0%, the standard error of measurement would be 1.6%.

TABLE 11.4

Standard Error of Measurement

s	r	$S_{Y \cdot X}$
5.6%	0.83	2.3%
5.6%	0.95	1.3%
4.0%	0.83	1.6%

* s = standard deviation; r = reliability coefficient for the test; $S_{Y \cdot X}$ = standard error of measurement

Remember the idea of the standard error of measurement when you are interpreting test scores. As indicated earlier, people sometimes have blind faith in measurements, particularly if the measurements seem scientific. With fat percentage estimation from skinfold thicknesses, for example, we need to keep in mind that error is connected not only with the skinfold measurements but also with the criterion that these measurements are predicting—that is, the body density values obtained from underwater weighing and the fat percentage determined from them. Yet we have observed people accepting as gospel that they have a certain amount of fat because someone measured a few skinfolds. Newspapers have reported that some athletes have only 1% fat, an impossibility from a physiological standpoint. Moreover, obtaining a predicted negative fat percentage from regression equations is possible. Please do not misunderstand. We are not condemning skinfold measurements. We are simply trying to emphasize that all measurements are susceptible to errors and that common sense, coupled with knowledge of the concept of standard error of measurement, can help us better understand and interpret the results of measurements. We revisit standard error in chapter 15 when we discuss errors associated with surveys.

Using Standard Scores to Compare Performance

Direct comparisons of scores are not meaningful without some point of reference. If Craig jumps 46 cm on the vertical jump and does 25 push-ups, which score is better? How can you compare centimeters and repetitions? If we know that the class mean for the vertical jump is 40 cm and that 20 is the mean for push-ups, we know that the performances of 46 cm and 25 push-ups are better than average, but how much better? Is one performance better than the other?

One way to compare the performances is to convert each score to a standard score. A standard score is a score expressed in terms of standard deviations from the mean. We now discuss how to determine standard scores by using z scores and T scales.

z Scores

The basic standard score is the *z* **score**. The z scale converts raw scores to units of standard deviation in which the mean is zero and a standard deviation is 1.0. The formula is the following:

$$z = (X - M) / s \tag{11.8}$$

Suppose the mean and standard deviation for vertical-jump scores are 40 cm and 6 cm, respectively, and for push-ups they are 20 and 5, respectively (see table 11.5). Thus, a score of 46 cm for the vertical jump is a z score of +1.00:

$$z = (46 - 40) / 6 = 6 / 6 = 1.00$$

A score of 25 push-ups is also a z score of +1.00:

$$z = (25 - 20) / 5 = 5 / 5 = 1.00$$

We see that Craig performed exactly the same on the two tests relative to the performance of others in the class. Both performances were 1 standard deviation above the mean. Similarly, scores of different students on the same test can be compared by z scores. If Grant jumps 37 cm, he has a z score of −0.5. If Mack jumps 44 cm, he has a z score of 0.67. All standard scores are based on the z score. But because z scores are expressed in decimals and can be positive or negative, they are not as easy to work with as some other scales.

z score—The basic standard score that converts raw scores to units of standard deviation in which the mean is zero and a standard deviation is 1.0.

TABLE 11.5

Means, Standard Deviations, and z-Scores

Participant	Task	Score	M	sd	z	T
Craig	Push-ups	25	20	5	+1.0	60
Craig	Vertical Jump	46	40	6	+1.0	60
Grant	Vertical Jump	37	40	6	-0.50	45
Mack	Vertical Jump	44	40	6	+0.67	56.7

T Scales

T **scale**—A type of standard score that sets the mean at 50 and standard deviation at 10 to remove the decimal found in *z* scores and to make all scores positive.

The *T* **scale** sets the mean at 50 and the standard deviation at 10. Hence, the formula is $T = 50 + 10z$ (see table 11.5). This removes the decimal and makes all the scores positive. A score 1 standard deviation above the mean ($z = 1.0$) is a *T* score of 60. A score 1 standard deviation below the mean ($z = -1.0$) is a *T* score of 40. Because more than 99% (99.73%) of the scores fall between $\pm 3s$, it is rare to have *T* scores below 20 ($z = -3.0$) or above 80 ($z = +3.0$). Table 11.6 shows the *z*-score distribution that some standardized tests use for different transformations of means and standard deviations.

The decision of which standard score to use depends on the nature of the research study and the extent of interpretation required for the test takers. In essence, then, it is a matter of choice in light of the use of the measures.

TABLE 11.6

Standardized Means and Standard Deviations of Well-Known Tests

Scale	M	s
Scholastic Assessment Test	1059	210
Stanford-Binet IQ	100	16
College Entrance Examination	500	100
National Teachers Examination	500	100
Wechsler IQ	100	15

Measuring Movement

Much of the research in physical activity obviously involves movement. The measurement of physical fitness has fascinated exercise physiologists and physical educators for years. The concept of health-related physical fitness is widely accepted as being represented by the components of cardiorespiratory endurance, muscular strength, muscular endurance, flexibility, and body composition. In addition, many research studies have examined other fitness parameters that are needed primarily for skilled performances, such as in athletics and dance. These components include power, speed, reaction time, agility, balance, kinesthetic perception, and coordination. Research in motor behavior generally deals with the acquisition and control of motor skills. Types of measurements include tests of basic movement patterns, sport skills, and controlled (often novel) laboratory tasks.

Research in biomechanics involves such measurements as high-speed cinematography, force transduction, and electromyography. Research in pedagogy typically involves observations of behavior in real-world settings, such as during physical education classes or sport participation. Video-recorded observations are often used to allow more precise analyses.

Researchers must defend the validity and reliability of all forms of measuring. In some respects, measuring movement is simpler and more straightforward than measuring cognitive or affective behaviors. For example, the amount of force a person can apply in a given movement can be measured accurately and directly. The distance a person can jump can be assessed with a measuring tape, and everyone accepts the result as a valid and reliable measure. On the other hand, cognitive and affective behaviors must usually be inferred from marks made on a sheet of paper or responses to a computerized task. Nevertheless, the measurement of movement is rarely a simple matter of measuring distance jumped. It is usually complex and often difficult to standardize. Each type of measurement has its own methodological difficulties, which pose problems for the researcher with regard to validity and reliability. We do not attempt a thorough description of measuring movement because much of this material is specific to your area of specialization (e.g., motor behavior, exercise, wellness). The focus of your graduate program is on many of these issues.

Measuring Written Responses

The measurement of written (and oral) responses is part of the methodology of numerous research studies in kinesiology. Research questions frequently deal with affective behavior, which includes attitudes, interests, emotional states, personality, and psychological traits.

Measuring Affective Behavior

Affective behavior includes attitudes, personality, anxiety, self-concept, social behavior, and sporting behavior. Many general measures of these constructs have been used over the years (e.g., *Cattell's 16 Personality Factor* [*PF*] *Questionnaire*, Spielberger's *State-Trait Anxiety Inventory*). Miller (2006) discusses the uses of affective measures for individuals and groups and explains various relevant affective constructs and associated measurement tools in physical education settings.

Attitude Inventories

A large number of attitude inventories have been developed. Undoubtedly, many test developers see a direct link between attitude and behavior. For example, if a person has a favorable attitude toward physical activities, that person will participate in such activities. Research, however, has seldom substantiated this link between attitude and behavior, although such a link seems logical.

Researchers usually try to locate an instrument that has already been validated and that is an accepted measure of attitude rather than construct a new one. Finding a published test that closely pertains to the research topic and that has scores validated with a similar sample is a problem. Another problem is that often the researcher wants to determine whether some treatment will bring about an attitude change. A source of invalidity discussed in chapter 18 is the reactive effects of testing. The pretest sensitizes the participant to the attitudes being surveyed, and this sensitization, rather than the treatment, may promote a change.

Another problem (mentioned in chapter 15) inherent in any self-report inventory is whether the person is truthful. It is usually evident what a given response to an item in an attitude inventory indicates. For example, a person may be asked to show their degree of agreement or disagreement with the statement "Regular exercise is an important part of my daily life." The person may perceive that the socially desirable response is to agree with the statement regardless of their true feelings. Some respondents deliberately distort their answers to appear good (or bad). Tests sometimes use so-called filler items to make the true purpose of the instrument less visible. For example, a test designer might include several unrelated items, such as "Going to the opera is a desirable social activity," to disguise the fact that the instrument is designed to measure attitude toward exercise. This issue is especially important when social desirability considerations may be biasing factors.

When a researcher seeks an attitude instrument to use in a study, validity and reliability must, of course, be primary considerations. Unfortunately, published attitude scales have not always been constructed scientifically, and limited information is provided about their validity and reliability. Reliability of methodology can be established easily. (We are not saying that attitude scales are easily made reliable.) Validity is usually the problem because of the failure to develop a satisfactory theoretical model for the attitude construct.

Personality Research

Many research studies in physical education and sport have attempted to explore the relationship between personality traits and aspects of athletic performance. Interest in this topic can be attributed to several factors, including the great deal of public attention focused on athletics. Athletes are obviously special people with regard to physical characteristics. Beyond that, however, is the hypothesis that athletes have certain personality traits that distinguish them from nonathletes.

One area of investigation is to identify personality traits that might be uniquely characteristic of athletes in different sports. For example, do people who gravitate toward vigorous contact sports differ in personality from those who prefer noncontact sports? That is, is there an American football type or a bowler type? Are superior athletes different from average athletes in certain personality traits? Can participation in competitive sport modify a person's personality structure? Moreover, do athletes within a sport differ on some traits that, if known, could point to different coaching strategies or perhaps be used to screen and predict which athletes will be difficult to coach or will lack certain qualities associated with success?

People with a strong interest in sports, such as former coaches and players, have been greatly attracted to the study of personality and athletics. But having a strong interest in athletics does not compensate for a lack of preparation and experience in psychological evaluation. Unfortunately, in some cases, personality-measuring instruments have been misused. Furthermore, sport psychologists have recognized the value of sport-specific measures of various behaviors and perceptions. Much of the impetus for this approach can be traced to Martens' *Sport Competition Anxiety Test* (Martens, 1977).

Some sport-specific measures relate to group cohesiveness (Carron, Widmeyer, & Brawley, 1985), intrinsic and extrinsic motivation (Weiss, Bredemeier, & Shewchuk, 1985), confidence (Vealey, 1986), and sport achievement orientation (Gill & Deeter, 1988). The rationale for the sport-specific approach is that general measures of achievement, anxiety, motivation, self-concept, social behavior, and fair play do not have high validity for sport situations. Studies in test development indicate sport psychologists' commitment to develop measures of affective behavior that are multidimensional and specific to the competitive environment.

Scales for Measurement

Likert-type scale—A type of closed question that requires the participant to respond by choosing one of several scaled responses; the intervals between items are assumed to be equal.

In measuring experiences, behavior, or perceptions, a variety of scales are used to quantify the responses. Two of the most commonly used are the **Likert-type scale** and the **semantic differential scale**.

Likert-Type Scale

The Likert-type scale is referred to in chapter 15 in connection with survey research techniques. It is usually a 5- or 7-point scale with assumed equal intervals between points. The Likert-type scale is used to assess the degree of agreement or disagreement with statements and is widely used in attitude inventories. An example of a Likert-type item follows:

I prefer quiet recreational activities such as chess, cards, or checkers rather than activities such as running, tennis, or basketball.

Strongly agree   *Agree*   *Undecided*   *Disagree*   *Strongly disagree*

A principal advantage of scaled responses is that they permit a wider choice of expression than responses such as *Always* or *Never,* or *Yes* or *No.* The use of five, seven, or more intervals may help increase the reliability of the instrument. For a consideration of the use of Likert-type and semantic differential questions using 5- and 7-point options in physical activity research, see Rhodes, Hunt Matheson, and Mark (2010).

Semantic Differential Scale

The semantic differential scale employs bipolar adjectives at each end of a 7-point scale. The respondent is asked to make judgments about certain concepts. The scale is based on the importance of language in reflecting a person's feelings. A sample item from a semantic differential scale is shown here:

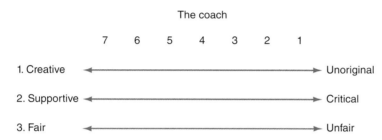

The 1- to 7-point scale between adjectives is scored, with 7 being the most positive judgment.

Rating Scales

Rating scales are sometimes used in research to allow participants to self-evaluate their perceptions. A self-rating scale concerning a person's perceived efforts during exercise that has been widely used in research is the **rating of perceived exertion (RPE)** scale by Borg (1962). The underlying rationale for the scale is to combine and integrate the many physiological indicators of exertion into a whole, or gestalt, of subjective feeling of physical effort. Borg quantified this feeling of perceived exertion into a scale with numbers ranging from 6 to 20, reflecting a range of exertion from *Very, very light* to *Very, very hard.*

Rating scales are also used to measure performance. For example, in a study that compares teaching strategies for diving, the dependent variable (diving skill) would most likely be derived from expert ratings because diving does not lend itself to objective skill tests. Thus, after the experimental treatments have been applied, people knowledgeable in diving rate all the participants on their diving skills. To do this in a systematic and structured manner, the raters need to have some kind of scale with which to assess skill levels in different parts or phases of the performance.

Types of Rating Scales

There are several kinds of rating scales, ranging from simple to complex, that use numerical ratings, checklists, or rankings; have verbal cues associated with numerical ratings; or require forced choices. Whatever the degree of complexity, however, practice in using the scale is imperative.

semantic differential scale—A scale used to measure affective behavior in which the respondent is asked to make judgments about certain concepts by choosing 1 of 7 intervals between bipolar adjectives.

rating scale—A measure of behavior that involves a subjective evaluation based on a checklist of criteria.

rating of perceived exertion (RPE)—A self-rating scale developed by Borg (1962) to measure a person's perceived effort during exercise.

When more than one judge is asked to rate performances, some common standards must be set. Training sessions with video-recorded performances of people of different ability levels are helpful in establishing standard frames of reference before judging the actual performances. Intertester agreement and reliability were discussed earlier in this chapter.

Rating Errors

leniency—The tendency for observers to be overly generous in performance ratings.

central tendency error—An error that results when the rater, avoiding the extremes of the scale, gives an inordinate number of ratings from the middle of the scale.

halo effect—A threat to internal validity wherein raters allow previous impressions or knowledge about certain people to influence all ratings of those people's behaviors.

proximity error—An error that results when a rater considers behaviors to be more nearly the same when they are listed close together on a scale than when they are separated by some distance.

observer bias error—An error that results when raters are influenced by their own characteristics and prejudices.

observer expectation error—An error that results when a rater sees evidence of certain expected behaviors and interprets observations in the expected direction.

Despite efforts to make ratings as objective as possible, pitfalls are inherent in the process. Recognized errors in rating include **leniency**, **central tendency**, the **halo effect**, **proximity**, **observer bias**, and **observer expectation**.

Leniency is the tendency for observers to be overgenerous in their ratings. This error is less likely to occur in research than in evaluating peers (e.g., coworkers). Thorough training of raters is the best means of reducing leniency.

Central tendency errors result from the inclination of the rater to give an inordinate number of ratings in the middle of the scale, thereby avoiding the extremes of the scale. Several reasons suggest the cause. Ego needs or status may play a part. For example, the judge is acting in the role of an expert and, perhaps unconsciously, may grade good performers as average to suggest that she is accustomed to seeing better performances. Sometimes, errors of central tendency are due to the observer wanting to leave room for better future performances. A common complaint in large gymnastics, diving, and skating competitions is that the performers scheduled early in the meet are scored lower for comparable performances than athletes scheduled later in the competition. Another central tendency error results when the judge avoids assigning very low scores, likely because he is reluctant to be too harsh. This kind of rating is, of course, a form of leniency.

The halo effect is the commonly observed tendency for a rater to allow previous impressions or knowledge about a certain person to influence all ratings of that person's behaviors. For example, knowing that a person excels in one or more activities, a judge may rate that person highly on all other activities. The *halo effect* perhaps is not the most appropriate term because negative impressions of a person tend to lead to lower ratings in subsequent performances.

Proximity errors are often the result of overdetailed rating scales, insufficient familiarity with the rating criteria, or both. Proximity errors manifest themselves when the rater associates behaviors that are listed in close proximity as being more nearly the same than behaviors that are separated (listed farther apart) on the scale. For example, if the qualities *Active* and *Friendly* are listed side by side on the scale, proximity errors result if raters tend to evaluate performers as more similar on those characteristics than if the two qualities were listed several lines apart on the rating scale. Of course, if the rater does not have adequate knowledge about all facets of the behavior, she may not be able to distinguish between behaviors that logically should be placed close together on the scale. Thus, the different types of behavior are rated the same.

Observer bias errors vary with the judge's own characteristics and prejudices. For example, a person who has a low regard for movement education may tend to rate students from such a program too low. Racial, sexual, and philosophical biases are potential sources of rating errors. Observer bias errors are directional because they are consistently too high or too low.

Observer expectation errors can operate in various ways, often stemming from other sources of errors such as the halo effect and observer bias. Observer expectations can contaminate the ratings because a person who expects certain behaviors is inclined to see evidence of those behaviors and interpret observations in the expected direction.

Research has demonstrated the powerful phenomenon of expectation. In classroom situations in which teachers are told that some children are gifted while others are slow learners, the teachers tend to treat the students accordingly, giving more attention and patience to the "gifted learners" and less time and attention to the "slow learners."

In the research setting, observer expectation errors are likely when the observer knows what the experimental hypotheses are and is thus inclined to watch for these outcomes more closely than if the observer were unaware of the expected outcomes. The double-blind experimental technique described in chapter 18 is useful for controlling expectation errors. In the double-blind technique, observers do not know which participants received which treatments. They also should not know which performances occurred during the pre- and posttests.

In summary, rating errors are always potentially present. The researcher must recognize and strive to eliminate or reduce them. One way to minimize rating errors is to define the behavior to be rated as objectively as possible. In other words, avoid having the observer make many value judgments. Another suggestion is to keep observers ignorant of the hypotheses and of which participants received which treatments. Bias and expectation can be reduced if the observer is given no information about the participants' achievements, intelligence, social status, and other characteristics. To achieve high levels of accuracy and interrater reliability, the most important precaution the researcher can take is to train observers adequately.

> Minimize rating error by defining the behavior in as specific, concrete, and objective terms as possible.

Measuring Knowledge

Obviously, the measurement of knowledge is a fundamental part of the educational aspects of physical education, exercise science, and sport science. But the construction of knowledge tests is relevant for research purposes as well. The procedures for establishing the validity and reliability of most pencil-and-paper measuring instruments used in research are similar to the procedures previously discussed. But we must also determine whether the difficulty of individual items and their ability to discriminate between levels of ability are functioning in the desired manner; that is, we must perform item analysis. (For a more thorough discussion of measurement methodology for knowledge tests, see Baumgartner et al., 2016.)

The purpose of **item analysis** is to determine which test items are suitable and which need to be rewritten or discarded. Two important facets of item analysis are determining the difficulty of the test items and determining their power to discriminate between levels of achievement.

item analysis—The process of evaluating knowledge test items for their suitability and ability to discriminate.

Analyzing Item Difficulty

Analyzing **item difficulty** is usually easy to accomplish. You simply divide the number of people who correctly answered the item by the number of people who responded to the item. For example, if 80 people answered an item and 60 answered it correctly, the item would have a difficulty index of .75 (60 / 80). The more difficult the item is, the lower its difficulty index is. For example, if only 8 of 80 answered an item correctly, the difficulty index is 8 / 80, or .10. Most test authorities recommend eliminating questions with difficulty indices below .10 or above .90. The best questions are those with difficulty indices around .50. Occasionally, a test maker may wish to set a specific difficulty index for screening purposes. For example, if only the top 30% of a group's applicants are to be chosen, the test maker could accomplish this by using questions with difficulty indices of .30. Questions that everyone answers correctly or that everyone misses provide no information about people's differences in norm-referenced measurement scales.

item difficulty—The analysis of each test item's difficulty level in a knowledge test, determined by dividing the number of people who correctly answered the item by the number of people who responded to the item.

Analyzing Item Discrimination

Index of discrimination (or item discrimination)—the degree to which test items discriminate between people who did well on the entire test and those who did poorly—is an important consideration in analyzing norm-referenced test items. The simplest way to compute an index of discrimination is to divide the completed tests into a high-scoring group and a low-scoring group and then use the following formula:

index of discrimination—The degree to which a test item discriminates between people who did well on the entire test and those who did poorly; also called *item discrimination*.

$$\text{Index of discrimination} = (n_H - n_L)/n \tag{11.9}$$

where n_H is the number of high scorers who answered the item correctly, n_L is the number of low scorers who answered the item correctly, and n is the total number in either the high or the low group. To illustrate, say we have 30 in the high group and 30 in the low group. If 20 of the high scorers answered an item correctly and 10 of the low scorers answered it correctly, the index of discrimination would be the following:

$$(20 - 10) / 30 = 10 / 30 = .33$$

Various percentages of high and low scorers, such as the upper and lower 25, 30, or 33%, are used in determining discrimination indices. The Flanagan method uses the upper and lower 27%. The proportion of each group answering each item correctly is calculated; then a table of normalized biserial coefficients (correlating the response to an individual item to the score on the total test) is consulted to obtain the item reliability coefficient. Thus, item reliability is the relationship between responses to each item and total performance on the test.

If approximately the same proportion of high scorers and low scorers answer an item correctly, the item is not discriminating. Most test makers strive for an index of discrimination of .20 or higher for each item. Obviously, a negative index of discrimination would be unacceptable. In fact, when this happens, the question needs to be examined closely to see whether something in the wording is throwing off the high scorers.

Item Response Theory

classical test theory (CTT)—A measurement theory built on the concept that observed scores are composed of a true score and an error score.

Most of the information concerning the validity, reliability, and item analysis presented thus far pertains to **classical test theory (CTT)**. Some radical changes have occurred recently in the study of the measurement of cognitive and affective behaviors. The advance that has received the most attention in the educational and psychological literature is **item response theory (IRT)**.

Characteristics of Item Response Theory

item response theory (IRT)—A theory that focuses on the characteristics of the test item and the examinee's response to the item as a means of determining the examinee's ability; also called *latent trait theory*.

In CTT, inferences are made about items and student abilities from *total* test score information, whereas item response theory (IRT), as the name implies, attempts to estimate examinees' abilities based on their responses to test items. CTT requires only a few assumptions about the observed scores and the true scores of people on a test. Group statistics regarding the particular test's total score for the total group being examined are used to make generalizations to an equivalent test and population. The estimate of error is assumed to be the same for all participants.

IRT does not limit measurements to one test. Multiple tests that measure the same trait can be used.

IRT is based on stronger assumptions than CTT is. The two major assumptions are unidimensionality and local independence. Unidimensionality refers to the measurement of a single ability or trait. However, this ability is not directly measurable, so IRT is sometimes referred to as *latent trait theory*. For an introduction to IRT and some of its applications, see Zhu and Yang (2016). According to Spray (1989), the real advantage of IRT is that the measurement of an examinee's ability from responses to test items is not limited to a particular test. Rather, it can be measured by any collection of test items that are considered to be measuring the same trait.

item characteristic curve (ICC)—A nonlinear regression curve for a test item that increases from left to right, indicating an increase in the probability of a correct response with increased ability, or latent trait.

In CTT, item difficulty is measured as a function of the total group. In IRT, item difficulty is fixed and can be assessed relative to an examinee's ability level. Thus, the probability that an examinee with a particular ability level will make a correct response to an item can be mathematically described by an **item characteristic curve (ICC)**. The ICC is a nonlinear regression for any item that increases from left to right, indicating an increase in the probability of a correct response with increased ability, or latent trait. **Parameter invariance** means that the item difficulty remains constant regardless of the group of examinees. The curve's steepness indicates the item's discriminating power. Therefore, the ICC can be used to analyze not only the item's discriminating power but also the item's difficulty and a so-called guessing parameter.

Application of Item Response Theory

Space limitations (and lack of knowledge by the authors) prohibit a detailed description and discussion of IRT. The concept is not simple, and complex computer programs are required. Item calibration and ability estimates need large sample sizes. The IRT model has been the subject of intense research in psychology and education for several years. It definitely has potential application for assessment problems in physical education, exercise science, and sport science.

Spray (1989) described several ways for which IRT can be used: **item banking**, **adaptive testing**, mastery testing, attitude assessment, and psychomotor assessment. Item banking is the creation of large pools of test items that are used to construct tests that have certain characteristics concerning the precision of estimating latent ability. Adaptive testing, some-times called *tailored testing*, refers to selecting items that best fit (are neither too difficult nor too easy for) each person's ability level. This function must be done on a computer by using items drawn from an item bank. Some of the widely used standardized tests, such as the Graduate Record Examination (GRE), use adaptive testing. IRT works similarly to computer games that vary the difficulty level based on the player's performance.

In criterion-referenced measurement, tests are constructed that use a cutoff score to show the proportion of items that should be answered correctly to represent mastery of the subject matter. IRT can be used to select the optimal number of items that yield the most precise indication of mastery for an examinee. In other words, people of different ability levels would require different numbers of items.

Item response theory has considerable potential application for assessing attitudes and other affective behaviors. Models of IRT have been proposed that estimate each respondent's attitude or trait parameter on an interval scale regardless of the ordinal nature of the scale. A score for each trait level is available for each category of each item. IRT models can also assess changes in attitude or other traits over time. To date, few affective measuring instruments have been constructed using IRT procedures.

The potential use of IRT for psychomotor assessment was postulated decades ago (Spray, 1987) but has not yet been implemented to any extent (Zhu & Yang, 2016). Motor perfor-mance tests differ from written tests regarding the number of items and trials. Also, some assumptions (particularly local independence) of IRT are not easily accommodated in psy-chomotor testing. Although researchers have occasionally used IRT for the development of physical education standards (Zhu, Fox, et al., 2011; Zhu, Rink, et al., 2011), it continues to be underutilized in the field of kinesiology (Zhu & Yang, 2016)

Summary

In this chapter, we discussed the concepts of validity and reliability of measurements and their application to research. Criterion validity (which includes both concurrent and predic-tive validity) and construct validity are two of the most popular methods of validating mea-sures used in research studies. One problem often identified regarding predictive measures is population or situation specificity. The value of using expectancy tables was discussed.

The topic of test reliability has prompted hundreds of studies and innumerable discus-sions among test theorists and researchers because a measure that does not yield consistent

parameter invari-ance—A postulate in item response theory that the item difficulty remains constant regard-less of the populations of examinees and that examinees' abilities should not change when a different set of test items is administered.

item banking—The creation of large pools of test items that can be used for constructing tests that have certain characteristics concerning the precision of estimat-ing latent ability.

adaptive testing—Selecting test items that best fit each person's ability level; also called *tailored testing*.

results cannot be valid. Classical test theory views reliability in terms of observed scores, true scores, and error scores, and the coefficient of reliability reflects the degree to which the measure is free of error variance. The rationale for using intraclass R instead of r for reliability and the computational procedures for intraclass R were presented. Various methods of estimating reliability were mentioned, such as stability, alternate forms, and internal consistency. An understanding of the concept of standard error of measurement is vital. Any test score should be viewed as only an estimate of a person's true score that probably falls within a range of scores.

Standard scores allow direct comparisons of scores on different tests using different types of scoring. The basic standard score is the z score, which interprets any score in terms of standard deviations from the mean. The T scale converts z-scores to a distribution with a mean of 50 and a standard deviation of 10.

Research studies in physical activity frequently use instruments involving written responses to measure knowledge or affective attributes such as attitudes, interests, emotional states, and psychological characteristics. More recently, sport- and exercise-specific affective measures have been developed.

Responses to items on these scales often use the Likert-type scale (a 5- or 7-point scale ranging from *Strongly agree* to *Strongly disagree*) and the semantic differential scale (a 7-point scale anchored at the extremes by bipolar adjectives). When using scales like these, researchers must be aware of problems such as rating errors, leniency, central tendency, the halo effect, proximity errors, observer bias errors, and observer expectation errors.

Knowledge testing requires item analysis techniques to ensure objective responses. The difficulty and discrimination of each test item must be established to determine the quality of the knowledge test.

Item response theory (IRT, also called *latent trait theory*) differs from classical test theory. It allows item difficulty to be fixed so that the participant's ability level can be determined. IRT is useful in tailoring tests for adaptive testing, mastery testing, attitude assessment, and psychomotor assessment.

✓ Check Your Understanding

1. Briefly describe two ways that evidence of construct validity could be shown for a motor performance test (such as throwing), a test of power, or a test of manipulative skill. How could criterion validity be shown in your example?

2. Find a study in a research journal that used a questionnaire (e.g., an attitude inventory). Describe how the author reported the validity and reliability of the instrument scores. What other technique could have been used to establish reliability or validity?

3. A girl received a score of 78 on a test for which the reliability is .85 and the standard deviation is 8. Interpret her score in terms of the range in which her true score would be expected to fall with 95% confidence.

PART III

Types of Research

Research is what I am doing when I don't know what I am doing.

—Wernher von Braun

Research may be divided into five basic categories: analytical, descriptive, experimental, qualitative, and mixed. In particular, researchers are beginning to plan studies across these basic categories because better and more complete descriptions and explanations of human behavior can be developed. These approaches are often called mixed methods, and chapter 20 addresses that topic.

Chapter 12 discusses sociohistorical research, which is a type of analytical research that answers questions through the use of past knowledge and events. Substantial changes have occurred in recent years in the reporting of sociohistorical research, particularly regarding the integration of the events of interest with other related events. In this chapter, David Wiggins and Daniel Mason provide an excellent overview of sociohistorical research methods.

Chapter 13 describes another type of analytical research that can be applied to physical activity: philosophical research. R. Scott Kretchmar and Tim Elcombe do a fine job of explaining and grouping philosophical research

methods and use examples to identify the strengths and weaknesses of each.

Chapter 14 presents research synthesis (specifically, meta-analysis), another type of analytical research that focuses on the shortcomings of the typical literature analysis. Meta-analysis is a useful method for analyzing a large body of literature. Although several other types of research are considered analytical, historical research, philosophical research, and meta-analysis are the most common and useful in the study of physical activity.

Chapter 15 discusses descriptive research and focuses on survey techniques. Then, other descriptive techniques, such as case studies, correlational studies, developmental studies, and observational studies, are presented in chapter 16. In general, descriptive research shows relationships between people, events, and performances as they currently exist.

Chapter 17, written by Duck-chul Lee and Angelique G. Brellenthin, provides methods in a relatively new type of descriptive research in our field: exercise epidemiology. Often, this approach uses large databases on exercise and health behaviors from various sources.

Chapter 18 introduces experimental research, which deals with future events or the establishment of cause and effect. Which independent variables can be manipulated to create change in the future in a certain dependent variable? After reviewing how difficult cause and effect are to establish, we divide the chapter according to the strengths of various designs: preexperimental, true experimental, and quasi-experimental. Our purpose is to demonstrate which designs and principles are best suited for controlling the sources of invalidity that threaten experimental research.

Chapter 19 presents qualitative research techniques, which are increasingly used in the study of physical activity. The assumptions underlying qualitative research differ from those used in research that adheres to the traditional scientific method. This difference does not mean that qualitative research is not science (i.e., systematic inquiry), but the techniques for acquiring and analyzing knowledge differ from the typical steps in the scientific method. This chapter discusses the differences in the quantitative and qualitative paradigms, procedures used in qualitative research, the interpretation of qualitative data, and theory construction.

Finally, in chapter 20, we present the mixed-methods approach to research. This approach combines one of the quantitative approaches (correlational, survey, experimental) with the qualitative model. The idea is that both approaches and types of data represent human behavior, and that achieving a full understanding of behavior is difficult without using both methods.

If you are a producer or a consumer of research in physical activity, you must understand accepted techniques for solving problems systematically. The nine chapters in part III provide the basic underpinnings of research planning. Many research types reported here are closely associated with the appropriate statistical analyses previously presented. We refer to the appropriate statistics as we discuss the types of research. Use the knowledge gained from the previous section to understand this application. You should learn about the relationship between the correct type of design and the appropriate statistical analysis as soon as possible.

12

Sociohistorical Process in Sport Studies

David K. Wiggins
George Mason University

Daniel S. Mason
University of Alberta in Edmonton

> **Often it does seem a pity that Noah and his party did not miss the boat.**
>
> —Mark Twain

In 1974, Marvin H. Eyler (p.74), one of the founders of the academic subdiscipline of sport history, contended that "the past can never be precisely replicated" but must be "reconstructed on the basis of evidence which has been selected from pre-suppositions." Furthermore, in contrast to science, he noted that history uses a "different standard of objectivity" and is concerned with knowledge that is "inferential and indirect" (Eyler, 1974, p. 74). "Historical research attempts to systematically recapture the complex nuances, the people, meanings, events, and even ideas of the past that have influenced and shaped the present" (Berg, 2001, pp. 210-211). Burke (1992, p. 2) offers yet another definition, suggesting that history is "the study of human societies in the plural, placing the emphasis on the differences between them and also on the changes which have taken place in each one over time." Quite simply, a history is an account of some event or series of events that has taken place in the past (Berg, 2001). To put historical research within the broader context of this book, perhaps the best justification for the use of historical methods is found in the following statement by Burke (1992, p. 139): "If we want to understand *why* social change takes place, it may be a good strategy to begin by examining *how* it takes place." Thus, the role of the historian is to uncover the evidence and analyze it in a way that provides new insights into social change over time.

This chapter examines the status of sport history as an academic subdiscipline and the major issues and methodological approaches associated with sport history, identifies exemplary research studies in sport history, and discusses future avenues of inquiry in the field. It is hoped that this review will provide a greater awareness of sport history as an academic subdiscipline and, more specifically, present the strengths and weaknesses of the research process in this field.

Development of the Discipline

At the beginning of the 1970s, there was unbridled optimism about the future of sport history; the first four years of that decade alone would witness the creation of two organizations, two journals, and three symposiums devoted to sport history (Barney & Seagrave, 2014; Berryman, 1973; Pope, 1997b; Struna, 1997). Early work, such as that of Eyler (1974), provided an initial blueprint for sport historians regarding the meaning of their work, methods for recovering information from the past and seeking historical truth, and a greater understanding of the development of sport.

Much of the momentum established in sport history during the early 1970s would be maintained over the next 25 years. Courses in various aspects of sport history at both the undergraduate and graduate levels were added to university curricula, and a number of research studies on sport history proliferated. Using a variety of methodologies and approaches, scholars examined what seemed to be every conceivable topic (Adelman, 1983; Baker, 1983; Guttmann, 1983; Hill, 1996; Kruger, 1990; Morrow, 1983; Park, 1983; Pope, 1997b; Struna, 1985, 1997, 2000; Walvin, 1984; Wiggins, 1986, 2000). Perhaps most significant, the number and quality of research outlets increased substantially. Besides *Sport History Review*, the *Journal of Sport History*, and the *International Journal of the History of Sport*, scholars could choose to publish their work in one of the journals from the parent discipline or in such sport studies–related journals as *Sport, Education and Society*; *Sporting Traditions*; *Sport in Society*; *Journal of Olympic Studies*; and *Olympika*. In addition to academic journals, both university and commercial presses added sport history titles to their publication lists. Importantly, university presses also created special series devoted to sport-related topics, including most notably the University of Illinois Press, Rutgers University Press, Syracuse University Press, University of Nebraska Press, The University of Arkansas Press, University of Tennessee Press, Temple University Press, and University of Texas Press (Schultz, 2017).

As Berryman, Struna, Pope, and others have pointed out, many important articles and books on aspects of sport history were published during the first half of the 20th century, including Frederic L. Paxson's "The Rise of Sport" in the *Mississippi Valley Historical Review* (1917), John A. Krout's *Annals of American Sport* (1929), Jennie Holliman's *American Sports, 1783-1835* (1931), Herbert Manchester's *Four Centuries of Sport in America, 1490-1890* (1931), and John R. Betts' "Organized Sport in Industrial America" (1951). The watershed for sport history, however, in regard to organizational structure and institutional support, occurred during the 1960s and early 1970s. Buoyed by the era's newfound interest in the history of the common person and everyday life (often called histories from below) and as part of the disciplinary movement pervading physical education at the time, a group of people interested in the topic held symposiums and founded organizations devoted to sport history.

Unfortunately, much of the enthusiasm for sport history has seemingly abated in recent decades (Booth, 1997, 1999, 2005; Nauright, 1999; Phillips, 1999, 2001; Ward, 2013). Although research studies in sport history continue to be produced and both individual and institutional memberships remain steady in sport history organizations, the field has gradually been marginalized as an academic subdiscipline. As a group, historians generally view sport as frivolous and unimportant in the grand scheme of things. The reasons for this attitude are impossible to explain with any certainty. Gorn and Oriard (1995, p. A52), in their appropriately titled and frequently cited essay "Taking Sports Seriously," noted that "athletes' bodies remain curiously off-limits" in cultural studies, history, and other disciplines and speculate that intellectuals are simply "uncomfortable with physicality" or are "playing out the long-standing faculty antagonism to the distorted priorities of universities with multimillion-dollar athletics programs."

The status of sport history is not much different in kinesiology departments, where academicians in biomechanics or exercise physiology, or even what is termed the *pedagogical sciences*, do not always see the relevance of sport history (Booth, 1997, 1999; Schultz, 2017). An interesting, but not all that surprising, response by some sport historians to this transformation in these fields has been to focus more directly on the history of the body and exercise rather than the institution of sport. Currently, a group of sport historians, as well as a group of sport sociologists, are more interested in examining the development of physical culture; representations of the human body; the interconnections of health, exercise, and medicine; and other such topics (Berryman & Park, 1992; Schultz, 2017; Todd, 1998; Vertinsky, 1990; Whorton, 1982).

However, the marginalization of sport history in kinesiology departments is not simply a result of the kinds of topics and phenomena explored. In fact, the relatively low status granted sport history in these departments is as much a result of the perceived limitations of its research techniques and methodological approaches as anything else (Booth, 1997). Scholars from various subdisciplines, either out of ignorance or deep philosophical differences, have been highly critical of the approaches taken by many sport historians. Some sport sociologists have railed against sport historians for their refusal to use theoretical models and adequately conceptualize their research data. In response, sport historians have criticized sport sociologists for their technical jargon, overemphasis on theory, and inability to grasp the complexities of the past (Phillips, 2001). In reality, there are elements of truth in the criticisms of both groups. Unfortunately both groups, but particularly sport historians, generally pay little attention to these criticisms and have been reluctant to reflect on methodological approaches and the meaning of their work. Although debatable, several scholars have surmised that historians avoid self-reflection because they are simply too busy collecting data and adding to the body of knowledge. It is hoped that this chapter provides a model for conducting sport history research that can at least partially reconcile some of the divisiveness within the field of sport studies and the opinions of some in other disciplines in kinesiology.

Theory and Sport History

Despite the issues identified in the preceding section, the field of sport history has much to offer to the realm of sport studies. In many respects, the problems within the discipline largely reflect a marginalization in other areas of sport studies and kinesiology or physical education programs. However, very influential and impactful scholars have used a variety of approaches to study the history of sport and the role sport has had within a broader social context. As noted earlier, one of the roles that sport historians play is in discovering and uncovering evidence that gives us a better understanding of how our world works today.

Burke (1992) identified two basic approaches to historical knowledge, which seem to work as ends of a continuum. He called them *particularizing* and *generalizing*. Others have used the terms *historical* and *theoretical*, and *splitting* and *lumping*. The difference between the approaches lies in the level and type of analysis used by the historian. For example, those who particularize are interested in minute details related to the past, whereas those who generalize are more interested in how things fit into a broad view. Similarly, many traditional historians have been interested in simply uncovering past events, whereas others are keener on discovering the broader theoretical implications of their research. Splitting can be described as chipping away at a theory like a sculptor by examining how theories hold against empirical data uncovered by the historian. In contrast, lumping consists of a complementary process of lumping together seemingly disparate facts, "working outwards from the particular to the general" (Burke, 1992, p. 130).

Looks like a potential historian—fitting pieces of the puzzle together.

© BSIP/Contributor

These two basic approaches shed some light on a movement within the history literature away from traditional histories that sought to examine the deeds of elite people (typically white and male), and toward understanding the social structures and processes of change that affected a given time period (Iggers, 1997). This can be described simply as a movement away from studying the limited and myopic realm of the great deeds of great men to more inclusive studies of what can be called *histories from below*. An example of this can be seen in the histories of Muhammad Ali. It could be argued that his athletic accomplishments (the great deeds of a great man) were not as important as the impact he had on millions of people in his generation (how his deeds affected those "from below"). In other words, how society was affected by his outspokenness and exploits during the turbulent period of the late 1960s could be considered of greater interest and importance than the number of knockouts he had in his career.

Relationship Between Theory and Method

It is important to note that Muhammad Ali would not have had such an impact were it not for his boxing ability. Thus, it is critical for historians to be aware of both splitting and lumping. A sport historian needs to be able to obtain evidence, critically ascertain its trustworthiness, and then understand how it adds to our greater understanding of the phenomenon being investigated. As this section will reveal, the way historians have sought to gain a greater understanding is directly related to the theoretical perspectives taken by researchers.

Sport historians have chosen a number of ways to undertake this scholarly process, and some have criticized sport historians for their reliance on several popular approaches. In this section, we review several analytical **paradigms**: structural functionalism and modernization, Marxism, and postmodernism. However, before we examine them in more detail, it is important to note that theory can never be applied to the past. What theory can do, on the other hand, is suggest new questions for historians to ask about their period, or new answers to familiar questions (Burke, 1992). The questions sport historians ask, and the

paradigm—An assumption about the nature of reality and the appropriate methodological approach in gathering and analyzing data.

conclusions they come to, are to a large degree influenced by the theories they use to draw broader inferences about the past. The more dominant approaches that have emerged in the field of sport history are outlined next.

Structural Functionalism and Modernization

One of the earliest and most enduring approaches sport historians used relates to social structures. Functionalism is characterized by an interest in social systems, institutionalization, specific periodization, and urbanization. Structural functionalism has been an especially popular approach among sport historians, particularly in the United States. Many historians during World War II and immediately thereafter adopted this approach, assuming that sport reflected America's superior way of life, democratic institutions, and unabated progress. Works of this genre include Foster Rhea Dulles' *A History of Recreation: America Learns to Play* (1965), Frederick W. Cozens and Florence Scovil Stumpf's *Sports in American Life* (1953), and John Rickards Betts' *America's Sporting Heritage, 1850-1950* (1974), the most well-known and influential survey of the three. Published posthumously by noted sport sociologist John Loy with assistance from Betts' son Richard and prominent sport historian Guy Lewis, the book was a reconstruction of Betts' 1951 dissertation "Organized Sport in Industrial America" (1951). The book, wonderfully researched and characterized by a combination of social and business history, provides a framework and organizational structure that has been used in subsequent monographs and sport history surveys. Betts discussed such topics as the technological revolution and the rise of sport, the Golden Age of Sport in the 1920s, the influence of the Great Depression and World War II on organized sport, and the interrelationships between sport and institutions in the fields of education, religion, and art. The structural-functionalist approach to the study of sport continues to dominate much of the work in the subdiscipline.

According to Burke (1992), functionalism's appeal to historians lies in the fact that it has compensated for the long-standing tendency of historians to explain too much of the past in terms of the actions and intentions of specific people. Instead, according to structural functionalists, the past is best explained by the broader social structures and forces exerted during a given time period. The structural-functionalist approach does have its critics, however. Sport history scholar Doug Booth, for one, believes the system has serious flaws. "Functionalist explanations in sport history," he noted, "typically gloss over disparities in sporting interests between different social and economic groups, exaggerates social stability and harmony, and ignores the inevitable losers and outsiders in local (community, regional, national) consensus building activities" (Booth, 2003, p. 11).

Certainly, one of the most influential analytical designs in sport history is the modernization paradigm first adopted by Allen Guttmann in his classic study *From Ritual to Record: The Nature of Modern Sports* (1978). According to Pope (1997b), modernization theory has developed as a way of investigating how society has evolved from a rural agrarian society to an urban and industrial society, at different times and in different places throughout the world. *From Ritual to Record* charts the formal structural characteristics of modern sports and the extent to which these characteristics were evident in sport during earlier periods in history. It was considered so influential that an entire issue of *Sport History Review* was devoted to a retrospective critique of the book (Booth, 2001; Brownell, 2001; Guttmann, 2001; Howell, 2001; Lippe, 2001). With its impressive list of diverse sources, astute historical and statistical comparisons, and cogent analysis and thoughtful questions, Guttmann's book is essentially a structural-functional analysis that outlines seven characteristics of modern sport: secularism, equality, rationalization, specialization, bureaucratization, quantification, and an obsession with records.

Melvin Adelman also used the modernization paradigm with much success in his pathbreaking study *A Sporting Time: New York City and the Rise of Modern Athletics, 1820-70*

(1986). The first book in the University of Illinois Press' sport and society series, *A Sporting Time* is one of the most frequently cited and important studies in sport history. Adelman, through painstaking research and perceptive analysis, examined the transformation of American sport from its premodern to modern forms by exploring early baseball; horse and harness racing; boxing; and such traditionally neglected activities as cricket, rowing, yachting, and billiards. Adelman challenged previous research by arguing that sport, particularly in New York City, evolved from an informal, unorganized activity to a more highly structured and organized phenomenon between 1820 and 1870 rather than during the late stages of the 19th century. The period from 1820 to 1870 witnessed the growth of commercialized athletics and evidence of a fledgling sport information system, the standardization of records and equipment, specialized player roles, and philosophical rationales for sport participation.

The modernization paradigm that Guttmann and Adelman adopted has received some criticism. In fact, in the aforementioned retrospective critique of *From Ritual to Record* in *Sport History Review*, several prominent scholars took turns poking holes in Guttmann's approach. Howell's criticism stemmed from the book's sweeping approach to history, and he argued that although it helped identify the process of change in sport over time, it was less successful in explaining the transformation of sport. "While many processes are identifiable in history," wrote Howell (2001, p. 13), "it is another thing altogether to understand history itself as a process. *From Ritual to Record* was written at a time of grand theorizing, when the grand narratives of development that postmodernist scholars now disdain were clearly in vogue." Booth generally had high praise for the book but claimed that Guttmann's work provides a "limited conceptualization of the process of modernization and the nature of modernity" and "lacks adequate political and dialectical analyses" (2001, p. 22). Pope (1997b, p. 5) also weighed in on the debate, noting that critics of modernization theory have "exposed modernization as a historical construct that distorted the past by emphasizing structure at the expense of human agency." In some ways, this criticism is similar to that leveled against supporters of structuralism.

Guttmann (2001, p. 5) countered these criticisms and others with his usual aplomb and erudition. He acknowledged that his modernization paradigm has limitations, including the fact that it can be

> **too abstract to account for every detail in history's vast panorama . . . can easily be misinterpreted to mean that the observed changes occurred as part of some uniform and inevitable process . . . [and] can be misused as a facile instrument of ethical judgment—as if modern ways were somehow a moral advance as well as a technological advance beyond traditional customs.**

He argued, however, that in spite of the obvious limitations of the modernization paradigm, no alternative model has been postulated that does a better job in chronicling the changes in sport over time and marking the differences between sport of yesteryear and that of today. Guttmann (2001, p. 6) wrote:

> **Despite the manifest possibility that the concept of modernization can be misunderstood and misrepresented, I continue to find it more useful than the postmodernism [to be described in more detail later in this chapter, and throughout this book] that is currently fashionable in what once were called the social sciences. An imperfect and admittedly selective account of the historical facts seems preferable to the startled discovery that there is no such thing as absolute truth.**

Recent decades, however, have witnessed the adoption of new and varied theoretical approaches and paradigms in sport history. Taking their cues from the vast changes taking place in historical scholarship during the 1970s and 1980s, sport historians increasingly turned to the analytical frameworks used in such fields as anthropology, cultural studies, and sociology to examine the development of sport. Although it was not always easy to discern specific patterns of inquiry, there was now a seeming emphasis in the subdiscipline on such concepts as community formation and power and the ways in which socially constructed notions of race, ethnicity, gender, class, and religion affected sport and vice versa. Increasingly, sport historians adopted the works of Max Weber, Norbert Elias, Clifford Geertz, and the like, to frame their analyses and interpretations (Pope, 1997b).

Marxism

Many sport historians have employed versions of Marxism to analyze the development of sport. Interested in the mode of production and the use of political and social power, Marxist sport historians have focused on sport in regard to social class aspirations and the way sport has been redefined by African Americans, women, workers, and other groups in society. Richard Gruneau (1983), the noted Canadian sociologist, has had a profound effect on the work of sport historians through his many publications on sport and Marxist cultural studies. Equally significant to sport historians have been the theoretical insights provided by such cultural Marxists as E.P. Thompson, Raymond Williams, Eric Hobsbawm, and Antonio Gramsci. Historian Colin Howell (1998, p. 101) correctly observed that "over the past few years, it has been difficult to pick up an issue of one of the major sport history journals and not find any article employing Gramsci's notion of **hegemony** [author's emphasis] or referring to notions of human agency."

hegemony—The cultural or political dominance or authority over others.

Howell might also have added that the theoretical approaches of Gramsci and other cultural Marxists have found their way in one form or another into some of the most important sport history monographs. Pope's *Patriotic Games: Sporting Traditions in the American Imagination, 1876-1926* (1997a) is one example of this. Using Eric Hobsbawm's concept of invented traditions, Pope examined the national sporting culture that was established in the United States between the 1876 centennial and the 1926 sesquicentennial. He illustrated how sport becomes a cultural transmitter of patriotic meaning, intertwined with Thanksgiving and Fourth of July holiday celebrations and military preparedness. Perhaps most important, Pope made clear that sport served as both a metaphorical activity and class drama and helped define public discourse on American hopes and dreams.

Pope's book is significant in that it is reflective of what Hill (1996) and Holt (1998) referred to as the *subtle shift in sport history from causation to meaning*, where scholars have been less concerned about the determinants of sport participation and more interested in the multiple and complex meanings sport has had for athletes, spectators, and other individuals, groups, and communities. Examples of this scholarship are numerous. John MacAloon focused on rituals and cultural symbols in his book *This Great Symbol: Pierre de Coubertin and the Origins of the Modern Olympic Games* (1981). Peter Levine argued in *Ellis Island to Ebbets Field: Sport and the American Jewish Experience* (1992) that sport served as a middle ground for American Jews who were striving to become full participants in American society while at the same time attempting to maintain their sense of ethnic pride and identity. Using an impressive list of source materials and wonderful stories of athletes and teams, Levine made clear that for East European Jewish immigrants and their children during the first half of the 20th century, sport was seen as an avenue of assimilation and a means to pursue the American dream. Elliott Gorn provided a fascinating analysis of the meaning of prizefighting in 19th-century America in his *The Manly Art: Bare-Knuckle Prize Fighting in America* (1986). Integrating the work of anthropologists, sociologists, folklorists, labor and social historians, and American and gender studies specialists, Gorn examined how prizefighting was transformed from a working-class ritual that symbolized physical prowess and toughness to a middle-class sport that illustrated dominant ideas of social mobility and meritocracy.

Postmodernism

Finally, it is important to recognize the presence of postmodern approaches to sport history. It is probably safe to say that most sport historians reject many of the basic tenets of the postmodernist approach, believing that it only results in a "dazzling kaleidoscope of impressions where anything can be construed to mean anything" and ultimately an anarchic nihilism (Holt, 1998, p. 18). As historian Georg Iggers (1997, p. 118) explained, "The basic idea of postmodern theory of historiography is the denial that historical writing actually refers to an actual historical past."

However, a few brave and adventurous souls in the field have employed versions of this approach in their research. Two studies that depend to some extent on postmodernist theory are Patricia Vertinsky's *The Eternally Wounded Woman: Women, Exercise, and Doctors in the Late Nineteenth Century* (1990) and Susan Cahn's *Coming on Strong: Gender and Sexuality in Twentieth-Century Women's Sport* (1994). Tina Parratt (1998) was absolutely correct when she noted that both of these works are indebted to postmodernism, particularly the notion that power and knowledge are the result of discursive practices.

The first and perhaps only genuine postmodernist study in sport history, however, is Synthia Sydnor's "A History of Synchronized Swimming" (1998). Published in a special issue of *Journal of Sport History* and titled "Sport History: Into the 21st Century," Sydnor's essay follows in the path of Marxist theorist Walter Benjamin in that it avoids such standard historical practices as recording facts and periodization to ponder synchronized swimming through an assemblage of symbols, images, and spectacles. Sydnor (1998, p. 260), designating her "essay as a modern text with postmodern concerns," did not analyze or explain or contextualize; rather she provided a number of evocative ideas through an "exhibition or performance" presented in a poetic and literary form. However, works such as Sydnor's remain very much in the minority in the field of sport history (Phillips, 2006).

An approach that is, in the words of Genevieve Rail (1998, p. x), "an amalgam of often purposely ambiguous and fluid ideas," postmodernism is variously thought of as a revolt against modernism and structuralism, a radical break from the past that leans toward a deconstruction of linguistic systems, a distaste for disciplinary boundaries, and a belief that the world is fragmented into many isolated and autonomous discourses that cannot be explained by any grand theory. Michel Foucault, one of the most prominent postmodernist thinkers, argues that truth is only a partial, localized version of reality and that power relations and meaning are the function of language and discourse (Rail, 1998).

All of the approaches described in this section have been shown to be viable ways to study the history of sport. However, each has shortcomings in helping us better understand the past. In a relatively old but still pertinent article, sport historian Colin Howell critiqued the use of the concept of hegemony in sport history. Although his criticisms were directed at one specific lens for historical research, in some respects, similar criticisms could also be leveled at the other approaches we have discussed:

> **Unfortunately, the idea of hegemony is often employed by the unimaginative as a formula to lend a patina of academic respectability to an academic paper or journal article. In turn, its overuse has made the concept appear hackneyed, especially when it is misconstrued as a more academically credible form of social-control analysis. One should keep in mind that the end of historical inquiry is not to confirm Gramsci, but rather to use his theoretical insights to better understand what happened. (Howell, 1998, p. 101)**

Perhaps the lesson for aspiring sport historians is to make sure they use an approach that will help the researcher more capably investigate the question at hand. In the meantime, as Howell warned, researchers must not get too caught up in an approach and use it simply to justify their own research. The most important issue at hand with regard to the theoretical approaches to the study of sport history is the fact that sporting activities, like social practices in general, have always tended to indicate, and at times been reinforced by, class distinctions (Booth & Loy, 1999). Within these class distinctions, it is also critical to identify issues and differences concerning gender, race, and other social concerns. For this reason, sport historians must try to gain a greater understanding of the broader social context of the periods they study, particularly in terms of the relationships between classes. Thus, sport history must be informed by social theories that seek to draw inferences from various contexts and time periods to better explain the role of sport in society.

Research Sources

Sport historians generally adhere to a specific and recognizable historical method. The first, and in some ways most important, task of the sport historian is to become conversant with the secondary literature. According to historians, there are two basic types of sources—primary and secondary. Primary sources are those that were created or witnessed by someone or something directly associated with the topic that is being investigated. For example, a primary source for a paper on an Olympic boxing match that took place in 1988 would be a person who was involved in the match itself, or a firsthand witness to the event. Primary source information can be obtained in a number of ways, such as interviewing a spectator, participant, or judge, or reading a newspaper account written by someone who was at the event (e.g., a sports reporter who covered the event and wrote an article that appeared in the next day's paper). Another way would be to watch video footage of the event. Primary sources are critical for sport historians because they bring them as close as possible to the event. Fortunately, advancements in data preservation and storage has allowed researchers more widespread access to primary sources. For example, some organizations and archives have digitally archived sources such as year-end reports, minutes of meetings, and other key sources dating back to the origins of modern sport. This has created more opportunities for scholars to quickly and inexpensively access sources, which makes the prospects of accessing primary sources far less daunting a task today (Osmond & Phillips, 2015).

Secondary sources are sources, such as books and articles, written by people who were not directly associated with the event or topic. In many cases, they are written by researchers who have done their own primary source–based research. However, the degree of filtering that results from having been examined by others makes secondary sources less valuable to historians. This is because someone else has analyzed the primary source material and come to their own independent conclusions.

At the outset of doing research, secondary sources are extremely important for the sport historian because they provide background information. Thus, when starting out, the historian usually explores the secondary literature. To locate secondary literature, sport historians traditionally turn first to a library catalog to find entries for book titles, subjects, and names of authors. They also depend on such resource aids as historical dictionaries, large general histories, specialized academic journals, biographical dictionaries, periodical indexes, bibliography of bibliographies, and current national bibliographies. Times have changed, however. In addition to the aforementioned tools, sport historians now conduct computer searches to locate secondary sources or contact colleagues on email lists to find secondary material (Atkinson, 1978; Gottschalk, 1969; D.H. Porter, 1981; Shafer, 1980). Databases such as SPORTDiscus also provide opportunities to search online for topics using keywords. In addition, other athletic bodies, such as the LA84 Foundation, have made journal and other sport sources available in full text through their websites.

Research Topics

According to Berg (2001, p. 213), "The major impetus in historical research, as with other data-collection strategies, is the collection of information and the interpretation or analysis of the data." Sport historians are fortunate that the number of improved search tools has coincided with a dramatic increase in both the quality and quantity of secondary works dealing with various aspects of sport history. Thought-provoking, creative, and imaginatively designed studies have been completed on numerous topics.

- For the pattern of sport in both rural and urban settings, see Adelman, 1986; Aiello, 2019; Berryman, 1982; Borish, 1996; Gems, 1997; Hardy, 1981, 1982; Liberti & Smith, 2017; Nathan, 2016; Norwood, 2018; Riess, 1989; R. Roberts, 2000; Somers, 1972; Uminowicz, 1984; Wilson & Wiggins, 2018

- For the history of baseball and other team and individual sports, see Davies, 2014; Dyreson & Schultz, 2015: Goldstein, 1989; Kirsch, 1989; Oriard, 1993, 2001; Riess, 1980; Riess, 2011; Sammons, 1988

- For the role of women in sport at the national and international levels of competition, see Anderson, 2017; Cahn 1994; Festle, 1996; Fields, 2005; Grundy & Shackelford, 2007; Guttmann, 1991; Mangan & Park, 1987; Schultz, 2014, 2018; Ware, 2014

- For the interconnections of organized sport, physical culture, and the medical profession, see Bachynski, 2019; Berryman & Park, 1992; Fair, 1999; H. Green, 1986; Todd, 1998; Verbrugge, 1988; Vertinsky, 1990; Whorton, 1982

- For the origin and evolution of the modern Olympic movement, see Barney, Wenn, & Martyn, 2002; Dyreson, 1998; Guttmann, 1984, 1992; Llewellyn, Gleaves & Wilson, 2015; MacAloon, 1981; Senn, 1999; Witherspoon, 2008; Young, 1996

- For the sport participation patterns of African Americans, Indigenous people, and those in other racial and ethnic groups, see Bass, 2002; Bloom, 2000; Davis, 2015; Demas, 2017; Gems, 2013; Hoberman, 1997; Kaliss, 2012; P. Levine, 1992; Mooney, 2014; Moore, 2017; Regalado, 1998; Riess, 1998; Ruck, 2011, 2018; Shropshire, 1996; Schultz, 2016; Thomas, 2017; Wiggins, 1997; Wiggins, 2018; Wiggins & Miller, 2003; Wiggins & Swanson, 2016; Forsyth & Giles, 2013; Osmond, 2019; Sikes, 2019; Phillips & Osmond, 2018

- For the conception of leisure and transformation of public recreation and organized sport at different times in history and among different classes and geographical locations, see Cavello, 1981; Hardy, 1990; Rosenzweig, 1983; Rosenzweig & Blackmar, 1992

- For the history of intercollegiate sport, see Austin, 2015; Ingrassia, 2012; Lester, 1995; Oriard, 1993, 2001, 2009; Smith, 1988, 2010

- For alternative sport forms, see Atencio et al., 2018; Laderman, 2014

The responsibility of sport historians is not simply to become familiar with this literature but to assess and analyze it in regard to the types of evidence used, generalizations drawn, and historical frameworks employed. A thorough and detailed understanding of the secondary literature assists scholars in formulating paradigms and asking good questions, which are at the heart of the historical process. It is precisely for this reason that the more experienced scholar has little difficulty selecting a topic and delimiting that topic to manageable proportions, whereas the fledgling scholar usually struggles to find a topic that will contribute to the body of knowledge and add to the research literature. Armed with a firm grasp of the secondary literature, the more experienced sport historian can usually establish appropriate themes and set parameters conducive to consistent and logical arguments. On the other hand, the less experienced members of the subdiscipline, equipped with a limited understanding of the secondary literature, often pick topics that are too broad, underestimating rather than overestimating the amount of source material available.

Novice sport historians may tend to choose topics that are too broad because they want to make their mark and are worried about choosing a topic that is too narrow and insignificant to be worthy of study. Peter Burke called historical examinations on a small, local scale *micro histories*. One thing to keep in mind when formulating a topic is that good primary sources that address small, obscure topics can shed light on broader social phenomena and should not be trivialized. As Burke (1992, p. 41) explained:

> **One might begin with the charge that the micro historians trivialize history by studying the biographies of unimportant people or the difficulties of small communities. . . . However, the aim of the micro historian is generally more intellectually ambitious than that. If they do not aspire to show the world in a grain of sand, these historians do claim to draw general conclusions from local data.**

For sport historians, then, a thorough grounding in the secondary literature is essential for formulating good questions, and these questions need not be on a grand scale. Scholars in the subdiscipline know full well, however, that good questions are also those of interest to the researcher, those that are answerable, and those that are tied to the primary evidence. Sport historians would be wise to listen to the advice of advisors and colleagues, but ultimately, they must address those questions that are of interest to them and not particularly to anyone else. A traditional yet still incredibly valuable strategy is to consider some basic questions. The first inquiries have to do with the phenomenon to be studied. Is the researcher most interested in examining sport? Athletics? Exercise? Leisure? Recreation? The next set of questions concern chronology. Is the researcher interested in a particular year? The Renaissance period? The late 19th century? The 1920s? The post–World War II period? Other questions have to do with geography. Is the researcher interested in Great Britain? Canada? The United States? Latin America? China? Finally, the researcher needs to ask questions about the individuals or groups to be studied. Is the researcher interested in elite athletes? Particular racial and ethnic groups? Ancient Greeks? Physical culturists? Health reformers?

Today, aspiring sport historians also have additional resources to use to refine their inquiries: their peers. Many topics of interest to sport historians have developed online networks of like-minded scholars who are very helpful in narrowing down topics, exchanging ideas, and directing researchers to secondary sources they might otherwise be unaware of. Ultimately, the sport historian must arrive at a final set of manageable questions that beg answers gleaned from primary sources.

Research Design

Once their topic questions have been finalized, sport historians put together research designs. Although they do not construct experiments or manipulate and control data, sport historians are similar to all scholars: They map strategies and develop research designs characterized by a systematic arrangement of questions and the approach necessary for answering those questions.

The two most commonly cited designs in sport history are descriptive history and analytical history. Descriptive history primarily addresses what happened in sport, and analytical history addresses the complexity of relationships and draws connections to determine how and why sport developed the way it did and the role it served in the lives of participants and spectators alike. Although the differences between the two basic research designs are not always discernible, most sport historians strive to complete analytical history, which has been most evident from the beginning (although some scholars in the subdiscipline would certainly disagree). As our earlier discussion of the need to link theory and method in historical research suggests, a reliance on descriptive histories may reinforce and rationalize the lack of respectability the field has within both sport studies and the parent discipline of history. As a result, we encourage aspiring sport historians to employ more analytical approaches to their research.

Regardless of the design chosen, all sport historians must be concerned with the collection of evidence. To answer their questions, sport historians seek as many primary sources or firsthand accounts as possible. Although recognizing the impossibility of locating all documents, and the inevitable incompleteness of the historical process, sport historians search archives, libraries, private collections, and a host of other repositories to find the primary sources necessary for gaining an understanding of the sporting past. Primary materials may be obtained through sources that are not necessarily considered first. For example, Iacovetta and Mitchinson (1998, p. 6) suggested exploring case files, which could include employment records, court proceedings, patient records, and psychiatrists' case histories. They explain that "historians have turned to case files because they offer us a rare window on human interaction and conflict. . . . These records can illuminate the ways in which dominant class, gender, and racial ideologies shaped official discourses and actions and relations between experts and clients." Thus, the aspiring sport historian in many ways has an opportunity to start doing historical research without many of the traditional restrictions to primary sources faced by sport historians of earlier years.

One consideration that bodes well for the sport historian is the greater ease of access to primary sources created by new technologies. As Cox and Salter (1998, p. 294) explained, "Technology will release the sport historian from unskilled, repetitive, time-consuming tasks to allow him or her to concentrate on the core activities of a sport historian—analysis, synthesizing, theorizing, and communicating information." For example, many newspapers, which historians typically research in microform format in libraries, are now available in pdf format through online databases. In many cases, these databases can be searched using keywords. So, rather than performing the time-consuming task of reading pages and pages of old newspapers on a microform reader, looking for articles concerned with a topic, researchers can now have a database pull up the appropriate newspaper pages on a given issue. An example is *The Toronto Star* Historical Newspaper Archive at www.thestar.com/about/archive-search.html.

Databases are just part of a host of digitized sources, tools, and social media platforms available to historians, including Wikipedia, blogs, Facebook, Twitter, YouTube, and Flickr. These sources are easy to access and provide information that is potentially valuable to historians in an assortment of ways. They do, however, pose methodological and philosophical questions for historians interested in assessing past events and analyzing changes over time. One question involves the accuracy and superficial nature of these materials. Another has to do with context, whether the selective digitalization process provides the relevant background information necessary for charting key events and providing historical narratives. In spite of these questions, historians, whether they embrace the digitized world or not, would certainly benefit from exploring it combined with standard print sources and archival material (Cohen & Rosenzweig, 2006; Evans & Rees, 2012; O'Malley & Rosenzweig, 1997; Osmond & Phillips, 2015; Rosenzweig, 2003).

Data Analysis and Interpretation

Having identified and located primary sources, the sport historian then exposes the sources to a rigorous two-step evaluative process known as external and internal criticism (Atkinson, 1978; Gottschalk, 1969; Porter, 1981; Shafer, 1980).

External Criticism

External criticism involves establishing the **authenticity** of the primary source. The sport historian, using tests similar to those employed by lawyers or detectives, must determine whether the chosen documents either are forged or have been tampered with since the time of the event. To differentiate between a forgery or misrepresentation and an authentic document, the sport historian must determine whether events, objects, people, or customs are correctly placed chronologically. Was a particular type of paper being used at the time the document was supposedly written? Could the alleged author of the document possibly have been at the location where it was supposedly written? Were the punctuation, spelling, proper names, and signatures in the document consistent with other contemporary writings? Does the alleged author of the document refer to events about which they could not possibly have known? Another common issue relates to authorship. For example, a book or an article may be credited to a person who might be considered a primary source (such as an autobiography). However, these types of books are often ghostwritten and therefore less reliable.

authenticity—The state of being true and legitimate.

Internal Criticism

Once authenticity has been established, the sport historian then assesses the credibility of documents through a process of internal criticism (Atkinson, 1978; Gottschalk, 1969; Porter, 1981; Shafer, 1980). A time-consuming and extremely important part of the historical method, internal criticism is concerned with determining who left documents and the relationship of the source of those documents to the event or events. In other words, what interests or biases might the author of the document have regarding the event? Internal criticism is also concerned with the wording of documents, particularly how words were defined and the way they were used during the historical period being studied.

Internal criticism is also concerned with the information included in and omitted from historical documents. Historians must be cognizant of the fact that all historical documents are only partial glimpses into the past and that some information is often either intentionally or unintentionally left out of written and oral accounts. This is precisely why scholars in sport history always seek corroboration by searching for the independent testimony of two or more trustworthy witnesses. If only one witness of an event is known to exist, scholars must seek other forms of corroboration, including the reputation of the author of the document, the absence of contradictions within the document, and the lack of anachronisms.

A good example of when internal criticism is critical is when relying on newspapers for data. According to Reah (1998), newspaper reports are not simply accounts of selected information on recent events. Reporters, subconsciously or not, put ideological spins on events, providing a skewed viewpoint. McNeil (2001) addressed this issue in a study of a worker's riot in Britain, in which he found that newspaper coverage was biased toward management. Thus, newspaper reports must be viewed with some skepticism because they tend to present the reader with aspects of news offered in a way that will influence the reader in a certain direction (Reah, 1998). Baker (1994, p. 530) wrote:

> **Journalists are actors and creators as well as reporters. They collect, select, and color the information they dispense to their customers; they frame the issues and assumptions from which cultural debates arise. Reporters thus shape opinion just as surely as do newspaper editors and columnists.**

With this in mind, the sport historian must try to gain an understanding of the potential biases that might influence the reporter. For example, it is likely that early coverage of professional sports teams in local newspapers favored the home team. This was due to civic boosterism and also to the very close ties early newspaper reporters had with their local sporting clubs. Thus, the historian cannot rely solely on information from such a source, because it represents only one viewpoint. The sport historian needs to seek other sources to corroborate facts or identify biases in the coverage. Perhaps the easiest way of doing this is to compare coverage in other newspapers (such as that of the visiting team's hometown), or to find other accounts from other sources (e.g., interviews with spectators).

In addition to establishing the credibility of documents, sport historians must carefully read the evidence to ascertain the information while keeping in mind the close interrelationship between empirical particulars and theory formation. Wiggins certainly discovered this in a study he conducted on Muhammad Ali's relationship to the Nation of Islam (Wiggins, 1995). *Muhammad Speaks*, the official organ of the Nation of Islam, provided important

Source Credibility

Every sport historian can probably relate personal stories about establishing the credibility of documents. David Wiggins, one of the authors of this chapter, found that his most challenging struggle with the problem of internal criticism took place while he was doing research on the role of leisure and recreation in the lives of enslaved people on plantations in the antebellum South (Wiggins, 1983a, 1983b). With this very sensitive and controversial subject, he was faced with the problem of how to reconcile the differing conceptions of enslaved people's leisure and recreation expressed by an enormously diverse group of witnesses who had their own biases and vantage points. He had access to British travel accounts located in the U.S. Library of Congress. British travelers, who visited the South in fairly large numbers between 1780 and 1860, varied greatly in their observational skills and represented such diverse professions and occupations as land scouts, missionaries, scientists, politicians, educators, and actors. White plantation owners provided another vantage point from which to view the institution. Wiggins used letters and plantation diaries from such repositories as the University of North Carolina at Chapel Hill's Southern Historical Collection and the Library of Congress to analyze a group of people who obviously had a strong allegiance to the institution of slavery and a reluctance to point out its many cruelties and horrors. The final group of witnesses to the cruel institution constituted the enslaved people themselves, who typically did not know how to read or write and frequently could recount their lives on southern plantations only after realizing their freedom. For their view of the institution, Wiggins turned to both published slave narratives and interviews conducted by the Works Progress Administration (WPA) during the Great Depression of the 1930s.

To assess the credibility of these divergent sources, Wiggins was forced to call on everything he had ever learned about the historical method. First, he had to constantly remind himself that what was most important was the credibility of statements made by authors rather than their overall credibility as witnesses. For example, plantation owners and their families, although generally providing a rather romantic view of slavery and refusing to acknowledge the insidiousness of the institution, were able to detail with some accuracy the play activities and games participated in by enslaved people because of their frequent contact and close proximity to them. Second, Wiggins had to be persistently aware of the great variability among witnesses in regard to the degree of attention they paid to the phenomenon being studied and the literary style they employed. For example, British travelers, many of whom were searching for large audiences and obviously seeking to appear omniscient, often recounted the leisure patterns and recreation of enslaved people based on nothing more than hearsay or tradition. Last, Wiggins had to be cognizant of the fact that many witnesses were far removed in time from the events they were describing. For example, many former enslaved people were at an advanced age when they recounted their days in slavery, a fact that not only tested their memory but also begged the question of whether their longevity was due to so-called "better" treatment than what other enslaved people experienced. In addition, they were often describing their childhood days, a time when their lives were perhaps less troubled by intense labor and horrors more characteristic of the institution of slavery as a whole.

details of Ali's membership in the organization but not a comprehensive view of the champion's religious beliefs. Ali's autobiography, *The Greatest: My Own Story*, provides insights into the champion's involvement with the Nation of Islam but virtually nothing about his strained and sometimes acrimonious relationship with both Elijah Muhammad and Malcolm X. Black newspapers covered Ali's relationship to the Nation of Islam in only a cursory fashion, seemingly either disinterested or unwilling to discuss the champion's involvement with the politically active and, some would say, racist organization. To piece together these outwardly incomplete sources and craft a portrayal of Ali's relationship to the Nation of Islam as accurately as possible, Wiggins had to pay particular attention to content, meaning, and historical context. This necessitated a close examination and comparison of written documents and a constant search for the biases that inevitably pervaded any discussion and analysis of Ali's involvement in and influence on the Nation of Islam. Wiggins also had to be speculative on occasion and draw historical generalizations that helped explain not only Ali's connection to the Nation of Islam but also how the great heavyweight champion fit into the larger Black Power movement and came to symbolize the tumultuous Vietnam Era.

Clearly, the sport historian's job is not as simple as having a good research topic grounded in the secondary literature.

> **The historian is still bound by his or her sources, and the critical apparatus with which he or she approaches them remains in many ways the same. Nevertheless, we view these sources more cautiously. We have become more aware of the extent to which they do not directly convey reality but are themselves narrative constructs that reconstruct these realities, not willy-nilly, but guided by scholarly findings and by a scholarly discourse. (Iggers, 1997, p. 10)**

In other words, today's sport historians should not be so bold as to believe that they can know exactly what happened in the past. Instead, they should attempt to understand what happened and how it fits into our broader understanding of the conditions that shaped both that time and the present. This understanding will certainly be influenced by the researcher's own experiences, the availability of sources, any biases of the resources, the underlying social dynamics of the time period studied (and the present because it affects the researcher), and the general epistemological and theoretical frameworks that guide the researcher and their subsequent research questions.

Research Findings

The limited information provided by each source is why all sport historians seek as much evidence as possible about their chosen topics. Once all the available evidence is collected and the sport historian has attempted to determine what it all means, the next job is to disseminate the findings to colleagues and, hopefully, a larger public. There is no specific formula for accomplishing these two tasks, but there are certain principles to which all good sport historians adhere when writing their final products. They present their evidence in as logical a fashion as possible, being careful to discuss as well as analyze and interpret the primary material within an overall theme. They present their evidence within the proper historical context, providing a meaningful framework for the data they have assembled. They must also consider the arrangement of historical data, recognizing the importance of establishing at least some form of framework that both reveals and clarifies the progression of events. And last, but certainly not least, they pay special attention to writing style and composition. Sport historians recognize, as do other scholars involved in qualitative approaches to the study of sport, that the way in which the researcher says something is just as crucial as the information they are imparting. Thus, the narrative, or how the researcher pieces together the data and results into a story, is absolutely critical. Words must convey exact meanings to avoid generalities and typification when possible, and to refrain from using artificial rhetoric. In many respects, because sport historians have adhered to this practice, writing in the field of sport history has remained some of the most accessible in the field of sport studies.

Exemplary Studies in Sport History

Now that we have reviewed the field of sport history and given an overview of how to go about doing research in this area, we now turn to reviewing several works that we believe exemplify the quality research that is being done in this subdiscipline. Three pathbreaking studies that address the complex meaning of sport and embody the best in scholarship are Rob Ruck's *Tropic of Football: The Long and Perilous Journey of Samoans to the NFL* (2018), Sean Dinces' *Bulls Markets: Chicago's Basketball Business and the New Inequality* (2018), and Maria J. Veri and Rita Liberti's *Gridiron Gourmet: Gender and Food at the Football Tailgate* (2019). Ruck's *Tropic of Football* provides a fascinating look into the presence of Americans of Samoan descent in arguably the most closely followed and popular of all commercialized sports in the United States. Like many of his previous works, including *Sandlot Seasons: Sport in Black Pittsburgh* (1987) and *Raceball: How the Major Leagues Colonized the Black and Latin Game* (2011), Ruck's book combines ethnography, history, sociology, and travelogue to tell the story of the growth of football among Samoans and how such a relatively high percentage of them made their way into the highest levels of college football and the National Football League. Ruck convincingly rejects the notion of biological determinism, attributing the enormous success of Samoans in football to the central tenets of Samoan culture, "the way of Samoa," which emphasizes hard work, competition, community, discipline, respect, tolerance of pain, and a Warrior ethos. There is much to recommend about the book, but perhaps what stands out most is how effective Ruck is in placing his topic in proper historical context, providing insights into the deeper meanings of Samoan life, and what Samoans have gained and lost from their devotion to football. What Ruck makes very clear is that in spite of the success of a select number of great Samoan football players, including the likes of Troy Polamalu and Junior Seau, there were literally hundreds of other boys who never came close to reaching the highest levels of the sport, with some of them undoubtedly at risk of contracting diabetes, obesity, kidney failure, and other health issues so highly prevalent in the community.

Sean Dinces' *Bull Markets* is decidedly different than Ruck's monograph, but no less impressive in concerning analyses, interpretations, and insights. Dinces relies on economic evidence in his analysis of the development of Chicago's United Center, which opened in 1995. It is not only a superb case study of urban sport, marked by convincing statistical and factual materials, but also a larger assessment of the so-called "new Gilded Age" playing out in the United States. *Bull Markets* examines the inimical effects that economic incentives can have on underserved communities. Like all good historians, Dinces depends on a wide variety of sources. By providing a thorough vetting of them, he makes a clear explanation of how a handful of disproportionately wealthy real estate developers, with the express consent of the municipal government, cunningly wrangled rents from local taxpayers, denying them the benefits they were promised in return for their capitulation to the new development. In fact, rather than providing alternate housing for the Near West Side population, the plan resulted instead in increasing rents and housing prices and putting more money into the hands of real estate developers and the municipal government. Importantly, Dinces also points out that the construction of the United Center is an example of how private-public urban sports are completed and accelerated, in this case, specifically by the arrival of Michael Jordan to the Bulls in 1984.

Finally, Maria J. Veri and Rita Liberti's *Gridiron Gourmet* is a highly creative, imaginative, and compelling study. Marked by the use of numerous kinds of sources, including cartoons, newspapers, cookbooks, television shows, and ethnographic and observational research, Veri and Liberti describe, among other things, the history of tailgating in the United States; elaborate cooking technologies and foods that are served; and the ways in which race, gender, and class play out on the blacktop. Perhaps most significant, Veri and Liberti astutely point out that tailgating initially emphasized feminine domesticity, but it would evolve into a hypermasculine activity by the 1970s. In fact, tailgating now provides men a "culinary cover" to carry out what was considered a traditional female activity—cooking—prior to and even after the big game. Enriching the book is the fact that Veri and Liberti bring themselves into the story, providing details and nuance to their analysis by conveying their interactions with tailgaters and observations during their visits to such football-crazed locations as Louisiana State University, Southern University, and the University of Tennessee.

Other studies that take more of a reflexivity approach are Dan Nathan's *Saying It's So: A Cultural History of the Black Sox Scandal* (2003); Peter Levine's *Ellis Island to Ebbetts Field: Sport and the American Jewish Experience* (1992); Douglas Hartmann's *Race, Culture, and the Revolt of the Black Athlete: The 1968 Olympic Protests and Their Aftermath* (2003); John Hoberman's *Darwin's Athletics: How Sport Has Damaged Black America and Preserved the Myth of Race* (1997); and Patricia Vertinsky and Sherry McKay's *Disciplining Bodies in the Gymnasium: Memory, Movement, Modernism* (2004). Of these studies, Levine's is the oldest, but it fits squarely into the reflexivity approach and indicates how important the present is to the study of the past. In the book's acknowledgments, Levine (1992, p. x) points out that historians "regardless of what they write about and however much they may seek to avoid it, always bring to their work their own values, interests, and beliefs. I am not uncomfortable with this fact." He then proceeds, through interviews, personal conversations and questions, and autobiographies of Jewish men between the ages of 60 and 80, to trace the involvement of Jews in American sport. Levine (1992, p. 7) ultimately concludes that participation in sport "both confirmed a meaningful Jewish identity while promoting assimilation and American acceptance." Perhaps even more important to Levine, however, was that the writing of the book "has been a chance for my father and myself to 'talk' in ways we never did when he was alive and for me to appreciate and accept what we had and didn't have together" and that hopefully the "story of the Jewish experience in American sport also has meaning for Jews today who debate whether or not Jewish life is still possible in the United States" (1992, pp. 275-276).

Summary

This chapter provided a brief overview of the subdiscipline of sport history, the theories underpinning research in this field, and a basic explanation of the process sport historians undergo to explore their research questions. In doing so, we have also brought to light some of the issues facing sport history as a field of study within the broader context of sport studies. We hope that we have identified the issues in a way that allows aspiring sport historians to continue undertaking interesting and worthwhile research projects that can contribute to this relevant and rewarding area of study.

Although it could be argued that the field of sport history has been marginalized to a certain degree, we would like to present several key points that we believe exemplify why sport history can and should continue to be an integral component of sport studies. As explained by Briesach (1994, p. 3):

> **Once we accept that human life is marked both by change as that which makes past, present, and future different from each other and continuity as that which links them together, we begin to understand why historians have played so central a role in Western civilization.**

We can only understand our present (and begin to ponder our future) with a greater understanding and appreciation of what occurred in the past. The process whereby sport historians have gone about achieving this understanding is, to a large degree, affected by a number of influences, including their epistemological and theoretical approaches, the context in which past events occurred, and the context in which they investigate their research questions. Thus, it can be argued that "historians are practitioners of an imperfect craft" (Iacovetta & Mitchinson, 1998, p. 13), and an awareness of this fact can only help them with their research endeavors.

We end this chapter with progressive statements from three prominent sport historians, Colin Howell, Steven Pope, and Patricia Vertinsky. According to Howell (2001, p. 16):

> **Any of you who know my work will be aware of my attempt to nudge sport historians away from their preoccupations with elite, urban, or metropolitan sporting experiences, away from sport at the national and international levels, to a study of hinterland localities and borderland regions.**

This comment reflects a desire to shift from the traditional acknowledgment and exploration of "great deeds of great men" toward a more inclusive historical paradigm. Similarly, Pope (1998, p. vii) notes:

> **Sport historians can no longer conceive of their works in a specialized, narrowly focused approach and expect to receive wide institutional approval. Historical case studies cannot be written in a scholarly vacuum; they must be relevant to contemporary issues and injustices.**

Finally, Vertinsky (2021, p. 130) notes the important role that sport historians play in kinesiology, the focus of this book:

> **Sport historians can help us to understand how and why ideas of balance have developed and shifted across time and cultural space in relation to the content and training best accomplished in kinesiology. They can illuminate how the transformative role of interdisciplinary collaboration, the contingency of knowledge development and creation, the political dimensions of the scientific and technological enterprise and the deep sociocultural situatedness of science and technology practices can all be productively brought to bear upon kinesiology teaching and research in the 21st century.**

With this in mind, we argue that the field of sport history remains an interesting, important, and rewarding field of study. By adhering to Howell's, Pope's, and Vertinsky's advice and following some of the basic research tools outlined in this chapter, aspiring sport historians may carve a significant niche in this foundational area of sport studies and produce work that signifies its continual relevance within the academy.

☑ Check Your Understanding

Find and read a historical study on some aspect of sport and physical activity.

1. Describe the research design.
2. Identify the primary sources.

13

Philosophical Research in Physical Activity

Tim Elcombe
Wilfrid Laurier University

Scott Kretchmar
Penn State University

> A great many people think they are thinking when they are merely rearranging their prejudices.
>
> —William James

To begin this chapter, consider how you might respond to the following questions:

1. Should athletes with certain advantages be excluded from competitions if those advantages are unearned and likely to directly translate into success?
2. Should we create as many categories in sport as necessary to ensure competition is fair?
3. How should the conditions that determine athlete eligibility (or relatedly, athlete categorization) be defined, and by whom?

These questions are fundamental to two significant and historically durable challenges that regularly arise in sporting contexts: Where (and how) do we draw the line between fair and unfair advantages; and On what grounds do we decide who may or may not compete in a certain category (if at all)? In the 21st century, these questions are best reflected in debates over athletes "gaining" advantages: for example, Oscar Pistorius, who competed in the Olympics using sophisticated prosthetics, or the justification for banning certain performance-enhancement practices in the Tour de France. Another example is the case of South African runner Caster Semenya. A women's division world champion middle-distance runner, Semenya races with naturally elevated testosterone levels due to differences of sex development (DSD). The governing body of track and field, World Athletics (formerly International Association of Athletics Federations, or IAAF), has decided this advantage is unfair to other female runners and created DSD regulations that exclude female athletes with testosterone levels above 5 nmol/L from competing in women's 400 m to 1-mile events—a policy upheld in 2020 by the equivalent of sport's Supreme Court, the Court of Arbitration for Sport (CAS). See the IAAF Eligibility Regulations for the Female Classification (DSD Regulations) sidebar for more information.

Last night, I went to a comedy and philosophy club. I laughed more than I thought.

IAAF Eligibility Regulations for the Female Classification (DSD Regulations)

In 2018, the IAAF (now known as World Athletics) enacted a new policy restricting the eligibility of athletes with differences of sex development (DSD) to compete in women's races. This policy was crafted largely in response to the domination of women's international middle-distance track events by South Africa's Caster Semenya—a runner socialized as a woman and with external female genitalia but born with internal testes and an XY chromosome sequence. Semenya's "intersex" characteristics result in testosterone levels well above the "normal" upper range for women (1.79 nmol/L) in the general population, yet below (or at the lower end of) the "normal" range for men (7.7-29.4 nmol/L). The new policy, simply stated, requires all athletes competing in IAAF-sanctioned women's races from the 400m to the 1 mile to exhibit testosterone levels below 5 nmol/L.

Semenya challenged the IAAF's new regulations at the CAS, arguing that the performance-enhancement research was flawed, and the policy violated her human rights. CAS ruled 2-1 in favor of the regulation in 2020, acknowledging the discrimination inherent in the policy but determining it necessary to protect women's sport. Athletes with DSD now must face the choice to take medical measures—drugs or surgery—to reduce their testosterone levels below the 5 nmol/L threshold or stop participating in women's competitions.

The response to CAS's ruling has been varied. Many of Semenya's competitors, discouraged by competing against athletes with the advantage of elevated testosterone levels, welcomed the decision. More broadly, many sporting officials viewed this as a win in the battle to "protect" women's sport. Others decried the sacrifice of human rights for sporting integrity and noted the potential negative impact on young women with DSD (in sport and beyond). In addition, the World Medical Association publicly urged physicians not to participate in the implementation of medical methods to reduce testosterone levels of women with genetic variations below 5 nmol/L for sporting rather than health reasons.

Throughout the chapter, we will refer to these important sport dilemmas: athlete categorization and limits to performance enhancement. Debates about athlete categorization and performance-enhancement bans not only relate to sport but also cut across the domains of inquiry we find in kinesiology, physical education, and sport and exercise science programs. Furthermore, these questions transcend sport and relate to broader deliberations about human rights, social justice, paternalism and individual health, and the role and limits of science—to name just a few. These issues, therefore, reflect the complexity of the kinds of questions we describe as "normative" and will be used to help us understand philosophical research methods—tools of inquiry important not just for philosophers but also for professionals of all kinds working in sport- or physical activity-related fields.

Let's consider some more questions, related to the ones posed at the start of the chapter, and in the context of the Caster Semenya case:

1. Should athletes who identify as female, have external female genitalia and naturally produce elevated levels of testosterone, be excluded from events where those advantages are believed to directly influence the outcome of competition?

2. Should we create new categories to ensure athletes with DSD can participate in IAAF events (similar to the Paralympic model) while preserving the competitive balance in the traditional women's division? Or should athletes with DSD simply be allowed to continue to compete in the women's division?

3. Do we need to treat athletes with DSD as "rare and exceptional" talents or as potential disruptions to the best version of competitive (women's) sport? On the other hand, should we worry less about the sporting implications of athletes with DSD and focus more on the wider social implications of categorization and restricting eligibility?

4. How should we address these questions? What information should we use? Who should be responsible for, or have the most influence on, final decisions regarding athletes' eligibility?

These are the kinds of questions we are tasked with answering using philosophical methods. In this chapter, we invite readers to develop both an appreciation for, and working knowledge of, philosophical research methods. The DSD regulations serve as a useful tool in the first half of the chapter to provide an example of real-world applications of the philosophical concepts we highlight, including the modern purposes of philosophy; the distinctions between objective, subjective, and normative inquiry; and differences between traditional areas of philosophical inquiry, including epistemology, metaphysics, and axiology. Along the way, you will be asked to participate in some exercises related to these philosophical domains and methods. Based on an article by Tamburrini (2000) that outlines a provocative position in the contemporary sports ethics debate over doping, you will be invited to stake out a nuanced and informed position on performance-enhancement bans. In this latter portion of the chapter, you will (with prompts from Professor Tamburrini's article) see the different ways a fourth area of philosophical inquiry—logical analysis—can be used to arrive at reasoned answers to difficult questions.

First, however, a few words of introduction are needed.

Unfortunately, some consider *philosophical research* a contradiction of terms. This viewpoint rests on the rise of empirical science, historic relationships of physical activity professions with medicine, and widespread contemporary doubts about the validity of reflective, reason-based procedures (Kretchmar, 1997, 2005). Some believe philosophy involves little more than establishing personal or professional "philosophies" or sharing opinions, albeit while using impressively long sentences and, sometimes, incomprehensible words to do so. Critics point out that philosophers themselves do not agree on the utility of their methods and the validity of their results. They suggest that nearly all philosophers today are necessarily **fallibilists**. That is, even the best and sharpest minds in philosophy admit (to varying degrees) that their evidence is often incomplete, and conclusions may be limited.

> **fallibilism**—A theory about limitations in seeking knowledge. It suggests that final and complete evidence is rarely, if ever, available for any truth claims. Because of this, the quest for knowledge is an ongoing process, and any specific claims are open to revision.

These criticisms and admissions have not proven fatal for philosophy or philosophers. In many ways, all researchers studying physical activity sail on the same boat. Most scientists, if honest, would acknowledge the fallibility of their own work. As we know all too well, many scientific "discoveries" or "findings" related to health, diet, safety, training, and sporting performance are regularly revisited and, just as regularly, revised. As we've already mentioned, scientific paradigms change. For example, Newtonian physics opens certain doors to understanding, but quantum and chaos theory provides a very different platform from which to view the world. New technologies change how and what things can be measured. For instance, the genome project was hardly conceivable 50 years ago. Research protocols evolve. Standards for verifiability or significance are contested and change from time to time. To illustrate, for some research, a p value of .01 is needed. For other research, it is .05. For yet other research, scientists disagree with one another over which p values should be required. In the case of Caster Semenya, there are debates and disagreements among scientists as to what levels of testosterone lead to a clear performance advantage in middle-distance running events—or, even more fundamentally, how to best use data to determine who "counts" as an eligible female athlete.

In short, the unavoidable limits of inquiry affect scholars across the subdisciplines involved in physical activity research. Philosophy is unique because it does not literally measure anything, gather data, or use p values. Relying heavily on reflection to address epistemological, metaphysical, axiological, and logical issues, philosophical inquiry faces distinctive challenges in justifying its methodologies and asking others to trust its conclusions. But once again, we will invite you to try them out rather than assume they do not work or, alternately, take it on faith from us that they can be effective.

Historical evidence suggests that, despite its intangible subject matter and fallibilistic nature, philosophical inquiry does real work in the real world. Philosophy remains an esteemed subject taught at virtually all colleges and universities around the world; best-selling books worldwide often deal with ethics, philosophy, wisdom, religion, and tips on how to live the good life. Casual conversations on the street corner or serious discussions in the halls of academia invariably turn to ethics—that is, to claims about how things should be, not how they are. Exercise scientists are well prepared to analyze the various mechanisms and particularities of performance optimization in classes and within their research. But students (and colleagues, friends, and community members) regularly want to talk about whether natural or acquired advantages are fair or unethical. Engaging informed and critical arguments against and in favor of eligibility criteria or doping bans requires a different kind of expertise and form of inquiry. Even books like this one that attempt to describe techniques for good research in a wholly objective and dispassionate way cannot avoid ethical issues. The editors, in fact, devote an entire chapter to ethical issues in research (see chapter 5).

At one time, virtually every physical education, exercise science, or kinesiology program taught philosophy and ethics. Today, many schools, for economic, pedagogical, or other reasons, consider their many science-based courses as not only necessary for physical activity professionals but also sufficient. We think this is unfortunate. While we agree that a full complement of science-based coursework is necessary, it is far from sufficient—a conclusion reached by many others (Fahlberg & Fahlberg, 1994; Glassford, 1987; *ICSSPE Bulletin,* no. 27, Fall, 1999; Kretchmar, 2005, 2007; Lawson, 1993; Sage, Dyreson, & Kretchmar, 2005; Anderson, 2001; Sheets-Johnstone, 1999). Developing a philosophical toolkit is important for more than professional philosophers—it is crucial for all who are invested in physical activity.

Identifying the Purposes of Philosophical Research

The purposes of philosophy have changed over time. In its earliest forms, for instance, in the Ancient Greek works of Plato and Aristotle, philosophy's role was to provide answers to virtually all of life's questions and offer solutions to most of its problems. Indeed, before the rise of science in the 15th and 16th centuries, most scholarship in the West was produced by philosophers working independently as secular thinkers, or as was often the case, under the influence and support of the Church. Philosophy's purposes reflected its expansive ambitions. These aspirations were seen to be both synoptic (dedicated to answering the so-called "big questions," pulling things together, and trying to make sense of the whole) and sufficient (considered independently capable of generating all-important truths). Over time, thinkers such as Copernicus, Kepler, Galileo, and Newton reconceptualized the world as a complex mechanism, a "self-governing machine," an entity to eventually be fully understood in geometrical and arithmetic, rather than grand philosophical, terms (Tarnas, 1991, p. 263).

With the evolution of the scientific method throughout the 17th century, philosophy needed to explain how and why reflective methodologies still capably provided avenues to important insights—that is, truths and plausible hypotheses inaccessible to physicists, chemists, and other scientists using their empirical methods. While its purposes continued to be synoptic and it still attempted to answer big questions related to the nature and purpose of human existence, philosophy now shared the stage with the many branches of natural science, the social sciences, and the other humanities like history and literature.

The revised purposes of philosophical research acknowledged a fundamental fact. Research today, including inquiry that is related to physical activity, comes in many shapes and forms. As one of the authors of this chapter has argued, inquiry best happens in three dimensions across disciplines (Elcombe, 2018). One dimension is the creation of abstract theory, often only comprehensible to a limited number of professional colleagues—think here of the incomprehensibility (to most) of theoretical physics or the complex and challenging philosophical ideas developed in Rawls' theory of distributive justice. A second dimension is applied inquiry. This can take place within a disciplinary framework as researchers apply concepts, theories, and methods to tangible problems. To maximize efficacy, researchers strip their studies of their real-world complexity, define and control dependent and independent variables, and proffer clear results. Studying the effects of testosterone on aerobic efficiency, for example, is best done in a lab to standardize the analysis and increase the experiment's validity and reliability. When we think of inquiry, we usually think of these two dimensions.

The third dimension, however, is necessarily "messy." This is where real-world problems are addressed in real time. Complexity and variance cannot be controlled, but problems demand working resolutions. In these situations, inquirers can work largely independently and later come together to attempt to fit their research puzzle pieces together. Or they can collaborate before designing research projects and possibly develop more effective studies because of this. The former has been identified as interdisciplinary collaboration, the latter as cross-disciplinary study (Kretchmar, 2005, 2008). The various chapters of this book, in effect, argue that no single discipline and no single methodology can go it alone. Rather, a combination of efforts and insights of researchers from any number of research silos will solve complex human problems.

Philosophy has both descriptive and prescriptive roles to play in helping resolve real human problems. Philosophers reflect on concepts from a broad perspective in the attempt to describe existence as accurately as possible. Reflection is critical; problems might exist with the current state of affairs, our beliefs, and our behaviors. When these issues arise, philosophers should not merely describe these problems but also recommend, counsel, urge, cajole, convince, and otherwise promote more culturally attuned values or more ethically justifiable behaviors. In other words, philosophers need to go beyond describing existence and initiate inquiry to prescribe how we *ought to* live and how we *should* act. Descriptive inquiry can be further divided into objective and subjective species (Morgan, 2007). Prescribed values and behavior fall within the purview of normative inquiry. In the following section, we discuss in further detail the various forms of inquiry.

Objective Inquiry

Objective inquiry is interested in describing concepts in a way that is independent of idiosyncratic experience, personal preference, or individual perspective. It is the most accepted and recognized form of inquiry and is shared by science and philosophy alike. It typically analyzes phenomena like health, fitness, and physical activity through the utilization of empirical methodologies: in particular, looking through microscopes at muscle tissue, collecting respiratory or blood pressure data, measuring testosterone levels, and employing statistical procedures to determine the strengths of possible causal relationships between group-based participation and rates of exercise adherence. Objective inquiry ideally elicits facts or laws that describe existence as accurately and universally as possible. For example, water freezes when it reaches a temperature of 0 °C is a universally accepted fact. Another example is the claim that competitions must be structured to allow participants of relative ability to measure differences in achievement. Generally regarded as logically required or, as philosophers would say, true *a priori*, it too might be identified as a fact about the world, or in this case, about how sport works.

Objective inquiry, as the theory of fallibilism suggests, seeks, but rarely achieves, impartial factual accuracy and macro-level generalizability. Most facts, particularly when related to human life, undergo revision and generate debate as ongoing empirical and reflective analyses reveal new evidence, challenge previous interpretations of the results, and lead to the emergence of alternate conceptual frameworks.

objective inquiry—An approach to producing universal knowledge that accurately describes reality by eliminating biases from language, history, socialization, or any other influences that would relativize truth claims to a time, place, or personal perspective.

Subjective Inquiry

subjective inquiry—A form of inquiry that takes personal experience and individual preferences seriously, speaking to the fact that different people are not entirely alike and that individual perspectives and personal backgrounds matter.

Subjective inquiry, like objective inquiry, emphasizes description. However, subjective inquiries seek to explain individual or small-group preferences by examining specifiable, micro-level variations and personal contexts. For instance, subjective inquiry may explore whether and why certain people prefer running to cycling, or chocolate ice cream to vanilla. A more critical application of subjective analysis might assess the quality of a dance performance or theater production. The validity and reliability of subjective inquiry depends entirely on the authenticity of beliefs articulated by the person(s) under study or providing the critical assessment, rather than the repeatability of the methods and generalizability of the results. A result is true in subjective inquiry if the person or small group presents an authentic expression of their own lived experience. It makes sense, in other words, that some people might prefer running to cycling or chocolate ice cream to vanilla even though it makes equal sense that others might prefer the opposite. While debate and disagreement still occurs in subjective inquiry, a lot of leeway must be given for individual preference claims. Consequently, subjective inquiry that explores the realm of opinion and preference is regularly considered less scientific than objective inquiry that trades in facts and neutrality.

Normative Inquiry

normative inquiry—A form of inquiry that seeks to challenge, reinforce, or apply a culture's norms, or ideal standards, which serve to recommend behavior, promote action, and indicate a way in which life should be lived.

Normative inquiry is the form of philosophical inquiry that moves from describing existing conditions to prescribing measures that may alter or improve the current status quo. It at once acknowledges why things are as they are while simultaneously responding to the widely held view that things could be different and better. The process of confirming, establishing, and revising norms requires constant analysis to ensure the beliefs and values of a culture or social phenomena (like sport) are best reflected in its practices. When norms are in place, they function as ideal standards to use when making evaluations about good and bad, right and wrong, or better and worse.

A crucial distinction between normative and subjective inquiry rests on the degree to which the implications of beliefs or behaviors extend beyond mostly innocuous individual preference. Choosing to wear a favored blue singlet instead of a red one in an 800 m track competition hardly matters to anyone except the runner. On the other hand, deciding to covertly use a banned substance in the same race deeply affects others. Crafting a sporting policy that excludes some women from participation in the race will affect even more people. In other words, purely subjective inquiry describes idiosyncratic preferences; normative inquiry seeks to prescribe justifiable positions.

Normative inquiry also differs from objective inquiry because its recommendations are not likely to generate agreement from all parties and for all time. For example, objective inquiry in science may "factually" conclude that a completely sedentary lifestyle increases the risk of type 2 diabetes. Objectively oriented philosophers may conclude with similar levels of consensus that sports include tests and contests that stand in a certain relationship to one another (see Philosophical Exercise 1). But it is improbable that everyone would agree with normative claims that sedentary people, as a consequence of their lifestyle choices, should be denied the same level of health care provided to physically active people; or, that in the realm of physical activity, the IAAF should not be concerned that Caster Semenya dominates middle-distance races, because other female competitors can still run personal best times even if they always lose.

Although philosophers may engage in both objective and subjective inquiry, normative inquiry is arguably the most important contribution of philosophical research to human life. Data about a phenomenon derived from the descriptive methodologies of objective inquiry stop short of prescribing what participants in a culture can (or should) do. A physiologist can provide objective facts about the effects of testosterone on aerobic performance. However,

Philosophical Exercise 1

Logical Relationships Between Sport Tests and Contests

In an article on Caster Semenya, Sigmund Loland (2020) identifies drama and uncertainty as important values in sport. He mentions them in the context of the Semenya case because he believes decisions on her eligibility might affect these values, specifically in relation to the drama of competition.

Loland's identification of drama in the contest is consistent with earlier research by Kretchmar (1975 & 2005), who argued that sport actually contains two sources of drama and uncertainty. One of them, as Loland noted, is the contest; for example, trying to reach the finish line before the other runners in the race. The drama and uncertainty here hinge on winning or losing the contest. The other source of drama and uncertainty is the test; for instance, attempting to complete an 800 m run in a certain time, which is something that can be done alone. The drama and uncertainty here would reside in passing or failing the test. The question at hand then is: How are these two sources of sporting drama and uncertainty related?

The following three Venn diagrams show possible relationships between tests (A) and contests (B). Which diagram do you think is the most accurate and why?

Diagram 1

- Some tests are not also contests (area A).
- Some contests are not also tests (area B).
- Some tests are also contests and vice versa (the overlapping area).

Diagram 2

- Some tests are not also contests (area A).
- All contests are also tests (area B).

Diagram 3

- Some contests are not also tests (area B).
- All tests are also contests (area A).

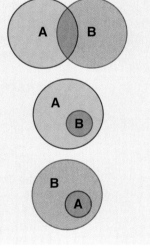

on their own, these facts do not provide an answer to the question, "Should female athletes, but not male athletes, with naturally elevated testosterone levels be excluded from competing in the 800 m in their respective gender categories at the Olympic Games?" Furthermore, the response to this question, no matter what that answer is specifically, has wide-ranging implications to all potential women's 800 m competitors, as well as all athletes born with natural physical advantages or people with DSD. Thus, normative inquiry goes beyond the description of preferences that subjective inquiry provides.

This is not to say objective and subjective descriptive analyses play no role in normative inquiry. Facts emerging from objective inquiry offer important evidence to consider when formulating normative positions (e.g., To what extent will elevated testosterone levels in women affect outcomes in an 800 m race?); subjective preferences inform normative inquiry of the host of values people hold relative to the issue under consideration (e.g., Would people prefer to see world record times challenged by rare and exceptional individual athletes or to watch slower but more closely contested 800 m races?). Philosophical research is needed not because empirical methodologies are ineffective but because they are, in isolation, insufficient. At the same time, philosophical inquiry, when dealing with existing (rather than abstract) problems, is similarly valuable yet insufficient and requires both objective and subjective inputs.

Philosophical Inquiry Continuum

Philosophers who tend to have more objective or subjective leanings when engaging in philosophical inquiry do so by employing reflective, rather than empirical, methods to explore their topics of interest. This distinction between objectively and subjectively oriented inquiry points to one key purpose of philosophy: the investigation of how we obtain knowledge and the degree of assurance we can claim that our findings are factual (or more broadly, what count as warrantable assertions). In some ways, this purpose undergirds normative inquiry, as well as metaphysics, axiology, and logic—all of which we explore in the following sections. This key purpose is foundational because extreme positions make a significant difference in the efficacy of philosophy itself. At one extreme are philosophers who are categorized as skeptics or relativists. They are radically subjective oriented and believe that no knowledge or truth exists "out there." Thus, they regard the role of philosophy as one that reinforces the validity of group preferences and different cultural perspectives. At the other extreme are philosophers who belong in the category of absolutists or realists. They are radically objective oriented and believe that reason is both powerful and capable of uncovering enduring knowledge and truths about human existence that provide insights on how people should live. Between the two extremes are the pragmatic philosophers—those who believe philosophical insights are valuable but limited. This ongoing debate is inherent to **epistemology**, or, in common terminology, the *theory of knowledge*.

epistemology—A branch of philosophy that addresses securing knowledge and understanding its status.

Epistemology

When you reflect on a difficult normative problem, such as the categorization of athletes with DSD or the validity of performance-enhancement bans (discussed in further detail in the upcoming section), you will have to decide where you stand on this matter of epistemology. Figure 13.1 shows three broad positions along a continuum from minimal to maximal confidence. We begin with those who place minimal faith in the power of philosophical reasoning and believe that objectivity is unattainable and truth finding is not philosophy's purpose.

Minimal Confidence: Skepticism, Nominalism, and Relativism

Philosophers at this end of the continuum are pessimistic about the validity of objective knowledge, or truth claims, for two fundamental reasons. The first has to do with the world and what is to be reflected on. Skeptics and nominalists doubt that the world is composed of neat and tidy objects simply "out there" waiting to be discovered. They assert that these objects have no firm nature; in other words, what appear as objects are instead internal constructions shaped by the human brain. These philosophers are nominalists because the identification of things in the world amounts to nothing more than an arbitrary process of naming chunks of reality. Thus, for skeptics and nominalists, a phenomenon like sport is not

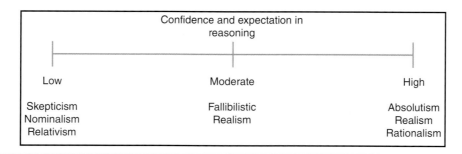

Figure 13.1 Epistemological confidence: from minimal to maximal.

simply one distinct thing among others; it is built and rebuilt by us, by those who perceive it. In other words, if we wanted sport to look different, we could change it immediately. Sport, from this viewpoint, is socially constructed, not found. Such constructions are subjectively controlled, therefore, objective truth about sport is unattainable.

The second has to do with oneself—that is, the power of reflection itself. Philosophers with minimal confidence in the power of reason, or relativists, doubt the validity of objectively oriented inquiry by pointing to themselves rather than the world. While a stable, objective world awaiting discovery might exist, relativists claim that we lack reliable access to it. Language, socialization, learning, idiosyncratic perspective, and all other manner of reflective contamination prevent us from trusting our powers of reasoning. Anyone's claims about sport, for instance, would be relative to their own perspective or upbringing. Once again, no truths or grand narratives about sport can be found—only authentic, personal, or small-group experiential accounts.

Maximal Confidence: Absolutism, Realism, and Rationalism

The opposite epistemological position is often practiced by absolutists, realists, and rationalists. These thinkers, unlike the skeptics and relativists, assume both physical and non-physical entities exist independently of the perceiver. Valid truth claims thus correspond to the way things really are in the world. In other words, it goes beyond what an individual person thinks about some phenomenon. There is a truth about the phenomenon that is the same for everyone for all time. Sport, from this perspective, has an essence in need of discovery and clarification—and that essence cannot be altered for an activity to still be considered sport. Consequently, realists and rationalists place a great deal of confidence in the power of logic, or reasoning, to solve philosophical problems. Rationalists perceive the world as one giant math problem. And just as mathematicians can be sure that $1 + 1 = 2$, more complex problems will be solved in the future with equal certainty. They believe there are right answers and, like many empirical scientists, argue that eventually these truths will be uncovered.

Moderate Confidence: Fallibilistic Realism and Pragmatism

Many philosophers today adopt a moderate position, believing that the world is reasonably durable even if it is not fixed. Pure objectivity and pure thoughtfulness are considered unhelpful abstractions. Because of this, philosophers need to return to their objects of inquiry, study them repeatedly, and thereby acknowledge both their dynamic qualities and inherent difficulties in knowing them completely. Sport is a more or less durable thing, but it may evolve along with the people who play it, and it is extremely complex. Likewise, reason is powerful, but there is no Archimedean or external vantage point from which to view something like sport. As noted in the introduction, most philosophers today acknowledge that reason is inherently and significantly limited. And because of this, philosophers are inherently fallible. While philosophers make progress, there will always be mistakes requiring correction, and there will always be more to be seen, described, and recommended.

As you can see in the first sections of this chapter, all inquirers should retain a degree of humility with respect to their methods and the applicability of their findings. But at the same time, they must have enough confidence in their research methodologies to make a positive and useful difference. These philosophers are often considered pragmatists for taking a moderate view of the power of philosophical analysis.

As a traditional branch of philosophy, some professional philosophers are interested in engaging in extended debates about epistemological positions. Most of us are not. Even so, different degrees of epistemological confidence will affect the interest we have in engaging in philosophical inquiry and the trust we put in any philosophical conclusions. Low confidence levels discourage philosophical reflection by portraying it largely as a waste of time. Low confidence levels also reduce the stock we place in philosophical conclusions. It would be hard to trust conclusions, after all, if we didn't trust the methods that produced them. High levels of confidence, on the other hand, have the opposite effects of encouraging reflection and increasing our trust in products of those reflections. But overconfidence in philosophical methods, like in other disciplines, may lead to unhelpful or even counterproductive conclusions.

Given the importance of epistemological attitudes, you will have two opportunities to determine your own level of epistemological confidence. In Philosophical Exercise 2, you will see if you can determine, in relationship to issues raised by the Caster Semenya case,

Philosophical Exercise 2

Are Sport Contests Fair?

Sports are said to be different than life in important ways. A person can be born with advantages that affect their future prospects—advantages based on genetic, economic, ethnic, gender, and citizenship lotteries, among others. Effects of these chance events are said to be unfair because they produce an unlevel playing field and because success is more often granted than earned. Moreover, in any contest between those who are lucky and those who are less fortunate, the outcome is easy to predict. Those who start life with an advantage will keep it.

Sport, argues Loland (2020), should not be like this. It should be designed to neutralize initial or inborn advantages and generate contests whose outcomes are earned, not granted. This is why Caster Semenya's case is so interesting. Her participation in certain women's events would seem to make the competition unfair . . . or at least significantly less fair. However, fairness is a function of many rules, restrictions, policies, and other strategies.

Consider the following steps taken by sport organizers to neutralize unearned advantage and ensure relatively even contests. Do they guarantee fair games or not? Are they applied consistently or not?

1. Using competitive divisions to neutralize biological differences and promote even contests (e.g., weight classes in boxing and wrestling, age-based categories for many sports)
2. Establishing competitive divisions to neutralize school size and promote even contests (e.g., Class AAA, AA, and A in high school sport leagues based on school population)
3. Writing rules that standardize legal preparations for contests (e.g., permitting Olympic athletes to choose their own coach and use preferred training facilities prior to the games)
4. Enforcing rules that neutralize in-game advantages (e.g., changing ends of the court or field at halftime, switching lanes in bowling)
5. Establishing rules that neutralize home-field advantage (e.g., home scheduling)
6. Using penalties to neutralize advantages gained from breaking the rules (e.g., yardage assessments in football for holding, no calls or penalties for flopping in basketball)

Based on your analysis these six fairness-promoting strategies, mark a spot on the scale below indicating your response to the following statement: Sport is constructed, organized, and conducted in ways that successfully ensure fairness:

(5)	(4)	(3)	(2)	(1)
Strongly agree	Moderately agree	Uncertain	Moderately disagree	Strongly disagree

What is the logical relationship between your judgment here and normative decisions on whether Semenya should be allowed to compete in women's competitions? If a high degree of unfairness already exists in sporting competitions, and if sport does not seem to be significantly harmed by it, and if it is virtually impossible to make sport perfectly fair, what follows regarding Semenya's case for inclusion? Do these factors strengthen or weaken her case . . . or perhaps have both effects?

what counts as fair competition. You will learn whether you can support the empirical claim made by some that sport effectively ensures fair play and whether answers to this question logically affect normative decisions about Semenya's eligibility to compete in women's competitions.

In the final section of this chapter, you will be guided through Tamburrini's arguments for eliminating current rules that ban the use of steroids and other performance enhancers. Tamburrini believes his arguments are persuasive; in other words, he has a reasonably high degree of epistemological confidence. You will need to see if his arguments strike you as forceful or not. If not, you should decide if other more valid arguments exist. If you agree with Tamburrini or counter with other persuasive arguments, you embody fairly high degrees of epistemological confidence. On the other hand, if you decide this whole issue boils down to nothing more than personal preferences and opinions, you exhibit attitudes of those with fairly low levels of epistemological confidence.

Metaphysics

Epistemological commitments directly influence how much confidence a person places on a second key branch of philosophical inquiry, **metaphysics**. The term *meta* refers to comprehensiveness or transcendence. Metaphysical inquiry, which traces back to Plato and Aristotle, is the process by which philosophers attempt to analyze the nature of things to transcend, or move beyond the physical, and to fully and comprehensively understand the essences. For example, if we asked you to close your eyes and generate an image of a "horse," everyone would be able to do so—but no two images would be the same or even fully describable. Yet we would all have a vision of what a horse is. Therefore, metaphysicians could argue that essential features that define the category "horsiness" exist even though we cannot feel or weigh an insight or idea.

A sport philosopher may, for instance, reflect on the conceptual features of sport in relation to games, or compare and contrast sporting activities to dance performances to offer clearer and more generalizable descriptions of differences between sport and art. In its traditional form, metaphysical analysis tends to be objectively oriented. Like the empirical scientist, metaphysicians carefully examine some phenomenon and try to describe it as accurately as possible. Unlike empirical scientists, however, they cannot use tools to measure, examine a response to some stimulus, or collect surveys to explain something like sport. Metaphysical inquiry relies on how compellingly philosophers articulate their reflections in relation to some reality that exists beyond the physical, which often results in disagreement on the essences under investigation.

The scope of this work runs the gamut from simple distinctions (e.g., contrasting sport from dance) to more complex and controversial theses (e.g., describing the nature of excellence or fairness in competitive events). In the case of Caster Semenya, the essence of "femaleness" is up for debate. In the past, female athletes were distinguished from male athletes based on observed biological (genitalia) differences. As scientists developed an understanding of genomes, XX or XY chromosomal profiles defined male and female distinctions. Current standards for sport governing bodies such as the IAAF employ testosterone levels as key indicators of femaleness or maleness. Standards for distinguishing male and female athletes continue to evolve, and this would seem to raise doubts about any metaphysical conclusions that identified male and female as fixed, binary gender categories.

Nonetheless, some metaphysicians with very high confidence levels would argue that essential differences do exist between genders. On their view, binary categories can be crafted through careful, objectively oriented analysis. To develop these categories, philosophers would reflectively consider what defines femaleness, and what distinguishes it from maleness. This is a similar process that you used in Philosophical Exercise 1 to identify differences between tests and contests. Without this metaphysical spadework, these philosophers

metaphysics—A branch of philosophy that addresses the nature of things.

would argue that using testosterone levels (or genetic composition) may be irrelevant for deciding who is or is not an eligible female athlete. Testosterone levels must be an objective measure connected to the essence of femaleness if it is to be a useful marker for eligibility. The metaphysical analysis, in other words, must come before the objective tool development.

Alternately, some philosophers reject metaphysics entirely, and instead focus their energies on challenging the categorization of the world. These anti-metaphysicians (subjectively oriented philosophers) might argue that gender and sex are complicated social concepts with a wide range of variance—therefore the creation of artificial and fixed categories based on a single factor is more problematic than useful. A pragmatist would likely agree that categorization is complicated but can still be useful when considered as family resemblances with fuzzy or blurry lines of demarcation open to constant revision (rather than as clear and distinct differences). Those who occupy this middle position might agree that categories related to sex or gender or both can be beneficial in sport, but such categories would need to be carefully crafted and adaptable to change to address the complex sporting and human issues at play.

Axiology

axiology—A branch of philosophy that addresses the value of things including proper behavior and the nature of the good life.

Philosophers also engage in a form of inquiry that combines subjective and objective approaches. In traditional philosophical language, this third branch of philosophy, **axiology**, focuses on inquiry related to values. Axiological work explores the values and meanings attached to (a) achievements, acquisitions, or states of affairs such as fitness, health, knowledge, and excellence (nonmoral value); (b) the display of appealing human qualities such as good sporting behavior (moral value); (c) the respect for the social tenets of fair play (ethical value); as well as (d) the dramatic or sensuous qualities of movement (aesthetic value). All four brands of axiology—theories of nonmoral, moral, ethical, and aesthetic values—take personal opinion and preference seriously. Thus they are necessarily subjective in nature. Yet all four types of reflection on value also aim at a degree of objectivity. For instance, while many different behaviors might promote fair play, not every behavior qualifies. While many elements of sporting experiences are beautiful, not everything can be described this way. Furthermore, some philosophers believe values are universal—the same for everyone if we dig deep enough. Alternatively, others argue values are always situated—relevant only to certain people in certain places at certain times. Thus, axiology involves an interesting blend of, and tension between, objective and subjective description.

This unique objective-subjective reflective approach in axiology allows philosophers to pose and answer questions empirical methodologies cannot (or do not) address. For instance, after chemists, physiologists, sociologists, psychologists, and historians gather and analyze data on exercise, questions still remain about the human meanings and values associated with it. Why should we exercise? To live longer? To live better? Both? If both, which is more important, the presence of more life per se or the existence of a certain quality of life? On what criteria would we determine quality of life? Would it be freedom, power, wealth, novelty, love, knowledge, security, adventure? Something else? Each person might respond differently to these questions with no "correct" answer in sight. Can people be "wrong" for exercising to extend their lives more so than to improve the daily quality of their existence? Is it "correct" to prefer cycling to running for the goal of developing aerobic capacities? Ambiguous responses to these questions show that we need to respect personal perspectives on the one hand and acknowledge the limited range of lifestyles and ethical choices that genuinely merit respect on the other.

Much philosophical ink has been spilled on logical problems of moving from "is" to "should." However, most modern philosophers believe that the mandate for action lies within the nature of the value or behavior that is being described with careful, axiological analyses. For instance, if our axiological reflections showed that cheating makes it difficult to determine

I'm working out my philosophy of number theory scoring for chasing and catching cats.

© Zharastudio/fotolia

how well an athlete played, and if this knowledge of how well that athlete played is on the list of noncontroversial sport values, it is clear that athletes should not cheat. If we value sport as a space where humans push the boundaries of possibility, then we are less likely to be concerned about advantaged athletes dominating lesser opponents.

The move from description to prescription, in other words, is not hard to negotiate.

Normative recommendations parallel the axiology areas for descriptive analysis listed in this section. Thus, *should* statements might accompany the following list of four value-related topics. Here is how the conversion from description to prescription would look:

- Because nonmoral values like excellence, knowledge, and happiness are good, people *should* seek to realize such nonmoral values in physical activity.
- Because moral values like prudence, determination, honesty, and respect are good, people *should* seek to realize moral values in physical activity.
- Because ethical values of playing fairly and attempting to avoid injuring an opponent are good, people *should* seek to employ and model these ethical values in physical activity.
- Because close, dramatic finishes are exciting and because sports like diving and figure skating include any number of beautiful physical movements, people *should* promote aesthetic values in physical activity.

If you have not already done so, you are encouraged to read an example of axiological argumentation: Tamburrini's article "What's Wrong with Doping?" (2000). This will also be helpful as you work through the next two sections of this chapter on locating and analyzing a research problem.

Locating a Research Problem

Philosophical problems are not difficult to find. They are literally everywhere. You cannot wake up in the morning without being confronted with potential philosophical issues. What is worth doing today? What is the good life? How should I go about my work or play? Why should I be ethical—particularly if everyone else is cutting corners and there is little chance of getting caught? Or, for the purposes of this text, what counts as valuable research? How

A Summary of Claudio Tamburrini's "What's Wrong With Doping?"

We highly recommend that you read Professor Tamburrini's chapter or perhaps access his related 2006 journal article. The following summary merely provides a quick and incomplete overview of his arguments against doping bans. But along the way we provide more information about Tamburrini's arguments against doping bans as we explore different ways of applying logic.

Tamburrini challenges the assumption that doping in sport is unethical. He alternatively argues that doping bans lack justification and should be revised or removed entirely. Tamburrini first distinguishes between doping substances and doping techniques. Doping substances are divided into three categories:

1. Harmful though legal (physician-prescribed substances)
2. Harmful and strictly illegal (e.g., cocaine)
3. Harmless and legal (e.g., diuretics and caffeine)

Doping techniques are categorized in two ways, blood doping and altitude training. Sports governing bodies ban certain performance-enhancement practices while permitting others, which leads to questions of why and how the lines between permissible and impermissible forms of performance enhancement are justified.

Tamburrini then offers three quick challenges to the justification of doping bans:

1. Bans arbitrarily restrain the development of sport.
2. Bans limit the development of knowledge about the harmful effects of doping.
3. Bans create uncertainty and transparency, raising doubt about the validity of any successful athletic endeavor now that doping happens in the shadows.

Tamburrini dedicates the bulk of his article to analyzing and countering four common arguments used to justify doping bans:

1. Doping is harmful to athletes.

    Lots (in fact, most) of sporting activities accept risk. There isn't enough aggregated evidence that shows social harm to justify bans.

2. Doping creates a pressure on others to dope (the coercion argument).

    No one is really forced to do what it might take to be among the small group of habitual winners.

    - Athletes are social role models.

    It is arbitrary to hold up athletes who participate in high-risk and unhealthy, elite sports activities as role models. There are far greater social concerns for impressionable young people than doping (and we should be having open conversations regardless).

    - Doping has negative effects on top and emerging amateur athletes.

    The most ambitious athletes should reap rewards (and accept risks) in the attempts to reach the top. There is no reason to ban harmless doping, and permitting [it] might lead to valuable knowledge that reduces the harms of doping.

    - Doping has harmful effects on youth and junior athletes.

    Minors should be protected from harmful substances, so only harmless or sterilized doping practices should be permitted.

3. Doping is unfair. Some athletes will gain unearned competitive advantages over opponents.

    CT: Sport is riddled with inequities; bans actually reinforce unearned advantages.

4. Doping undermines the nature of sporting competitions (the essentialist argument).

    - Doping deprives sport of its excitement.

    CT: There is no compelling evidence on the tensive quality of sport that shows doping causes any such deprivation.

    - Doping is incompatible with the idea of contest between persons.

    Doping is not a magic bullet. The suggestion that a majority of athletes are opposed to doping on moral grounds (value based) is just not supported.

does one conduct research in an ethical manner? Normative, epistemological, metaphysical, and axiological questions lie beneath every rock in the landscape.

In the domain of physical activity, the situation is the same. One cannot deal with health, disease, fitness, movement, sport, games, and the like, without any number of philosophical issues presenting themselves. Debates over classification are one example and referred to regularly in the previous sections. Tamburrini identifies another enduring issue challenging humans throughout the history of game playing—namely, the methods that should be allowed and disallowed in our pursuit of success and victory.

Athletes have always attempted to find ways to improve their performances. The quest for higher, faster, and stronger seems built into sport itself. One cannot study competitive activity historically without finding athletes pushing the performance envelope—whether ethically, unethically, or by means difficult to categorize as clearly one or the other.

A first step in locating a research problem is ensuring that the question at hand is, in fact, worthy of examination. If answers are obvious, if no good arguments can be rallied on one side or the other of the issue, then the problem is probably a pseudo-problem and not worth examining. This would be the case with the following two ethical questions in the area of performance enhancement:

1. Should very young children be allowed to make their own decisions on whether to use illegal performance enhancers or not?

2. Should ambitious parents be free to force their underage children to use steroids for purposes of promoting sporting success?

Both of these questions have fairly obvious answers. We want children to enjoy sport without taking dangerous risks. We also expect parents to exercise good judgment on behalf of their young children. It would be difficult to rally an argument on the affirmative side for either of these issues. These questions, in other words, constitute an unsuccessful attempt to locate a research problem. Tamburrini was aware of this and thus limited his discussion to the issue of performance enhancement among informed adult athletes. The question becomes more complex in this context because we do, in fact, allow adults to make decisions about their lives as long as those choices do not negatively affect others. We permit informed adults to use dangerous tobacco and alcohol products, for instance. Thus, a potentially interesting normative question arises: Why allow smoking and alcohol consumption and not the use of steroids?

On closer inspection, we see that the issue of deregulating bans on certain performance enhancers actually taps into a durable dilemma in the history of modern philosophy. It pits libertarians wanting to maximize human freedoms, on the one hand, against conservatives or communitarian philosophers who, for a variety of reasons, believe that limitations are warranted. You should be able to feel the tension here. You value your freedoms and undoubtedly resist the heavy hand of paternalism whether from your parents, the government, the church, or some other source. You resist unwelcome intrusions of others telling you how to live your life. On the other hand, you benefit from many social arrangements that put restrictions on yourself and others—for example, the requirement to pay taxes, the need (in some countries) to serve in the military, or the importance of obeying noise ordinances. Thus, on reflection, you can see that this debate goes beyond sport. It is about broader issues related to the limits of human freedom—particularly among informed, consenting adults. This further suggests that Tamburrini had hold of a good research problem.

A second step in locating a philosophical research problem is more technical. Analogous to the limitations and delimitations accompanying some empirical study, philosophers need to clarify the issue under their reflective microscope. Tamburrini did this early in his article by identifying three kinds of doping: harmful but legal, harmful and illegal, and both harmless and legal. Anabolic steroids, the kind of enhancement that lies at the center of much debate over the bans, fall into the first category. Steroids are legal in the sense that doctors and other health professionals can prescribe steroids for medical purposes; yet they cause harm when taken without professional supervision and in large doses. Tamburrini also drew an important distinction between banned substances (such as steroids) and banned techniques (such as blood doping, in which a person's own blood is withdrawn and then injected back into the body prior to competition). These clarifications—the three categories of doping substances and the distinction between substances and techniques—all come into play in his argument.

Tamburrini took two additional steps to locate and establish his research problem. He identified three reasons for eliminating the ban (with a fourth—the preservation and promotion of human freedom—underlying all three) and considered four prominent arguments used by supporters of the ban. (These arguments will be identified in the analysis that follows.) This set the stage, as it were, for the philosophical debate to begin—after all, there would be little rationale for Tamburrini's essay if conflicting arguments were difficult to find and if the use of steroids was unrelated to the promotion of good and the avoidance of harm.

The final preliminary step was to identify his thesis. Tamburrini indicated that he would focus on the four arguments supporting the current prohibition and attempt to show that they lack validity. He wrote, "I will scrutinize these [four, pro-ban] arguments to support the claim that the ban on doping is not justified and should therefore be lifted" (2006, p. 202). He chose this strategy because of the "widespread intuition" that doping is "inherently wrong." That is, many take the four arguments used to support the ban—whether intuitively or reflectively—as conclusive. Tamburrini wanted to challenge that assumption.

> Good philosophical research identifies a good problem, clarifies the problem, and establishes a clear purpose or thesis.

Analyzing a Research Problem

This is where much of the hard work requiring good logic, clear thinking, unbiased description, attention to detail, and other factors related to solid philosophical practice begins. We call this work hard because the criticisms of extreme skeptics and relativists—those arguing that persuasive justifications for certain perspectives do not exist—can be difficult to overcome. But once again, we want you to walk through these methods to see their utility. Even if they cannot provide final answers, they may point us in the right direction.

This section provides examples of four kinds of logic that philosophers use—inductive, deductive, descriptive, and speculative-critical. All four approaches to correct reasoning can be applied to Tamburrini's project and are implicit in his reasoning. Even though Tamburrini did not do so, we highlight them to show how they work.

Inductive Logic

> inductive logic—Reasoning that moves from observation of a few samples or exemplars to a statement of general conclusions. It typically asks: What commonalities exist in the samples or exemplars being studied? (See figure 2.1 in chapter 2.)

Inductive logic is reasoning that moves from a limited number of specific observations to general conclusions about the phenomenon (or object) under consideration. It relies on intelligent discernment to identify common elements or similarities in the samples.

A major issue underlying doping bans is the question of paternalism. When and under what conditions can second parties, such as sports governing bodies, justifiably limit an athlete's freedom? One obvious answer is that externally imposed prohibitions are appropriate when a behavior may bring serious harm to that athlete and fellow competitors. Safety rules in sport are common . . . and commonly accepted! This logic seems to provide a very strong counterargument to Tamburrini's case. However, he can use inductive logic to provide a rejoinder. He can identify a limited number of activities considered dangerous yet permitted, such as skydiving, deep-sea diving, climbing very high mountains, playing tackle (American) football, chewing tobacco, riding motorcycles, and eating high-fat meats or high-calorie desserts.

With this list in hand, a philosopher can look for commonalities or general conclusions regarding their similarities, such as the following:

- They are all dangerous to the point of causing debilitating injury or illness or even premature death.
- All of them are legal. (No paternalistic agency prohibits participation by informed adults.)
- All of them are (or can be) enjoyable. Many people engage in them simply because they serve as a source of pleasure.

These inductively generated insights led Tamburrini to a couple of interesting questions and a potentially persuasive conclusion: What if athletes found that taking steroids enhances their enjoyment of sport? And what if the risks were no greater than those incurred by others involved in this list of dangerous but permissible activities? The conclusion would point to a potentially indefensible inconsistency. Society tolerates any number of high-risk behaviors in the name of individual freedom and the pursuit of personal pleasure or happiness. So, too, should it be with using steroids. Consistency requires it.

You would be right to question the force of this argument because any number of additional reasons may support the prohibition, reasons that may trump the "tolerance for risk" and the "pursuit of pleasure" arguments. But inductive logic proves to be a reliable tool in pointing out an apparent inconsistency in how we deal with risk. It raises important questions about double standards and, possibly, a kind of irrationality underlying doping prohibitions. Even use of the term *doping* suggests something illegal, surreptitious, and otherwise distasteful. Why not call it a technology or simply a performance enhancer so that more objective and dispassionate reviews of its status can take place?

Inductive logic has both strengths and weaknesses. For one, the development of the original list is crucial. If items are not representative of the phenomenon at hand (here it was "dangerous but permitted activities"), then the conclusions will likely be wrong or misleading. Also, the process of identifying generalities can prove challenging. If too obvious (e.g., all of these activities involve people), the inductive generalization may be true but uninteresting and lacking helpful insight. A generality might also be wrong if the list of particulars is incomplete. A philosopher thinking only of skydiving, riding motorcycles, and white water rafting might consider permissible risks to be only those that threaten external harm to the person. This would eliminate steroids that are used inside the body. A more expansive list of dangerous but permitted activities that involve ingestion or internal threats to well-being (chewing tobacco and eating high-calorie desserts) would reduce chances of making this error.

Philosophers in general, but particularly those taking the cautious middle epistemological position we identified as fallibilistic realism, test their conclusions derived from inductive reasoning repeatedly. This is the case because neither scientists nor philosophers can be certain that the items scrutinized the first time around are representative of those that were not examined. As noted, the previous sample, in other words, might be flawed or contaminated and lead to unwarranted conclusions. Or the way the sample was tested might have introduced errors. Scientists therefore test and retest in attempts to replicate previous findings. Philosophers reflect inductively on a set of items; then reconsider the items selected and reflect some more. Progress is made, but certainty is rarely achieved.

Deductive Logic

Deductive logic is a companion technique of inductive reasoning, and many philosophers adroitly and spontaneously intermix the two. Deduction requires intellectual movement in the opposite direction from induction. Whereas inductive reasoning starts working from particulars to construct abstractions or generalities, deductive reasoning starts with general claims to see what particulars follow.

deductive logic—Reasoning that moves from general claims or premises to particular conclusions. (See figure 2.2 in chapter 2.)

The initial general claims or premises used in deduction are of two sorts. The first are statements of fact often phrased as *Because such and such is true, then it follows that . . . .* But premises can also be hypothetical: *If such and such were true, then it follows that . . . .* We will take a look at a deductive line of reasoning using hypothetically stated premises.

Those supporting the current bans offer one standard justification, the so-called *coercion argument.* Those employing this argument might agree with the libertarian contention that society should err on the side of freedom even when a person chooses to take extreme or unnecessary risks. However, they would disagree that people should be allowed to coerce others into similar actions. Because steroids confer considerable competitive advantages (so the argument goes), athletes otherwise inclined to avoid such risks would feel coerced to accept them. This coerciveness is what some doping ban defenders deem unacceptable and consequently what serves to justify paternalistic prohibition. Tamburrini, however, advanced a series of counterarguments that can be put in deductive form:

Premise 1: If it is true that high-risk behaviors occur in business (e.g., working all night, taking pep pills to stay awake), and

Premise 2: If it is true that such behaviors confer a clear competitive advantage on those willing to take such risks and are, for this reason, potentially coercive for fellow workers, and

Premise 3: If it is true that many fellow workers still choose to resist coercive pressures by not adopting such behaviors, then

It follows that high-risk, advantage-conferring behaviors in business are not (strongly) coercive.

Would lawyers, accountants, or small business owners accept a mandatory 40-hour week with fines imposed for working overtime? Unlikely. Therefore, Tamburrini can then ask: Why should it be any different in sport? Just like people working in offices, athletes can resist coercive pressures and simply choose not to dope. Moreover, it could be argued that those willing to take the risks (in business or in sport) should reap the benefits and suffer the consequences of such actions. Justice is being served.

Tamburrini was aware of another argument, one that hinges on a difference by degree. Working all night on the job may confer *some* advantage, but this may also lead to burnout, inefficiency, increased mistakes due to fatigue and the like. The advantage, in other words, may not be significant. And besides, relatively few workers might risk compromising their own health and family life (if they have one), thus making those working inordinately long days statistical outliers. Prudent workers therefore will be compared with a large number of other prudent workers (apples compared to apples). They will not be at a disadvantage. Thus, for both reasons (justice is served and unfair advantages are mitigated), prohibiting voluntary overtime in the workplace is not needed.

Sport is different, however, because advantages conferred by steroids are clear and decisive. In the case of the Tour de France, ban supporters claimed that clean athletes simply could not compete against those using illegal enhancements (justice is absent). Also, in this environment, it was commonly known that almost everyone doped. The coercive pressures were enormous. If one wanted to compete, one had to dope. Thus a few prudent cyclists who refused enhancements would be compared to an overwhelmingly large number of imprudent cyclists who used them (apples compared to oranges). This is patently unfair.

Tamburrini presented the following rejoinder. Again, this could be put in a deductive format.

Premise 1: If athletes choosing not to dope are less successful competitively, yet

Premise 2: If less successful athletes still have access to a number of athletic benefits, then it follows that athletes choosing not to dope still have access to a number of athletic benefits.

Tamburrini speculated on two scenarios in which non-doping athletes plausibly could receive benefits even if bans were lifted. The first suggests that the national and international levels of interest in athletics is so great that even second-tier athletes (those choosing not to dope) can have successful professional careers. The second argument points to the perceived superior moral position maintained by those who do not dope. Many fans may prefer a doping-free version of the activity (perhaps a separate division), and thus support the athletic aspirations of individuals and groups choosing, on moral grounds, not to partake in certain performance-enhancement practices.

These arguments may strike you as a little far-fetched. In a world in which winning seems valued over all else and fans regularly ignore poor behavior exhibited by favorite players or teams, Tamburrini's scenarios may seem unrealistic. However, the deductive logic, as far as it goes, is sound. The two premises necessarily generate the conclusion, and this is perhaps the greatest strength of deduction. As much as any other method, it aims at certainty. St. Thomas Aquinas used it to good effect during the Middle Ages to argue that if certain premises about God's nature, or God's will, or claims of Holy Scripture were true, then important theological conclusions necessarily followed. This is much like a geometric proof and thus favored by those with high degrees of confidence in the powers of reason and strong tendencies toward objectively oriented epistemological realism.

However, two important weaknesses must be addressed. The first relates to the validity of the premises. If any of the premises are false, the deductive conclusion is likely false as well. You might question, for instance, Tamburrini's second premise that athletic benefits will remain available to elite athletes choosing not to dope. If that premise is wrong, the argument fails. Thus, one of the best ways to attack a deductive conclusion is to raise questions about one or more of the premises resulting in that conclusion.

Of course, the deduction itself can be faulty. The conclusion can introduce terms or concepts not included in any of the premises. See if you can pick out the missing premise in the following line of reasoning:

Premise 1: If it is true that Armstrong doped, and

Premise 2: If it is true that doping is both illegal and unethical, then

It follows that Armstrong should be stripped of all victories gained while he was doping.

This conclusion, although it may seem reasonable enough, is not deductively valid. It introduces concepts not included in either of the premises—concepts about grounds for stripping medals or victories. Thus, a third premise is required, one that states, "If it is true that all illegal and unethical conduct warrants the vacation of victories," then the conclusion follows. The logical chain of reasoning from the fact of Armstrong's doping to the stripping of his titles has been connected.

Descriptive Logic

Descriptive logic provides a direct and disarmingly simple way to conduct philosophical research. Philosophers using this method describe phenomena in the context of lived or hypothetical experience. To be sure, they manipulate that experience adroitly, as will be described shortly, but descriptive techniques rely on the fidelity of observations and resultant descriptions.

Descriptive philosophers prefer to conduct their research in less roundabout ways than those provided by inductive and deductive logic. They do not line up a series of particulars and ask themselves what they all have in common (induction), nor do they begin with a set of givens or premises to see what follows (deduction). They simply describe what they see reflectively when examining a phenomenon.

The issue of fairness, one that is central to the doping debate, can be addressed descriptively. We can reflect, for instance, on elite downhill skiers from Norway, who happen to

descriptive logic—Reasoning that describes lived experiences and relies on descriptions of observations.

be identical twins separated at birth, now vying for top world ranking. We wonder if their competition is fair. It would seem that it is . . . or at least largely so. They both play by the rules, so neither gains an advantage by cheating. Moreover, raised in nearby towns, they had (let us say) equal amounts of winter weather as well as access to the same equipment, the same coaches, and the same mountainous slopes during their formative years while learning to ski. And, perhaps most important of all, they share the same genes. On this particular afternoon, weather conditions are the same for both. Multiple runs lessen the possible effects of chance or luck, and the order in which they ski is reversed each time to neutralize advantages from slope conditions. The competition, as much as humanly possible, seems to pit the merits of one sister against the merits of the other. We intuit this as a level playing field and describe the activity as a fair competition.

But then we vary the situation. We now picture two genetically unrelated downhill competitors who grew up under very different conditions—one in Norway as before, but the second one in a desert environment, from a relatively poor family, in possession of only inferior equipment, with inadequate coaching, and finally, with very limited access to good downhill training facilities. In spite of this, she has become reasonably accomplished and now competes against a Norwegian woman, a person with all the aforementioned advantages. Do we still intuit this as a fair competition?

In a way, we do. Both are following the rules. Nobody is gaining an advantage by cheating. Both will be skiing down the same slope. However, in another important sense, we would have to describe it as very unfair. The merits of the two skiers are not the only things being tested. Their backgrounds are very much in play. The Norwegian woman will probably win, but she can hardly take credit for her many advantages. We might even speculate that sport is not designed to measure the kinds of things that produced the very different skills of these women—benefits of birth, geography, weather, wealth, and other such factors that seem beside the point. Sport is supposed to show who *deserves* top ranking, who *earned* the victory. Thus, this reflective variation allows us to describe this competition as unfair as well as identify the kinds of factors that make it so.

These kinds of descriptive or intuitive realizations led Tamburrini to draw a surprising conclusion. He suggested that if international sporting organizations were genuinely interested in promoting fair play, they would allow selective doping to level the playing field and better test the true merits of those competing. It would amount to a kind of high-tech handicapping designed to nullify unearned advantage. In our scenario, only the second skier from the desert climate would be permitted to dope to equalize the competition. Once again, this may strike you as far-fetched, but Tamburrini, like many good philosophers, followed the logic where it led him. After all, we handicap in a number of sports (e.g., in horse racing by putting weight on faster horses, in golf by giving strokes to the lesser competitor). This further suggests we not dismiss his idea out of hand.

The processes of looking carefully and describing accurately require considerable skill, and like all skills, they can be performed well or poorly. Because human beings can be inattentive, often jump to conclusions, regularly fail to notice details, take things for granted, and often confuse parts and wholes, it is not particularly easy to produce accurate, insightful, and careful description. When done well, however, descriptive reasoning can be extremely useful and enlightening.

Descriptive reasoning highlights at least one advantage philosophers enjoy over their scientific colleagues. Philosophers can vary their subject matter at will. In the preceding example, we did not have to find a skier from the desert, and we did not have to conduct an interview. We could simply create the new scenario and describe it. If we did our work well, our conclusions would map onto the real world in effective ways. In this case, we could say with a high degree of assurance that they would. This is the case because we see many sport organizations trying, as best they can, to neutralize irrelevant factors that could influence performance. They implement weight classes in sports in which crude physical mass can create advantages. They standardize equipment to ensure that the contest is less likely to turn on the merits of the equipment manufacturer. They create multiple categories in the Paralympics to address different functional capacities. And we learned that, at least in principle, selective doping could promote rather than detract from fair play.

Descriptive methodologies place a great deal of confidence in the capacity of reason to portray existence accurately. It is important to recognize that this confidence resides in traits and skills of human intelligence depending on the level of description attempted. If description is intended to depict an actual lived experience—with its real perceptions, feelings, ups, and downs—then we must rely on good memory, honesty, attention to detail, an unwillingness to embellish facts, and so on. This subjective level of description (similar to that used in qualitative methodologies) in effect involves making claims about what someone really experienced, and such portrayals can vary between fact and fiction.

A second, more objective level of descriptive work, such as the one modeled earlier, is disinterested in the actual subjective experience itself—for instance, whether the skier from the desert actually experienced her competition as fair or unfair, and what the content of that lived experience was. At this more abstract level of descriptive work, the concern is with the nature of experience *in principle*. For example, what accounts for experiences of unfairness? What is it for a contest to be encountered as unfair? This level of description places confidence in the power of reason to notice important differences when varying the subject matter as we did when we imagined two scenarios of competitive skiing.

The danger with this second type of descriptive methodology is not poor memory, dishonesty, or tendencies to exaggerate events, but rather people's internal biases that contaminate the ability to see clearly. For instance, you might have been raised in a conservative religious household. Or a family member might have fallen prey to drug addiction. In either case, you might not be able to reflect on the issue of drug taking in sport objectively and dispassionately. You might not be able to see how and why doping could enhance fairness.

> Descriptive methodologies rely on reason's capacity to reliably portray reality, whether depicting an actual experience or working on distinctions at a more abstract level.

Speculative and Critical Forms of Logic

Speculation can be thought of as an extension of both descriptive and prescriptive philosophy. It is a philosophical method that raises possibilities that cannot be substantiated or otherwise forcefully defended. **Critical logic** is the other side of the coin. It is a philosophical method that deconstructs or debunks philosophical analyses of many types, particularly those that are speculative. Critical reasoning attempts to uncover the biases shaping philosophical conclusions. Some critical philosophy regards all attempts to find the truth as futile or otherwise misguided.

Speculative philosophy attempts to answer questions without applying simple logic or using clear description. In this sense, speculative conclusions lack supporting evidence. It does not follow, however, that speculation requires no argumentation whatsoever. Again, speculating can be done well or poorly. Even though speculative conclusions cannot be proven or demonstrated, they can vary in their degree of plausibility.

Tamburrini addressed one speculative claim about sport and took a more or less neutral position on its validity. It has been speculated that much of the value and charm of sport lies in its uncertainty, the attempt in sport to pull something off without any assurances of success. In other words, if sport became predictable, it would lose much of its appeal, and quite likely, fewer people would want to play or watch it. It could be further speculated that this uncertainty has two roots—the test and the contest, as we proposed in Philosophical Exercise 1. The test is the game problem—such as running 4 miles (6 km) in a short period of time. This produces the uncertainty of passing or not passing the test, of getting an A, a B, a C, or a worse grade as a result of our efforts. We also evaluate against ourselves: Can I do better than I did yesterday? Can I set a new personal record? This delicious uncertainty

speculation—Claims about subjects, such as values, that are significant to humans but cannot be forcefully defended.

critical logic—An activity intended to show reasoning errors, faulty claims, and misplaced beliefs.

draws us to the track, the golf course, the bowling alleys—wherever we can face this tension or uncertainty. Once again, this species of uncertainty is generated by taking a test.

A second source of uncertainty lies in the contest: Regardless of how well I do on the test today, will it be good enough to beat you? This produces a dichotomous contest result of victory or defeat, but in actuality it is a graduated comparison extending from no difference (a tie game) to an extreme difference in two directions (a lopsided victory or crushing defeat). Regardless, some delightful drama and uncertainty are present here too (Kretchmar & Elcombe, 2007). Will I win or lose? If I lost to you the last time by a wide margin, can I keep it closer today?

Other philosophers speculate that factors in addition to or other than drama account for much of the value of sport. In other words, different plausible answers to sport's value present themselves seemingly without any way to adjudicate between them. Thus, speculative philosophy leaves room for subjective preference. Some may play sport for the dramatic tensions it includes. Others may play sport for the knowledge they can gain about their personal abilities. This is like preferences for vanilla and chocolate ice cream. We have to respect these unique tastes or opinions.

However, Tamburrini was not willing to give too much credence to subjective preference. He believed that there is at least a degree of truth to claims about the importance of uncertainty. Thus, he took pains to show how doping does not threaten this quality of sport. He provided several reasons for claiming that doping and uncertainty are compatible. You will have to read these arguments presented in his article to see if you agree.

This raises a point made earlier in the chapter. Descriptive metaphysics and prescriptive normative inquiry are often interrelated. To know how we should act in X, we need to know what X is, how it works, and what its values are. Tamburrini acknowledged this relationship by first doing some speculative metaphysics—establishing that uncertainty is part of the nature of sport. He then moved on to an axiological description, determining that uncertainty is valuable. He concluded by making a normative recommendation: Bans should be lifted because doping does not threaten the value of uncertainty.

The advantage of speculative philosophy is that it broaches subjects with deep human significance and the potential to markedly change human lives. Wars, for instance, have been fought over a way of life grounded in the perceived superiority of some values in relation to others—say, democracy over some form of totalitarianism. Accordingly, given that people mold their lives around values, it may well be important to do research on intangibles, even if knockdown arguments are not available.

The research principles here would be twofold. First, it is better to know a little (even if it is inconclusive or incomplete) than to know nothing at all. Second, one can be unable to prove a claim and still be right about it. Some speculate that a spiritual reality of some sort exists, and they provide evidence, they believe, pointing in that direction. Others disagree and present their own evidence suggesting nothing is real but atoms and void. Neither side can present knockdown arguments. Yet one side is still right. Or perhaps, in some hard-to-imagine way, both of them are right.

Disadvantages also exist. Narrow, selfish, manipulative, and even mean philosophies can masquerade as legitimate speculative statements about reality. Either purposely or unintentionally, some people spew groundless opinions, biases, and self-serving and status-quo-preserving visions of life in the name of philosophy. Even arguments intended to show the plausibility and attractiveness of certain claims might only play to the biased predispositions of the audience. For instance, claims about the superiority of sport when conceptualized as a mutual quest for excellence may seem persuasive in a capitalistic, achievement-oriented culture in which most people are raised to be all they can be. In another society, such arguments for excellence might seem odd or misplaced.

In some ways, Tamburrini's unusual defense of doping can be read as critical philosophy. It debunks the metaphysical superstructures, or grand narratives, built up around sport.

Although it does not move significantly in a political direction, it raises questions related to power and politics. Who benefits by maintaining the status quo of banning drugs? Why would sporting organizations such as the International Olympic Committee make a show of enforcing doping regulations? When some folks argue that doping violates the nature of sport, why should their version of sport be accepted as definitive or absolute?

At one time in history, and to a certain extent still today, amateur competition was regarded as the only true form of sport. Critical sociologists and philosophers debunked that myth and exposed it as a very limited and self-serving ideological view of sport that favored certain elite classes. Tamburrini may have accomplished a similar end by exposing the preeminence of drug-free competition as a myth, one that supports various commercial interests. Or at least he has tried to do so.

Critical logic is based on skeptical attitudes toward the power of reason and therefore seems unduly negative. Some critical philosophy reads like an exercise in tearing things down, not building them back up. However, exposing myths and errant thinking is no trivial activity. Exposing myths and attacking stereotypes about gender, ethnicity, religious organizations, sexual orientation, or certain privileged versions of sport serve an important purpose. It is far better to question what is held to be true than perpetuate a lie or tacitly support errant beliefs.

This is the dilemma that Tamburrini's challenging article addresses. Are doping bans a mistake? Or are they grounded in defensible concerns? Hopefully, you will be able to interact with the position he developed and add further arguments to either support or question his conclusions. A bibliography is provided at the end of the chapter for those who wish to look at what a number of philosophers have written about this issue.

We also introduced the Caster Semenya case earlier in this chapter. It offers a second option for class debate or analysis using the same tools of logic reviewed in this section. In addition to information provided at the start of this chapter and ideas contained in the two Philosophical Exercises, the previously-cited article by Sigmund Loland (2020), as well as the series of articles in the same journal issue reacting to Professor Loland's position on the IAAF's DSD Regulation, can be used to provide important pro and con arguments for allowing Semenya to compete in women's track events.

Summary

In some ways, philosophizing well is like performing a motor action skillfully. And like many acts that require expertise, confidence grows as skill improves. Within broad boundaries of orthodoxy, different strategies and styles work for different people. They allow us to get in touch with our preferences. This is the important subjective side of philosophy, the part that respects a range of likes and dislikes that vary from one person to another, and from one group of people to another. Different strategies also allow us to appreciate the descriptive and prescriptive purposes of philosophy—the metaphysics and axiology of description and the normative recommendations of prescription. As many philosophers have noted, the constant subtext of philosophy is the reminder that the world can be different . . . and better.

☑ Check Your Understanding

1. Have the purposes of philosophy remained constant across history? If not, how have they changed?

2. What is the difference between objective philosophy and subjective philosophy? Why are both important? Why does much traditional philosophy favor objective analyses?

3. What is the difference between metaphysics and axiology? Why are both identified as descriptive aspects of philosophy? Identify the four subcategories of axiology and describe their differences.

4. What is the difference between metaphysics and axiology, on the one hand, and normative philosophy on the other? See if you can identify three normative claims, one that is highly subjective, one that falls between the subjective and objective ends of the continuum, and one that is highly objective—that is, one that would be difficult to refute.

5. Why is epistemology such a foundational aspect of philosophy? Which of the three epistemological positions did you find most comfortable . . . and why?

6. What counts as a good philosophical problem? Why would Tamburrini's question count as a good problem? Why are definitions important? What role does a thesis play in philosophical research?

7. Should Caster Semenya be required to compete in a separate category of women who share her genetic makeup? Is categorization the only strategy (or one of the most important strategies) for promoting fair competition?

8. How do the logics of deduction, induction, description, speculation, and critical philosophy work? What are their strengths and weaknesses?

Appendix: Using Philosophical Research Methods

Philosophical Exercise 1: Logical Relationships Between Sport-Related Projects

One of the easiest ways to determine relationships using Venn diagrams is to look for examples that support or contradict the relationships depicted. Using this method, you should have been able to quickly dismiss diagram 3. It indicates that every test must also be a contest. But we know that isn't true because we can find multiple examples of test-taking that do not involve attempts to beat others. One of them was used in this very exercise—namely, golfing alone. We also try to solve crossword puzzles, sudoku puzzles, math problems, riddles, and any number of other tests without attempting to beat anybody else. Making a flaky crust for an apple pie, driving under snowy conditions, and threading a needle are all tests that need not be turned into competitions. Testing activity, in other words, is both ubiquitous and intelligible on its own. Thus, we need a Venn diagram unlike diagram 3 that has an area of testing that falls outside the contesting circle.

Both diagrams 1 and 2 satisfy this requirement. Yet there is a big difference between the two. Diagram 1 indicates that contesting can exist in the absence of testing, whereas diagram 2 indicates the opposite—that all contests include tests.

In attempting to choose between these two, you may have had trouble coming up with examples that support diagram 1—that is, examples of contests that did not include a test. It may have even struck you that attempting to compete in the absence of any challenge or test does not make much sense. If so, you should have been leaning in the direction of diagram 2.

The word *contest* itself provides a clue to the merits of this selection. To *con-test* means "to bear witness together," particularly in disputes. In contests, two or more parties testify to their abilities to solve a common testing problem—getting golf balls into holes in fewer strokes, scoring more runs, lifting heavier weights, running faster, and so on. Without attempts to solve some testing problem, there would be nothing to testify or bear witness to.

Word derivations can be helpful in solving puzzles like this one, but philosophers want stronger evidence, as we previously argued in terms of playing games. A reflective lab experiment can help us verify this tentative conclusion about contests requiring tests. Picture an activity that everyone can do 100% of the time—such as picking up a pencil lying on a desk. Then picture yourself challenging someone to a contest over this act of simply picking up a pencil. It should strike you that this wouldn't make any sense. You can't compete over something that everyone can easily do. Without a valid test that can be solved to varying degrees, no way exists to differentiate winners and losers. Everyone would always score 100%.

Likewise, picture an activity that nobody can do, ever—such as jumping to the moon. Then picture yourself challenging someone to a contest over jumping to the moon. Again, this doesn't make any sense. You can't compete over something that nobody can do. Jumping to the moon per se cannot be solved to varying degrees. It cannot be a race in which one person jumps successfully before others. It simply cannot be done. Again, there would be no way to differentiate winners and losers. Everybody in such a contest would always tie with a score of 0.

Contesting then makes sense only in the presence of a valid test—that is, some activity that is neither gratuitously easy nor impossibly difficult for the people participating in it. Contesting in the absence of a shared challenge, in short, makes no sense. It would amount to an illogical attempt to show superiority where no ways to show superiority are available. Contests therefore are grounded in (valid) tests.

Philosophical Exercise 2: Are Sport Contests Fair?

It probably occurred to you there is no easy answer to this question. In other words, there is no right or wrong place on the scale for your answer. Many people, when working on this exercise, tend to realize there is more unfairness in sport than they previously noticed. Thus, claims that sport is a very special place of fairness and justice may appear to be exaggerations at best. Sporting structures, organizations, and regulations certainly promote fairness but are never fully successful in achieving it. In other words, sport is a domain in which many inequities are reduced but never entirely eliminated. A person could reasonably expect to be treated more fairly in sport than in life, but even in sport, competition is not entirely fair.

A review of the following six techniques cited in the exercise for promoting fairness can quickly show that fairness is a desirable goal but not one that is easily realized. Insufficiencies of these strategies are italicized.

1. Using competitive divisions to neutralize biological differences and promote even contests (e.g., weight classes in boxing, wrestling). *Problem: Classes are ranges, and a person at the bottom of a range would be at a disadvantage competing against someone at the top of that range. Also, biological differences are only one kind of difference between competitors. What about psychological and social differences? Moreover, if weight classes make boxing fairer, why don't organizations have height classifications that make basketball fairer for short people?*

2. Establishing competitive divisions to neutralize school size and promote even contests (e.g., Class AAA, AA, and A in high school sport leagues based on school population). *Problem: As with divisions in strategy 1, a school at the bottom of a range would be at a disadvantage playing against a school at the top of that range. Moreover, school size is only a rough and thus imperfect way to measure athletic potential of a school district. Wealth, traditions, history, quality of facilities, random variation, and other factors play into competitive potential.*

3. Writing rules that standardize legal preparations for contests (e.g., permitting Olympic athletes to choose their own coach and use preferred training facilities prior to the games). *Problem: Enabling legislation is always suspect because it advantages those who can access the stipulated allowances—for instance, wealthy athletes who can afford the best coaches, the best equipment, and the best training facilities.*

4. Enforcing rules that neutralize in-game advantages (e.g., changing ends of the court or field at halftime; switching lanes in bowling). *Problem: This only approximates fairness. Weather conditions, the game's score, degree of fatigue, and other factors are different when the time comes to switch ends of the field. Those who go first or, with the flip of a coin, get the first choice, often have an advantage that cannot be entirely erased.*

5. Establishing rules that neutralize home-field advantage (e.g., home scheduling). *Problem: One school's or team's home field advantage is rarely the same as another school's or team's home field advantage. For example, some basketball facilities are notoriously difficult venues for visiting teams. Others are not.*

6. Using penalties to neutralize advantages gained from breaking the rules (e.g., yardage assessments in football for holding; no calls or penalties for flopping in basketball). *Problem: Occasionally, officials will blow a call that unfairly costs some team a game they rightfully should have won. More commonly, penalties either overcompensate or undercompensate for illegal advantages gained. Those who commit so-called "smart fouls" take advantage of the inherent unfairness in some penalties that undercompensate the team disadvantaged by the rule breaking.*

What then is the logical relationship between these partially successful attempts to promote competitive fairness and close outcomes? One of them is consistency. If certain kinds of unfair behavior are currently allowed, and if they do not ruin competition or prevent close finishes, why make an exception of Semenya? Consistency coupled with absence of harm requires that she be allowed to compete.

This is not the only conclusion that can be reached. It could be argued that the degree of unfairness that currently exists in sport is unfortunate. For practical reasons, most of it cannot be eliminated. But we can establish competitive categories to eliminate unearned biological advantages. We have already done so with weight classes in wrestling. Thus, for purposes of reducing manageable levels of unfairness, Semenya should not be allowed to compete in certain women's races.

Selected Bibliography on Ethics of Performance Enhancement

Brown, M. (1980). Drugs, ethics, and sport. *Journal of the Philosophy of Sport, 7,* 15-23.

Burke, M., & Roberts, T. (1997). Drugs in sport: An issue of morality or sentimentality? *Journal of the Philosophy of Sport, 24,* 99-113.

Gardner, R. (1989). On performance-enhancing substances and the unfair advantage argument. *Journal of the Philosophy of Sport, 16,* 59-73.

Holowchak, A. (2002). Ergogenic aids and the limits of human performance in sport: Ethical issues and aesthetic considerations. *Journal of the Philosophy of Sport, 29,* 75-86.

Kretchmar, S. (Fall, 1999). The ethics of performance-enhancing substances in sport. *Bulletin: International Council of Sport Science and Physical Education, 27,* 19-21.

Kretchmar, S., & Elcombe, T. (2007). In defense of competition and winning. In W. Morgan (Ed.), *Ethics in sport* (2nd ed., pp. 181-194). Champaign, IL: Human Kinetics.

Lopez Frias, J. (2014). The challenges of modern sport to ethics: From doping to cyborgs. *Journal of the Philosophy of Sport, 41* (3), 400-419.

Petersen, T. S., & Kristensen, J. K. (2009). Should athletes be allowed to use all kinds of performance-enhancing drugs? A critical note on Claudio M. Tamburrini. *Journal of the Philosophy of Sport, 36,* 88-98.

Schneider, A. J., & Butcher, R. B. (1993). Why Olympic athletes should avoid the use and seek the elimination of performance-enhancing substances and practices for the Olympic Games. *Journal of the Philosophy of Sport, 20*(1), 64-81.

Simon, R. (1985). Good competition and drug-enhanced performance. *Journal of the Philosophy of Sport, 11,* 6-13.

Simon, R. (2004). *Fair play: The ethics of sport* (2nd ed., pp. 69-90). Boulder, CO: Westview.

Tamburrini, C. M. (2000). What's wrong with doping? In T. Tansjo and C. Tamburrini (Eds.) *Values in sport: Elitism, nationalism, and gender equality and the scientific manufacture of winners* (pp. 200-216). London: Spon Press.

Tamburrini, C. M. (2006). Are doping sanctions justified? A moral relativistic view. *Sport in Society, 9*(2), 199-211.

Selected Bibliography on the Ethics of Gender Classification

Camporesi, S. (2016). Ethics of regulating competition of women with hyperandrogenism. *Journal of Clinical Sports Medicine, 35*(2), 293-301.

Camporesi, S. (2019). When does an advantage become unfair? Empirical and normative concerns in Semenya's case. *Journal of Medical Ethics, 45*(11), 700-704.

English, C. (2017). Toward sport reform: Hegemonic masculinity and reconceptualizing competition. *Journal of the Philosophy of Sport, 44*(2), 183-198.

Loland, S. (2020). Caster Semenya, athlete classification, and fair equality of opportunity in sport. *Journal of Medical Ethics, 46,* 584-590.

Sullivan, C. F. (2011). Gender verification and gender policies in elite sport: Eligibility and 'fair play.' *Journal of Sport and Social Issues, 35*(4), 400-419.

Teetzel, C. (2014). The onus of inclusivity: Sport policy and the enforcement of women's category in sport. *Journal of the Philosophy of Sport, 41*(1), 13-127.

14

Research Synthesis
(Meta-Analysis and Systematic Reviews)

Within the last few chapters, we have used 10 puns hoping to make you laugh. No pun in ten did.

—Jerry R. Thomas

An analysis of the literature is a part of all types of research. The scholar is always aware of past events and how they influence current research. Sometimes, however, the literature review stands by itself as a research paper that analyzes, evaluates, and integrates the published literature. A term used to describe this is *research synthesis*. As mentioned in chapter 2, many journals consist entirely of literature review papers, and some research journals publish review papers occasionally.

All the procedures discussed in detail in chapter 2 apply to research synthesis. This chapter addresses using the literature to draw empirical and theoretical conclusions rather than to document the need for a particular research problem. A good research synthesis results in several tangible conclusions and should spark interest in future directions for research or to affect practice. Sometimes, a research synthesis leads to a revision to or the proposal of a theory. The point is that a research synthesis is not simply a summary of the related literature; it is a logical type of research that leads to valid conclusions, hypothesis evaluations, and the revision and proposal of theory.

> A research synthesis leads to valid conclusions, evaluates hypotheses, and possibly leads to the proposal of a theory or of future directions for research. It is more than a summary of related literature, which are generally not publishable.

The approach to a research synthesis is like the approach to any other type of research. The researcher must clearly specify the procedures that are to be followed. Unfortunately, the literature review paper seldom specifies the procedures that the author used. Thus the basis for the many decisions made about individual papers is usually unknown to the reader. Of course, an objective evaluation of a literature review then becomes nearly impossible. Questions that are important yet usually unanswered in the typical review of literature include the following:

- How thorough was the literature search? Did it include a computer search and hand search? In a computer search, what descriptors were used? Which journals were searched? Were theses and dissertations searched and included? Were presentations included? Was the data from the same research study included if presented in multiple publications?

- On what basis were studies included or excluded from the written review? Were theses and dissertations arbitrarily excluded? Did the author make decisions about inclusion or exclusion based on the perceived internal validity of the research, on sample size, on research design, or on appropriate statistical analysis?

systematic review—A technique of literature review that contains a definitive methodology, analyzes methods, rigor, and strength; typically includes a flowchart showing inclusion and exclusion of articles and a table summarizing findings of included articles.

meta-analysis—A technique of literature review that contains a definitive methodology and quantifies the results of various studies to a standard metric that allows the use of statistical techniques as a means of analysis.

- How did the author arrive at a particular conclusion? Was it based on the number of studies supporting or refuting the conclusion (i.e., vote counting)? Were these studies weighted differentially according to sample size, the meaningfulness of the results, the quality of the journal, and the internal validity of the study?

Many more questions could be asked about the decisions made in the typical literature review paper because good research involves a systematic method of problem-solving. But in most literature reviews, the author's systematic method remains unknown to the reader and thus prohibits an objective evaluation of these decisions.

In recent years, several attempts have been made to solve the problems associated with literature reviews. **Systematic review** and **meta-analysis** are two techniques that use clearly defined research questions and include a priori decisions that guide the research—factors that distinguish them from other approaches to literature review. The purpose of this chapter is to present the procedures used in meta-analysis and systematic review.

Purpose of Research Synthesis

Like any research procedure, research synthesis involves selecting an important problem to address. Two critical distinctions between a literature review and research synthesis are (a) the level of intention defined by the procedures and (b) the transparency of the process. Table 14.1 delineates the steps in meta-analysis and systematic review. Both techniques share the same steps at the beginning and end, but they also diverge to arrive at conclusions.

TABLE 14.1

Comparison of the Steps for Meta-Analysis and Systematic Review

Meta-analysis	Systematic review
Identify a problem.	Identify a problem.
Perform a literature search by specified means.	Perform a literature search by specified means.
Review identified studies to determine inclusion or exclusion.	A research team of two or more work independently to determine inclusion or exclusion of studies.
Carefully read and evaluate to identify and code important study characteristics.	Research team members extract data independently.
Calculate effect sizes.	Resolve differences in both data inclusion and exclusion as well as extraction.
Apply appropriate statistical techniques.	Present process in a flowchart and data in an appropriate table.
Report all these steps and the outcomes in a review paper.	Report all these steps and the outcomes in a review paper.

Using Meta-Analysis to Synthesize Research

Since Glass introduced meta-analysis in 1976, thousands of meta-analyses have been published, especially in the social, behavioral, and medical sciences. At a national conference in 1986 that was sponsored by the U.S. National Institutes of Health, Workshop on Methodological Issues in Overviews of Randomized Clinical Trials, participants focused considerable attention on meta-analysis for health-related and medical research. The *Handbook of Research Synthesis*, edited by Cooper and Hedges (1994), is still one of the most useful sources of information for planning and conducting meta-analytical reviews.

In this section, we describe the steps of meta-analysis in greater detail. The first step is to report a definitive methodology concerning the decisions in a literature analysis. The next step involves quantifying the results of various studies to a standard metric called *effect size* (ES). This allows the use of statistical techniques as a means of analysis. Of course, one of the major problems in a literature review paper is the number of studies that must be considered. To some extent, analyzing all these studies is like trying to make sense of all the data points collected on a group of participants. However, within a study, using statistical techniques reduces the data to make them understandable. The procedures of meta-analysis are similar. The findings within individual studies are considered the data points to use in a statistical analysis of the findings of many studies.

How, then, can findings based on different designs, data collection techniques, dependent variables, and statistical analyses be compared? Glass (1976) addressed this issue by using the estimate effect size (ES, or the symbol Δ). Note that we discussed this general concept in chapters 7 and 9 as a way of judging the meaningfulness of group differences. ES is determined by the following formula:

$$ES = (M_E - M_C)/s_C \tag{14.1}$$

where M_E is the mean of the experimental group, M_C is the mean of the control group, and s_C is the standard deviation of the control group. Note that this formula places the difference between the experimental and control groups in control group standard deviation units. For example, if $M_E = 15, M_C = 12, and S_C = 5, then ES = (15 - 12)/5 = 0.60$. The experimental group's performance exceeded the control group's performance by 0.60 standard deviations. If this were done across several studies addressing a common problem, the findings of the studies would be in a common metric, ES (Δ), which could be compared. The mean and standard deviations of ES can be calculated from several studies. This allows a statement about the average ES of a particular type of treatment.

Suppose we want to know whether a particular treatment affects males and females differently. In searching the literature, we find 15 studies on males comparing the treatment effects and 12 studies on females. We calculate an ES for each of the 27 studies and the mean (and standard deviation) of the ES for males ($n = 15$) and females ($n = 12$). If the ESs were distributed normally, an independent t test could then be used to see whether the average ES differed for males and females. If the t values were significant and the average ES for the females were greater, we could conclude that the treatment had more effect on females than on males. Cooper and Hedges (1994) provide considerable detail on the methods of meta-analysis, including literature search strategies, ways to calculate ES from the statistics reported in various studies, suggestions for how and what to code from studies, and examples of meta-analysis usage.

Certainly, meta-analysis is not the answer to all the problems associated with research synthesis. But Glass provided an objective way to evaluate the literature. Advances (Hedges, 1981, 1982a, 1982b; Hedges & Olkin, 1980) in studying the statistical properties of ES have contributed considerably to the appropriate use of meta-analysis. The text by Hedges and Olkin (1985) provides a complete accounting of the procedures and statistical analyses appropriate for meta-analysis. (Other books have appeared on meta-analysis advocating slightly different procedures, e.g., Hunter & Schmidt, 2004.) A tutorial by Thomas and French (1986) is an overview of Hedges and colleagues' techniques and includes examples from the study of physical activity. Some examples of meta-analyses that have appeared in the physical activity literature follow.

Examples of Meta-Analysis

Meta-analyses have been published since 1976, including many about physical activity. A brief overview of some of the more recent studies shows the value of meta-analysis.

Mann, Williams, Ward, and Janelle (2007) calculated 388 effect sizes across five dependent measures (response accuracy, response time, fixation duration, number of fixations, and quiet eye duration) to provide practical and theoretical implications for sport expertise.

They assigned study variables to three categories: sport type, research paradigm, and stimulus presentation. Sport type was coded as *interceptive*, *strategic*, or *other*. Results indicated that experts were faster and more accurate in decision-making than athletes who were less skilled. Experts demonstrated different eye movement patterns than those with less skill. Meta-analysis allowed a comparison of the research presentation types—video, static, and field. The field type produced the greatest differences favoring experts. This finding should influence future research and explain some previous findings; it also suggests the importance of ecological validity when designing studies to determine the expert advantage. Unfortunately, field experiments are difficult to control; on the other hand, video presentations are easier to control and better than static presentations. The following three examples illustrate the value of the meta-analysis approach.

Schieffer and Thomas (2012) evaluated the findings from 12 school-based intervention programs in which data were reported on both intervention and control schools using components of the coordinated school health model. The benefits from interventions were generally small except for health knowledge. Outcomes were better when more components of the model were included and when interventions were longer. However, they reported problems with some of the designs and analyses.

Gordon, Tucker, Burke, and Carron (2013) examined the effectiveness of physical activity interventions on preschool children. Using effect sizes from 15 studies, they found that interventions had small to moderate effects on general physical activity (ES = 0.44). They reported that most children were not meeting general physical activity guidelines but that the interventions did increase general physical activity.

Kang, Summers, and Cauraugh (2015) completed a meta-analysis on 17 studies using direct current stimulation to improve motor function after a stroke. Of particular interest was long-term retention. Their conclusion was that transcranial stimulation produced robust benefits in stroke recovery.

> A properly applied and interpreted meta-analysis will reveal the underlying principles for a large number of studies.

Meta-analysis, when applied appropriately and interpreted carefully, offers a means of reducing a large quantity of studies to underlying principles. These principles can become the basis for program development, future research, and theory testing, as well as practical applications, such as practice and training.

Methodological Considerations

Although meta-analysis has been a widely used technique in the last 30 years, early on, there was considerable controversy about its validity (e.g., Carlberg et al., 1984; Slavin, 1984a, 1984b). Among the most severe criticisms of meta-analysis is that it combines findings from studies representing different measurement scales, methodologies, and experimental designs. This mixing of apples and oranges was compounded by the results of early meta-analyses using the methods of Glass (1977), which tended to reveal few differences in ESs, even between studies in which internal validity and methodological control clearly varied. Hedges (1981, 1982a, 1982b) and Hedges and Olkin (1983, 1985) extended Glass' original work in 1977 and proposed a new set of techniques and statistical tests to address the following questions and criticisms of meta-analysis:

- How should the process of deciding the variables to code and the systematic coding be organized?
- What should be used as the standard deviation when calculating an ES?
- Because sample ESs are biased estimators of the population of ESs, how can this bias be corrected?
- Should ESs be weighted for their sample size?
- Are all ESs in a sample from the same population of ESs? This is the apples and oranges issue: Is the sample of ESs homogeneous?

- What are appropriate statistical tests for analyzing ESs?
- If a sample of ESs includes outliers, how can they be identified?

In the following sections, we address each of these questions, summarize the theoretical basis for the statistical procedures introduced by Hedges (1981, 1982a, 1982b) and Hedges and Olkin (1983, 1985), and suggest applications of these techniques in the study of physical activity. Rosenthal (1994) has an excellent chapter (16) that provides all the formulas for conducting a meta-analysis when data are normally distributed.

Deciding What to Code

One of the most difficult yet most important tasks in meta-analysis is choosing variables to code and developing a scheme for coding them. Stock (1994) provided a useful discussion of this topic. In particular, the meta-analyst must realize the trade-off between the number and importance of items to code and the time required for coding. Of course, the more information that is coded about a study, the more time is required to develop a coding scheme and do the coding. On the other hand, omitting potentially important items to code also creates major problems because the scholar then has to go back through all of the studies to pick up these omitted items.

> Being familiar with the relevant literature and proficient with meta-analysis procedures helps in the planning of coding.

The best way to be successful in selecting items to code and creating a coding scheme is to know the theoretical and empirical literature on which you are doing the meta-analysis and to understand meta-analysis procedures. Stock (1994) suggested the following two important considerations for planning the coding for a meta-analysis—selecting the items to code and organizing them.

The meta-analyst looks for variables to code that are likely to influence (or be related to) the effect sizes that will be calculated. For example, Mann, Williams, Ward, and Janelle (2007) coded three moderators in cognitive-perceptual research on experts: sport type, research paradigm, and stimulus presentation. We previously listed the levels for sport type and stimulus presentation. Research paradigm includes the following levels: anticipation, decision-making, eye movement, recall, recognition, spatial awareness, task performance, and temporal awareness. The moderators and levels were selected a priori based on Mann's and colleagues' knowledge of the literature and then coded as part of each study included in the meta-analysis.

Stock (1994) provided a generalized example of a coding form. An important issue here is how to code variables, or how to assign numbers to characteristics to be coded. Sometimes specific information is maintained; for example, if the average age of participants from each study is important, this average age can be entered as the code. In other instances, variables are grouped into categories for coding; for example, categories of level of expertise could be coded as 1 = internationally ranked athletes, 2 = nationally ranked athletes, 3 = college or university athletes, and 4 = high school athletes. Other important issues include the training of coders and establishing and maintaining intra- and intercoder reliability.

Choosing the Standard Deviation for the Effect Size

Originally, Glass and colleagues (1981) proposed the use of the control group's standard deviation as the most appropriate measure of group variability. From an intuitive perspective, the control group's variability represents the "normal" variation in an untreated population. The control group's standard deviation also has the advantage of assigning equal ESs to equal treatment means when a study contains two or more experimental groups that have heterogeneous variances. Therefore, the control group's standard deviation serves as a common standard metric from which treatment differences can be compared.

In most cases, estimates of group variance are homogeneous. Hedges (1981) argued that a pooled estimate of the variance provides a more precise estimate of the population variance. (The square root of the pooled estimate of the variance is the pooled estimate of the standard deviation.) One advantage of pooling variances is an increase in the degrees of freedom associated with the estimate of variance. The pooled estimate Hedges (1981) suggested is given in equation 14.2. Note that the variance of each group is weighted by the sample size of that group in a way that is similar to the procedure used in t tests with unequal ns. The formula for the pooled estimate of standard deviation (s_p) is

$$s_P = \sqrt{\frac{(n_E - 1)S_E^2 + (n_C - 1)S_C^2}{n_E + n_C - 2}} \tag{14.2}$$

where n_E is the sample size of the experimental group (group 1), n_C is the sample size of the control group (group 2), s_E^2 is the variance of the experimental group (group 1), and s_C^2 is the variance of the control group (group 2).

Many studies involve tests of effects of categorical variables (e.g., race and sex) where there is no control condition. Early meta-analyses used an average of the standard deviations of the groups compared (Glass & Smith, 1979; Hyde, 1981; Smith, 1980). Hedges and Olkin (1985) suggested using the weighted pooled standard deviation (s_p, equation 14.2) as the estimate of the standard deviation (for an example using sex differences in motor performance, see Thomas & French, 1985).

In some cases, the variance for the groups is heterogeneous. Glass and colleagues (1981) showed that the ESs are biased when the group variances are unequal. Because parametric tests are based on the assumption of equal variances, ESs calculated from t and ANOVA may be biased if the group variances are unequal.

Researchers should evaluate whether heterogeneous variances are a common phenomenon in the area of research for the meta-analysis. We offer these suggestions: When the ES compares an experimental group and a control group and the group variances are unequal, use the standard deviation of the control group to calculate the ES in all studies. If the ES compares two groups (such as different ages or sexes) in which no clear control condition is present, we believe that the weighted pooled estimate (equation 14.2) suggested by Hedges (1981) is the best choice.

Calculating Effect Sizes for Within-Subjects Designs

Often, a researcher wants to calculate an effect size for a within-subjects design (or repeated-measures effect), usually between a pretest and posttest for a treatment effect (i.e., experimental group). The appropriate formula for this effect size is to use the pretest standard deviation in the denominator of equation 14.1 (Looney, Feltz, & VanVleet, 1994). This represents the best source of untreated variance against which to standardize differences between the pre- and posttest means.

Using Effect Size as an Estimator of Treatment Effects

Hedges (1981) provided a theoretical and structural model for the use of ES as an estimator of treatment effects. An individual ES may be viewed as a sample statistic that estimates the population of possible treatment effects within a given experiment.

Hedges (1981) also demonstrated that ESs are positively biased in small samples; the bias is 20% or less, however, when the sample size exceeds 20. A virtually unbiased estimate of ES can be obtained by multiplying the ES by the correction factor given in the following formula:

$$C = 1 - \frac{3}{4m - 9} \tag{14.3}$$

where $m = n_E + n_C - 2$ when a pooled estimate is used as the standard deviation, $m = n_1 + n_2 - 2$ if a pooled estimate is used in a categorical model, or $m = n_C - 1$

when the control-group standard deviation (or pretest standard deviation in the case of a within-subjects ES) is used.

Each ES should be corrected before averaging or further analysis. If ESs are not corrected before averaging, the average of even a large number of ESs remains biased and simply estimates an incorrect value more precisely (Hedges, 1981).

Although the individual ES estimate can be corrected for bias, the variability associated with the estimate remains a function of the sample size. Hedges (1981) showed that the variance of an individual ES can be directly calculated from the following formula:

$$\mathrm{var(ES}_i) = \frac{n_\mathrm{E} + n_\mathrm{C}}{n_\mathrm{E} n_\mathrm{C}} + \frac{ES_i^2}{2(n_\mathrm{E} + n_\mathrm{C})} \tag{14.4}$$

where n_E is the sample size of the experimental group (group 1), n_C is the sample size of the control group (group 2), and ES_i is the estimate of ES. Note that the variability associated with the sample statistic, or ES, is a function of the value of the ES and the sample size. An ES based on a large sample has a smaller variance than an ES based on a small sample. Therefore, ESs based on large samples are better estimates of the population parameter of treatment effects. Hedges (1981) and Hedges and Olkin (1985) suggested that each ES should be weighted by the reciprocal of its variance, that is, 1 / variance (as estimated from equation 14.4). Therefore, ESs that are more precise receive more weight in each analysis.

Testing Homogeneity

Meta-analysis has been criticized for combining studies with various measurement scales, designs, and methodologies. No appropriate test existed to determine whether all the ESs were estimating the same population treatment effect until Hedges (1982b) introduced his test for homogeneity. The homogeneity statistic, H, is specifically designed to test the null hypothesis, $H_0: ES_1 = ES_2 = ... = ES_i$. This is equivalent to saying that all ESs tested come from the same population of ESs.

The statistic H is the weighted sum of squared deviations of ESs from the overall weighted mean. The contribution of each ES to the overall mean is weighted by the reciprocal of its variance (equation 14.4). Effect sizes with smaller variances receive more weight in the calculation of the overall mean. Under the null hypothesis, H has a chi-square distribution with $N - 1$ degrees of freedom, where N equals the number of ESs.

When the null hypothesis is not rejected, all ESs are similar and represent a similar measure of treatment effectiveness. In this case, the researcher should report that the homogeneity statistic indicates that ESs are homogeneous and use the weighted mean ES with confidence intervals for interpretation. If the null hypothesis is rejected, ESs are not homogeneous and do not represent a similar measure of treatment effectiveness, or grouping. Two methods have been proposed by Hedges (1982b) and Hedges and Olkin (1983) to examine explanatory models for ESs. We look at these models next.

Analyzing Variance and Weighted Regression

The first method to fit an explanatory model to ES data is analogous to ANOVA, in which the sum of squares for the total statistic is partitioned into a sum of squares between groups of ESs (H_B) and a sum of squares within groups of ESs (H_W). Each sum of squares can be tested as a chi square with $k - 1$ df for H_B and $N - k - 1$ df for H_W (where N is the number of ESs and k is the number of groups). Therefore, a test can be conducted for between-group differences (H_B), and a test can be conducted to determine whether all ESs within a group are homogeneous (H_W). Further discussion of the categorical model is presented in Hedges (1982b) and Hedges and Olkin (1985).

The second method proposed by Hedges and Olkin (1983) to fit an explanatory model to ES data is a weighted regression technique. We have chosen to present a more detailed discussion of the regression techniques for several reasons. First, one of the meta-analyses discussed previously in exercise and sport used regression techniques to analyze ES (Thomas & French, 1985). Second, it is common for a continuous variable to influence ES. Third, often, more than one study characteristic influences ES, especially when there are many ESs. The regression procedures can accommodate a larger number of variables in the analysis without inflating the alpha level of the statistical test. Conducting many tests using the categorical or ANOVA-like procedure results in alpha inflation, so the researcher would need to report the experiment-wise error rate or adjust the alpha level using the Bonferroni technique (see chapter 9). Fourth, categorical variables can easily be dummy- or effect-coded and entered into the regression procedures. For example, published versus unpublished papers can be dummy-coded (1 or 0) and entered into the regression. Thus, conducting a separate analysis for categorical variables is not necessary.

Each ES is weighted by the reciprocal of its variance in the weighted regression technique. Most standard statistical packages (e.g., SAS, SPSS, BIOMED) have an option to perform weighted regression. Thus, the computations can be done easily on most computers.

The sum of squares total for the regression is equivalent to the homogeneity statistic, H. It is partitioned into a sum of squares regression and a sum of squares error. The **sum of squares** for regression yields a test of the variance due to the predictor variables (H_R). The H_R is tested as a chi square with $df = p$, where p equals the number of predictor variables. The sum of squares for error (H_E) provides a test of model specifications, or whether the ESs are homogeneous when the variance caused by the predictors is removed. The H_E is tested as a chi square with $df = N - p - 1$, where N equals the number of ESs and p equals the number of predictor variables. The test of model specification evaluates the deviation of the ESs from the regression model. A nonsignificant test of model specification indicates that the ESs do not deviate substantially from the regression model, whereas a significant test for model specification shows that one or more ESs deviate substantially from the regression model. Ideally, the researcher wants the H_R for regression to be significant and the H_E for model specification to be nonsignificant.

When the test for model specifications is significant, one or more ESs do not follow the specified regression. Often, the ESs that do not follow the same pattern may suggest other characteristics that can be added to the model (e.g., published versus unpublished studies). Moreover, some ESs may represent outliers (unrepresentative scores). In either case, the use of outlier techniques is helpful in identifying these ESs.

sum of squares—A measure of the variability of scores; the sum of the squared deviations from the mean of the scores.

Testing for Outliers

Hedges and Olkin (1985) outlined procedures to identify outliers in categorical and regression models. We use an example from a regression model (for more information concerning the categorical model, see chapter 7 in Hedges & Olkin, 1985).

Outliers in regression models can be identified by examining the residuals of the regression equation. The absolute values of the residuals are standardized to z scores by subtracting the mean and dividing by the standard deviation. Effect sizes with standardized residuals larger than 2 are often examined as potential outliers because they fall outside 95% of the distribution.

According to Hedges and Olkin (1985), each ES makes a contribution to the regression model. The outliers identified by standardized residuals may vary, depending on which ESs are in the model at the time. Hedges and Olkin (1985) suggested computing the standardized residuals multiple times with a different ES deleted from the model each time. For example, if you have 10 ESs, standardized residuals would be computed 10 times with a different ES deleted from the model each time. But some of our preliminary work suggested that if more than 20 ESs are included in the regression model, going through all possible combinations of deleting ESs from the model may be of limited value. Simply calculate the residuals with

Just because things are on the same path does not mean they fit together in one analysis.

© Andrew Kazmierski/Fotolia

all the ESs in the model, standardize the residuals, and evaluate whether ESs with a z score larger than 2 might be outliers.

Alternatively, if the ESs are not normally distributed (as Hedges & Olkin, 1985, reported is often the case), the rank-order procedures described in chapter 10 (and developed in greater detail by Thomas, Nelson, & Thomas, 1999) are effective if ESs are changed to ranks before statistical analysis.

Accounting for Publication Bias

Journals have a tendency to accept papers that report significant findings. Thus, there may be some number of studies on any topic (including one on which a meta-analysis is being done) that have not been submitted to or published in scholarly journals. This problem has been labeled the file-drawer effect.

An important question in a meta-analysis is how many of these studies are sitting around in file drawers. The meta-analyst should find every study possible, but how many studies have been done that have not been reported? Hedges and Olkin (1985) suggested a technique for estimating how many unpublished studies with no significant effect on the variable of interest would be needed to reduce the mean ES to nonsignificance.

$$K_0 = [K(d_{ES\ mean} - d_{ES\ trivial})]/d_{ES\ trivial} \qquad (14.5)$$

where K_0 is number of studies needed to produce a trivial ES, K is the number of studies in the meta-analysis, $d_{ES\ mean}$ is mean of all the ESs in the meta-analysis, and $d_{ES\ trivial}$ is an estimate of a trivial ES.

For example, if there were 75 studies in the meta-analysis (K), if the mean ES ($d_{ES\ mean}$) was 0.73, and if a trivial effect size ($d_{ES\ trivial}$) was established as 0.15, to estimate the number of file-drawer studies needed to produce a trivial ES (K_0), we solve equation 14.5:

$$K_0 = [75(0.73 - 0.15)] / 0.15 = 290$$

Thus, there would need to be an inordinate number (290) of unpublished studies on this topic to reduce the average ES of 0.73 in the study to a trivial ES of 0.15.

Other Considerations

Sometimes all the information needed to calculate an ES is not available in a research report. Rosenthal (1994) provided an excellent discussion, formulas, and examples of how to estimate effect sizes from other statistics that are often provided in individual studies. On occasion, however, only the statement that two groups do not differ significantly is reported. What should the meta-analyst do? To drop the study is to bias the results of the meta-analysis toward significant differences. One can insert a zero ES, using the rationale that no significant differences means that the differences did not reliably differ from zero. Using many zero ESs may cause problems in the analysis of homogeneity by making the studies look too consistent. Hedges (1984) suggested a correction factor that adjusts for this problem. Regardless of the solution selected, the meta-analyst must cope with this issue and select a logical solution.

Presenting Effect Size Data

Graphic displays are extremely useful when presenting effect size data. According to Light, Singer, and Willett (1994), graphs enable the reader to see important characteristics of relationships in the ES data. In particular, graphic displays of ES data are useful for showing sources of variation in effect size displaying the center, spread, and shape of the distribution; such displays include stem-and-leaf plots, schematic plots (box and whiskers), and funnel graphs (effect sizes plotted against sample sizes). Graphic displays of ES data are also useful for sorting effect sizes by important study characteristics, such as treatment intensity, age, and sex. These displays should also reflect the variability of effect sizes for sorted characteristics (e.g., 95% confidence intervals).

Light, Singer, and Willett (1994, pp. 451–452) listed five valuable concepts for graphic displays of meta-analytic data (of course, these concepts apply equally well to studies reporting original data):

1. The spirit of good graphics can even be captured by a table of raw ESs, provided that they are presented in a purposeful way (e.g., by magnitude or important study characteristics).

2. Graphic displays should be used to emphasize the big picture (e.g., make the main points clear).

3. Graphics are especially valuable for communicating findings from a meta-analysis when each individual study outcome is accompanied by a confidence interval (this demonstrates not only the variability of the effect sizes of interest but also whether they are different from the null value).

4. Graphic displays should encourage the eye to compare different pieces of information.

5. Stem-and-leaf plots and schematic plots are especially efficient for presenting the big picture that emerges in a research synthesis.

Using Systematic Review for Research Synthesis

The purpose of a systematic literature review (SLR or systematic review) is to synthesize evidence within a given predefined area to address specific research questions. Similar to meta-analysis, searches are conducted to find published research with predetermined criteria for inclusion and exclusion. This distinguishes the systematic review (SLR) from the literature review (LR) used to identify and introduce a research topic as described in chapter 2. Robinson and Lowe (2015) identify eight points of comparison:

1. SLRs have a specific question, whereas LRs often have no specific question and cover several topics.

2. SLRs use targeted search of specific databases, whereas LRs are not systematic and are often random.

3. SLRs extract data, while LRs present "take home messages."

4. SLRs include 10-50 papers, whereas LRs include more papers.

5. SLRs analyze methods, rigor, and strength of the research; LRs are the writers' interpretations.

6. SLRs include a chart of included studies, but LRs are narratives.

7. SLRs are publishable, but LRs are not publishable.

8. SLRs produce actions based on evidence, while LRs are informed by evidence in the papers reviewed.

We think a strength of the systematic review is thoroughness. This is a result of two characteristics of systematic reviews, reduced bias and increased transparency. Using two or more reviewers who determine inclusion and exclusion and data extraction stand as one control for bias (Robinson & Lowe, 2015). Transparency is achieved by clearly stating the method at the outset and carefully following, reporting, and documenting the method (Pati & Lorusso, 2018).

Systematic reviews are useful to practitioners as the number of studies and volume of information in health care and health science explodes (Pati & Lorusso, 2018). A well-done systematic review provides ready access to and collection of multiple research publications on a single topic. Extensive guidance for conducting systematic reviews of health care interventions is available from the *Cochrane Handbook for Systematic Reviews of Interventions* (Higgins, Thomas, Chandler, Cumpston, Li, Page, & Welch, 2019). Pati and Lorusso (2018) advocate for PRISMA (Preferred Reporting Items for Systematic Reviews and Meta-Analysis). PRISMA was a collaborative effort to enhance rigor and quality of systematic reviews of literature. While Cochrane and PRISMA were developed by and for health care, the method in each is informative for other fields including kinesiology.

Examples of Systematic Reviews

Logan, Webster, Getchell, Pfeiffer, and Robinson (2015) conducted a systematic review examining the impact of fundamental motor skills on physical activity in children. There were seven inclusion criteria. Children were typically developing and between ages 3 and 13 years, fundamental motor skills (FMS) were measured using process instrument, physical activity (PA) was measured, and the relationship between FMS and PA was statistically analyzed. Thirteen studies met the criteria. Two researchers extracted data from the studies with 93% agreement. They included a figure delineating the literature search and a table summarizing the findings of the included studies. The relationship between FMS and PA was low to moderate in early childhood (4 studies) and adolescence (2 studies), and low to high in middle to late childhood (7 studies). This systematic review found varying levels of support for the two dominate models of motor development and suggested areas for future research.

Stathokostas, Little, Vandervoort, and Paterson (2012) examined the impact of flexibility training on functional outcomes in healthy adults over 65 years of age. Functional outcomes were a critical interest of the Canadian government and to all who plan to age gracefully.

Twenty-two studies met the criteria for inclusion. They determined that flexibility training had a positive impact on range of motion but were unable to determine whether that training had a positive impact on functional outcomes.

Alcantara, Sanchez-Delgado, Martinez-Tellez, Labayen, and Ruiz (2019) conducted a systematic review to determine the impact of cow's milk on muscle function and exercise performance. The study followed PRISMA guidelines and was registered with PROSPERO (an international preregistration for SLRs). The search retrieved 7,708 articles of which 11 met the inclusion criteria. The authors deemed the studies generally "suboptimal." The results varied widely; thus, the authors were unable to draw conclusions other than more research is necessary. While this seems inconclusive, the systematic review demonstrates an important point. Results of studies should be reliable as demonstrated through replication, or at a minimum, similar results across studies. The authors made specific suggestions for higher-quality future research, for example, double-blind studies. Our point is that systematic reviews provide information not available when examining one published study with significant findings.

Conducting Systematic Reviews

Research begins with a good question, so an SLR must also do the same. To frame a well-founded question, one must have a grasp of the literature: Are there enough studies to warrant research synthesis? The answer to a meaningful research question will affect the field. The impact may be to underscore what is currently known, influence practice, or suggest future research. Key dependent and independent variables must be specified for the search as well as other factors, such as theses and dissertations, English-only journals, years of publication, and a selection of databases to be searched. Graduate student research is by nature a team event—minimally, the student, advisor, and committee. For systematic reviews, team participation is required because more than one person will review the literature. Further, your institution's library staff are valuable assets, particularly in selecting and using databases. Preregistration of the study occurs at this point with either PROSPERO or Open Science Framework (Gunnell, Poitras, & Tod, 2020). Now the work begins with the four parts of the systematic review; identification, screening, eligibility, and inclusion (Moher, Liberati, Tetzlaff, Altman, and the PRISMA Group, 2009). These steps will be presented in a flowchart. During the process, modifications are often made to the protocol. Reviewers are better able to judge the appropriateness of modifications when the initial protocols are clearly written (and preregistered) and modifications explained in the final product.

- *Identification:* At this point, we count all studies found in the database search. Additional studies may also be identified by other means, such as combing the reference lists and dissertation abstracts. We count these separately from those we identified in our initial database search, and then write a clear statement of the process we've used. This includes noting when we ceased searching the database. Considering the amount of time it takes to screen studies and extract data, study identification is much like a two-step process. First, we gather and process the majority of studies and then we do a second search to identify studies published since our previous searches. This last search should occur as near to completion of the work or submission for publication as possible (Gunnell et al., 2019).

- *Screening:* Titles and abstracts typically provide the information needed for screening. Some studies will appear that are clearly not related to your question; for example, you are interested in humans but a title indicates the subjects are canine. Of course, the same study may appear in more than one database, so this is the time to remove any duplicates and report the number of duplicates you've removed. You will also have to determine how to handle the same study being published, usually in parts or parsed differently, then published separately.

- *Eligibility:* The best practice is having at least two people review the studies to determine eligibility. Another approach is to have a second reviewer examine 20% of the articles (Pati & Lorusso, 2018). Here we report the agreement of reviewers in all cases.

- *Inclusion:* In this last step, we report the number of studies from which we will extract data based on a full article review. These studies are reported in the table with the critical review.

The results from identification, screening, eligibility and inclusion are presented in a flowchart. The charts will look different but clearly present the following:

- How the literature was searched
- What and when exclusions were made
- The number of exclusions

Two examples of these charts are presented in figures 14.1 and 14.2. We previously described the Logan et al. (2015) systematic review that looked at the relationship between fundamental motor skills and physical activity. McBain, Shrier, Schultz, Meeuwisse, Klügl, Garza, and Matheson (2012) examined interventions to reduce sport injuries that measured biomechanical and physiological outcomes. They noted a change in study design from cross-over primarily examining equipment design to randomized control trials in recent years. Further, they noted that few studies attempted to identify factors leading to injury.

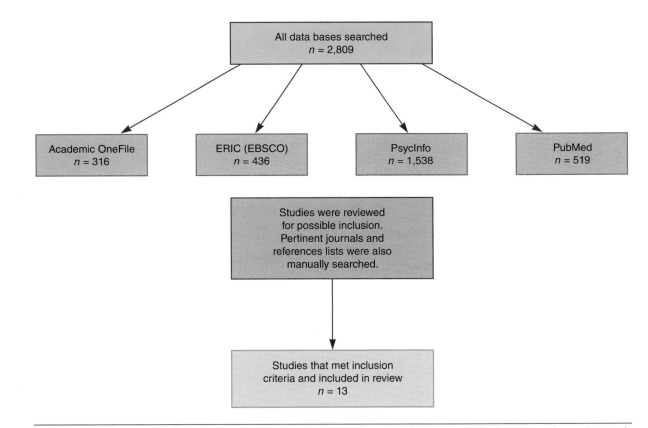

Figure 14.1   Example description of the process used to identify studies from varying databases and determine study inclusion.

Reprinted by permission from S. Logan, E. Webster, N. Getchell, et al., "Relationship Between Fundamental Motor Skill Competence and Physical Activity During Childhood and Adolescence: A Systematic Review," *Kinesiology Review* 4 (2015): 416-426.

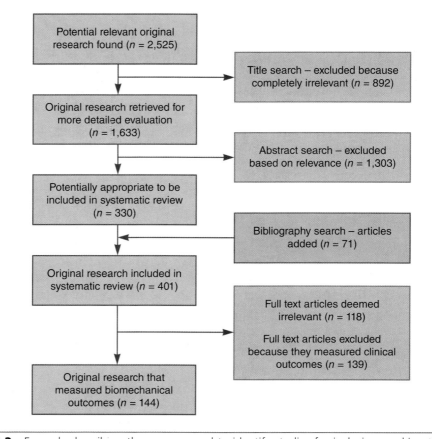

Figure 14.2   Example describing the process used to identify studies for inclusion working through various components of the study.

Reprinted from K. McBain et al., "Prevention of Sports Injury I: A Systematic Review of Applied Biomechanics and Physiology Outcomes Research," *British Journal of Sports Medicine* 46 (2012): 169-173.

Data extraction and critical analysis focus on the research question. A table is created based on the data extracted from included studies. The table includes author name, year of the publication, pertinent information about the publication, description of the inclusion variables, outcomes, assessment of robustness, quality, risk bias, and other information as appropriate. We previously described the method used by Stathokostas et al. (2012). Here we present one section of the table quantifying the 22 studies (table 14.2). To assess quality, Stathokostas and colleagues used a checklist modified from other sources with a high score of 24. In the example, the study earned 19 of the possible 24 points for quality.

The discussion summarizes the findings based on the research question, starting with the most important finding first. The discussion should include publication bias, limitations, and conclusions (Moher et al., 2009). As you can see, a systematic review has similar organization to other research reports (see chapter 21).

TABLE 14.2

An Example Table Used for Data Analysis

Publication country/study type	Objective	Population	Methods	Outcomes	Comments and conclusions	Quality
Christiansen 2008 [22]. USA. RCT. **Focus** Hips and ankles	To examine the effects of a hip and ankle static stretching program on freely chosen gate speed of healthy, community-dwelling older people not active in exercise.	$n = 37$ Age: 72 ± 4.7 yr. **Intervention group:** $n = 18$. Age: 72 ± 4.7 yr. 3 male, 15 female. **Control group:** $n = 19$. Age: 72 ± 5.0 yr. 5 male, 14 female. Independently living. **Inclusion criteria:** In desired age range, healthy, no joint or musculoskeletal pain that limited movement in the past month, no diagnosed gait or balance disorder, no falls history, has not participated in a formal exercise program during the previous 6 months, and has not used an assistive device for walking.	Pre-post 8 weeks. **Intervention group:** Hip and ankle stretching, 2 static stretches held for 45 seconds and repeated 3 times, alternating sides; total 9 minutes (540 s)/session. Stretches are standing calf stretch and standing hip flexor stretch. **Control group:** Ensured no changes in current physical activity. **Assessments:** *Passive ROM:* goniometric measurements of hip (hip extensions based on Thomas test position) and ankle. *Gait:* shoes on; two walking speeds. **Analysis:** Independent t tests or chi-square tests, ICC model 2 and form 1, 2-factor repeated-measures ANOVA, separate repeated-measures ANOVA, paired t tests, and Bonferroni adjustment.	85% compliance. *Gait:* significant increase in freely-chosen gate speed for intervention group (1.23 to 1.30 m s⁻¹, +0.7 m s⁻¹, and $p = 0.016$ versus control (no change). *Joint motion:* peak hip extension and knee flexion increased significantly (59.7 to 66.5°) in intervention group, with no change in control group (56.2 to 56.1°). Significant increase in intervention group (7.8 to 11.3°) for passive dorsiflexion, with no change in the control group. *Other:* No significant changes in stride length or joint angular displacement.	Evidence is provided from the results that joint motion is a modifiable impairment that can be effectively targeted for older people with simple, static stretching home-based intervention.	19/24

Adapted from Stathokostas et al. (2012).

Summary

A research synthesis is not just a summary of studies arranged in some kind of sequence. It should be structured, analytical, and critical and lead to specific conclusions. Meta-analysis and systematic review use a definitive methodology that quantifies results and allows critical analysis or statistical techniques to be used as a means of analysis.

The research synthesis explicitly details how the search was done, the sources used, the choices made regarding inclusion or exclusion, the coding of study characteristics, and the analytical procedures used. The basis of a meta-analysis is the effect size, which transforms differences between experimental and control groups' performances to a common metric, which is expressed in standard deviation units. The basis of systematic review is an a priori protocol for identification, screening, eligibility, and inclusion that reduces bias and increases transparency, thereby leading to critical analysis.

Research synthesis offers a means of reducing large quantities of studies to underlying principles. There is available guidance for conducting a review from Cochran and PRISMA including a system for preregistering planned research all with international input. This demonstrates not only the importance of research synthesis but also the efforts to improve the quality of research of this type.

✓ Check Your Understanding

1. Select a possible problem (e.g., sex differences in running speed, the effects of weight training on males and females, or the influence of training at 60% versus 80% of $\dot{V}O_2max$). Find five studies that compare the characteristics that you select (e.g., males and females). Make sure each study has the means and standard deviations for the variables of interest. Calculate the ES for each study and correct it for bias. Calculate the average ES for the five studies and their pooled standard deviations. Interpret this finding.

2. Select a possible problem (e.g., optimal practice organization for learning a motor skill, ACL injuries by sport and sex, audience influence on sport performance). Identify three databases and search for publications in each. Record the number of unique (not duplicate) titles that are pertinent to your problem.

15

Surveys

Most surveys represent the average opinion of people who don't know.

—Unknown

Descriptive research is a study of status and is widely used in education and the behavioral sciences. Its value is based on the premise that problems can be solved and practices improved through objective and thorough description. The most common descriptive research method is the **survey**. The survey is generally broad in scope. The researcher usually seeks to determine present practices (or opinions) of a population. The survey is used in the fields of education, psychology, sociology, and physical activity. The questionnaire, the Delphi method, the personal interview, and the normative survey are the main types of surveys.

Questionnaires

The **questionnaire** and the **interview** are essentially the same except for the method of questioning. Questionnaires comprise specific questions that participants usually answer online or on paper, whereas interviews are usually conducted orally. As noted later in this chapter and in chapter 19, interviews for qualitative research are generally open ended and may have different goals than the interview-based survey. The procedures for developing questionnaire and interview items are similar. Consequently, much of the discussion regarding the steps in the construction of the questionnaire also pertains to the interview. Similarly, much of this advice is the same for either a paper survey or one conducted electronically on the Internet. The method of delivering the survey may depend on the sample—electronic if participants are geographically dispersed and paper if it is easier to gain direct access to the participants in person, although much survey research is now conducted online.

Researchers use the questionnaire to obtain information by asking participants to respond to questions rather than by observing their behavior. The obvious limitation of the questionnaire is that the results consist simply of what people say they do or what they say they believe or like or dislike. But certain information can be obtained only in this manner, so the questionnaire must be planned and prepared carefully to ensure the most valid results. The survey research process includes eight steps, and we discuss each in this section, which

survey—A technique of descriptive research that seeks to determine present practices or opinions of a population; it can take the form of a questionnaire, interview, or normative survey.

questionnaire—A paper-and-pencil or electronic survey used in descriptive research in which information is obtained by asking participants to respond to questions rather than by observing their behavior.

interview—A survey technique similar to the questionnaire except that participants are questioned and respond orally rather than in writing.

covers both online and paper-based surveys that use the postal service for sending and returning the survey.

Determining the Objectives

This step may seem too obvious to mention, yet countless questionnaires have been prepared without clearly defined objectives. In fact, poor planning may account for the low esteem in which survey research is sometimes held. The investigator must have a clear understanding of what information is needed and how each item will be analyzed. As with any research, the analysis is determined in the planning phase of the study, not after the data have been gathered.

The researcher must decide on the specific purposes of the questionnaire: What information is wanted? How will the responses be analyzed? Will they be described by merely listing the percentages of participants who responded in certain ways, or will the responses of one group be compared with those of another?

One of the most common mistakes in constructing a questionnaire is not specifying the variables to be analyzed. In some cases, when investigators fail to list the variables, they ask questions unrelated to the objectives. In other cases, investigators forget to ask pertinent questions.

Delimiting the Sample

Most researchers who use questionnaires have in mind a specific population to be sampled. Obviously, the participants selected must be the ones who have the answers to the questions. In other words, the investigator must know who can supply what information. If information about policy decisions is desired, the respondents should be those involved in making such decisions.

Sometimes the source used in selecting the sample is inadequate. For example, some professional associations are focused on very narrow groups while others are less focused in their membership. A study examining strength and conditioning practices in college athletes would find the CSCCa (Collegiate Strength and Conditioning Coaches Association) a good source of participants because that association serves college and university professional strength and conditioning coaches. However, if your question was related to senior citizen strength and conditioning, CSCCa would not be an appropriate choice. Determining your target participants and how you will contact them is critical. Unless some screening mechanism is used regarding place of employment, many incomplete questionnaires will be returned because the questions were not applicable.

The representativeness of the sample is an important consideration. Stratified random sampling, as discussed in chapter 6, is sometimes used. If a researcher is studying fitness programs for persons at least 65 years of age, the majority of the programs should be community based (e.g., YMCA, health club) because 78% of people over age 65 live in the community. So, 78 programs surveyed should be community-based and 22 facility-based. However, if the target age increases to 75 years and older, the number of facility-based programs would increase as the number of people living in facilities increases with age.

The selection of the sample should be based on the variables specified to be studied. This affects the generalizability of the results. If an investigator specifies that the study deals with just one sex, age level, or one institution, then the population is narrowly defined, and it may be easy to select a representative sample of that population. But the generalizations that can be made from the results are also restricted to that population. On the other hand, if the researcher is aiming the questionnaire at everyone in a population (e.g., strength and conditioning coaches or all fitness instructors), then the generalizability is enhanced, but the sampling procedures become more difficult. The representativeness of the sample is more important than its size.

Although a questionnaire is limited by what respondents say they do, believe, or prefer, it is the only technique that can gather certain information.

Accordion to a recent survey, replacing words with the names of musical instruments in a sentence often goes undetected.

In some instances, surveys address populations that are difficult to sample in the ways noted earlier. For example, identifying all the possible participants for a national study of physical education teachers may be impossible. Obtaining a listing from every state or school district is impractical, and each may have records that are either more or less current. In these types of studies, once the type of participant is identified, the researcher works to get the largest and most diverse sample size possible. This procedure occurs in a number of ways and is discussed further in the section on online surveys.

> In survey research, representativeness is more important than sample size.

Sampling Error

Generally, news reports of local or national polls about some issue or about an upcoming election include a statement about the sample size and sampling error, such as "The poll involved 1,022 Americans, and the results were subject to a sampling error of plus or minus 3.1 percentage points." How do the pollsters arrive at that error range? We discussed standard error in earlier chapters. Standard error of sampling can be expressed as

$$SE = \sqrt{\text{variance}/n} \tag{15.1}$$

where SE is the standard error of sampling and n is the size of the sample. We know that variance can be defined as the sum of the squared deviations from the sample mean divided by n. In this case, however, variance is calculated as a function of proportions; that is, the percentage of a sample that has a certain characteristic or gives a certain response. Variance of a proportion is calculated as $p(1-p)$, where p is the proportion that has a characteristic or that responds a certain way and $1-p$ is the proportion that does not have the characteristic or that responds differently. These proportions could be of many things, such as males and females, voters and nonvoters, smokers and nonsmokers, or those who believe one way versus those who believe differently, as is common in surveys. The formula can be written as follows:

$$SE = \sqrt{p(1-p)/n} \tag{15.2}$$

If we do not know the proportions (or expected proportions), $p = 0.50$ is used. This proportion results in the largest error. Let us go back to our example of the survey of 1,022 Americans with the 3.1% sampling error. Using equation 15.2, we have this:

$$SE = \sqrt{.50(1 - .50)/1022} = .0156, \text{ or } 1.6\% \tag{15.3}$$

But remember that standard errors are interpreted as standard deviations in that ±1 SE includes roughly 68% of the population; to say it another way, we are only 68% confident that our sample is representative. So, most polls use a 95% confidence interval, which we know is ±1.96 SE. Hence, our finalized formula should be as follows:

$$SE = 1.96\sqrt{p(1-p)/n} \tag{15.4}$$

$$= 1.96\sqrt{.50(1 - .50)/1022}$$

$$= 3.1\%$$

Sample Size

The size of the sample needed is an important consideration from two standpoints: (a) for adequately representing a population and (b) for practical considerations of time and cost. Certain formulas can be used to determine the adequacy of sample size (SurveyMonkey, 2020). These formulas involve probability levels and the amount of sampling error deemed acceptable. The practical considerations of time and cost need attention in the planning phase of the study. While the cost does not change when using an online survey, if using a paper survey, you should sit down with a calculator and tabulate the costs of printing, initial mailing (which includes self-addressed, stamped return envelopes), follow-ups, scoring, and data analysis. Sometimes the costs are so substantial that a sponsoring agency or grant must be found to subsidize the study; otherwise, the project must be narrowed or moved to an online platform—or abandoned entirely. Time is also important concerning participant availability, possible seasonal influences, and deadlines. See Fink (2003, 2005) and Fowler (2009) for discussions on survey sampling, including the calculation of sample size.

Constructing the Questionnaire

The notion that constructing a questionnaire is easy is a fallacy. Questions are not simply "thought up." Anyone who prepares a questionnaire and asks someone to read it soon discovers it is not an easy task after all. The questions that were clear and concise to the writer may be confusing and ambiguous to the respondent.

One of the most valuable guidelines for writing questions is to ask yourself continually what specific objective each question is measuring. Then, ask how you are going to analyze the response. While you are writing questions, it is a good idea to prepare a blank table that includes the categories of responses, comparisons, and other breakdowns of data analysis so that you can readily determine exactly how each item will be handled and how each will contribute to the objectives of the study. Next, you must select the format for the questions, some examples of which follow.

Open-Ended Questions

open-ended question— A category of question in questionnaires and interviews that allows the respondent considerable latitude to express feelings and expand on ideas.

closed question—A category of question in questionnaires and interviews that requires a specific response and that often takes the form of rankings, scaled items, or categorical responses.

ranking—A type of closed question that forces the respondent to place responses in a rank order according to some criterion.

Open-ended questions such as "How do you like your job?" or "What aspects of your job do you like?" may be the easiest to write. Such questions allow the respondent considerable latitude to express feelings and expand on ideas. But several drawbacks to open-ended questions usually make them less desirable than **closed questions**. For example, most respondents do not like open-ended questions, which require more time to answer than closed questions do. For that matter, most people do not like questionnaires because they think that surveys encroach on their time. Another drawback is limited control over the nature of the response: The respondent often rambles and strays from the question. In addition, such responses are difficult to synthesize and to group into categories for interpretation. Although open-ended items can yield valuable information, they are hard to analyze by any means other than simple description.

Sometimes open-ended questions are used to construct closed questions. Student course evaluations are often developed by having students list all the things they like and dislike about a course. From such lists, closed questions are constructed by categorizing the open-ended responses.

Closed Questions

Closed questions come in a variety of forms. A few of the more commonly used closed questions are rankings, scaled items (some measurement scales were covered in chapter 11), and categorical responses.

A **ranking** forces the respondent to place responses in a rank order according to some criterion. As a result, value judgments are made, and the rankings can be summed and analyzed quantitatively. Here is an example of a rank-order response question:

Rank the following activities regarding how you like to spend leisure time. Use numbers 1 through 5, with 1 being the *most preferred* and 5 the *least preferred.*

_____ **Reading**

_____ **Watching television**

_____ **Arts and crafts**

_____ **Vigorous sports such as tennis and racquetball**

_____ **Mild exercise activity such as walking**

Scaled items are one of the most commonly used types of closed questions. Participants are asked to indicate the strength of their agreement or disagreement with some statement or to cite the relative frequency of some behavior, as in the following example:

> **Indicate the frequency with which you are involved in committee meetings and assignments during the academic year.**
>
> *Rarely*   *Sometimes*   *Often*   *Frequently*

scaled item—A type of closed question that requires participants to indicate the strength of their agreement or disagreement with some statement or the relative frequency of some behavior.

A Likert-type scale is a scale with three to nine responses, in which the intervals between responses are assumed to be equal:

> **In a required physical education program, students should be required to take at least one dance class.**
>
> *Strongly agree*   *Agree*   *Undecided*   *Disagree*   *Strongly disagree*

The difference between *Strongly agree* and *Agree* is considered equivalent to the difference between *Disagree* and *Strongly disagree*. Following are examples of words and phrases that can be used in scaled responses: *Excellent, Good, Fair, Poor,* and *Very poor*; *Very important, Important, Not very important,* and *Of no importance.*

Categorical responses offer the respondent only two choices. Usually, the responses are *Yes* and *No* or *Agree* and *Disagree*. An obvious limitation of categorical responses is the lack of other options such as *Sometimes* or *Undecided*. Categorical responses do not require as much time to administer as scaled responses do, but they also do not provide as much information about the respondent's degree of agreement or the frequency of the behavior.

categorical response— A type of closed question that offers only two responses, such as *Yes* or *No*.

Sometimes, questions in questionnaires are keyed to the responses of other items. For example, a question might ask whether the respondent's institution offers a doctoral degree. Respondents who answer *Yes* are directed to answer subsequent questions about the doctoral program. Those who answer *No* are directed either to stop or to move on to the next section.

Gall, Gall, and Borg (2006) offered the following rules for the construction of questionnaire items:

- The items must be clearly worded so that they have the same meaning for all respondents. Avoid words that have no precise meaning, such as *usually, most,* and *generally.*

- Use short questions rather than long questions because short questions are easier to understand.

- Do not use items that have two or more separate ideas in the same question, as in this example: "Although everyone should learn how to swim, passing a swimming test should not be a requirement for graduation from college." This item cannot really be answered by a person who agrees with the first part of the sentence but not the second, or vice versa. Another example is "Does your department require an entrance examination for master's and doctoral students?" This question is confusing because the department may have such an examination for doctoral students but not for master's students, or vice versa. If the response choices are only *Yes* and *No,* a *No* response would indicate no examination for either program, and a *Yes* response would mean exams for both. This subject should be divided into two questions.

- Avoid using negative items, such as "Physical education should not be taught by coaches." Negative items are often confusing, and the negative word is sometimes overlooked, causing respondents to answer in exactly the opposite way from what they intended.

- Avoid technical language and jargon. Attempt to achieve clarity and the same meaning for everyone.

- Be careful not to bias the answer or lead the respondent to answer in a certain way.

Sometimes questions are stated in such a way that the person knows what the "right" response is. For example, you would know what response was expected if you encountered this question: "Because teachers work so hard, shouldn't they receive higher salaries?" The same advice applies to the problem of threatening questions. A respondent who perceives certain items to be threatening will probably not return the questionnaire. A questionnaire for physical therapists on client success, for example, may be viewed as threatening because poor outcomes would indicate that the therapist is not doing a good job rather than the client failing to follow the protocol. Thus, a therapist who feels threatened either may not return the questionnaire or may give responses that seem to be "right" answers.

Results of experiments on surveys show that minor changes in wording, or even in ordering the alternatives, can cause differences in responses. At a seminar at Washington State University, Don Dillman, a recognized authority on the survey method, reported results of experiments on the effects of question form and survey mode on survey results. As an illustration of some of the variations studied, when students were asked how many hours they studied per day, 70% said more than 2.5 hours, but when the same question was asked another way, only 23% indicated more than 2.5 hours.

Appearance and Design

Finally, the entire appearance and format of the questionnaire can have a significant bearing on the return rate. Questionnaires that appear to be poorly organized and poorly prepared are likely to be clicked out of or "filed" in the wastebasket. Remember, many people have negative attitudes toward questionnaires, so anything the researcher can do to overcome this negative attitude enhances the likelihood that the questionnaire will be answered. Some suggestions are merely cosmetic, such as the use of alternating colors for questions. Even small enhancements, such as grouping related items or using an easy-to-read font, may pay big dividends.

The questionnaire should provide the name, address, and email of the investigator. The instructions for answering the questions must be clear and complete, and examples should be provided for any items that are expected to be difficult to understand.

The first few questions should be easy to answer; respondents are more likely to start answering easy questions and are more apt to complete the questionnaire after they are committed. A poor strategy is to begin with questions that require considerable thought or time to gather information. Every effort should be made to make difficult questions as convenient to answer as possible. For example, questions that ask for enrollment figures, the size of a faculty, or the number of graduate assistants can often ask for responses in ranges (e.g., 1-10, 11-20, 21-30). First, you will probably group the responses for analysis purposes anyway. Second, the respondent can often answer range-type questions without having to consult the records or at least can supply the answers more quickly. An even more basic question to ask yourself is Do I really need that information? or Does it pertain to my objectives? Unfortunately, some investigators simply ask for information off the top of their heads with no consideration for the time required to supply the answer or the relevance of the information.

Generally, shorter questionnaires have higher response rates and more validity than longer ones do. The issue is not more data but better data (Punch, 2003). An analysis of 98 questionnaire studies by Gall et al. (2006) showed that, on average, each page added to a questionnaire reduced the number of returns by 0.5%, and this also likely is the case for online surveys. Because many people are reluctant to complete questionnaires, the cover letter (discussed later) and the size and appearance of the questionnaire are crucial. A lengthy questionnaire requiring voluminous information will likely be put aside until later if not discarded immediately.

Conducting the Pilot Study

A pilot study is recommended for any type of research but is imperative for a survey. In fact, the designer of a questionnaire may be well advised to do two pilot studies. The first trial run consists simply of asking a few colleagues or acquaintances to read over the questionnaire. These people can provide valuable critiques about the format, content, expression and importance of items, and feasibility of adding or deleting questions.

After revising the questionnaire in accordance with the criticisms obtained in the first trial, respondents who are a part of the intended population are selected for the second pilot study. The questionnaire is administered, and the results are subjected to item analysis (discussed in chapter 11). In some questionnaires, correlations may be determined between scores on each item and the whole test to see whether the items are measuring what they are intended to measure. Responses are always examined to determine whether the items seem clear and appropriate. First, questions that are answered the same way by all respondents need to be evaluated; they probably lack discrimination. Unexpected responses may indicate that the questions are poorly worded. Some rewording and other changes might also be necessary if the participants, who might be sensitive to some questions, do not respond to them. Furthermore, the pilot study determines whether the instructions are adequate. Another value of the pilot study relates to determining the length of the survey. The researcher should record how long it takes the average participant to complete the survey.

A trial run of the analysis of results should always be accomplished in the pilot study. The researcher can see whether the items can be analyzed in a meaningful way and then ascertain whether some changes are warranted for easier analysis. This is one of the most profitable outcomes of the pilot study. Of course, if substantial changes are mandated by the results of the pilot study, another pilot study is recommended to determine the questionnaire's state of readiness for distribution.

Writing the Cover Letter

Unquestionably, the success of the initial mailing depends largely on the effectiveness of the **cover letter** that accompanies the questionnaire. If the cover letter explains the purposes and importance of the survey in a succinct and professional manner (and if the purposes are worthy of study), the respondent will likely become interested in the problem and be inclined to cooperate.

An effective cover letter should also assure respondents that their privacy and anonymity will be maintained. Furthermore, the cover letter should make an appeal for the respondent's cooperation. Some subtle flattery about the respondent's professional status and the importance of his or her response may be desirable. This effort must be done tactfully, however, and only when appropriate. In this situation, the person's name and address should appear on

cover letter—The letter attached to a survey that explains the purposes and importance of the survey.

the cover letter. The letters should look as if they were individually typed, even when a word processor is used. People will be insulted if it is suggested that they have been handpicked for their expertise and valued opinions when the letters are addressed "Dear Occupant."

If the survey is endorsed by recognized agencies, associations, or institutions, specify this in the cover letter. If possible, use the organization's or institution's stationery. Respondents will be much more cooperative if some respected person or organization is supporting the study. In addition, acknowledge whether financial support is being given and by whom. Identify yourself by name and position. If the study is part of your thesis or dissertation, give your advisor's name. Having the department chair or the dean of the college endorse the letter may be advantageous. To increase the response rate, contact individuals by email or telephone, asking for their participation in the survey. If the purposes of the study are worthwhile, offering the respondent a summary of the results is usually effective to enhance participation. Before making this offer, give ample consideration to how much work and money such a summary will entail. If you do offer to provide a summary, be sure to follow through on your promise.

Most universities require informed consent by those completing surveys. The survey topic and university procedures may determine the type of consent required (i.e., passive or active) and the method by which participants provide consent (e.g., by checking a box on an online survey or sending a signed consent in a separate envelope when a sensitive topic is being addressed). As the study is being designed and institutional review board documents are being prepared, determine what type of consent is required and how to make this process efficient for the participants.

Because a questionnaire imposes on a person's time, strategies involving rewards and incentives are sometimes used in an effort to amuse or involve the respondent. Some questionnaires include money (perhaps a dollar) as a token of appreciation. This incentive may appeal to a person's integrity and evoke cooperation. Then again, you may just be out a dollar. The disadvantage is that inflation plays havoc with the effects of such rewards. A quarter may have been effective years ago, whereas one or more dollars may now be required to elicit the same sense of guilt for not responding. Research results concerning the effectiveness of monetary and gift incentives have been mixed (Agarwal et al., 2016; Becker, Moser, & Glauser, 2019), so there is no guarantee of their effectiveness—and certainly are not as effective as a well-designed and piloted questionnaire.

The cover letter should request that the questionnaire be completed by a certain date. (This information should also be specified on the questionnaire.) When establishing the date for completion, consider such things as the respondents' schedules, responsibilities, and vacations. Be reasonable about the time that you allow the person to respond, but do not give the respondent too much time because of the tendency of ignoring emails that get further down in an inbox. A couple of weeks is ample time for the respondent to answer.

The appearance of the cover letter for both online and paper surveys is just as important as the appearance of the questionnaire. Grammatical errors, misspelled words, and improper spacing and format give the respondent the impression that the author does not attach much importance to details and that the study will probably be poorly done.

Researchers have tried a number of subtle and tactful approaches to establish rapport with respondents. Some have tried a solemn appeal to the monumental importance of the survey, some have attempted casual humor, and others have tried a folksy approach. Obviously, the success of any approach depends on the skill of the writer and the receptiveness of the reader. Any of these attempts can backfire.

A number of years ago, one of this book's previous authors (Jack Nelson) received from his department head a note to which a letter like this was attached:

Dear Dr. _____:

It has been said: "There are two kinds of information, knowing it or knowing where to find it." I asked three men you know, Doctors Eeney, Meeney, and Moe, a certain question, and each said this: "Sorry. I don't know his name." But they came back with this helpful suggestion: write to you.

Now for the question: What is the name of the person in your department or school who is in charge of your graduate program in physical education? . . .

Cordially yours,

Harry Homespun

The department head answered the letter, naming Nelson as head of the graduate program; shortly thereafter, this letter arrived:

Dear Dr. Nelson:

Your good work in administering a graduate program is well known throughout our profession. Even though there are many knotty problems, everyone wants to upgrade the resources for graduate studies. But how? Because of your scholarly approach, my associates Doctors Eeney, Meeney, and Moe suggested that I write to you.

It has been said, "If you want to get something important done, ask a busy person." My associates tell me you are busier than a bird dog in tall grass. They also say that you have a high regard for excellence. . . . If you could find it convenient to return your checked copy on or before May 1, you might help reduce the cultural lag.

Cordially yours,

Harry Homespun

We have, of course, omitted the main parts of the letters concerning what was being studied and what the instructions specified, but those parts were generally well done.

About a year later, another questionnaire arrived from the same university with essentially the same folksy approach, including the analogy of a bird dog in tall grass. It is safe to assume that these two graduate students had the same research methods course at the same institution. Most likely, in the discussion of the survey method, some examples of cover letters were given with different approaches. Professors Eeney, Meeney, and Moe would undoubtedly be embarrassed if they knew that the two students had written such nearly identical letters.

In most respects, the cover letters contained the essential information, and the topics were worthwhile. The authors were simply too heavy handed in their attempts at a down-home approach, and their efforts to flatter the respondent regarding his "well-known" expertise lacked subtlety.

The Sample Email Cover Letter sidebar shows a cover letter for a questionnaire that deals with a potentially threatening topic. The letter effectively explains to the respondent how confidentiality will be ensured. It also focuses on the importance of the topic and explains how the process will work. Bourque and Fielder (2003) offered helpful suggestions regarding what should be included in cover letters.

Sample Email Cover Letter

Dear _____:

I am a biomechanics doctoral student conducting a study for my dissertation titled *Perceptions of Teachers of Biomechanics on the Physics Knowledge of Undergraduate Kinesiology Students*. As a part of this study, I will compare the responses of biomechanics teachers in programs where college physics is required with programs where physics is not required. My experience as a graduate assistant in biomechanics laboratories suggests that this information is important. Often students struggle with the course material and the support materials available may assume a physics background not required of all students. Finally, this has curricular implications; do most programs require physics? How do teachers compensate when physics is not required? I hope you will agree to participate in this study.

Your participation will require approximately 20 minutes to complete an online survey. If you would like to participate, all you need to do is click on the link below and it will take you to a site where you can begin. Before starting the survey, you will be informed of your rights and asked to give consent to participate. Included in the introductory material is a description of how the information you provide will be kept completely confidential. You will be anonymous, and at no time will your answers be identified with your name or any other information that could identify you. In addition, the only data that will be reported are means and other summary scores. Neither you nor your university will be identified in this report, the dissertation, or any papers that are published once the dissertation is complete. We have identified a stratified random sample based on the American Kinesiology Association program types (e.g., undergraduate only, MS, or PhD) and I am contacting you based on your university contact information on the website.

I hope you will agree to participate in this study. I would be happy to answer any questions. You can respond to this email or call me at 212-555-5555.

Thank you for your time and for considering this request. Your help will enable me to conduct a strong study that will provide valuable results for our field.

Sincerely,
Ronald Smith
Graduate Student

Click Here to Participate in This Study

Sending the Questionnaire

You must take care in choosing the time for the initial emailing or mailing. Such considerations include holidays, vacations, and especially busy times of the year for the respondents. If using a paper survey, a self-addressed, stamped envelope should be included. It is almost an insult to expect respondents not only to answer your questionnaire but also to provide their own envelope and stamp to mail it back to you.

Other matters regarding the distribution of the questionnaire, such as establishing the date to be returned, have been covered in previous sections. As we noted earlier, the initial mailing of a paper questionnaire represents a substantial cost to the sender. Securing a sponsor to underwrite or defray expenses and using bulk-mailing services are important considerations, and it is likely an electronic survey would be much less expensive and yield greater response rates.

Following Up

This should not come as a big shock, but it is unlikely that you will get 100% return on the initial request for participation. A follow-up is nearly always needed, and this task can be done in many ways. One is to wait about two weeks after the initial request and then send a reminder to all people in the sample, stressing the importance of their participation (and apologizing if they have already completed the questionnaire). Approximately two weeks from the first reminder, another reminder with the link for the questionnaire or, if mailed, another copy of the questionnaire and another self-addressed, stamped envelope should be sent to those who have not responded. This approach takes time and may be expensive, of course, but follow-ups are effective. In a number of cases, the person has forgotten to respond, and a mere reminder will prompt completion. Other nonrespondents who had not initially planned to return the questionnaire may be influenced by the researcher's efforts in reiterating the significance of the study and the importance of the person's input.

The follow-up letter should be tactful. The person should not be chastised for failing to respond. The best approach is to write as though the person would have responded had it not been for some oversight or mistake on your part (see the Sample Follow-Up Letter sidebar).

A lack of response is especially likely when the questionnaire deals with a sensitive area. For example, surveys about program offerings and grading practices are often not returned by schools with inadequate programs and poor grading practices. Thus, the obtained responses are apt to be biased in favor of the better programs and better teachers. Regardless of the nature of the survey, the results from a small return rate (e.g., 10 to 20%) cannot be given much credibility. People who have a particular interest in the topic being surveyed are more likely to respond than those who are less interested. These respondents are self-selected, and the responses are almost invariably biased in ways that are directly related to the purposes of the research (Fowler, 2009).

> Survey results with a small return rate have less credibility because respondents are more likely to be those who are interested in the topic. This pattern almost always produces a bias directly related to the purposes of the research.

In fact, in most studies in which more than 20% of the questionnaires are not completed, it is recommended that a sample of the nonrespondents be surveyed. Of course, this task is not easy and may be impossible if you are recruiting participants through professional organizations or government agencies that will not provide direct contact information. If people have ignored the initial mailing and one or two follow-ups, the chances are not good that they will respond to another request, but it is worth a try. The preferred technique is to select a small number (e.g., 5 to 10%) of the nonrespondents randomly by using the table of random numbers (this technique is sometimes called *double-dipping*). Then make contact either by telephone or email. After the responses have been obtained, comparisons are made between the answers of the nonrespondents and the answers of the people who responded initially. If the responses are similar, you can assume that the nonrespondents

Sample Follow-Up Letter

Dear _____:

I sent an email with a link to a questionnaire regarding criteria for planning a recruitment campaign to you a few weeks ago and have not heard from you. As you can appreciate, it is important that we obtain response from everyone possible inasmuch as only a few select individuals were contacted. Our facility is planning a new recruitment campaign based on the results of this study, so it is of vital concern to the health of our program.

The email was sent during the summer when you may have been away from your office. I have included the link below and it would be most helpful if you could take 15 to 20 minutes to give your opinions on the information requested.

Thank you so much for your cooperation.

Sincerely yours,

Director of Fitness and Health Services

are not different from those who replied. If there are differences, you should either try to get a greater percentage of the nonrespondents or be sure these differences are noted and discussed in the research report. In addition, if a professional organization emailed your requests for participation, it may be impossible to determine an exact return rate. This too should be noted in the research report.

To follow up on nonrespondents for mailed questionnaires, you have to know who has not responded. Keeping records of those who have and have not responded may seem to fly in the face of guaranteed anonymity. However, this is not a difficult task. An identifying number written on the questionnaire or return envelope can be used. Of course, it is recommended that you inform the respondents about the use of the number in your cover letter. You might offer an explanation such as the following:

> **Your responses will be strictly confidential. The questionnaire has an identification number for mailing purposes only. With this numbering system, I can check your name off the mailing list when your questionnaire is returned. Your name will never be placed on the questionnaire.**

Another approach, again for paper surveys, which works well, is to send a separate postcard (self-addressed and stamped) with the questionnaire. The card contains an identifying number and is to be mailed back separately when the person returns the questionnaire. The card simply states that the person is sending the questionnaire and therefore that you do not need to send any more reminders. This procedure ensures anonymity and yet informs you that the person has responded (Fowler, 2009).

A good return rate is essential. We frequently read journal and news reports of surveys in which something like 2,000 surveys were mailed and 600 were returned. The number of returns may sound respectable, but these 600 people represent only a 30% return rate. They are not a random sample. They are self-selected, and their responses may be quite different from those of the 1,400 who did not respond.

Analyzing the Results and Preparing the Report

These last two steps are discussed in chapter 21, which deals with the results and discussion sections of the research report. The main consideration is that the method of analysis should be chosen in the planning phase of the study. Many questionnaires are analyzed merely by tallying the responses and reporting the percentage of the respondents who answered one way and the percentage who answered another way. Often, not much in the way of meaningful interpretation can be gained if only a simple tally is given. For example, when the researcher states simply that 18% of the respondents strongly agreed with some statement, 29% agreed, 26% disagreed, 17% strongly disagreed, and 10% had no opinion, the reader's reaction may be, "So what?" Questionnaires, like all surveys, must be designed and analyzed with the same care and scientific insight as experimental studies.

Additional Considerations for Online Surveys

As noted throughout the previous section, much of the information pertaining to surveys is applicable to both online and paper surveys. The care in determining the objectives, constructing and designing the questionnaire, writing a cover letter, and communicating with participants is the same. And, of course, conducting a pilot study is equally important. The use of electronic surveys, however, has some different demands.

After you have finalized the objectives and are in the process of writing questions, you should determine which survey software is available on your campus. For most online questionnaires the survey is placed on the Internet after it is constructed. These software programs are relatively easy to use. Most have fees associated with their use, but your university is likely to have a site license that allows you to get a user name for Internet surveys or use it on a department computer without cost. If you have more than one choice about which program to use, talk with those at the computer help desk who specialize in this software. The best choice is the software that is easiest to use and allows you to construct the survey, post it, and get the results in the most efficient way possible.

After you determine which survey software to use, we highly recommend attending an orientation session on campus or completing the online tutorial. Either approach will alert you to the capabilities that the software has to offer and help you learn how to insert questions, change appearance and design, and prepare data for analysis. A few hours invested here will likely save many hours as you develop and pilot-test the survey.

After you have written the questions and had them reviewed by advisors and colleagues, you can construct the online survey. If you have become familiar with the software, this step should be relatively straightforward. Most electronic surveys, as with paper surveys, begin with an introduction of the purpose and provide directions for completing the survey; for example, "Click *next page* after you have competed the questions," or "You cannot go back and complete questions if you already have submitted a page." Consent for participating in the study appears early in the survey and should be worded according to the stipulations of the institutional review board.

Care is needed in designing a survey that is visually appealing and easy to use (Fink, 2016). Most survey software permits an infinite array of color and design options. Choose these carefully because people may react differently to similar color schemes and layouts. One of us was recently asked to complete an online survey that alternated between a light blue and a dark blue color with each question. Although this design was appealing from an artistic viewpoint, it was nearly impossible to read the questions in dark blue because the text was posted in black. All decisions about design should be made to enhance the number of participants who complete the survey.

As you design the cover letter, use the information from the previous section. This letter will direct the participant to the link to complete the survey. Check and double-check that the link in the email letter is correct. If the link is not correct, participants may be lost and may not respond to a second request if they wasted their time trying to connect the first time.

The pilot study should be conducted in the same way that the final survey is conducted. A good approach is to have colleagues and your major professor respond to the email request and complete the survey before you use a larger group. This step allows you to correct formatting and design issues, as well as make sure the link to the survey works and that the survey is accessible from a variety of locations. This step, which often can be conducted in a few days, may identify problems that would create havoc to the study's time frame if a second, larger pilot study has to be conducted. After the initial pilot study with colleagues, and then again with the larger pilot study, modify the questions and adjust the design and layout of the survey as needed. It always is helpful to download the data to make certain it is a useful format. If it is not, you will want to make changes so you can go directly from downloading the survey data to data analysis.

If you are using a sampling procedure, confirm that the email addresses for participants are correct before sending them anything. As noted earlier, for many studies, it is not possible to use sampling techniques, so requests for participation are made in cooperation with professional groups and government agencies and through requests on often-visited websites. Arranging to get help with requesting participation and deciding how to conduct follow-ups should be started early. Determine who in the organization will be responsible for giving permission and who will send out email requests. Work with that person to make the process efficient and as pleasant as possible. The final sample size will be determined by who is solicited to participate, and this may be contingent on how much others want to help you.

After all follow-ups have been sent and you are ready to begin the analysis, download the data file with all the completed surveys. Most survey software programs have a number of format and file type options for saving the data. As we noted earlier, the pilot is an ideal time to determine which formats and file types will enhance subsequent analysis. You generally will have no need to reconfigure anything to use SPSS or another statistical program—if you plan ahead.

Delphi Method

Delphi survey method—A survey method that uses a series of questionnaires in such a way that the respondents (usually experts) reach a consensus about the subject.

round—A stage of the Delphi survey method in which respondents are asked for their opinions and evaluations of various issues, goals, and so on.

The **Delphi survey method** uses questionnaires but in a different manner than the typical survey. The Delphi method uses a series of questionnaires in such a way that the respondents finally reach a consensus about the topic. It is basically a method of using expert opinion to help make decisions about practices, needs, and goals.

The procedures include the selection of experts, or informed people, who are to respond to the series of questionnaires. A set of statements or questions is prepared for consideration. Each stage in the Delphi method is a **round**, the first of which is mostly exploratory. The respondents are asked for their opinions on various issues, goals, and so on. Open-ended questions may be included to allow the participants to express their views and opinions.

The questionnaire is then revised as a result of the first round and sent to the respondents, asking them to reconsider their answers in light of the analysis of all respondents to the first questionnaire. In subsequent rounds, the respondents are given summaries of previous results and asked to revise their responses, if appropriate. Consensus about the issue is finally achieved through the series of rounds of analysis and subsequent considered judgments. Anonymity is a prominent feature of the Delphi method, and the consensus of recognized experts in the field provides a viable means of confronting important issues. For example, the Delphi method is used sometimes to determine curricular content for programs, to decide on the most important objectives of a program, or to choose a solution to a problem.

Personal Interviews

As mentioned earlier, the steps for the interview and the questionnaire are basically the same. This section focuses only on the differences.

Preparing for the Interview

The most obvious difference between the questionnaire and the interview is in the gathering of the data. In this respect, the interview is more valid because the responses are apt to be more reliable. In addition, the percentage of returns is much greater.

Participants should be selected using the same sampling techniques used for a questionnaire. Generally, the interview uses smaller samples, especially when a graduate student is doing the survey. Cooperation must be secured by contacting the participants selected for interviewing. If some refuse to be interviewed, you must consider possible bias to the results, as with nonrespondents in a questionnaire study.

Conducting an effective interview takes a great deal of preparation. Graduate students sometimes have the impression that anyone can do it. Because the procedures used for the questionnaire are followed in preparing the items, you must be very familiar with them. Carefully rehearse the interview techniques. One source of invalidity is that interviewers tend to improve with experience, and thus the results of earlier interviews may differ from those of interviews conducted later in the study. A pilot study is extremely important. Make sure the vocabulary level is appropriate and the questions are equally meaningful given the ages and educational backgrounds of the participants. Use careful planning in organizing the questions and visual aids to provide a comfortable flow of presentation and transitions from question to question.

Training is required in making initial contact and presenting the oral "cover letter" by phone or videoconference. At the meeting, establish rapport and help the person feel at ease. If you are audio-recording the interview, obtain permission. If you are not recording, have an efficient system of coding the responses without consuming too much time and appearing to be taking dictation. You must not inject your own bias into the conversation and certainly should not argue with the respondent. Although you hold many advantages over the questionnaire with regard to the flexibility of the questioning, it introduces the risk that the respondent will stray from the questions and get off the subject. You must tactfully keep the respondent from rambling, and doing this requires skill.

The key to getting good information is to ask good questions. Some surveys that employ in-person interviews are quantitative, and the questions may be similar to those in an online or paper questionnaire. Other surveys deal with qualitative data. In these studies, standardization is less of a concern, and more emphasis is given to description. Good interviewers in qualitative surveys do not ask yes-or-no questions, they do not ask multiple questions disguised as a single question, and they try to avoid inserting their own points of view. Merriam (2007) defined four major categories of questions: hypothetical, devil's advocate, ideal position, and interpretive. Here are some examples of each:

- Hypothetical: "Let's suppose that this is my first day of student teaching. What would it be like?"
- Devil's advocate: "Some people say that professional educational courses are of little value for the student-teaching experience. What would you say to them?"
- Ideal position: "What do you think the ideal student-teaching preparation program would be like?"
- Interpretive: "Would you say that the student-teaching experience is different from what you expected?"

You may recognize these approaches from interviews that you have seen on television. The interview has the following advantages over the mailed or emailed questionnaire:

- The interview is more adaptable. Questions can be rephrased, and clarification can be sought through follow-up questions.
- The interview is more versatile regarding the personality and receptiveness of the respondent.
- The interviewer can observe how the person responds and can thus achieve greater insight into the sensitivity of the topic and the intensity of the respondent's feelings. This feature can add considerably to the validity of the results, because respondents' tendency to avoid sensitive topics is one of the greatest threats to validity in questionnaire studies.
- Because each person is contacted before the interview, interviews have a greater rate of return. Moreover, people tend to be more willing to talk than to fill out a questionnaire. A certain amount of ego is involved because a person who is interviewed feels more flattered than one who receives only a list of questions.
- An advantage of in-person interviews is that visual aids, such as flash cards, can be used to simplify long questions and explain the response lists.

A variety of resources provide information to help prepare interview questions and perfect techniques for qualitative interviews. Books by Patton (2014) and Merriam (2007) provide general advice. More specific advice is found in books by Rubin and Rubin (2020) and Seidman (2019). The references you select and use will be determined by the research questions and the type of interview that will yield the best responses to them.

A potential problem in interviews is losing the data. If you audio-record your interviews and back them up appropriately, this problem may be less consequential because you will have a transcript and can go back to the recording if questions arise. Because of confidentiality restrictions in some studies, recordings will not be used and names may not be placed on the interview data instrument. Consequently, the researcher may later become confused about which data belong to which respondents and what each respondent said. Use several identifiers for each interview, and practice taking notes during pilot studies. Oishi (2003) provides an excellent coverage of the dos and don'ts in planning and conducting the personal interview.

Conducting a Telephone or Videoconference Interview

Interviewing by telephone is becoming more prevalent. Telephone interviewing has a number of advantages over face-to-face interviewing:

- Telephone interviewing or videoconferencing is less expensive than traveling to visit respondents. Researchers can conduct these interviews in significantly less time than in-person interviews.

- The interviewer can work from a central location, which facilitates the monitoring and the quality control of the interviews and provides a better opportunity for using computer-assisted interviewing techniques (Gall et al., 2006).

- Many people are more easily reached by telephone than by personal visit.

- The telephone interview enables the researcher to reach a wide geographical area; geography is a limitation of the personal interview. This advantage also can increase the validity of the sampling.

© Belinda Pretorius/fotolia

Surveys are sometimes the average opinion of people who don't know.

- Some evidence indicates that people respond more candidly to sensitive questions over the telephone than they do in personal interviews, in which the presence of the interviewer may inhibit some responses.

Using a computer in telephone interviewing eases data collection and analysis. For example, a computer can display the questions for the interviewer to ask, to which the interviewer types the responses obtained. Each response triggers the next question, so the interviewer does not have to turn pages and is less likely to ask inappropriate questions. Responses are stored for analysis (thus reducing scoring errors), and the stored responses can then be accessed for statistical analysis.

Among the disadvantages of telephone surveys is that obtaining a representative sample is increasingly difficult because of cell phones, answering machines, caller ID, and call-blocking devices. Telemarketing has greatly soured the public on telephone interviews.

An email sent prior to calling can be effective in securing cooperation, thereby increasing the response rate. The drawback is that you must have the names, email addresses, and other contact information of the target population. In some telephone surveys, the researcher uses random-digit dialing (RDD) to select the sample. In this method, a table of random numbers can be used to select the four numbers following the three-number exchange. Both listed and unlisted numbers can be reached using this method.

Much time is consumed in callbacks because of no answer, voicemail, unwanted people answering the phone, busy signals, and language barriers. The researcher usually must hire other people because of the many calls that must be made. An important source of bias concerns the time when telephone or videoconference surveys are done. If conducted during the day, the sample is biased toward the people most likely to be home: homemakers, remote workers, the unemployed, and the retired. Thus, interviewers usually must work after 5:00 p.m. In addition, it is estimated that one-third of the people contacted are not home on the initial call, which means that callbacks are necessary (sometimes as many as 10 or 12!).

> The timing of a telephone call can bias the results. Calling before 5:00 p.m. may skew the sample toward homemakers, remote workers, the unemployed, and the retired.

Commonly cited weaknesses of telephone surveys include the sample being limited to people who have either landline or mobile phones and the fact that many people do not answer their phone if they do not recognize the number of the person calling. If these are factors, a telephone survey is not recommended.

Bourque and Fielder (2003) gave a thorough description of the relative advantages, disadvantages, and recommended methodology of the telephone survey.

Normative Surveys

The **normative survey** is not described in most research methods textbooks. As implied by the name, this method involves establishing norms for abilities, performances, beliefs, and attitudes. A cross-sectional approach is used: Samples of people of various ages, sexes, and other classifications are selected and measured. The steps in the normative survey are generally the same as in the questionnaire, the difference being the way the data are collected. Rather than asking questions, the researcher selects the most appropriate tests to measure the desired performances or abilities, such as the components of physical fitness.

> **normative survey—** A survey method that involves establishing norms for abilities, performances, beliefs, and attitudes.

In any normative survey, the test must be administered in a rigidly standardized manner. Deviations in the way measurements are taken give meaningless results. The researcher collects and analyzes the data from the survey by some norming method, such as percentiles, t scores, or stanines and then constructs norms for the categories of age, sex, and so on.

SHAPE America (formerly known as AAHPERD) has sponsored several normative surveys. Probably the most notable was the Youth Fitness Test (see American Association for Health, Physical Education, and Recreation, 1958), conducted in response to the furor caused by the results of the Kraus-Weber test (Kraus & Hirschland, 1954), which had revealed that U.S. children were inferior to European children in minimum muscular fitness. The Youth

Fitness Test was originally given to 8,500 boys and girls in a nationwide sample. Follow-up testing was done in 1965 and 1975.

In the 1958 AAHPER normative survey, a committee determined a seven-item motor fitness test battery. The University of Michigan Survey Research Center selected a representative national sample of boys and girls in grades 5 through 12 and then requested the cooperation of each school. They prepared directions for giving the test items and selected and trained physical education teachers in various parts of the country to administer and supervise the testing.

In addition, AAHPERD conducted a sport skills testing project, which established norms for boys and girls of various ages in a number of sports skills. The NHANES National Youth Fitness Survey (2012) collected data on physical activity and fitness for a large group of children.

Sometimes comparisons are made between the norms of different populations. In other studies, the major purpose is simply to establish norms. The primary drawbacks of any normative survey occur in test selection and the standardization of testing procedures. With many test batteries, there is a danger of generalizing based on the test results. For example, when only one test item is used to measure a component (such as strength), making unwarranted generalizations is possible.

The standardization of testing procedures is essential for establishing norms. But when a normative survey involves many testers from different parts of the country, problems with standardization become a potential source of measurement error. Published test descriptions simply cannot address all the aspects of test administration and the ways of handling the many problems of interpreting test protocol that arise. Extensive training of testers is the answer, but this is often logistically impossible.

Summary

The most common descriptive research technique is the survey, which includes questionnaires and interviews. The two are largely similar except for the method of asking questions. Surveys can be conducted online or on paper, and similar steps are required for getting good results for both methods. The questionnaire is a valuable tool for obtaining information over a wide geographical area. A good cover letter is important in getting cooperation. One or two follow-ups are often necessary to secure an adequate response. Personal interviews usually yield more valid data because of the personal contact and the opportunity to make sure that respondents understand the questions. Telephone and videoconference interviews are becoming increasingly popular. They have most of the advantages of personal interviews and the flexibility to involve more participants over a larger geographical area. The Delphi survey method uses a series of questionnaires in such a way that the respondents eventually reach a consensus about the topic. It is frequently used to survey expert opinion to make decisions about practices, needs, and goals. The normative survey is designed to obtain norms for abilities, performances, beliefs, and attitudes. The AAHPERD fitness tests developed over the years are examples of normative surveys. In all surveys, representative sampling is extremely important.

☑ Check Your Understanding

1. Find a thesis or dissertation that used a questionnaire to gather data. Briefly summarize the methodology, such as the procedures used in constructing and administering the questionnaire, selecting participants, and following up.

2. Locate a study that used interviews for gathering data. What type of interview was completed, face-to-face, phone, or other? Which type of question was used (e.g., hypothetical, devil's advocate, etc.)?

16

Other Descriptive Research Methods

Analyzing humor is like dissecting a frog. Few people are interested and the frog dies of it.

—E.B. White (1899-1985)

Forms of the survey method of descriptive research were discussed in chapter 15. In this chapter, we address several other descriptive research techniques. One is developmental research. Through cross-sectional and longitudinal studies, researchers investigate the interaction of growth and maturation and of learning and performance variables. The case study, in which the researcher gathers a large amount of information about one or a few participants, is widely used in a number of fields. Through observational research, the researcher obtains quantitative and qualitative data about people and situations. Some research studies employ unobtrusive methods in which the participant is not aware of being studied. Correlational studies determine and analyze the interrelationships of variables and generate predictions.

Developmental Research

Developmental research is the study of changes in behaviors across years. Although much of the developmental research has focused on infancy, childhood, and adolescence, research on senior citizens and even across the total human life span is increasingly common.

Longitudinal and Cross-Sectional Designs

Developmental research focuses on cross-age comparisons. For example, researchers can compare how far children can jump when they are six, eight, and then 10 years old, or they can compare adults who are 45, 55, and 65 years old on their knowledge of the effects of obesity on life expectancy. The major distinction between the two basic approaches in developmental studies is whether researchers follow the same participants over time (longitudinal design) or select different participants at each age level (cross-sectional design).

Longitudinal studies are powerful because the changes in behavior across the time span of interest are seen in the same people. Longitudinal designs, however, are time-consuming. A longitudinal study of children's jumping performances at ages six, eight, and 10 obviously requires five years to complete. That choice of design is probably not wise for a master's

developmental research—The study of changes in behaviors across years.

longitudinal studies—Research in which the same participants are studied over a period of years.

thesis or a doctoral dissertation where the student does not have years to complete the study. Longitudinal designs have additional problems besides the time required to complete them. First, over the several years of the study, some children are likely to move away when parents change jobs or change schools when school districts are rezoned. In longitudinal studies of senior citizens, some may die over the course of the study. The problem is not knowing whether the sample characteristics remain the same when participants are lost. For example, when children are lost from the sample because parents change jobs, is the sample then composed of children of lower socioeconomic levels because the more affluent parents moved? Furthermore, if obesity is related to longevity, are older people more likely to be less obese and consequently have increased knowledge because the more obese people with less knowledge have died? Thus, knowledge about obesity may not change from ages 45 to 65; rather, the sample may change.

Another problem with longitudinal designs is that as participants become increasingly familiar with the test items, these items may cause changes in behavior. The knowledge inventory on obesity may prompt the participants to seek information about obesity, thus changing their knowledge, attitudes, and behaviors. Therefore, the next time they complete the knowledge inventory, they have gained knowledge. This knowledge gain, however, is the result of having been exposed to the test earlier and might not have occurred without that exposure.

cross-sectional studies—Research in which samples of participants from different age groups are selected to assess the effects of maturation.

cohort problem—A problem in cross-sectional design concerning whether all the age groups are really from the same population.

Cross-sectional studies usually require less time to carry out than longitudinal studies. Cross-sectional studies test several age groups (e.g., six, eight, and 10) at the same point in time. Although these studies are more time efficient than longitudinal studies, a limitation, the **cohort problem**, exists: Are all the age groups really from the same population (group of cohorts)? Asked another way, are the environmental circumstances that affect jumping performance for six-year-olds the same today as when the 10-year-olds were six, or have physical education programs improved over this four-year span so that six-year-olds receive more instruction and practice in jumping than their 10-year-old peers previously did at that age? If the latter is true, then we are not looking at the development of jumping performance but rather at some uninterpretable interaction between normal development and the effects of instruction. The cohort problem exists in all cross-sectional studies.

Examples of longitudinal developmental studies are Johnson et al. (2021) who studied for over a 10-year period changes in brain structure and cognitive and motor symptoms of Huntington's disease carriers and control subjects and Lau, Dowda, McIver, and Pate (2017) who studied changes in physical activity during the school day of children in 5th, 6th, and 7th grades. In a cross-sectional study, Möhring and colleagues (2021) examined children between 8 and 13 years old to assess their dual-task abilities (motor and cognitive) to study executive function. Each of these studies suffers from the specific defects associated with the type of developmental design. Both Johnson and colleagues (2021) and Lau et al. (2017) lost participants over the several years of the study. For example, participants dropped from 1,083 in the first year to 958 in the final year of data collection in the study by Lau et al. Further, Möhring and colleagues concluded that their study would have been stronger if they could have examined changes within each child.

Although both longitudinal and cross-sectional designs have some problems, they are the only means available to study development. Thus, both are necessary and important types of research. While these two designs are considered descriptive research, either can also be experimental research (chapter 18); that is, an independent variable can be manipulated within an age group. Bellows and colleagues (2017) implemented a community-based intervention in a longitudinal study, to assess fundamental motor skills and determine the impact of an intervention during Headstart preschool one and two years later. The control group did not participate in the intervention.

Methodological Problems of Developmental Research

Whether the developmental research is longitudinal or cross-sectional, several methodological problems exist (for a more detailed discussion, see Thomas, 1984). We address four examples in this section: unrepresentative scores, unclear semantics, lack of reliability, and statistical problems.

Unrepresentative Scores

One of the most common problems in developmental research is unrepresentative scores, or outliers, which occur in all research but are particularly problematic at the extremes of developmental research (children and senior citizens). Outliers frequently result from shorter attention spans, distraction, and a lack of motivation to perform the task. The best way to handle unrepresentative scores is to do the following:

- Conduct the testing situation within a reasonable time frame to account for attention span.
- Set up the testing situation so that distractions do not occur.
- Be aware of what an unrepresentative score is and retest when one occurs.

The last thing a researcher wants is to use unrepresentative scores. Therefore, the developmental researcher should expect and plan to handle any undetected outliers from the testing stage when studying the distribution of the data. There are several ways to test for these extreme and unrepresentative scores. One example is examining scores from the subgroup's mean that exceed standard deviations.

Unclear Semantics

Selecting the words to use in explaining a task to children in various age groups can be a formidable challenge. If the researcher is not careful, older children will perform better than younger children primarily because they grasp the idea of what to do more quickly. Although the standard rule in good research is to give identical instructions to all the participants, researchers must bend this rule for developmental studies with children. They must explain the test in a way that the participants can understand. Researchers must then obtain tangible evidence that children of various ages understand the test before conducting it. One approach could be having children demonstrate the activity to some criterion level of performance before collecting the data.

Lack of Reliability

When researchers obtain a performance score for a child, they must ensure that it is a reliable one; that is, if the child is tested again, the performance score should be about the same. Obtaining reliable performance is frequently a problem when testing younger children for many of the same reasons that outliers occur. Of course, making sure that the child understands the task must be the first consideration, and maintaining motivation the second. A task made to be fun and enjoyable is more likely to elicit a consistent performance. Researchers can accomplish this by using cartoon figures, encouragement, rewards, and other forms of motivation as well as performing frequent reliability checks during testing sessions (for appropriate techniques, see chapter 11).

Statistical Problems

It is worth mentioning statistical problems as our final example of common issues that researchers encounter in developmental research. A frequent means of making cross-age comparisons is ANOVA (see chapter 9), which assumes that the groups being compared have equal variances (spread of scores about the mean). However, researchers often violate this assumption in making cross-age comparisons. Depending on the nature of the task, older children may have considerably larger or smaller variances than younger children. It is useful to be aware of this potential issue and some of the solutions. In particular, pilot work using the tasks of interest in the research should provide insight into this problem.

Protecting Participants

In chapter 5, we discussed the protection of human participants in research. Of course, this protection also pertains to children. Parents or guardians must grant permission for minors to participate in research. Researchers should obtain this permission as they do for adults, except that they give the explanation and consent forms to the parents or guardians. If minors are old enough to understand the methodology, researchers should also obtain their consent. This means explaining the purpose of the research in terms minors can understand. Most public and private schools have their own requirements for approval of research studies.

The normal sequence of events for obtaining permission is as follows:

1. Planning the research
2. Acquiring the approval of your university's committee for the protection of human participants
3. Locating and gaining the approval of the school system, the school involved, and the teachers
4. Obtaining the approval of parents and, when appropriate, children

You can see that this step requires a good deal of paperwork. Thus, beginning the process well before you plan to begin data collection is essential.

Retrospective Developmental Studies

Another approach to developmental studies is to look backward in time rather than forward. Sometimes data have been collected on a group of people over many years, and the researcher can use that data to evaluate development. For example, Thomas and Thomas (1999) discovered that two professional athletes were from the same Iowa town and that the same physical education teachers had taught these two athletes from grades K through 6. They interviewed the physical education teachers about characteristics of these professional athletes as they progressed through the elementary grades.

The Thomas and Thomas (1999) study was qualitative, but collecting quantitative data post hoc is also possible. For example, the Professional Golfers Association (PGA) collects performance data each year on players in all tournaments. Vermilio, Thomas, Thomas, and Morrow (2013) used this data for a study of the performance of the same professional golfers over a six-year period. Of course, they first had to identify players who competed over this time period, thus excluding young players just starting on the tour and older players who may have dropped off the tour. If you go to the PGA website where we obtained the data used in this study, you can find data on professional golfers over the past several years. Using this sort of data, a researcher could analyze quantitative changes in performance of individual or groups of golfers during the period for which data are available.

Case Studies

In the **case study**, the researcher strives for in-depth understanding of a single situation or phenomenon. This technique is used in many fields, including anthropology, clinical psychology, sociology, medicine, political science, speech pathology, and educational areas such as disciplinary problems and reading difficulties. It has been widely used in the health sciences and to some extent in exercise science, sport science, and physical education.

The case study is a form of descriptive research. Whereas the survey method obtains a rather limited amount of information about many participants, the case study gathers a large amount of information about one or just a few participants. Although the study consists of a rigorous, detailed examination of a single case, the underlying assumption is that the case is representative of many other such cases. Consequently, a greater understanding about similar cases is achieved. This is not to say, however, that the purpose of case studies is to make generalizations. On the contrary, drawing inferences about a population from a case study is not justifiable. However, the findings of a number of case studies may play a part in the inductive reasoning involved in the development of a theory.

The case study is not confined to the study of an individual but can be used in research involving programs, institutions, organizations, political structures, communities, and situations. The case study is used in qualitative research to deal with critical problems of practice and to extend the knowledge base of the various aspects of education, physical education, exercise science, and sport science (qualitative research is discussed in chapter 19).

Information about case study methodology is difficult to find. As Merriam (2007) observed, material on case study research strategies can be found everywhere and nowhere. Methodological material on case study research is scattered about in journal articles, conference proceedings, and research reports of the many fields that use this form of research. Frustration in locating substantive material about case study research in an educational setting prompted Merriam (1988) to write her interesting and informative text, *Case Study Research in Education*. Yin (2017) presented a comprehensive discussion on design and methods for case study research in the social sciences.

> **case study**—A form of descriptive research in which a single case is studied in depth to reach a greater understanding about other similar cases.

Types of Case Studies

In many ways, case study research is similar to other forms of research. It involves identifying the problem, collecting data, and analyzing and reporting results. As in other research techniques, the approach and the analysis depend on the nature of the research problem. The case study enables a more in-depth, holistic approach to the problem than may be possible with survey studies. Case studies can be descriptive, interpretive, or evaluative as described in the next section.

Descriptive Studies

A descriptive case study presents a detailed picture of the phenomenon; however, it does not attempt to test or build theoretical models. Some descriptive case studies are historical in nature, carried out for the purpose of achieving a better understanding of the present status. Descriptive case studies frequently serve as an initial step or database for subsequent comparative research and theory building (Merriam, 1988, 2007).

Interpretive Studies

Interpretive case studies also employ description, but the major focus is interpreting the data to classify and conceptualize the information and perhaps theorize about the phenomena. For example, a researcher might use the case study approach to gain a better understanding of the cognitive processes involved in sports.

Evaluative Studies

Evaluative case studies also involve description and interpretation, but the primary purpose is to use the data to evaluate the merit of some practice, program, movement, or event. The efficacy of this type of case study relies on the competence of the researcher to use the available information to make judgments (Yin, 2017).

Case Study Participants

The selection of participants in a case study depends, of course, on the problem being studied. The individual (or case) may be a person (e.g., student, teacher, coach), a program (e.g., Little League baseball), an institution (e.g., a one-room school), a project, or a concept (e.g., mainstreaming). In most studies, random sampling is not used because the purpose of a case study is not to estimate a population value but to select cases from which one can learn the most, often called *purposive sampling*, or sometimes, *criterion-based sampling*. The researcher establishes criteria necessary to include in the study and then finds a sample that meets the criteria. Criteria can include age, years of experience, evidence of level of expertise, and situation and environment. The case may be one classroom that meets certain criteria or a state that is involved in a specific program.

Characteristics of the Case Study

The case study involves the collection and analysis of information from many sources. In some respects, the case study has some of the same features found in historical research. Although it consists of the intensive study of a single unit, the ultimate worth of a case study may be that it provides insight and knowledge of a general nature for improved practices. The generalizability of a case study is ultimately related to what the reader is trying to learn from it (Yin, 2017). The case study approach is probably most frequently used to understand why something has gone wrong.

Gathering and Analyzing Data

The case study is flexible concerning the amount and type of data that are gathered and the procedures used in gathering the data. Thus, the steps in the methodology are not distinct or uniform for all case studies.

Data for case studies can be interviews, observations, or documents. It is not uncommon for a case study to employ all three types of data. A case study involving an elite athlete, for example, may include interviews with the athlete, coach(es), and teammates. The researcher may systematically observe the athlete in practice and during games. Documents could include medical examination reports, physical performance test scores, game statistics, psychological test scores, and media. As noted previously, some case studies are mainly descriptive, but others focus on interpretation or are evaluative. Some case studies propose and test hypotheses, while others attempt to build theories through inductive processes.

According to Yin (2017), the analysis of case study data is one of the least developed and most difficult aspects of conducting case studies. Yin maintained that too often, the researcher begins a case study without the slightest notion about how to analyze the data. It is a formidable task because of the nature of the data and the massive amount of information. Merriam (2007) states that analysis continues and intensifies after the data have been collected. The data must be sorted, categorized, and interpreted. As in any research, the ultimate value of a study rests on the researcher's insight, sensitivity, and integrity because the researcher is the primary instrument in the collection and analysis of the data in the case study. This attribute is both a strength and a weakness. It is a weakness if the researcher fails to use the appropriate sources of information. The researcher can also be guilty of either oversimplifying the situation or exaggerating the actual state of affairs (Guba & Lincoln, 1981). On the other hand, a competent researcher can use the case study to provide a thorough, holistic account of a complex problem.

Applying Case Study Research in Physical Activity

There are several case studies in physical activity settings. For example, Kolodziejczyk and colleagues (2021) studied four Croatian soccer players and how games in over-time affected player performance in the next game during World Cup play. Another interesting study by Jiang, Huang, and Fisher (2019) examined how people used an urban park in Beijing based on the day's air quality. The study by van Munster, Lieberman, and Grenier (2019) analyzed how teachers differentiated instruction for students with disabilities. Yan and Cardinal's 2013 study was on physical activity in Chinese female graduate students identifying barriers (e.g., time, low self-efficacy) and facilitators (e.g., social support, changing perception of femininity). This information could lead to interventions to increase physical activity among this population.

One of the principal advantages of the case study approach is that it can be fruitful in formulating new ideas and hypotheses about problem areas, especially areas for which there is no clear-cut structure or model. The researcher selects the case study method because of the nature of the research questions being asked. The case study, when used effectively, can play an important role in contributing to knowledge in our field.

> Case studies can help to formulate new ideas and hypotheses, especially for areas that lack established, clear-cut structures or models.

Observational Research

A variety of research endeavors use observation, a descriptive method of researching certain problems, to provide a means of collecting data. In the questionnaire and interview techniques, the researcher relies on self-reports about how the participant behaves or what the participant believes. A weakness of self-reports is that people may not be candid about what they really do or feel and may give what they perceive to be socially desirable responses. An alternative descriptive research technique is for the researcher to observe people's behavior and qualitatively or quantitatively analyze these observations. Some researchers claim that this technique yields an increased accuracy of data. There are, of course, several limitations to observational research.

Basic considerations in observational research include the behaviors that will be observed, who will be observed, where the observations will be conducted, and how many observations will be made. Many other considerations are connected to these basic ones. Depending on the problem and the setting, each investigation has unique procedures, which limits our discussion to only the basic considerations.

What Behaviors Will Be Observed?

Deciding what behaviors will be observed relates to the statement of the problem and to the operational definitions. For example, a study on teacher effectiveness must have clearly defined observational measures of teacher effectiveness. Definite behaviors must be observed, such as the extent to which the teacher asks students questions. Some other aspects of teacher effectiveness include giving individual attention, demonstrating skills, dressing appropriately for activities, and starting class on time. In determining the behaviors to observe, the researcher must also limit the scope of the observations to make the study manageable.

> Have you observed someone in an elevator pushing the floor button repeatedly? Does pushing the button make the elevator arrive more quickly or does it make the button pusher feel better?

Whom Will Be Observed?

As with any study, the population from which samples will be drawn must be determined. Will the study focus only on elementary school teachers? Which grades? Will the study include only physical education specialists, or will it also include classroom teachers who teach physical education? Another question concerns the number of participants to be observed. Will the study include observations of students in addition to teachers? In other words, the researcher must describe the participants of the study precisely.

Where Will Observations Be Conducted?

The setting for the observations must be considered, in addition to the basic considerations of the size of the sample and the geographical area. Will the setting be unnatural or natural?

Hmmm, I'm not sure how to describe this.

© Eric Isselée/fotolia

Using an unnatural setting means bringing the participant to a laboratory, room, or other locale for observation.

An unnatural setting offers some advantages in terms of control and freedom from distractions. For example, a one-way mirror is advantageous for observation because it removes the influence of the observer on the behavior of the participant. That the participants' behavior is affected by the presence of an observer is also shown in classroom situations. When the observer first arrives, the students (and perhaps the teacher) are curious about the observer's presence. Consequently, they may behave differently than they would if the observer were not there. The teacher may also act differently, possibly by perceiving the observer as a threat or being aware of the purpose of the observations. In any case, the researcher should not make observations on the initial visit. Allowing the students to become gradually accustomed to the observer's presence is best.

Whether a participant will be observed alone or in a group is also related to the setting. In a natural setting, such as the playground or classroom, the child may behave more typically, but additional extraneous influences on behaviors are likely to be present.

How Many Observations Will Be Made?

Many factors determine how many observations to make. First are the operationally defined behaviors in question and the time constraints of the study itself. For example, if you are studying the amount of activity or participation of students in a physical education class, you must consider several things. In the planning phase of the study, you must decide the type of activity unit and the number of units encompassed in the study. The number of observations for each activity depends on the particular stage of learning in the unit (e.g., introductory, practice, and playing stages). This must be specified in the operational definitions, of course, but both the length of the unit and the subsequent length of each stage within each unit play major roles in determining the number of observations that are feasible.

Another factor to consider is the number of observers. If only one person is doing the observations, either the number of people being observed or the number of observations per person (or both) are restricted. Attempting to generalize typical behavior from observing a few individuals on a few occasions is hazardous.

Some types of behavior may not be manifested frequently. Fair play, aggression, leadership, and other traits (as operationally defined) are not readily observable because of the lack of opportunity to display such traits (among other things). The occasion must present itself, and the elements of the situation must materialize so that the participant has the opportunity to react. Consequently, the number of observations is bound to be extremely limited if left to chance occurrence. On the other hand, situations that are contrived to provoke certain behavior are often unsuccessful because of their artificiality.

We cannot say how many observations are necessary; we can only warn against too few. We recommend a combination of feasibility and measurement considerations. This question is readdressed in the discussion on scoring and evaluating observations.

When Will Observations Be Made?

You can easily understand that all the basic considerations being discussed are related and overlap. The determination of when to make the observations includes decisions about the time of day, day of the week, phase in the learning experience, season, and other time factors.

In our previous example of observing the amount of student activity in a physical education course, different results would be expected if the observations were made at the beginning of the unit rather than at the unit's end. As previously mentioned, allowing students to become accustomed to the situation so that the observational procedures do not interfere with normal activity is another consideration. In observing student teachers, for example, differences would certainly be expected if some were observed at the beginning of their student-teaching experience and others were observed at the end.

Graduate students encounter major problems dedicating enough time in observational research, which makes it difficult to make a sufficient number of observations that yield reliable results. Furthermore, graduate students usually must gather the data by themselves, making it much more time-consuming than it would be if other observers were available.

How Will Observations Be Scored and Evaluated?

Researchers employ a number of techniques for recording observational data. Methods that employ computers and other computer-assisted event recordings have alleviated much of the technical problems that have long plagued observational research. Following are some of the more commonly used procedures for recording observational data:

- Narrative, or continual recording
- Tallying, or frequency counting
- Interval method
- Duration method

In the **narrative**, or *continual-recording*, **method**, researchers record in a series of sentences observed occurrences as they happen. This method of recording is the slowest and least efficient. The observer must be able to select the most important information to record because not everything that occurs in a given situation can be recorded. Probably the best use of this technique is in helping to develop more efficient recording instruments. The researcher first uses the continuous method and then develops categories for future recording from the narrative.

The **tallying**, or *frequency-counting*, **method** involves recording each occurrence of a certain behavior. The researcher must clearly define the behavior and make the frequency counts within a certain period, such as the number of occurrences in a 10- or 30-minute session.

The **interval method** is used when the researcher wishes to record whether the behavior in question occurs in a certain interval of time. This method is useful when counting individual occurrences is difficult. The Academic Learning Time in Physical Education (ALT-PE) is a

narrative method—A method of recording in observational research in which researchers describe their observations as they occur; also called *continual-recording method*.

tallying method—A method in observational research in which researchers record each occurrence of a clearly defined behavior within a certain period; also called *frequency-counting method*.

interval method—A recording method in observational research that is used when counting individual occurrences is difficult; the researcher records whether the behavior in question occurs in a certain interval of time.

commonly used observational instrument developed by Siedentop and graduate students at Ohio State University for use in physical education (Siedentop, Birdwell, & Metzler, 1979; Siedentop, Trousignant, & Parker, 1982). The instrument entails time sampling during which the researcher observes and codes a child's activities for a specified period. Blocks walked method (BWM) was a reliable method to observe physical activity in various neighborhoods (Suminski, Petosa, & Stevens, 2006). Observers recorded people walking, cycling, and walking with a dog among other categories. The goal was to assess the type and amount of physical activity in specific city blocks.

The **duration method** involves timed behavior. The researcher uses a stopwatch or other timing device to record how much time a person spends engaged in a particular behavior. A number of studies have used this method in observing student time on and off task. In the previous example about the amount of activity in a physical education class, a researcher could simply record the amount of time a student spends in actual participation or the amount of time spent standing in line or waiting to perform. The researcher usually observes a student for a given unit of time (e.g., a class period), starting and stopping a stopwatch as the behavior starts and stops to record cumulative time on (or off) task.

duration method—A recording method in observational research in which the researcher uses a stopwatch or other timing device to record how much time a participant spends engaged in a particular behavior.

Using Video for Observation

Video is a potentially valuable instrument for observational research. Its greatest advantage is that the researcher need not worry about recording observations when the behavior is occurring. Furthermore, it allows the researcher to observe a number of people at one time. For example, teachers and students can be observed simultaneously, which is difficult in normal observational techniques. In addition, the researcher can both replay a video as needed for evaluating the behavior and retain it as a permanent record.

The use of video does have a disadvantage: The presence of a video-recording device may also alter behavior to the extent that the participants do not behave normally. But if this disadvantage can be resolved, video can be effective for observational research.

Weaknesses of Observational Research

Problems and limitations of observational research include the following:

• A primary danger in observational research lies in the study's operational definitions. The behaviors must be carefully defined and observable. Consequently, the actions may be so restricted that they do not depict the critical behavior. For example, teacher effectiveness encompasses many behaviors, and to observe only the number of times the teacher asked questions or gave individual attention may be an inadequate sample of effectiveness.

• Using observation forms effectively requires much practice. Inadequate training therefore represents a major pitfall in this form of research. Attempting to observe many things can also be difficult. Often, the observation form is just too ambitious for one person to use.

• Certain behaviors cannot be evaluated as finely as some observation forms dictate. A common mistake is to ask the observer to make discriminations that are too precise, thus reducing the reliability of the ratings.

• The observer's presence usually affects participants' behavior. Therefore, the researcher has the added responsibilities of not only being aware of this likelihood but also finding ways to reduce the disturbance as much as possible.

• Generally, having more than one observer greatly expedites observational research. Failure to use more than one observer results in decreased efficiency and objectivity.

Unobtrusive Research Techniques

Gathering information about people is not simply limited to the use of questionnaires, case studies, and direct observation. Webb, Campbell, Schwartz, and Sechrest (1966) first discussed a variety of approaches that they termed **unobtrusive measures**. Examples they mentioned include the replacement rate for floor tiles around museum exhibits as a measure of the relative popularity of exhibits and the shrinking diameter of a circle of seated children as a measure of the degree of fear caused by telling ghost stories. Researchers have measured boredom by the amount of fidgeting movements in an audience. The rate of library withdrawals of fiction and nonfiction books has been studied to determine the influence of television in communities. Researchers have also noted that children demonstrate their interest in Christmas by the size of their drawings of Santa Claus and the amount of distortion in the figures.

In some of the preceding examples, the experimenter was not present when the data were being produced. In other situations, the researcher is present but acts in a nonreactive manner (i.e., the subjects are not aware that the researcher is gathering data). For example, a researcher in psychology measured the degree of acceptance of strangers among delinquent boys by measuring the distance maintained between a delinquent boy and a new boy to whom the researcher introduced the delinquent participant. Sometimes, the researcher intervenes to speed up the action or force the data but in a manner that attracts little attention to the method. In a classic study of the cathartic effect of activity on aggression, Ryan (1970) had an accomplice behave in an obnoxious manner and then measured the amount of electric shock the participants administered to that accomplice and to innocent bystanders. Other researchers have intervened by causing participants to fail or succeed so that they could observe responses to winning and losing in competitive situations.

People are filmed regularly by security cameras without consent or knowledge; they have no expectation of privacy when walking or playing in public. However, the ethical issue of invasion of privacy arises in some forms of unobtrusive measures. Informed-consent compliance has placed considerable restraints on certain research practices, such as those that involve entrapment and experiments that aim to induce heightened anxiety.

unobtrusive measures—Measures of behavior taken of people who are unaware that the researcher is gathering data.

Unobtrusive measures can raise ethical concerns regarding privacy and informed consent.

Correlational Research

Correlational research is descriptive in that it explores relationships that exist between variables. Sometimes researchers make predictions based on the relationships, but correlation cannot determine cause and effect. The basic difference between experimental research and correlational research is that the latter does not cause something to happen. Correlational research involves no manipulation of variables or no administration of experimental treatments. The basic design of correlational research is to collect data on two or more variables on the same people and determine the interrelationships of the variables. But, of course, the researcher should have a sound rationale for exploring the relationships.

We discussed correlational techniques in chapter 8 by providing examples of situations that lend themselves to correlational research. The two main purposes for doing a correlational study are to analyze the interrelationships of variables and to predict.

correlational research—Research that explores interrelationships of variables and sometimes also involves the prediction of a criterion variable.

Steps in Correlational Research

The steps in a correlational study are similar to those used in other research methods. The problem is first defined and delimited. The selection of the variables to be correlated is of critical importance. Many studies have failed in this regard. Regardless of how sophisticated the statistical analysis may be, the statistical technique can deal only with the variables that are entered; as the saying goes, "Garbage in, garbage out." The validity of a study that seeks to identify basic components or factors of fitness hinges on the identification of the variables to be analyzed. A researcher who wishes to discriminate between starters and substitutes in a sport faces the crucial task of determining which physiological or psychological variables are the important determinants of success. The researcher must lean heavily on past research in defining and delimiting the problem.

Researchers select participants from the pertinent population by using recommended sampling procedures. The magnitude and even the direction of a correlation coefficient can vary greatly, depending on the sample used. Remember that correlations show only the degree of relationship between variables, not the cause of the relationship. Consequently, because of other contributing factors, a researcher might obtain a correlation of .90 between two variables in a sample of young children but a correlation of .10 between the same variables in adults, or vice versa. Chapter 8 presented examples of how factors, such as age, can influence certain relationships.

Another aspect of correlation is that the size of the correlation coefficient depends, to a considerable extent, on the spread of the scores. A sample that is fairly homogeneous in certain traits seldom yields a high correlation between variables associated with those traits. For example, the correlation between a distance run and maximal oxygen consumption with a sample of elite track athletes is almost invariably low because the athletes are so similar; the variability is insufficient to permit a high correlation because the scores on the two measures are too uniform. If some less-trained runners were included in the sample, the size of the correlation coefficient would increase dramatically.

In prediction studies, the sample must be representative of the study's target population. One of the major drawbacks to prediction studies is that the prediction formulas are often sample specific, which means that the accuracy of a formula is greatest (or maybe only acceptable) when it is applied to the particular sample on whom it was developed. We discussed this shrinkage phenomenon as well as cross-validation, which is used to counteract shrinkage, in chapter 11.

The collection of data requires the same careful attention to detail and standardization that all research designs do. A variety of methods for collecting data can be used, such as physical performance tests, anthropometric measurements, pencil-and-paper inventories, questionnaires, and observational techniques. The scores must be quantified, however, to be correlated.

Analysis of the data can be performed by a number of statistical techniques. Sometimes the researcher wishes to use simple correlation or multiple correlation to study how variables, either by themselves or in a linear composite of variables, are associated with some criterion performance or behavior. Factor analysis is a data-reduction method that helps determine whether interrelationships of a number of variables can be reduced to small combinations of factors or common components. Structural modeling is a technique used to test some theoretical model about causal relationships of three or more variables. (See chapter 8 for a discussion on simple correlation, factor analysis, and structural modeling.)

Prediction studies usually employ multiple regression or similar techniques because the accuracy of predicting some criterion behavior is nearly always improved by using more than one predictor variable. Discriminant function analysis is a technique used to predict group membership; logistic regression is a similar technique to predict the probability of an occurrence; and canonical correlation is a method for predicting a combination of several criterion variables from several predictor variables. (Multiple regression and canonical correlation are discussed in chapter 8, and discriminant function analysis in chapter 9.)

Limitations of Correlational Research

Limitations of correlational research include those of both planning and analysis. We have already pointed out the importance of identifying pertinent variables and selecting proper tests to measure those variables. Hypotheses should be based on previous research and theoretical considerations; simply correlating a set of measurements to see what happens is not an effective way to approach research. Selecting an inadequate measure as a criterion in a prediction study is a common weakness in correlational research. For example, a criterion of success in some endeavor is often difficult to define operationally and may be more difficult to measure reliably.

> The easier a criterion is to define operationally and measure reliably, the more effective correlational research will be.

Summary

Descriptive research encompasses many techniques. In this chapter, we described the basic procedures and strengths and weaknesses of five techniques in the category of descriptive research.

Developmental research seeks to study growth measures and changes in behavior over a period of years. In a longitudinal design, the same people are followed over time. When participants of different ages are sampled, the design is cross-sectional. Of paramount importance in developmental research is whether the participants represent their populations and their performances. A developmental study may involve experimental treatment in which the researcher attempts to determine the interaction of some treatment with age. Developmental studies may also be done in a post hoc manner in which the researcher evaluates data that have been collected over a number of years.

In a case study, the researcher attempts to gather a lot of information about one or a few cases. An in-depth study of a single case can result in a greater understanding of other similar cases. Case studies can be descriptive, interpretive, or evaluative. One of the principal advantages of the case study approach is its potential to lead to the formulation of ideas and hypotheses about problem areas.

Observational research is a descriptive technique that involves the qualitative and quantitative analysis of observed behaviors. Unlike the survey method, which relies on self-reports about how a person behaves, observational research attempts to study what a person actually does. Behavior is usually coded by what occurs and when, how often, and how long. Standardized observation instruments such as the ALT-PE and the BWM are frequently used. Researchers often video-record their participants to store the observations for later analysis.

Some studies use unobtrusive measures in which the participants are unaware that the researcher is gathering data. For example, instead of (or in addition to) asking how much a person smokes, the researcher might count the cigarette butts in an ashtray after a certain period. Although the methods can be interesting and innovative, they may be constrained by ethical considerations such as invasion of privacy and entrapment.

Correlational research examines interrelationships of variables. Sometimes the relationships are used for prediction. Although correlations are often used with experimental research, the study of relationships is descriptive in that it does not involve the manipulation of variables. A major pitfall in correlational research is assuming that because variables are related, one causes another.

✓ Check Your Understanding

1. Locate a cross-sectional study and a longitudinal study in the literature and answer the following questions about each:
   a. Is the study descriptive or experimental?
   b. What age levels are studied?
   c. What are the independent and the dependent variables?
   d. What statistics are used to make cross-age comparisons?
2. Write an abstract of a case study found in the literature. Indicate the problem, the sources of information used, and the findings.
3. Find an observational study in the literature and outline the methodology.
4. Write a brief abstract of a correlational research study.

17

Physical Activity Epidemiology Research

Duck-chul Lee
Iowa State University

Angelique G. Brellenthin
Iowa State University

> Those who think they have not time for bodily exercise will sooner or later have to find time for illness.
>
> —Edward Stanley

The emergence of the mid-20th-century heart disease epidemic fueled a number of observational epidemiological studies of large scale aimed at identifying the determinants of heart disease to undertake preventive measures to improve the public's health. A series of observational studies was initiated between the late 1940s and 1960s. A number of them were particularly important to the development of physical activity epidemiology because they were the first to develop methods for measuring physical activity and conduct systematic studies of the link between physical activity and a life-threatening disease. These studies, to name a few (and the important people associated with each), include the Framingham Heart Study, London busmen and British civil servants study (Jeremy Morris), Tecumseh Community Health Study (Henry Montoye), the College Alumni Health Study (Ralph Paffenbarger), and the Aerobics Center Longitudinal Study (Steven Blair).

U.S. National Physical Activity Guidelines and Plan

By the early 1980s, researchers had begun to amass solid evidence demonstrating that low levels of physical activity were associated with increased risk of heart disease and overall mortality. Because heart disease was (and still is) the leading cause of death in the United States and the countries of the European Union, physical activity emerged as an important public health concern. Accordingly, the U.S. Public Health Service initiated surveillance programs to quantify the leisure-time physical activity patterns in the U.S. population in the mid-1980s. These programs demonstrated that more than 60% of U.S. adults were not physically active

We wish to acknowledge the efforts of the previous authors of this chapter, Barbara E. Ainsworth and Charles E. Matthews.

on a regular basis. In the 1990s, epidemiological evidence continued to accumulate, showing that low levels of physical activity were associated with increased risk for a number of important health conditions, including all-cause and cause-specific mortality, cardiovascular disease, type 2 diabetes, osteoporosis, some forms of cancer, and mental health and quality-of-life problems (U.S. Department of Health and Human Services, 1996). Major initiatives are currently underway to understand the best way to change the behavior of the population at both the individual and community levels.

One such initiative is Healthy People 2030, a plan the U.S. Department of Health and Human Services (HHS) developed in 1979 and has updated every 10 years since. The initiative includes a section on physical activity in addition to other health behaviors such as nutrition and healthy eating, tobacco use, drug and alcohol use, and sleep (U.S. Department of Health and Human Services, 2020). The physical activity objectives for 2030 focus on reducing inactivity and increasing physical activity to levels recommended in the U.S. Physical Activity Guidelines. Another well-documented and recognized initiative is the U.S. National Physical Activity Plan that complements Healthy People 2030 by recommending detailed and evidence-based policies, programs, and initiatives designed specifically to promote physical activity in these nine societal sectors (National Physical Activity Plan Alliance, 2016):

1. Business and industry
2. Community recreation, fitness, and parks
3. Education
4. Faith-based settings
5. Health care
6. Mass media
7. Public health
8. Sport
9. Transportation, land use, and community design

The Plan was originally released in 2010 and updated in 2016 by the National Physical Activity Plan Alliance (NPAPA), a nonprofit organization with a close partnership with HHS. The Plan is based on a socioecological model of health behavior that physical activity is determined by various factors at the personal, family, institutional, community, and policy levels. Each sector includes broad strategies to promote physical activity in the U.S. population, and each strategy outlines specific tactics for communities, organizations, and agencies. Some examples of specific tactics include "Create or enhance access to places for employees to engage in physical activity before, during, and after work hours;" "Conduct periodic worksite-based health screenings that measure physical activity and fitness levels of workers," under the business and industry sector; and under the education sector, "Support adoption of school design strategies to support active transport" and "Support adoption of policies requiring that students at all levels be given physical activity breaks during the school day."

The U.S. Physical Activity Guidelines provide specific recommendations for the types and amounts of physical activity for individuals. However, the Plan provides recommendations for policy and practices, specifically in nine societal sectors in the population, to further support and facilitate the U.S. Physical Activity Guidelines. Figure 17.1 depicts the percent of adults by various demographic subgroups who meet the aerobic physical activity guidelines relative to the Healthy People 2030 target.

2018 U.S. Physical Activity Guidelines

Adults should move more and sit less throughout the day.

Adults should do at least 150 minutes (2 hours and 30 minutes) to 300 minutes (5 hours) a week of moderate-intensity, or 75 minutes (1 hour and 15 minutes) to 150 minutes (2 hours and 30 minutes) a week of vigorous-intensity aerobic activity.

Preferably, aerobic activity should be spread throughout the week.

Additional health benefits are gained beyond the recommended amounts of 300 minutes of moderate-intensity or 150 minutes of vigorous-intensity physical activity a week.

Adults should do muscle-strengthening activities that involve all major muscle groups on 2 or more days a week.

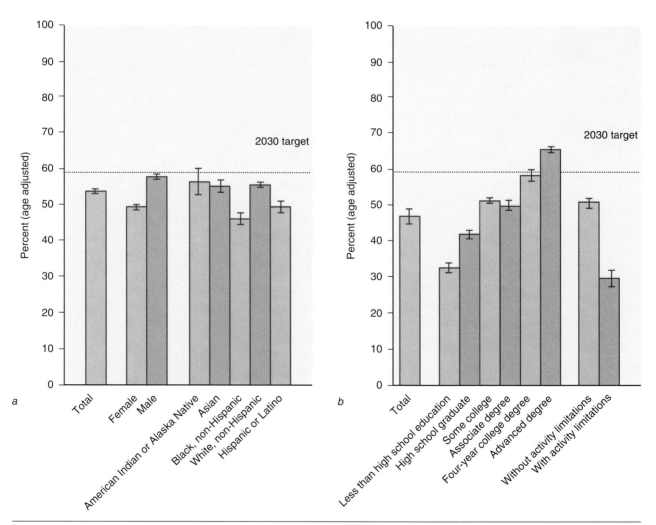

Figure 17.1 Percentage of U.S. adults meeting aerobic physical activity guidelines of ≥150 minutes per week of moderate-intensity or ≥75 minutes per week of vigorous-intensity physical activity, or an equivalent combination, by sex, age, race and ethnicity, and education in 2018. I = standard errors. Dotted line = Healthy People 2030 target of 59.2% of adults meeting the aerobic physical activity guidelines.

Data from National Center for Health Statistics, National Health Interview Survey (2018).

risk factor—An exposure that has been found to be a determinant of a disease outcome or health behavior.

This small history of physical activity epidemiology mirrors the major goals of epidemiological inquiry that we describe in the remainder of this chapter. Epidemiological methods are used to provide the scientific backbone for public health endeavors, including (a) quantifying the magnitude of health problems, (b) identifying factors that cause disease (i.e., **risk factors**), (c) providing quantitative guidance for the allocation of public health resources, and (d) monitoring the effectiveness of prevention strategies using population-wide surveillance programs (Caspersen, 1989). In this chapter, we define epidemiology, describe methods of measuring physical activity in epidemiological studies, describe epidemiological study designs, consider threats to the validity of these designs, and provide an outline for reading and interpreting an epidemiological study.

Observational Versus Experimental Research

Observational research considers how existing differences (e.g., in physical activity, diets) affect disease outcomes.

A key to understanding epidemiological methods lies in considering the difference between observational and experimental research. It has been said that an epidemiologist is most happy when conditions conspire to produce circumstances much like those of a true experiment (Rothman, 1986).

Observational research uses existing differences in factors that may cause disease within a population, such as physical activity, dietary habits, or smoking. Some people in a population choose to be physically active, whereas others do not. Epidemiologists use these naturally occurring differences in a population to observe, and therefore understand, the effect of these differences on specific disease **outcomes**.

outcome—The dependent variable in an analysis.

A primary reason epidemiological research has become such an important discipline is that it is virtually the only way to obtain a quantitative understanding of the health risks of many behaviors. This circumstance arises from the fact that it would be unethical to conduct true experimental research on health behaviors such as physical inactivity. For example, researchers could not randomize a group of people to either a completely sedentary lifestyle or an active lifestyle, wait 5 to 10 years, and then see how much death and heart disease resulted from physical inactivity. Although an experimental study such as this would unequivocally demonstrate that physical inactivity is a causal factor in the development of heart disease, the ethical problems of the experiment are obvious. Thus, in **physical activity epidemiology**, both long-term observational research, such as cohort studies on physical activity and health, and well-designed experimental research, such as controlled trials of exercise that are randomized, are important since they complement each other.

physical activity epidemiology—The study of the distribution and determinants of physical activity, its associations with health-related outcomes, and the application of this study to disease prevention and health promotion.

What Is Physical Activity Epidemiology?

Epidemiology has been defined as "the study of the distribution and determinants of health-related states or events in specified populations, and the application of this study to the control of health problems" (Last, 1988, p. 141). Caspersen defined physical activity epidemiology as a two-part process (Caspersen, 1989, p. 425):

1. First, as a science, "it studies the association of physical activity, as a health-related behavior, with disease and other health outcomes; the distribution and the determinants of physical activity behavior(s); and the interrelationship of physical activity with other behaviors."

2. Second, as a practice, "it applies that knowledge to the prevention and control of disease and the promotion of health."

Based on these definitions and concepts, physical activity epidemiology can be concisely defined as "the study of the distribution and determinants of physical activity, its associations

Components of Epidemiological Research

Distribution

Frequency—prevalence, incidence, mortality rate

Patterns—person, place, time

Determinants

Defined characteristics—associated with change in health

Application

Translation—knowledge to practice

with health-related outcomes, and the application of this study to disease prevention and health promotion." Physical activity can be either an exposure or outcome depending on the study design and purpose. When studying the associations of physical activity with disease and other health outcomes, it is an exposure. However, when studying the determinants of physical activity, it is an outcome. From both clinical and public health perspectives, physical activity epidemiology is an important growing area of science and practice for the prevention of disease and health promotion.

Distribution

The distribution of disease relates to the *frequency* and *patterns* of disease occurrence in a population. Frequency, or how often the disease occurs, is typically measured as the prevalence, incidence, or mortality rate of a disease. Disease prevalence refers to the number of people in a given population that have a disease at a particular point in time. The frequency of disease occurrence may also be calculated as the rate of new disease or health events, as *incidence* or *mortality rates* (i.e., new cases or new deaths from a specific disease within a specified period). For example, the mortality rates for cardiovascular disease were 219.4 per 100,000 U.S. adults in 2017 (Virani et al., 2020).

A presentation of the frequency of disease occurrence relative to the number in the population of interest enables a comparison of prevalence, incidence, or mortality rate across populations. For example, the prevalence of cardiovascular disease is about 50% higher for men than it is for women in the U.S. (Virani et al., 2020).

Evaluating the basic *patterns* of disease occurrence within a specified population is often useful for developing hypotheses about risk factors for the disease. Patterns of disease occurrence refers to characteristics related to *person*, *place*, and *time*. Personal characteristics include demographic factors such as age, sex, and socioeconomic status. Characteristics of place include geographic differences, urban–rural variation, and, particularly important in the history of physical activity epidemiology, differences in occupation. Historically, differences in occupational classifications enabled researchers to make crude comparisons of occupational physical activity levels and were used in some of the first epidemiological studies of the relationship between physical activity and heart disease.

Time of disease occurrence refers to annual, seasonal, or daily patterns of occurrence. Quantification of temporal changes in disease rates often leads to hypotheses that generate a more detailed examination of the factors that caused this change. A good example of this was the observation that the incidence of upper-respiratory infections (i.e., common colds) in runners increased in the 14-day period after running an ultramarathon (56K) (Peters & Bateman, 1983). Of the 141 marathoners studied, 47 (33%) came down with a cold after

exposure—Factors (variables) in epidemiological studies that are tested for their relationship with the outcome of interest.

the race, whereas only 19 (15%) of the 124 people who did not run the marathon had a cold during the same period. Thus, the incidence of colds was more than twice as high among the runners, apparently because of the **exposure** of the marathon event. This observation led to intensive research in the area of exercise immunology and ultimately resulted in a greater understanding of the relationship between exercise and the immune system (Nieman, 1994).

In physical activity epidemiology, distribution can also refer to the prevalence of meeting the physical activity guidelines (as opposed to the occurrence of disease) in a certain population, area, or time. For example, in 1,922 community-recruited adults (mean age 65) with or at high risk for knee osteoarthritis, more than 50% of men and nearly 80% of women failed to meet the U.S. Physical Activity Guidelines, defined as at least 150 weekly minutes of moderate-to-vigorous-intensity physical activity measured by accelerometers, which suggests a sex difference in physical activity in this population (Chang, Song, Lee, Chang, Semanik, & Dunlop, 2020).

Determinants

determinant—A factor that changes a characteristic; often called *risk factor*.

A **determinant** is "any factor, whether event, characteristic, or other definable entity, that brings about change in a health condition, or other defined characteristic" (Last, 1988, p. 500). In physical activity research, the goals are usually to test the hypothesis that activity is or is not a determinant for a particular disease outcome, or to identify the determinant of physical activity behaviors such as sex, age, race, income, occupation, and environment. Determinants of disease are often called *risk factors* because they increase a person's risk for disease. Epidemiological studies have been instrumental in identifying risk factors for heart disease, including obesity, high blood pressure, high LDL cholesterol, low HDL cholesterol, and physical inactivity. Epidemiological studies have also identified that women are less likely than men and older individuals are less likely than younger individuals to meet the physical activity guidelines. The identification and surveillance of specific determinants of a particular disease or physical activity behaviors allow for targeted health promotion campaigns that present this new health knowledge to the public.

Associations

The associations of physical activity with various health-related outcomes (e.g., cardiometabolic diseases, cancer, mental health, cognition, quality of life, mortality) are the central part of physical activity epidemiology. Findings from these studies provide important data (e.g., how much of the risk of experiencing a heart attack can be reduced in physically active compared to inactive individuals) to develop effective public health strategies and policies such as the U.S. Physical Activity Guidelines. The associations of physical activity with health outcomes are commonly expressed as odds ratios or hazard ratios from observational studies and the differences in the primary outcomes (e.g., body weight, blood pressure) before and after exercise intervention from randomized controlled trials. These epidemiological measures and study designs are described in detail with examples in the following sections.

Application

The application of the established understanding of the causal factors related to disease is a major goal of public health. Thus, after epidemiologists have identified the cause of disease, health educators interact with communities to make them healthier places to live.

Other names for the application of research in a community setting are *translation* and *dissemination*. Successful public health dissemination strategies are those that engage the community using a variety of methods, from motivating people to change their behaviors to affecting public policy. This is referred to as the *ecological model for health promotion* (McLeroy, Bibeau, Steckler, & Glanz, 1988). Public health disseminators translate knowledge from epidemiological studies to help increase physical activity among individuals and within social groups and community organizations. Strategies are also used to affect the community environment (e.g., by building walking trails) and encourage policy makers to enact legislation or appropriate funds to enable people to lead physically active lives.

Definitions of Physical Activity and Its Components

Because physical activity can be described in many ways, definitions are necessary to increase measurement consistency and reduce variability across studies. *Physical activity* is a global term that is defined as "any bodily movement produced by the contraction of skeletal muscle that substantially increases energy expenditure" (U.S. Department of Health and Human Services, 2018). Physical activity includes all forms of movement done in occupational, exercise, home and family care, transportation, and leisure settings. *Exercise* is a form of physical activity that is planned, structured, repetitive, and performed with the goal of improving health or fitness (U.S. Department of Health and Human Services, 2018). All exercise is physical activity, but not all physical activity is exercise. *Physical fitness* is the ability to carry out daily tasks with vigor and alertness, without undue fatigue, and with ample energy to enjoy leisure-time pursuits and respond to emergencies (U.S. Department of Health and Human Services, 2018). Types of physical fitness are cardiorespiratory (endurance or aerobic power), musculoskeletal, flexibility, balance, and speed of movement.

The two most common types of physical activity are aerobic physical activity and muscle-strengthening physical activity on which the U.S. Physical Activity Guidelines have specific and separate recommendations (U.S. Department of Health and Human Services, 2018). *Aerobic physical activity* is the activity in which the body's large muscles move in a rhythmic manner for a sustained period of time, which is also called *endurance* or *cardio activity* such as brisk walking, running, swimming, and bicycling. *Muscle-strengthening physical activity* is physical activity including exercise that increases skeletal muscle strength, power, endurance, and mass, such as strength training, weight lifting, or resistance training.

When physical activity is measured in epidemiological studies, the frequency, duration, and intensity of physical activities performed are studied in relation to their associations with health.

- *Frequency* relates to how often a person does activity. For example, in the U.S. Physical Activity Guidelines, at least two days per week of muscle-strengthening activities are recommended. In addition, figure 17.2 shows an epidemiological example of all-cause and cardiovascular mortality risk by leisure-time running in 55,137 adults aged 18 to 100 years (mean age 44) with an average 15-year follow-up from the Aerobics Center Longitudinal Study (Lee, Pate, Lavie, Sui, Church, & Blair, 2014). In this prospective cohort study, one exposure was weekly frequency of running, and the study suggested that even one to two times per week of running was sufficient to provide significant mortality benefits.

- *Duration* refers to how long a person does an activity as hours or minutes in any one session. Duration is used in combination with frequency to calculate weekly amount of physical activity. For example, the U.S. Physical Activity Guidelines recommend doing moderate-intensity aerobic activity for at least 150 minutes per week, which is calculated as duration multiplied by frequency (e.g., 150 minutes could be achieved as 30 minutes of activity per session by five sessions per week). In figure 17.2, Lee and colleagues (2014) presented reduced mortality risks in runners across total weekly running time in minutes compared to non-runners as an example of how duration is used in physical activity epidemiology. They also presented mortality risk across total weekly distance of running in miles as another example of the amount of total weekly physical activity. In muscle-strengthening

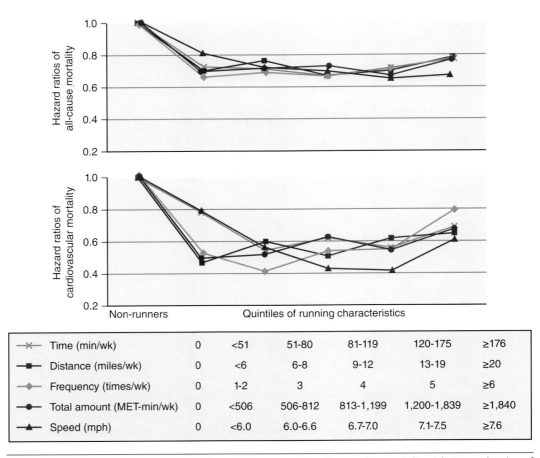

		Non-runners	Quintiles of running characteristics			
──✳──	Time (min/wk)	0	<51	51-80	81-119	120-175	≥176
──■──	Distance (miles/wk)	0	<6	6-8	9-12	13-19	≥20
──◆──	Frequency (times/wk)	0	1-2	3	4	5	≥6
──●──	Total amount (MET-min/wk)	0	<506	506-812	813-1,199	1,200-1,839	≥1,840
──▲──	Speed (mph)	0	<6.0	6.0-6.6	6.7-7.0	7.1-7.5	≥7.6

Figure 17.2   Leisure-time running reduced all-cause and cardiovascular mortality risk. Hazard ratios of all-cause and cardiovascular mortality by running characteristics (weekly running time, distance, frequency, total amount, and speed). Participants were classified into six groups: non-runners (reference group) and five quintiles of each running characteristic. All hazard ratios were adjusted for baseline age (years), sex, examination year, smoking status (never, former, or current), alcohol consumption (heavy drinker or not), other physical activities except running (0, 1 to 499, or ≤500 MET-minutes per week), and parental history of cardiovascular disease (yes or no). All p values for hazard ratios across running characteristics were <0.05 for all-cause and cardiovascular mortality except for running frequency of ≤6 times per week ($p = 0.11$) and speed of <6.0 miles per hour, or 10 km, ($p = 0.10$) for cardiovascular mortality. MET = metabolic equivalent.

Reprinted by permission from D.C. Lee et al., "Leisure-Time Running Reduces All-Cause and Cardiovascular Mortality Risk," *Journal of the American College of Cardiology* 64, no. 5 (2014): 472-481.

activity, sets and repetitions are used to define how many times a person does a muscle-strengthening activity such as number of times a person lifts a weight or performs a push-up. The combination of sets and repetitions is comparable to the duration component from the aerobic activity definition.

• *Intensity* refers to the magnitude of the effort required to perform an activity or exercise. In muscle-strengthening activity, intensity is often expressed as the amount of weight lifted or moved relative to the maximum weight a person is able to lift. Intensity may be expressed in absolute or relative terms. The recommended unit for intensity in absolute terms is the metabolic equivalent (MET), which is defined as "the ratio of the activity's metabolic rate to the resting metabolic rate, without considering the physiological capacity of the individual." One MET is approximately equal to 3.5 ml · kg^{-1} · min^{-1} of oxygen consumption, or about 1 kcal· kg^{-1} · hr^{-1} of energy expenditure for a 60 kg (132 lb) person.

To provide consistency in assigning intensity levels to activities, the Compendium of Physical Activities, which provides MET intensities for over 500 activities, was developed

(Ainsworth et al., 2011). On an absolute scale, aerobic activity intensity is generally classified as sedentary (≥1.5), light (1.6 to 2.9), moderate (3.0 to 5.9), vigorous (6.0 to 8.9), or high (≤9.0) based on MET values in epidemiological settings (Norton, Norton, & Sadgrove, 2010; U.S. Department of Health and Human Services, 2018). Absolute intensity of activity can also be expressed differently as shown in figure 17.2 using running speed in miles per hour on mortality risk (Lee et al., 2014). Expressing physical activity intensity in relative terms allows for the adjustment of difficulty based on individual differences (e.g., on cardiorespiratory fitness). Relative intensity measures include the percentage of maximal oxygen uptake (% $\dot{V}O_2$max) or maximal heart rate (% max heart rate) and rating of perceived exertion (RPE). A difficulty in using relative intensities in epidemiological studies is that comparison of intensity levels for similar activities is unequal between studies. Thus, it is preferable to express physical activity intensity in relative terms when measuring activity for a given person, such as in prescribing exercise programs on an individual basis. On a relative scale, aerobic activity intensity is generally classified as sedentary (<20%), light (20 to 39%), moderate (40 to 59%), vigorous (60 to 84%), or high (≥85%) based on the percentage of heart rate reserve (HRR), which is used in the U.S. Physical Activity Guidelines (Norton et al., 2010; U.S. Department of Health and Human Services, 2018). These percentage classification cut-points based on HRR are also similar to the percentage of maximal oxygen uptake (% $\dot{V}O_2$max). The current guidelines focus more on moderate- to vigorous-intensity aerobic activities that have been most often studied in the past. However, recent studies have investigated the ends of the physical activity spectrum including sedentary and light-intensity physical activities such as sitting, standing, and slow walking as well as high-intensity activity such as high-intensity interval training (HIIT) on various health outcomes (e.g., cardiometabolic disease and its risk factors, cancer, mortality). While any amount of physical activity generally provides health benefits compared to no activity, only physical activity that is at least of a moderate intensity counts toward meeting the aerobic physical activity guidelines.

• *Dose* refers to the combination of the frequency, intensity, and duration of aerobic physical activity and is expressed as kcal per day, or MET-hours per day. There is no widely accepted, singular metric to characterize the dose of muscle-strengthening activities at this time. In figure 17.2, Lee and colleagues (2014) presented total weekly amount of running in MET-minutes per week using frequency, intensity, and duration of running, which was calculated as the MET value for a given speed (intensity) multiplied by the weekly running time (duration times frequency). They found that runners had approximately 30 to 50% lower risk of all-cause and cardiovascular mortality, compared to non-runners. However, the greatest reduction in mortality risk was found between non-runners and runners with the lowest dose of running.

Assessment of Physical Activity

Physical activity can be measured using a variety of methods ranging from direct measurement of the amount of heat a body produces during activity to asking people to rate how active they recall being during the past week or year. From the 1950s to the 1980s, job titles were used to classify physical activity patterns in epidemiological studies involving occupational physical activity. However, with the changing profile of the labor market, occupational titles no longer reflect the physical requirements of a job, eliminating the use of job titles to classify occupational energy expenditure (Montoye, Kemper, Saris, & Washburn, 1996).

Because of the exceptionally large number of people in many epidemiological studies, self-administered questionnaires or brief interviews are often used to capture activity spent at work and at home and in exercise, transportation, and leisure settings. Questionnaires can be classified as global, short recall, and quantitative histories depending on the length and complexity of the items.

• *Global* questionnaires are instruments of one to four items that present a general classification of a person's habitual activity patterns. Global surveys are most accurate for classifying people according to their level of vigorous-intensity physical activity. Because these surveys may take little time to administer (less than two minutes), they are preferred for use in epidemiological studies.

• *Short recall* questionnaires generally have 5 to 15 items and reflect recent physical activity patterns (during the past week or month). They are effective for classifying people into categories of activity, such as inactive, insufficiently active, or highly active based on health guidelines and recommendations for minimal levels of activity. Short recall questionnaires may take from 5 to 15 minutes to complete and are recommended for surveillance activities and observational epidemiological studies designed to assess the prevalence of adults and children who obtain national recommendations for physical activity and health.

• *Quantitative history* questionnaires are detailed instruments that have from 15 to 60 items and reflect the intensity, frequency, and duration of activity patterns in various categories, such as occupation, household, sports and conditioning, transportation, family care, and leisure activities. Quantitative history questionnaires allow researchers to obtain detailed information about physical activity energy expenditure and patterns of activity from the past month to year. Because of their length and complexity, however, they may take from 15 to 30 minutes to complete and are usually interviewer administered. Quantitative history questionnaires are appropriate for studies designed to examine issues of dose–response in heterogeneous populations with a wide variety of physical activity patterns.

One of the most commonly used questionnaires is the International Physical Activity Questionnaire (IPAQ), and there are short and long forms (versions) using either telephone

© germanskydive110/fotolia

Understanding the "risk factors" in research is important.

or self-administered methods (www.ipaq.ki.se). The seven-item short form is an example of a short recall questionnaire including four sections on vigorous activity, moderate activity, walking, and sitting. The 27-item long form is an example of a quantitative history questionnaire including five activity domains of occupation, transportation, housework, leisure time, and sitting. The IPAQ has been developed in different languages through extensive reliability and validity testing across 12 countries and provides data processing and a scoring protocol to calculate total MET-minutes per week and identify individuals who meet the U.S. Physical Activity Guidelines. Therefore, the IPAQ can be used to compare physical activity data internationally and between studies.

Since their introduction as an objective measure of free-living physical activity in the early 1980s, waist-mounted activity monitors (accelerometers and pedometers) have become a staple of the physical activity assessment repertoire. They have been used extensively in the validation of self-reported physical activity surveys, as outcome measures of physical activity in intervention studies, and in research designed to identify the psychosocial and environmental correlates of physical activity behaviors. The great advantage of objective physical activity measures is that they overcome some of the limitations inherent in self-report methods that rely on information reported by participants. With activity monitors, there are far fewer reporting errors or errors introduced by interviewers, and real-time data collection and automated data reduction can provide a rich description of the activity profiles of people and populations. After the logistics of monitor delivery and retrieval are overcome, activity monitors offer a relatively simple and efficient method of measurement that is suitable for small clinical studies and epidemiological studies of intermediate size (e.g., fewer than 5,000).

The ActiGraph is a small, battery-operated accelerometer-based motion sensor commonly worn on the waist. It provides a computerized record of the intensity and duration of ambulatory movement, presented as movement counts. The ActiGraph accelerometer is used in the U.S. National Health and Nutrition Examination Surveys to augment physical activity questionnaires in a subsample of participants. Minute-by-minute activity count data for the ActiGraph accelerometer on two days of monitoring for a middle-aged woman participating in a physical activity intervention study are presented in figure 17.3. Activity count levels above 1,950 counts per minute reflect purposeful moderate-intensity walking in the range of 3 to 4 miles per hour (4.8 to 6.4 km per h) (Freedson, Melanson, & Sirard, 1998). Panel

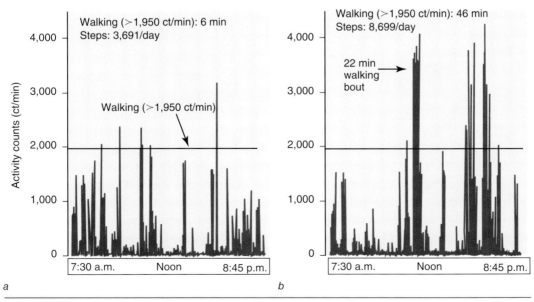

Figure 17.3   ActiGraph output and interpretation: *(a)* sedentary day, *(b)* active day.

a depicts a relatively sedentary day with few extended bouts of walking. Data summary for this day indicated that only 6 minutes of walking and 3,691 steps were accumulated. In contrast, panel *b* represents a more active day in which there was a walking bout (22 min) in the middle of the day and several additional shorter episodes of activity later in the day, eventually accumulating 46 walking minutes and 8,699 steps over the course of the day. Objective information such as this can be used to track changes in activity patterns in response to an intervention or to describe differences in the activity profiles of different populations using the same objective standard. A number of other accelerometer-based activity monitors that are comparable to the ActiGraph devices are commercially available.

Pedometers also have been used in epidemiological studies to measure the amount of accumulated steps in free-living settings. Pedometers are used extensively in health promotion programs to provide feedback to participants about their activity levels or in community trials to measure the effect of walking behavior interventions. For example, to quantify levels of walking among residents in a small southern community, Tudor-Locke, Williams, Reis, and Pluto (2004) mailed pedometers to 209 U.S. adults residing at randomly selected households who agreed to wear the pedometer for seven consecutive days and record their accumulated steps in a logbook each day. The mean step counts were $5,932 \pm 3,664$ steps per day. Steps per day were higher in men ($7,192 \pm 3,596$) than in women ($5,210 \pm 3,518$), higher in whites ($6,628 \pm 3,375$) than in other racial and ethnic groups ($4,792 \pm 3,874$), and inversely related to body mass index (normal weight = $7,029 \pm 2,857$; overweight = $5,813 \pm 3,441$; obese = $4,618 \pm 3,359$).

In general, physical activity questionnaires are more feasible, especially in larger observational studies, and objective physical activity measures are more accurate and valid (e.g., direct calorimetry for energy expenditure in a smaller mechanistic study). Epidemiologists usually value feasibility over accuracy while physiologists usually value accuracy over feasibility. A major limitation in using questionnaires is recall bias that people, especially children or older adults with cognitive impairment, do not accurately remember their previous physical activity and tend to overreport their activities. Questionnaires also require active participation of study participants to fill out the questionnaires. However, questionnaires are inexpensive and generally easy to administer. Researchers can also identify the type of physical activity (e.g., walking, bicycling, tennis, weight lifting) and detailed contextual information about where (e.g., home, work, school) and when (e.g., leisure time, transportation) the physical activity occurred, all of which may better inform the design of behavioral interventions or physical activity policies. Major limitations in objective measures of physical activity include the high cost to purchase activity monitors, difficulties in identifying the type of activities, lack of contextual information, considerable time and effort to download and process the data, and compliance issues including that participants should wear it for several days. However, activity monitors (e.g., smart watches and smartphone apps) are becoming more affordable, popular, and accurate for both researchers and the public, although the frequent technological updates and new commercial versions of these monitors makes it difficult to compare the results between studies over time. Furthermore, most objective physical activity assessments are for aerobic activities, and there is no feasible objective measure of free-living lifestyle muscle-strengthening activities. Other objective measures for physical activity are cardiorespiratory fitness and muscular strength, often used to reflect recent aerobic and muscle-strengthening activities, respectively. An important consideration when using physical fitness measures to represent recent physical activity is that they are also influenced by other factors such as age, sex, and genetics, which should be addressed in the design of epidemiological studies. The best method of physical activity assessment depends on research priorities such as study size, population, purpose, and outcome measures.

Epidemiological Study Designs

Initial hypotheses are developed and tested using observational study designs such as cross-sectional, case-control, and **cohort** studies. Mortality or disease prevalence or incidence is often the dependent variable for the investigations. These designs have been used for the last 50 years, and a solid body of evidence has been amassed to support their basic methods.

After observational epidemiological studies have consistently demonstrated an exposure–disease link, in conjunction with supportive laboratory evidence, experimental studies can then be initiated to test, in a rigorous experimental design, the validity of the observational findings. The design of choice for experimental epidemiological studies is the randomized controlled trial. Outcomes may be mortality, disease incidence, or an intermediate end point such as blood cholesterol levels or blood pressure. In the case of physical activity intervention research, the dependent outcome variable could be a person's or community's physical activity level.

Sometimes it is neither practical nor ethical to conduct a randomized controlled trial to test the validity of observational findings. For example, given the strong associations between smoking and lung cancer from observational studies, it would be unethical and impractical to randomize participants to either smoking or nonsmoking control groups for several years to investigate the long-term effects of smoking on cancer mortality. In cases like these, there are criteria or guidelines that, if met, strengthen the inference that an observed association between an exposure and outcome is causal. These guidelines called *Hill's criteria* were first outlined by epidemiologist and statistician Austin Bradford Hill in 1965 as part of the process that led to the Surgeon General's Report on Smoking and Health (Hill, 1965). While Hill outlined nine criteria in his seminal report, the six most well-established and agreed-upon criteria are as follows:

1. Strength of the association: The rate of the outcome is greater in the exposed group compared with the non-exposed group. The larger the association between the exposure and outcome, the more likely it is to be causal.
2. Consistency: The association between the exposure and outcome is observed regardless of other factors such as sex, age, race, investigator, or methods.
3. Temporality: The exposure precedes the outcome with appropriate delay to allow the outcome to occur (e.g., disease progression).
4. Biological gradient: Greater amounts or degrees of the exposure are associated with a higher rate of the outcome. Also known as *dose–response*.
5. Plausibility: The observed association has a plausible underlying mechanism that is consistent with existing biological knowledge.
6. Experiment: Manipulating the exposure through experimentation (e.g., randomized controlled trials) should affect the rate of the outcome.

Figure 17.4 from a study by Lee and colleagues (2010) investigating the associations between cardiorespiratory fitness (exposure) and all-cause and cardiovascular disease mortality (outcome) provides a good example of meeting several of Hill's criteria. Compared with the least-fit group, the risks of mortality were significantly lower in the groups with higher fitness (strength of the association). This association was similar for men and women (consistency). In this prospective cohort study, cardiorespiratory fitness was assessed prior to mortality ascertainment in adults without cardiovascular disease or cancer at baseline (temporality). There was a significant linear trend between higher fitness levels and lower risk of mortality (biological gradient and dose–response). There are plausible biological

cohort—A group of individuals who are followed over a period of time.

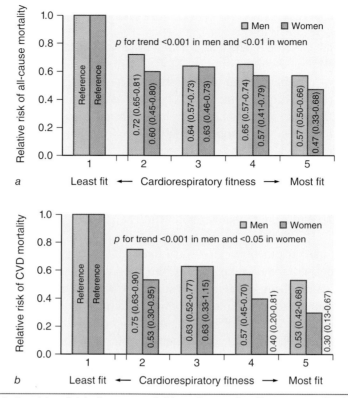

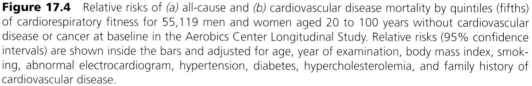

Figure 17.4 Relative risks of *(a)* all-cause and *(b)* cardiovascular disease mortality by quintiles (fifths) of cardiorespiratory fitness for 55,119 men and women aged 20 to 100 years without cardiovascular disease or cancer at baseline in the Aerobics Center Longitudinal Study. Relative risks (95% confidence intervals) are shown inside the bars and adjusted for age, year of examination, body mass index, smoking, abnormal electrocardiogram, hypertension, diabetes, hypercholesterolemia, and family history of cardiovascular disease.

Reprinted by permission from D.C. Lee, E.G. Artero, X. Sui, and S.N. Blair, "Mortality Trends in the General Population: The Importance of Cardiorespiratory Fitness," *Journal of Psychopharmacology* 24, no. 11 (2010): 27-35.

mechanisms between higher fitness and lower risk of death including better cardiac and pulmonary function. Finally, the observed association between cardiorespiratory fitness and mortality has been supported by randomized controlled trials showing that those who improve their fitness through aerobic exercise have reductions in several cardiovascular disease risk factors like blood pressure (experiment).

Cross-Sectional Studies

The cross-sectional design is perhaps the most frequently conducted type of study examining the relationship between physical activity and health outcomes. Cross-sectional studies examine the associations of physical activity with both intermediate end points (e.g., cardiovascular disease risk factors) and hard end points of disease (e.g., ischemic stroke). In fact, a great deal of the initial work examining the effect of physical activity on cardiovascular disease risk factors (e.g., high blood cholesterol or blood pressure levels) used simple cross-sectional designs. Researchers simply assembled two groups of people, highly active people (usually athletes) and sedentary control participants, and then measured the cardiovascular risk factors of interest. A simple test of the mean differences between groups formed the analysis. Because these studies typically measure the outcome variables at a single point in time, they are relatively simple and easy to conduct.

An example of a cross-sectional study shows the associations between self-reported physical activity and disability in a nationally-representative sample of 8,309 adults who were at least 50 years old (Delaney, Warren, Kinslow, de Heer, & Ganley, 2019). The outcome variable was participants reporting at least one disability, defined as "some difficulty," "much difficulty," or being "unable to perform" mobility tasks (e.g., walking up 10 steps), activities of daily living (e.g., dressing yourself), or social engagements (e.g., going out to the movies). Participants were asked the frequency and duration of moderate- and vigorous-intensity aerobic physical activities, and the total dose of physical activity was used to categorize them as meeting or not meeting the aerobic physical activity guidelines of at least 150 minutes of activity with moderate-to-vigorous intensity, or an equivalent combination, per week. Results showed a significant dose–response relationship between different levels of meeting the physical activity guidelines and the odds of reporting one or more disability as shown in figure 17.5. Participants who reported no physical activity had two times higher odds of reporting a disability compared with those who met double the physical activity guidelines.

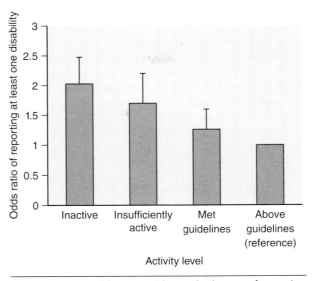

Figure 17.5 Odds ratios with standard errors of reporting at least one disability by levels of meeting the U.S. Physical Activity Guidelines. Inactive ($n = 2,912$) = no physical activity; Insufficiently active ($n = 1,360$) = 10 to 149 minutes of moderate or 10 to 74 minutes of vigorous activity per week; Met guidelines ($n = 1,028$) = 150 to 299 minutes of moderate or 75 to 149 minutes of vigorous activity; Exceeded guidelines ($n = 3,009$) = 300+ minutes of moderate or 150+ minutes of vigorous activity. Adjusted for age, sex, race, education, body mass index, arthritis, myocardial infarction, stroke, depression, and diabetes.

Adapted by permission from M. Delaney et al. "Association and Dose-Response Relationship of Self-Reported Physical Activity and Disability Among Adults =50 Years: National Health and Nutrition Examination Survey, 2011-2016," *Journal of Aging and Physical Activity* 28 (2019): 434-441.

Cross-Sectional Ecological Studies

Ecological studies are typically cross-sectional in nature and use existing data sources for both exposure and disease outcomes to compare and contrast rates of disease by specific characteristics of an entire population. Typical data sources in these studies are census data, vital statistics records (for countries, states, counties), employment records, and national figures for health-related information such as food consumption.

For example, Morris, Heady, Rattle, Roberts, and Parks (1953) reported on a series of studies that used occupational classifications as a surrogate measure of physical activity levels. Perhaps the most famous comparisons were among London busmen working on double-decker buses. Conductors, who had to walk up and down the bus stairs all day, every day, were compared with the bus drivers, who sat and drove the bus all day, every day. Drivers consistently had heart disease mortality rates that were twice as high as those of conductors. Also included in this seminal work was an ecological study of heart disease mortality and occupational physical activity. The investigators tabulated heart disease mortality rates by age and occupational activity level among Englishmen and Welshmen using existing public health statistics for the years 1930 to 1932. The retabulation and classification of these data by occupational types into heavy-, intermediate-, and light-activity categories provided a scientifically crude, but ultimately valid, understanding of the effect of physical activity on heart disease mortality.

Figure 17.6 shows a graded inverse relationship between occupational physical activity and heart disease mortality in each age group. That is, men who were engaged in heavy occupational activity had the lowest death rates, and men in the light-activity occupations had the highest death rates, regardless of age. Men in intermediate-activity categories had death rates that were in between. A clear increase in heart disease mortality with increasing age was also observed.

Advantages of Cross-Sectional Studies

An advantage of many cross-sectional studies is that the physiological differences of individuals (rather than groups) can be compared, and with planning, investigators can control for factors that could potentially confound the relationship of interest. Confounding factors can be controlled using statistical methods when the size of the study is adequate, or by matching the two comparison groups on an important factor. For example, to control for the confounding effect of body fat on the physical activity–blood cholesterol relationship, one could recruit highly active and sedentary people with similar levels of body fat (e.g., ± 2%). Comparisons of the two groups would then be independent of the effects of body fat. Cross-sectional studies can also be conducted in large samples in a cost-effective manner in a relatively short amount of time since they do not follow participants over time.

Disadvantages of Cross-Sectional Studies

The major limitation of the cross-sectional study design is that the study outcome and exposure are measured at the same point in time. Therefore, it is impossible to know whether physical activity exposure was actually responsible for the effects observed. In the previous

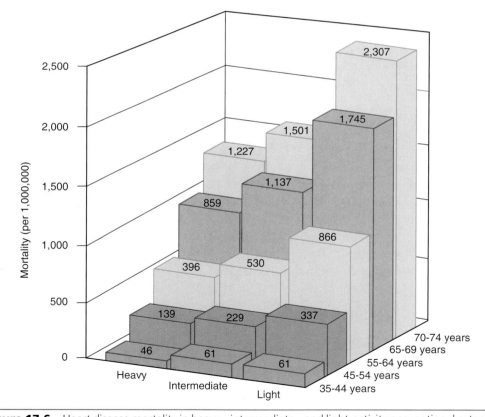

Figure 17.6   Heart disease mortality in heavy-, intermediate-, and light-activity occupational categories by age group among men: England and Wales, 1930 to 1932.

Reprinted by permission from J.N. Morris et al., "Coronary Heart-Disease and Physical Activity of Work," *The Lancet* 262, no. 6796 (1953):1111-1120.

example of meeting the physical activity guidelines and presence of disability, it is equally possible that individuals have disabilities because they are less active or they are less active because they have disabilities. That is, cross-sectional study designs do not allow for definitive conclusions about cause and effect because the temporal sequence (timing) of the relationship between the outcome and the exposure is not known.

Another disadvantage of ecological designs in particular is that the level of analysis is a population group (e.g., those with a certain occupation) rather than an individual. Therefore, there is no way to link individual physical activity levels to a specific heart disease outcome. A final and often overlooked limitation of cross-sectional studies is that a lack of association in a cross-sectional study may not mean that no longitudinal relationship exists between the two factors being examined. For example, on a cross-sectional basis, there can be little or no relationship between dietary fat intake and blood cholesterol levels. In longitudinal studies, however, there is a clear, positive relationship between fat intake and cholesterol levels (Jacobs, Anderson, & Blackburn, 1979). Conversely, cross-sectional studies have also been known to produce spurious results. For example, ecological studies first conducted in the 1970s examining the relationship between dietary fat intake and breast cancer mortality revealed strong linear relationships between this exposure and breast cancer. Women in countries whose citizens ate more fat had higher breast cancer mortality rates. Since that time, however, a large body of research using methodologically superior study designs has been unable to demonstrate the strong link between fat intake and breast cancer observed in the early ecological studies (Willett, 1990). Accordingly, cross-sectional studies and ecological studies should be considered only a first line of investigation aimed at the development of testable hypotheses.

In summary, cross-sectional studies are useful for developing and crudely testing initial hypotheses about exposure–disease relationships. More definitive, scientifically sound results, however, can come only from a more advanced observational study design such as a cohort study or a case-control study, or from an experimental study design such as the randomized controlled trial.

To understand cohort studies and case-control studies more thoroughly, we must consider a simplified model of the natural history of chronic disease (see figure 17.7). Chronic diseases, such as heart disease, osteoporosis, and cancer, can take 10 to 40 years to develop into a problem

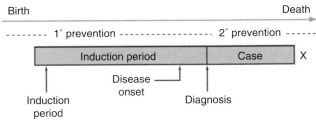

Figure 17.7 Natural history of chronic disease.

sufficient to be diagnosed clinically. Physical activity epidemiologists are typically interested in how exposure to physical activity early in the disease process alters the course of the natural history of the disease. The disease course could be extended by preventive exposures (an active lifestyle) or hastened by adverse exposures (a sedentary lifestyle).

The induction period of a disease is conceptualized as the period between the point at which the exposure of interest alters the physiology of the disease process and the point at which the disease is clinically diagnosed. The clinical diagnosis always occurs later than the disease onset. After being clinically diagnosed with the disease, a person becomes an incident case. Primary prevention (1°) refers to prevention of the first occurrence of the disease, and secondary prevention (2°) refers to prevention of disease recurrence. The usual goal of cohort and case-control studies is to quantify the effect of exposures that occur during the induction period on eventual disease risk. That is, the goal is to identify predictors of the disease outcome during the early portions of the natural history of the disease.

Cohort Studies

The terms *follow-up studies*, *prospective studies*, and *longitudinal studies* have all been used to describe the **cohort study design**. In this design, a large disease-free population is defined, and an assessment of relevant exposures is obtained. Baseline data are used to categorize the cohort into levels of exposure (e.g., low, medium, high). After this baseline assessment has been made, the follow-up period begins. Because chronic diseases such as colon and breast cancer are relatively rare, the follow-up period can last from as little as 1 to 2 years to more than 20 years. At the end of the follow-up period, the number of people within the cohort who died or who were diagnosed with the disease outcome of interest during follow-up is tabulated.

The analysis of cohort studies is relatively simple. Because we are interested in how various levels of baseline exposure predict disease occurrence, the basic analysis consists simply of calculating disease rates for the levels of exposure. For example, the mortality rates among people reporting regular exercise would be compared with the mortality rates of people who do not exercise. In more refined analyses, mortality rates for three to five levels of physical activity may be calculated, and comparisons between the low activity level and each of the higher activity levels could be made.

Disease rates in cohort studies are often expressed relative to person-years of follow-up. A person-year represents one year of observation for one person during the follow-up period. For example, in figure 17.8, five people were observed from time 0 to time 5 (6 full years). A participant's person-year contribution to the cohort begins when they enter follow-up (at time 0 for the top three people, at time 1 for the fourth person, at time 2 for the fifth person). Follow-up ends for each person when the person has an

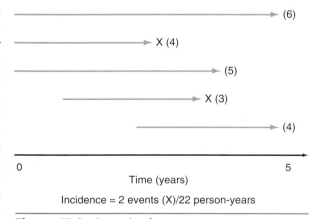

Figure 17.8   Example of person-years.

event (X), when the person is lost for follow-up, or when the follow-up period ends. In the example, each participant's person-year contribution to the cohort is in parentheses, and the summary measure of the incidence rate is 2 events (X) / 22 person-years of observation.

The summary measures of the exposure–disease relationship in cohort studies are typically expressed in absolute or relative terms. Differences in disease rates between exposure groups expressed on an absolute basis can be obtained by simple subtraction of rates. More frequently, estimates of the exposure–disease relationship are expressed as the **relative risk**. Relative risks are calculated as the ratio of the referent category to various exposure levels. The null value, or no effect, of the relative risk is 1.0. Values below 1.0 indicate reduced risk, whereas values above 1.0 indicate increased risk.

A landmark cohort study in the physical activity epidemiology literature is the College Alumni Health Study. Ralph S. Paffenbarger Jr. and colleagues initiated this study by enrolling 16,936 male Harvard alumni aged 35 to 74 years in 1962 and 1966. At baseline, participants completed mailed questionnaires that included questions about their daily physical activities. Using simple questions about walking, stair climbing, and exercise activities, they categorized participants into six levels of physical activity energy expenditure (i.e., <500, 500 to 999, 1,000 to 1,999, 2,000 to 2,999, 3,000 to 3,999, and ≤4,000 kcal per week). In a follow-up assessment 6 to 10 years later (1972), it was determined that 572 men had experienced their first heart attack (Paffenbarger, Wing, & Hyde, 1978). Rates of fatal, nonfatal, and all first heart attacks were compared in each of the baseline physical activity exposure categories. These data, presented in figure 17.9, clearly indicated that the rates for each category of

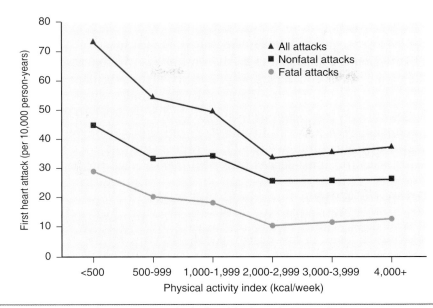

Figure 17.9   Heart attack rates versus physical activity in male Harvard alumni.

Adapted by permission from R.S. Paffenbarger, Jr., A.L. Wing, and R.T. Hyde, "Physical Activity as an Index of Heart Attack Risk in College Alumni,"*American Journal of Epidemiology* 142, no. 9 (1995): 889-903.

heart attack were inversely related to physical activity index levels, up to about 2,000 kcal per week, where the heart attack rates reached a plateau or increased slightly.

Stratified comparisons of these data by age group, smoking status, blood pressure level, and adiposity revealed that these potentially confounding factors did not account for this relationship between physical activity and first heart attacks. Thus, the results strongly suggested that the effect of physical activity on heart attack risk was independent of many important factors that could potentially confound this relationship. The summary relative risk estimate of 1.64 was obtained by dividing the heart attack rates for less active men (<2,000 kcal per week, 57.9 attacks per 10,000 person-years) by the rates for highly active men (≤2,000 kcal per week, 35.3 attacks per 10,000 person-years). That is, less active men had a 64% increased risk of having a heart attack relative to highly active men (Paffenbarger et al., 1978).

Advantages of Cohort Studies

A distinct advantage of the prospective cohort is that the temporal sequence between exposure and outcome is clearly defined. Exposure assessments are obtained before disease onset, so the timing of the exposure in the natural history of the disease is in the correct sequence. This feature is the primary strength of cohort studies.

Cohort studies are also good for rare exposures. For example, if you wanted to study the effects of a rare exposure such as cycling, you could specifically recruit a cohort of cyclists into your study to ensure an adequate number of people in short-, medium-, and long-distance categories. If you recruited for such a study using a random sample of the U.S. adult population, you would have only 3% of your study population with your exposure of interest (Woods, 2017).

Cohort studies are good for understanding the multiple effects of a single exposure. For example, within the Harvard alumni cohort study mentioned earlier, the effects of physical activity on a diverse number of health outcomes were examined, including lung, colon, prostate, and pancreatic cancers, as well as the risk of first heart attack, stroke, hypertension, depression, suicide, Parkinson's disease, and cardiovascular and all-cause mortality.

Disadvantages of Cohort Studies

Cohort studies are difficult and costly to conduct because of the challenges inherent in keeping track of large numbers of people over long periods of time (about 40 years in the case of the College Alumni Health Study). Loss to follow-up can be problematic because large numbers of losses can result in biased estimates of the exposure–disease relationship.

A second limitation of cohort studies is that some disease outcomes are sufficiently rare that even an extremely large cohort may not produce enough cases for meaningful analyses. For example, it took only 6 to 10 years for the Harvard alumni study ($N = 16,936$) to accumulate 572 cases of first heart attack (or, an incidence of 42.2 per 10,000 men per year). By comparison, in a cohort study of the effects of physical activity on the rarer outcome of breast cancer, 14 years of follow-up in 25,624 women were required to accumulate 351 cases of breast cancer (or an incidence of 9.8 per 10,000 women per year).

Case-Control Studies

case-control study—A study that involves the identification of causally related factors for disease outcome in populations with and without disease.

odds ratio—A measure of the exposure–disease relationship typically employed in case-control studies.

The **case-control study**, like the cohort study, aims to identify factors that are causally related to disease outcome. In this design, people with disease (i.e., cases) and without disease (i.e., controls) are recruited into the study over the same period. Frequently, the disease-free control participants are selected to match cases with respect to potentially important confounding factors, such as age, sex, and ethnicity. Both cases and controls are then asked about their exposures to potential causal factors using in-person interviews or self-administered questionnaires. Thus, the objective of this retrospective exposure assessment in case-control studies is to identify factors that influenced the natural history of the disease during its induction period (i.e., before disease onset). The analysis of case-control data contrasts the exposure history of the cases to that of controls and typically expresses this relationship as an **odds ratio (OR)**. The odds ratio is interpreted as an estimate of the relative risk that would have been calculated in the study group if a cohort study had been conducted. Like the relative risk, the null value of the odds ratio is 1.0. Values above 1.0 typically indicate increased risk, and values less than 1.0 indicate reduced risk.

The case-control epidemiological design has been particularly important in studying the relationship between physical activity and relatively rare diseases, such as cancer. For example, breast cancer is the second leading cancer diagnosis, after skin cancer, among U.S. women, accounting for 250,520 cases in 2017 (U.S. Department of Health and Human Services, 2020). Clearly, it is a major public health concern, yet the relatively low incidence of the disease among women in general (125 cases per 100,000 women) makes it a challenging outcome to investigate using the cohort design.

An interesting case-control study examining the relationship between lifetime physical activity and endometrial cancer, another somewhat rare cancer among women, was published by John and colleagues (2010). Investigators examined 472 cases with endometrial cancer and 443 controls who were free of the disease. Participants were 35 to 79 years old and women in the control group were matched to cases according to age and race or ethnicity. Upon recruitment, each woman was interviewed about the domain, type, duration, frequency, and intensity of her physical activity participation at different ages over her lifetime and a number of additional endometrial cancer risk factors. Physical activities were assigned a MET value to compute average energy expenditure of physical activity during four age periods (12 to 19, 20 to 39, and 40+ years old). Using these data, women were categorized into tertiles (i.e., thirds) of average lifetime weekly physical activity exposure.

Figure 17.10 was adapted from the John et al. article and presents multivariate (adjusted) odds ratios for this association. Women who performed at least 91.9 MET-hours of total physical activity per week were at a significantly lower risk of having endometrial cancer compared to the least active women who performed less than 43.2 MET-hours of physical activity (reference group) because the value of the odds ratios is below 1.0 and the 95% confidence interval does not cross 1.0.

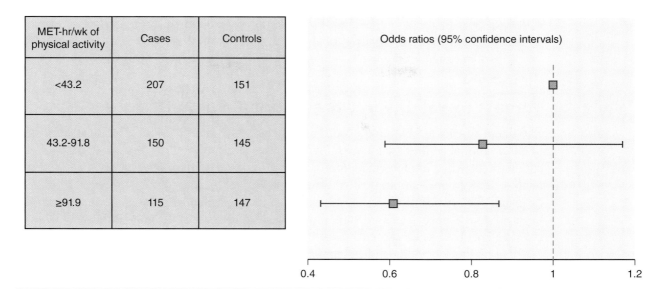

MET-hr/wk of physical activity	Cases	Controls
<43.2	207	151
43.2-91.8	150	145
≥91.9	115	147

Figure 17.10   Odds ratios for endometrial cancer by tertiles of lifetime physical activity.

Based on data from John, Koo, and Horn-Ross (2010).

Advantages of Case-Control Studies

The case-control study design can provide valid estimates of exposure–disease relationships in a shorter time and with less monetary expense in comparison to the cohort design.

They are particularly efficient for investigating rare diseases because they avoid the need for extremely large populations or long follow-up periods to accumulate a useful number of cases for analysis.

They also enable hypothesis testing for multiple exposures for a single disease outcome.

They can be used to acquire exposure information in greater detail because more resources can be used to gather this information as a result of the smaller scale of most case-control studies.

Disadvantages of Case-Control Studies

The major disadvantages of case-control studies are that exposure information is obtained after the disease has been diagnosed and that recruiting an appropriate control group can be challenging. The exposure information and the resulting odds ratios may be adversely affected by problems in each of these areas. First, people who have just been diagnosed with a major disease that is often life threatening may recall their exposures differently than control participants do simply because of their recent diagnosis. Bias of this type has been called **recall bias**. For example, in a case-control study examining the physical activity–heart disease relationship, participants with disease (cases) may report that they were less active than they truly were because it is well known that exercise is protective against heart disease. This recall bias would increase the apparent difference in physical activity levels between the controls who reported their exposure without a recall bias. In this example, recall bias would bias odds ratios away from the null value of 1.0 and suggest that a stronger relationship existed than was actually present.

Second, for the contrast of exposure histories between cases and controls to provide valid odds ratios, control participants need to be representative of the population from which the cases were obtained. This requirement means that the control group does not have to be generalizable to the general population, just to the population that produced the case series.

> **recall bias**—Systematic errors introduced by differences in recall accuracy between comparison groups (i.e., between cases and controls).

If the exposure histories of the control participants in the study differ importantly from the unknowable exposure distribution of the cohort from which the cases arose, then **selection biases** may cause odds ratios that are not valid estimates of the true exposure–disease relationship.

Threats to Validity in Observational Study Designs

The major threats to obtaining a valid understanding of the exposure–disease relationship in observational studies are methodological biases and confounding factors. Epidemiological methods, when implemented appropriately, limit the effects of biases and confounding factors. **Bias** in epidemiological research typically refers to systematic deviations of risk estimates (i.e., odds ratios or relative risk) from their true value. Recall and selection biases have previously been discussed in the case-control studies section. Other important biases to consider are observer biases and biases that result from exposure misclassification. Observer bias can be introduced in studies that use interviewers to collect exposure information but do not blind the interviewers to the participants' disease. A result of this lack of blinding can be a systematic bias introduced if the interviewer collects exposure data differently from cases and controls.

Biases that result from the misclassification of physical activity exposure levels often arise from measurement errors introduced when the exposures are measured. Because physical activities are often measured rather crudely, or are assessed at only one point in a person's life, the measure of exposure may not adequately reflect the person's true physical activity exposure during the induction period of the disease of interest. Accordingly, the data collected may not clearly describe the exact relationship between the exposure and the study outcome. If this misclassification of exposure is randomly distributed in the study population, these errors will result in attenuation in the measures of effect. That is, an actual effect on the disease outcome by the exposure could be less pronounced than it actually is, or the effect may be missed completely. This type of bias is usually referred to as *bias toward the null* (i.e., no effect).

A **confounding factor** in epidemiological studies refers to an additional factor that obscures the observable relationship between an exposure and the outcome. Confounding factors must be associated with both the exposure and the outcome and are identifiable by their effect on the measure of the association. That is, changes in odds ratios or relative risk estimates between unadjusted and adjusted analyses suggest the presence of confounding factors. Confounding factors can be controlled using statistical methods such as testing for interactions between variables or stratification. For example, if smoking was a confounding factor in the relationship between physical activity and blood cholesterol levels, its confounding effect could be controlled by examining the relationship in smokers and nonsmokers separately.

There are other various threats to validity in observational studies including additional sources of bias and types of confounding relationships that are well beyond the scope of this chapter. Interested readers are encouraged to consult standard epidemiology texts.

Experimental Designs With Randomized Trials

Regular physical activity increases cardiorespiratory fitness, promotes weight loss, and if done at regular intervals, reduces the risks for heart attack, colon cancer, and premature mortality (U.S. Department of Health and Human Services, 2018). Experimental designs allow researchers to identify the effects of a specific intervention on a health outcome in a group of people (experimental group) while simultaneously monitoring changes in the same health outcome among people not receiving the intervention (control or comparison group). Randomized trials that are focused on changing health at the individual level are referred to as **clinical trials**. Because denying treatment that is known to cause health benefits is illegal, randomized trials for physical activity often provide a form of physical activity to the comparison group that is not expected to have the same effects as the intervention treatment provided to the intervention group, or that compares different types of treatment.

An example of a randomized trial involving a physical activity intervention is Project Active, which compares the effects of a personal, lifestyle physical activity intervention with a traditional, structured exercise intervention. The 235 participants were initially sedentary and apparently healthy adults who were randomized into one of two types of intervention groups: lifestyle or structured. The hypothesis was that there would be no difference in the physical activity and cardiorespiratory fitness between the two groups at the end of the treatment. Both groups were instructed in programmatically different, but physiologically equivalent, methods to increase their levels of daily physical activity (Kohl, Dunn, Marcus, & Blair, 1998). Results showed increases in physical activity and cardiorespiratory fitness at the end of the six-month intensive intervention, but these changes were similar between the two intervention styles (Dunn, Garcia, Marcus, Kampert, Kohl, & Blair 1998).

Randomized trials that are focused on changing behaviors in communities are referred to as **community trials**. The rationale for community-level interventions are as follows: (a) targeting everyone may prevent more cases of disease than targeting just high-risk people; (b) environmental modifications may be easier to accomplish than large-scale, voluntary behavior change; (c) risk-related behaviors are socially influenced; (d) community interventions reach people in their native habitat; and (e) community interventions can be logistically simpler and less costly on a per-person basis.

community trials—Randomized trials focused on changing behaviors in communities.

In community trials, whole communities are randomized to receive multiple treatments in the form of mass media campaigns, school-based programs, point-of-decision prompts (e.g., posting signs to encourage individuals to use the stairs instead of the elevator), and related activities. The comparison (i.e., control) communities do not receive the intervention but often share several demographic characteristics with the treatment community and serve to show the effect of naturally occurring changes in community behaviors over time. Data collection and analysis include monitoring the extent to which an intervention is implemented as intended and the effects of the intervention on community behaviors (process evaluation), individual-level changes in behavior and health outcomes (individual evaluation), and changes in the environment influenced by the intervention (community-level indicator evaluation). Data analysis must account for community- and individual-level variations in behaviors. An advantage of community trials is that they allow researchers to see the population changes in behaviors as a result of interventions. But this benefit is tempered by the difficulty of changing individual behaviors and the amount of time that may be required (sometimes many years) to see changes in community behaviors.

A famous example of a community trial is the Stanford Five-City Project (Winkleby, Taylor, Jatulis, & Fortmann, 1996) wherein intervention activities designed to reduce cardiovascular disease risk factors were delivered to two communities from 1979 to 1985. The physical activity interventions were targeted toward increasing moderate- and vigorous-intensity physical activity in intervention communities. The interventions used electronic and print media, individual and community activities, and school-based functions. Results showed modest but statistically significant changes in physical activity in the intervention communities. Men were more likely to participate in vigorous-intensity activities, and women reported spending more time in moderate-intensity activities. Similar changes were not observed in the three control communities.

Reading and Interpreting a Physical Activity Epidemiological Study

Because epidemiological studies may involve a variety of study designs and large numbers of people, it is a good idea to follow a guideline when reading and interpreting a research paper. The example used in this section is an article titled "Associations of Light, Moderate,

and Vigorous Intensity Physical Activity With Longevity: The Harvard Alumni Health Study"
(Lee & Paffenbarger, 2000), which is a prospective cohort study.

Introduction

The introduction should provide a global overview of the rationale for the study, highlighting
major research that has shown associations (or lack of associations) between the exposure and
outcomes and explaining why this health concern is worthy of study. The rationale should
also address the biological mechanism for the association between the physical activity and
a health outcome. Research questions should reflect findings and questions arising from
earlier studies and should state the study's purpose.

> *Example introduction.* **The authors recognized the growing acceptance that a healthy
> lifestyle includes regular physical activity. However, they noted the inconsistent findings
> about the optimal intensity of physical activity needed to reduce premature mortality.
> The purpose of the paper was to present associations between bouts of light-, moderate-,
> and vigorous-intensity activity and mortality among men enrolled in the Harvard Alumni
> Health Study.**

Materials and Methods

This section should include information about the study design, study population, variables,
and analytical methods.

The *study design* should be appropriate to answer the research questions. To describe
events in a population (descriptive studies), either an ecological or a cross-sectional study
design is appropriate. To identify causal associations that reflect changes in health status
across time, however, prospective cohort studies or experimental study designs are needed.
Case-control studies are useful to examine exposure and outcome relationships in small
study samples and provide only limited evidence of the existence of a temporal relationship
between the exposure and outcome.

The *study population* should be appropriate to answer a study's research questions. Writers should describe their *sample size* to include how many participants were enrolled in the
study, completed the tasks as assigned, and were included in the data analysis. The writers
also should provide an explanation of differences in the sample sizes from the start to the
end of the study (e.g., how many participants completed follow-up measures). This is an
important issue for avoiding biased studies. The methods used to recruit the participants
and inclusion or exclusion criteria should be listed.

Characteristics of the study population should include participants' age, sex, race, income,
education, or other factors related to the research questions. Is the study population representative of a group used for generalization?

The *study variables* should be described for the exposure (physical activity), outcome
(e.g., mortality), and potential confounders (e.g., smoking). Writers should identify how the
exposure was measured and scored and whether data are available that address the validity and reliability of the measures. The *outcome* should be identified by a case definition
(e.g., International Classification of Diseases-version 10 codes) and a description of how
data were obtained, whether from hospital records, death certificates, physical exams, or
self-reports. Disease outcomes should have a case definition that the authors use to guide
the researchers in the classification of the outcome status. *Potential confounders* should be
identified with the data collection methods for each variable described. Writers should also
describe the quality control measures used to ensure accuracy in the data collection process.

Methods used for *data analyses* should reflect the research questions and hypotheses
tested in the study.

Example study population. Participants were described as men who matriculated as undergraduates at Harvard University between 1916 and 1950. Inclusion criteria included having returned a survey sent in 1977 (n = 17,835). Exclusion criteria were the presence of physician-diagnosed cardiovascular disease, cancer, chronic obstructive pulmonary disease (n = 3,706), or failure to identify information about potential covariate variables on the survey (n = 644). The sample used in the analysis was 13,485 men with a mean age of 57.5 ± 8.9 years.

Example exposure. Physical activity was measured with a mailed survey that had participants recall the number of stairs climbed, city blocks walked, and sports and recreational activities performed in the past week. A physical activity index (PAI) was expressed as a composite of these activities in kilojoules per week. The survey has a one-month reliability of r = .72 and a correlation with physical activity records of r = .65. MET levels associated with intensity levels were as follows: light: <4 METs; moderate: 4 to 6 METs; vigorous: >6 METs.

Example outcome. Identification of participants' deaths was obtained from the National Death Index. The causes of death were determined from death certificates.

Example potential confounders. Confounder variables were obtained from self-reported information recorded on the mailed health survey. Variables included age, body mass index, cigarette smoking, alcohol consumption, and early (<65 years) parental death.

Example data analysis. The association between total energy expenditure in kilojoules per week (kJ/wk) and mortality was examined using Cox proportional hazards models.

Results

The text should present a global overview of the results and refer the reader to the tables and figures that show the data. Tables and figures should be self-explanatory and identify the confounding variables controlled in the analyses in a table footnote and figure legend. Statistical comparisons should present p values or confidence intervals around parameter estimates (e.g., odds ratio, relative risk) and dose–responses by presenting comparisons of the data for multiple levels.

Example results. A total of 2,539 men died between 1977 and 1992. Compared with men who expended <4,200 kilojoules per week, those who expended more energy through regular physical activity reduced their risk for mortality by 20 to 27%. Trends for associations between type and intensity of activity and mortality were significant at p < .05 for distance walked, stairs climbed, and moderate- and vigorous-intensity physical activity (see figure 17.11). Light-intensity physical activity was not associated with decreases in mortality rate.

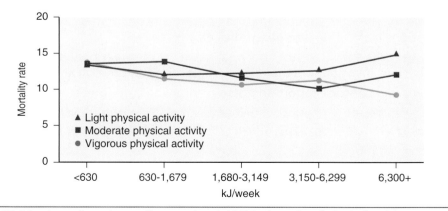

Figure 17.11 Age-adjusted mortality rates (per 1,000) by intensity of physical activity.

Reprinted by permission from I.M. Lee and R.S. Paffenbarger, Jr., "Associations of Light, Moderate, and Vigorous Intensity Physical Activity with Longevity. The Harvard Alumni Health Study," *American Journal of Epidemiology* 151 (2000): 293-299.

Discussion and Conclusions

In the discussion, writers should highlight the main study findings, compare and contrast the results with those of similar studies, and relate the findings to data that highlight the magnitude of the problem. The strengths (e.g., large sample size, prospective study design) and limitations (e.g., self-reported physical activity, limited generalizability to well-educated men) of the study should be identified and clearly described. Sources of bias that may have influenced the study results should be addressed as potential weaknesses of the study. Potential biological mechanisms between the exposure and outcome variables should be explained (e.g., how physical activity reduces the risk of premature mortality). The criteria for causality (e.g., temporality, strength of association, consistency, biological gradient– or dose–response, and biological plausibility) should be discussed as a prelude to determining whether the associations and hypotheses tested support a causal association between the physical activity exposure and a health outcome.

> ***Example discussion and conclusion.*** **The authors concluded that greater energy expenditure is associated with increased longevity. Participation in moderate-intensity activities showed a trend toward lower mortality rates, and participation in vigorous-intensity activities showed a strong association with lower mortality rates. Participation in light-intensity activities was unrelated to mortality rates. Biological mechanisms for the association are supported by other studies showing lower blood pressure, glucose, and insulin levels; higher HDL cholesterol; and enhanced cardiac function and hemostatic factors with regular moderate- and vigorous-intensity physical activity. Strengths of the study include a detailed physical activity assessment and a long follow-up period. Limitations include the possibility of recall bias in using a self-administered physical activity survey and the inability to account for diseases not measured on the survey. The public health impact of the study showed benefits of moderate- and vigorous-intensity physical activity in reducing risks for mortality among men.**

Summary

This chapter provided an overview of the research methods used in the study of physical activity epidemiology. Epidemiological studies identify the distribution and determinants of physical activity, investigate its associations with health or disease in population groups, and apply the research findings to reduce the burden of disease in communities.

☑ Check Your Understanding

1. Describe the differences between observational and experimental research.
2. Describe the six criteria that help strengthen the inference that exposure-outcome associations from observational studies may be causal.
3. Briefly describe the strengths and weaknesses of observational and experimental study designs in examining the relationships between physical activity and health.
4. Compare and contrast cohort and case-control study designs.
5. Write a review of a physical activity research study published in the *American Journal of Epidemiology*, the *Journal of Clinical Epidemiology*, the *International Journal of Epidemiology*, or *Epidemiology*.
6. Compare and contrast the ways epidemiologists study the distribution and determinants of physical activity, its associations with health-related outcomes, and its application to address health issues.

18

Experimental and Quasi-Experimental Research

> **Life is trying things to see if they work.**
>
> —Ray Bradbury

A simple idea underpins science: Trust but verify. To verify, good science must be open to challenge from experimental studies. In modern science, we do not verify often enough. For example, in 2012, scientists from a biotech firm could reproduce only 6 of 53 important studies from cancer research; another drug company group could verify only 25% of 67 important papers. Many of these problems are a result of journals publishing mostly papers that find important differences. In fact, negative results (no differences) account for 14% of published papers (*Briefing*, 2013). Although we like to think of scientific research as self-correcting, without the replication of experiments, it may not be. Thus, in this chapter we discuss experimental research and its design, but we certainly advocate for efforts to verify previously found important results.

Experimental research attempts to establish cause-and-effect relationships. That is, an independent variable is manipulated to judge its effect on a dependent variable. But the process of establishing cause and effect is a difficult one. Three criteria must be present to establish cause and effect:

1. *The cause must precede the effect.* For example, the starting gun in a race precedes the runners' beginning the race; the runners' beginning does not cause the starting gun to go off.

2. *The cause and effect must be correlated with each other.* As we have already discussed, just because two variables are correlated does not mean one causes the other; cause and effect, however, cannot exist unless two variables are correlated.

3. *The correlation between cause and effect cannot be explained by another variable.* Recall from chapter 1 that the relationship between the academic performance of elementary school children and shoe size was explained by a third variable, age.

We may think of cause and effect in terms of necessary and sufficient conditions (Krathwohl, 1993). For example, if the condition is necessary and sufficient to produce the effect, then it is the cause. But alternative situations exist as well (see the Examples of Cause and Effect in Golf sidebar):

- *Necessary but not sufficient.* Some related condition likely produces the effect.
- *Sufficient but not necessary.* Some alternative condition is likely the cause.
- *Neither necessary nor sufficient.* Some contributing condition is likely the cause.

Also remember that statistical techniques can only reject the null hypothesis (establish that groups are significantly different) and identify the percentage of variance in the dependent variable accounted for by the independent variable or the effect size. This makes statistics not sufficient for establishing cause and effect. Only by applying logical thinking to well-designed experiments can cause and effect be established. This process demonstrates that no other reasonable explanation exists for the changes in the dependent variable except the manipulation of the independent variable. The application of this logic is made possible by the following:

- Selection of a good theoretical framework
- Use of appropriate participants
- Application of an appropriate experimental design
- Proper selection and control of the independent variable (treatment)
- Appropriate selection and measurement of the dependent variable
- Use of the correct statistical model and analysis
- Correct interpretation of the results

> **Establishing cause and effect requires logical thinking applied to well-designed experiments.**

In this chapter, we discuss experimental designs by explaining how you can recognize and control sources of invalidity and threats to both internal and external validity. We also explain several types of experimental designs. Before going any further into this chapter, review the following terms (introduced in chapters 1 and 3), which we use throughout the discussion:

- *Independent variable*
- *Dependent variable*
- *Categorical variable*
- *Control variable*
- *Extraneous variable*

Sources of Invalidity

All the design types we discuss have strengths and weaknesses that pose threats to the validity of the research. The importance of validity was stated well by Campbell and Stanley (1963):

> **Fundamental . . . is a distinction between internal validity and external validity. Internal validity is the basic minimum without which any experiment is uninterpretable: Did in fact the experimental treatments make a difference in this specific experimental instance? External validity asks the question of generalizability: To what populations, settings, or treatment variables can this effect be generalized? (p. 5)**

Both internal validity and external validity are important in experiments, but they are frequently at odds in research planning and design. Gaining internal validity involves controlling all variables so that the researcher can eliminate all rival hypotheses as explanations for the outcomes observed. Yet in controlling and constraining the research setting to gain internal validity, the researcher places the generalization (external validity) of the findings in jeopardy. In studies with strong internal validity, the answer to the question of to whom, what, or where the findings can be generalized may be very uncertain. This is because in

Examples of Cause and Effect in Golf

1. No matter how bad your last shot was, the worst is yet to come. This condition does not expire until after the 18th hole.
2. Your best round of golf will be followed almost immediately by your worst round ever. (The probability increases based on the number of people you tell about your best round.)
3. Brand-new golf balls are water magnetic. (The magnetic effect increases with the price of the ball.)
4. The higher your handicap, the more qualified you deem yourself to be an instructor.
5. Every par-three hole has a secret desire to humiliate golfers. (The shorter the hole, the greater its desire.)
6. Electric golf carts always fail at the point farthest from the clubhouse.
7. The person you most hate to lose to will always be the one who beats you.
8. The last three holes of a round will automatically adjust your score to what it really should be.
9. Golf balls never bounce off trees back into play. (If one does, the groundskeeper will remove that tree before your next round.)
10. All vows taken on a golf course are valid only until sundown.

ecologically valid settings (perceived by the participants as intended by the researchers), not everything is controlled and operated in the same way as in the controlled laboratory context. Thus, the researcher must decide whether it is more important to be certain that the manipulation of the independent variable caused the observed changes in the dependent variable or to be able to generalize the results to other populations, settings, and so on. We cannot provide an easy answer to that question, which is often debated at scientific meetings and in the published literature.

To expect any experiment to meet all research design considerations is unreasonable. A more realistic approach is to identify the specific goals and limitations of the research effort. Is internal validity or external validity the more important issue? After that is decided, the researcher can plan the research with one type of validity as the major focus while maintaining as much of the other type of validity as possible. Another recourse is to plan a series of experiments in which the first experiment has strong internal validity even at the expense of external validity. If the first experiment confirms that changes in the dependent measure are the result of manipulating the independent variable, subsequent experiments can be designed to increase external validity even at the expense of internal validity. This approach allows evaluation of the treatment in settings more like the real world. (Review the related discussion of basic and applied research in chapter 1 and Christina, 1989.)

Threats to Internal Validity

Campbell and Stanley (1963) identified eight threats to the internal validity of experiments, and Rosenthal (1966) identified a ninth:

1. History—events occurring during the experiment that are not part of the treatment
2. Maturation—processes within the participants that operate as a result of time passing (e.g., aging, fatigue, hunger)
3. Testing—the effects of one test on subsequent administrations of the same test

4. Instrumentation—changes in instrument calibration, including lack of agreement within and between observers

5. Statistical regression—the fact that groups selected based on extreme scores are not as extreme on subsequent testing

6. Selection bias—choosing comparison groups in a nonrandom manner

7. Experimental mortality—loss of participants from comparison groups for nonrandom reasons

8. Selection–maturation interaction—the passage of time affecting one group but not the other in nonequivalent group designs

9. Expectancy—experimenters' or testers' anticipation that certain participants will perform better

If these threats are uncontrolled, the change in the dependent variable may be difficult to attribute to the manipulation of the independent variable.

History

A history threat to internal validity means that some unintended event occurred during the treatment period. For example, in a study evaluating the effects of a two-year community-based walking program on blood pressure, the fact that a pandemic occurred during year two of the intervention would constitute a history threat to internal validity. The pandemic is likely to produce benefits for some participants because they did not work and had more time to walk, while other participants were reluctant to participate because of the pandemic. Therefore, it would be difficult to separate the impact of the program from the pandemic and to extrapolate benefits to a time without a pandemic.

Maturation

Maturation as a threat to internal validity is most often associated with aging. This threat occurs frequently in designs in which one group is tested on several occasions over a long period. Elementary school physical education teachers frequently encounter this source of invalidity when they give a physical fitness test in the early fall and again in the late spring. The children nearly always do better in the spring. The teacher would like to claim that the physical education program was the cause. Unfortunately, maturation is a plausible rival hypothesis for the observed increase; that is, the children have grown larger and stronger and thus probably run faster, jump higher, and throw farther.

Testing

A testing threat is the effect that taking a test once has on taking it again. For example, if athletes are administered a multiple-choice test to evaluate their knowledge about steroids today and again two days later, they will do better the second time even though no treatment intervened. Taking the test once helps in taking it again. The same effect is present in physical performance tests, particularly if the participants are not allowed to practice the test a few times. If a group of seniors take a balance test one week and again the following week, they will generally do better the second time. They learned something from performing the test the first time.

Instrumentation

Instrumentation is a problem frequently faced in exercise science research. Suppose the researcher uses a spring-loaded device to measure strength. Unless the spring is calibrated regularly, it will decrease in tension with use. Thus, the same amount of applied force produces increased readings of strength compared with earlier readings. Instrumentation problems also occur in research using observers. Unless observers are trained and regularly checked, the same observer's ratings may systematically vary across time or participants (a phenomenon called *observer drift*), or different observers may not rate the same performance in the same way.

Statistical Regression

Statistical regression can occur when groups are not randomly formed but are selected based on an extreme score on some measure. For example, if someone rates the behavior of a group of children on a playground on an activity scale (very active to very inactive) and two groups are formed—one of very active children and one of very inactive children—statistical regression is likely to occur when the children are next observed on the playground. The children who were very active will be less active (although still active), and the very inactive children will be more active. In other words, both groups will regress (move from the extremes) toward the overall average. This phenomenon reflects only the fact that participants' scores tend to vary about their average performance (estimated true score). Extreme scores may simply reflect the observation of a performance on the high (or low) side of the participant's typical performance. The next performance is usually not as extreme. Thus, when average scores of extreme groups are compared from one time to the next, the high group appears to get worse, whereas the low group appears to get better. Statistical regression is a particular problem in studies that attempt to compare extreme groups selected on some characteristic, such as highly anxious, fit, or skilled participants versus not very anxious, fit, or skilled participants.

Selection Bias

Selection biases occur when groups are formed on some basis other than random assignment. Thus, when treatments are administered, because the groups were different to begin with, always present is the rival hypothesis that any differences found are due to initial selection biases rather than the treatments. Showing that the groups were not different on the dependent variable at the beginning of the study does not overcome this shortcoming. Any number of other unmeasured variables on which the groups differ might explain the treatment effect. Gall, Borg, and Gall (1996) asked important questions that apply to selection (or sampling) bias:

> *Did the study use volunteers?* **Use of volunteers is common in research in the study of physical activity. Yet volunteers are often not representative of anyone but other volunteers. They may differ considerably in motivation for the experimental task and setting from those who don't volunteer.**

> *Are participants extremely nonrepresentative of the population?* **Often we are unable to select participants at random for our studies, but it is useful if we at least believe (and can demonstrate) that they represent some larger group from our culture.**

Recall the discussion of sampling in chapter 6, particularly the concept of a good enough sample.

Experimental Mortality

Experimental mortality refers to the loss of participants. Even when groups are randomly formed, this threat to internal validity can occur. Participants may remain in an experimental group taking part in a fitness program because it is fun, whereas participants in the control group become bored, lose interest, and drop out of the study. Of course, the opposite can

You must keep participants' attention when doing interviews for research.

occur, too. Participants may drop out of an experimental group because the treatment is too difficult or time-consuming.

Selection–Maturation Interaction

A selection–maturation interaction occurs only in designs in which one group is selected because of some specific characteristic, whereas the other group lacks this characteristic. An example is a study of the differences between employees in two technology companies; the employees from the established company are the experimental group and receive a fitness program while the employees in the start-up company are the control group. One company is established and the other is a new start-up. Most of the employees at the start-up have just finished college while the other company has a range of employee experience and therefore the start-up has younger employees. Therefore, it would be difficult to determine whether the fitness program or the fitness program combined with the participants' advanced age produced the observed changes.

Expectancy

expectancy—A threat to internal validity in which the researcher anticipates that certain behaviors or results will occur.

One additional threat to internal validity not mentioned by Campbell and Stanley (1963) has been identified. **Expectancy** (Rosenthal, 1966) refers to experimenters' or testers' anticipation that certain participants will perform better. This effect, although usually unconscious on the part of the experimenters, occurs where participants or experimental conditions are clearly labeled. For example, testers may rate skilled participants better than unskilled participants, regardless of treatment. This effect is also evident in observational studies in which the observers rate posttest performance better than pretest performance because they expect change. If the experimental and control groups are identified, observers may rate the experimental group better than the control group even before any treatment occurs. The expectancy effect may influence the participants, too. For example, in a sport study, coaches may actually cause poorer performance by substitutes (compared with starters) because the substitutes realize that the coach treats them differently (e.g., the coach may show less concern about incorrect practice trials).

Any of these nine threats to internal validity may reduce the researcher's ability to claim that the manipulation of the independent variable produced the changes in the dependent variable. We discuss experimental designs and how they control (or fail to control) the threats to internal validity later in this chapter.

> There are nine threats to internal validity that can undermine the claim of a causal relationship between two variables.

Threats to External Validity

Campbell and Stanley (1963), in their classic volume, identified four threats to external validity, or the ability to generalize results to other participants, settings, measures, and so on:

1. Reactive or interactive effects of testing—The pretest may make the participant more aware of or sensitive to the upcoming treatment. As a result, the treatment is less effective without the pretest.

2. Interaction of selection bias and experimental treatment—When a group is selected on some characteristic, the treatment may work only on groups possessing that characteristic.

3. Reactive effects of experimental treatment—Treatments that are effective in constrained situations (e.g., laboratories) may not be effective in less constrained settings (more like the real world).

4. Multiple-treatment interference—When participants receive more than one treatment, the effects of previous treatments may influence subsequent ones.

Reactive or Interactive Effects of Testing

Reactive or interactive effects of testing may be a problem in any design with a pretest. Suppose a fitness program is the experimental treatment. If a physical fitness test is administered to the sample first, the participants in the experimental group might realize that their levels of fitness are low and be particularly motivated to follow the prescribed program closely. In an untested population, however, the program might not be as effective because the participants might be unaware of their low levels of physical fitness.

Interaction of Selection Bias and Experimental Treatment

The interaction of selection bias and the experimental treatment may prohibit the generalization of the results to participants lacking the particular characteristics (bias) of the sample. For example, a drug education program might be effective in changing the attitudes of first-year undergraduates toward drug use. This same program would probably lack effectiveness for third-year medical students because they would be much more familiar with drugs and their appropriate uses.

Reactive Effects of Experimental Arrangements

Reactive effects of experimental arrangements means that the experimental treatment may not be generalizable to real-world situations. These reactive effects are a persistent problem for laboratory-based research (e.g., in exercise physiology, biomechanics, motor behavior). In such research, is the researcher investigating an effect, process, or outcome that is specific to the laboratory and cannot be generalized to other settings? We have referred to this earlier as a problem of ecological validity. For example, in a study employing high-speed cinematography, the skill to be filmed must be performed in a certain place, and joints must be marked for later analysis. Is the skill performed in the same way during sport participation?

A specific type of reactive behavior, the **Hawthorne effect** (Brown, 1954), refers to the fact that participants' performances change when attention is paid to them. This effect may be a threat to both internal and external validity because it is likely to produce better treatment effects and reduce the ability to generalize the results.

Multiple-Treatment Interference

Multiple-treatment interference is most frequently a problem when the same participants are exposed to more than one level of the treatment. Suppose participants are going to learn to move to the hitting position in volleyball using a lead step or a crossover step. We want to know which step gets the players in a good hitting position more quickly. If the players are taught both types of steps, learning one might interfere with (or enhance) learning the other. Thus, the researcher's ability to generalize the findings may be confounded by the use of multiple treatments. A better design might be to have two separate groups, each of which learns one of the techniques.

The ability to generalize findings from research to other participants or situations is a question of random sampling (or at least good enough sampling) more than any other. Do the participants, treatments, tests, and situations represent any larger populations? Although a few of the experimental designs discussed later control certain threats to external validity, usually researchers control these threats through the way they select the samples, treatments, situations, and tests.

Controlling Threats to Internal Validity

Threats to internal and external validity are controlled in different ways by applying specific techniques. In this section, we describe useful approaches to solving threats to internal validity in the design of experiments. Many threats to internal validity are controlled by making the participants in the experimental and control groups as alike as possible. This objective is most often accomplished by randomly assigning participants to groups.

Randomization

As mentioned in chapter 6, randomization allows the assumption that the groups do not differ at the beginning of the experiment. The randomization process controls for history up to the point of the experiment; that is, the researcher can assume that past events are equally distributed among groups. This approach does not control for history effects during the experiment if experimental and control participants are treated at different times or places. The researcher must try to prevent an event (besides the treatment) from occurring in one group but not in the other groups.

Randomization also controls for maturation because the passage of time is equivalent in all groups. Statistical regression, selection biases, and selection–maturation interaction are controlled because they occur only when groups are not randomly formed.

Ways other than random assignment of participants to groups are sometimes used to control threats to internal validity. The matched-pair technique matches pairs of participants who are equal on some characteristic and then randomly assigns each to a different group. The researcher might want very tight control on previous experience in strength training. Thus, participants would be matched on this characteristic and then randomly assigned to the experimental and control groups.

> **Hawthorne effect**—A phenomenon in which participants' performances change when attention is paid to them, which is likely to reduce the ability to generalize the results.

A matched-group technique may also be used. This method involves nonrandom assignment of participants to experimental and control groups so that the group means are equivalent on some variable. This procedure is generally regarded as unacceptable because the groups may not be equivalent on other unmeasured variables that could affect the outcome of the research.

In within-subjects designs, the participants are used as their own controls. This means that each participant receives both the experimental and the control treatment. In this type of design, the order of treatments should be counterbalanced; that is, half the participants should receive the experimental treatment first and then the control, and the other half should receive the control treatment first and then the experimental treatment. If there are three levels of the independent variable (1 = control, 2 = experimental treatment A, 3 = experimental treatment B), participants should be randomly assigned to one of the six possible treatment orders: 1-2-3, 1-3-2, 2-1-3, 2-3-1, 3-1-2, 3-2-1. If the number of treatment levels administered to the same participants is larger than three, experimenters can assign a random order of the treatments to each participant rather than counterbalance treatments.

Placebos, Blind Setups, and Double-Blind Setups

Other ways of controlling threats to internal validity include placebos, blind setups, and double-blind setups. A **placebo** is used to evaluate whether the observed effect is produced by the treatment or is a psychological effect. Frequently, a control condition is used in which participants receive the same attention from and interaction with the experimenter, but the treatment administered does not relate to performance on the dependent variable.

A **blind setup** is a study in which participants do not know whether they are receiving the experimental treatment or the control treatment. In a **double-blind setup**, neither the participant nor the tester knows which treatment the participants are receiving. A triple-blind test has also been reported (Gastel & Day, 2016): The participants do not know what they are getting, the experimenters do not know what they are giving, and the investigators do not know what they are doing. We can only hope the triple-blind technique finds limited use in our field.

All these techniques (except the triple-blind technique) are useful in controlling the Hawthorne, expectancy, and halo effects, as well as what we call the **Avis effect** (a recent version of the John Henry effect), or the fact that participants in the control group may try harder simply because they are in the control group.

A good example of the use of these techniques for controlling psychological effects is the use of steroids to build strength in athletes. A number of studies were done to evaluate the effects of steroids. To combat the fact that athletes may become stronger because they think they should when using steroids, a placebo (a pill that looks just like the steroid) was used. In a blind study, the athletes did not know whether they received the placebo or the steroid. In a double-blind study, the athlete, the person dispensing the steroids or placebos, and the testers did not know which group received the steroids. Unfortunately, these procedures did not work well because taking large quantities of steroids made the athletes' urine smell bad. Thus, the athletes knew whether they were receiving the steroid or the placebo.

placebo—A method of controlling a threat to internal validity in which a control group receives a false treatment while the experimental group receives the real treatment.

blind setup—A method of controlling a threat to internal validity in which participants do not know whether they are receiving the experimental or control treatment.

double-blind setup—A method of controlling a threat to internal validity in which neither the participant nor the experimenter knows which treatment the participant is receiving.

Avis effect—A threat to internal validity wherein participants in the control group try harder just because they are in the control group.

Uncontrolled Threats to Internal Validity

Three threats to internal validity remain uncontrolled by the randomization process: reactive or interactive effects, instrumentation, and experimental mortality.

Reactive or Interactive Effects

Reactive or interactive effects of testing can be controlled only by eliminating the pretest. These effects can be evaluated, however, by two of the designs discussed later in this chapter: pretest–posttest randomized groups and the Solomon four-group design.

Instrumentation

Instrumentation problems cannot be controlled or evaluated by any design. Only the experimenter can control this threat to internal validity. In chapter 11, we went into some detail on techniques for controlling the instrumentation threat to develop valid and reliable tests. Of particular significance is test reliability. Whether the measurement is obtained from a laboratory device (e.g., an oxygen analyzer), a motor performance test (e.g., a standing long jump), an attitude-rating scale (e.g., on feelings about drug use), an observer (e.g., coding the percentage of time a child is active), a knowledge test (e.g., on basketball strategy), or a survey (e.g., on available sport facilities), the answer must be consistent. Controlling this threat frequently involves the assessment of test reliability across situations, between and within testers or observers, and within participants. The validity of the instrument must also be established to control for instrumentation problems: Does it measure what it was intended to measure? The total process of establishing appropriate instrumentation for research is called *psychometrics*.

One final point about instrumentation involves the halo effect. As we discussed in chapter 11, this effect occurs in ratings of the same person on several skills. Raters seeing a skilled performance on one task are likely to rate the participant higher on subsequent tasks, regardless of the level of skill displayed. In effect, the skilled behavior has "rubbed off," or created a halo, on later performances. An order effect may also occur in observation. Gymnastics and swimming judges often rate earlier performers lower to save room on the rating scale for better performers. Because gymnastics and swimming coaches know this, they always place better performers later in the event order.

Experimental Mortality

Experimental mortality is not controlled by any type of experimental design. The experimenter can control this threat only by ensuring that participants are not lost (at all, if possible) from groups. Many problems of participant retention can be handled in advance of the research by carefully explaining the research to the participants and emphasizing the need for them to follow through with the project. (During the experiment itself, begging, pleading, and crying sometimes work.)

Controlling Threats to External Validity

External validity is generally controlled by selecting the participants, treatments, experimental situation, and tests to represent some larger population. Of course, random selection (or good enough sampling) is the key to controlling most threats to external validity. Remember that elements other than the participants may be randomly selected. For example, the levels of treatment can be randomly selected from the possible levels, experimental situations can be selected from possible situations, and the dependent variable (test) can be randomly selected from a pool of potential dependent variables.

As previously noted, ecological validity is the ability to generalize the situation. Although a particular treatment's results can be generalized to a larger group if the sample is representative, this generalization may apply only to the specific situation in the experiment. If the experiment is conducted under controlled laboratory conditions, then the findings may apply only under controlled laboratory conditions. Frequently, the experimenter hopes that the findings can be generalized to real-world exercise, sport, industrial, or instructional settings. Whether the outcomes can be generalized in this way depends largely on how the participants perceive the study, and this influences the way participants respond to study characteristics. The questions of interest are: Does the study have enough characteristics of real-world settings that participants respond as if they were in the real world? and Is ecological validity present? These questions are not easy to answer; many scholars advocate that more research in physical activity settings be conducted in field settings.

The reactive or interactive effects of testing can be evaluated by the Solomon four-group design. Interaction of selection biases and the experimental treatment is controlled by the random selection of participants. The reactive effects of experimental arrangements can be controlled only by the researcher (this is again the issue of ecological validity). Multiple-treatment interference can be partially controlled by counterbalancing or randomly ordering the treatments among participants. But the researcher can control whether the treatments still interfere only through knowledge about the treatment rather than the type of experimental design.

Types of Designs

This section, much of which is taken from Campbell and Stanley (1963) and the updated volumes by Cook and Campbell (1979) and Shadish, Cook, and Campbell (2002), is divided into three categories: preexperimental designs, true experimental designs, and quasi-experimental designs. We use the following notation:

- Each line indicates a group of participants.
- R signifies the random assignment of participants to groups.
- O signifies an observation or a test.
- T signifies that a treatment is applied. A blank space in a line where a T appears on another line means that the group is a control.
- A dotted line between groups means that the groups are used intact rather than being randomly formed.
- Subscripts indicate either the order of observations and treatments (appearance on the same line) or observations of different groups or different treatments (appearance on different lines). For example, if terms T_1 and T_2 appear on different lines, they refer to different treatments; if terms T_1 and T_2 appear on the same line, they indicate that the treatment is administered more than once to the same group.

Preexperimental Design

This section discusses **preexperimental design**, a research design that comprises three types of studies that control few of the sources of invalidity. None of the study types has random assignment of participants to groups.

One-Shot Study

In a one-shot study design, a group of participants receives a treatment followed by a test to evaluate the treatment:

$$T \quad O$$

This design fails all the tests of good research. All that can be said is that at a certain point in time, this group of participants performed at a certain level. In no way can the level of performance (O) be attributed to the treatment (T).

One-Group Pretest–Posttest Design

The one-group pretest–posttest design, although very weak, is better than the one-shot design. At least we can observe whether any change in performance has occurred:

$$O_1 \quad T \quad O_2$$

preexperimental design—A type of research design that controls few of the sources of invalidity and does not include random assignments of participants to groups; comprises three types of study: one-shot study, one-group pretest–posttest design, and static group comparison.

If O_2 is better than O_1, we can say that the participants improved. For example, let's imagine that Bill Biceps, a qualified exercise instructor, conducted an exercise test at a health club. Participants then trained three days per week for 40 minutes each day at 70% of their estimated $\dot{V}O_2$max for 12 weeks. After the training period, participants retook the exercise test and significantly improved their scores. Can Mr. Biceps conclude that the exercise program caused the changes that he observed in the exercise test performance? Unfortunately, this design does not allow us to say why the participants improved. Certainly, the improvement could be due to the treatment, but it could also be due to history. Some event other than the treatment (T) may have occurred between the pretest (O_1) and the posttest (O_2); the participants may have exercised at home on the other days. Maturation is a rival hypothesis. The participants may have gotten better (or worse) just as a result of the passage of time. Testing is a rival hypothesis; the increase at O_2 may be the result only of experience with the test at O_1. If the group being tested had been selected for some specific reason, then any of the threats involving selection bias could occur. This design is most frequently analyzed by the dependent t test to evaluate whether a significant change occurred between O_1 and O_2.

Static Group Comparison

The static group-comparison design compares two groups, one of which receives the treatment and one of which does not:

$$\begin{array}{cc} T & O_1 \\ \hline & O_2 \end{array}$$

In this case, we do not know whether the groups were not equivalent when the study began, as the dotted line between the groups indicates. This means that the groups were selected intact rather than having been randomly formed. We are thus unable to determine whether any differences between O_1 and O_2 are because of T or only because the groups differed initially. This design is subject to invalidity because of selection biases and the selection–maturation interaction. A t test for independent groups is used to evaluate whether O_1 and O_2 differ significantly. But even if they do differ, the difference cannot be attributed to T.

As you can see, the three studies that make up preexperimental design research are not valid methods of answering research questions (see table 18.1). They do not represent experiments because the change in the dependent variable cannot be attributed to the manipulation of the independent variable. You will not encounter these preexperimental designs in research journals, and we hope that you will not find (or produce) theses and dissertations using these designs. The preexperimental designs represent much wasted effort because little or nothing can be concluded from the findings. If you submit studies using these designs to research journals, you are likely to receive rejection letters similar to one Snoopy (from the *Peanuts* comic strip) received: "Dear Researcher, Thank you for submitting your paper to our research journal. To save time, we are enclosing two rejection letters—one for this paper and one for the next one you send."

True Experimental Designs

true experimental design—Any design used in experimental research in which groups are randomly formed and that controls most sources of invalidity.

The designs discussed in this section are called **true experimental designs** because the groups are randomly formed, allowing the assumption that they were equivalent at the beginning of the research. This controls for past (but not present) history, maturation (which should occur equally in the groups), testing, and all sources of invalidity that are based on nonequivalence of groups (statistical regression, selection biases, and selection–maturation interaction). But only the experimenter can make sure that nothing happens to one group (besides the treatment) and not the other (present history), that scores on the dependent measure do not

TABLE 18.1

Preexperimental Designs and Their Control of the Threats to Validity

Validity threat	One-shot study	One-group pretest and posttest	Static group
Internal			
History	–	–	+
Maturation	–	–	?
Testing		–	+
Instrumentation		–	
Statistical regression	?	+	
Selection	–	+	–
Experimental mortality	–	–	
Selection × maturation		–	–
Expectancy	?	?	?
External			
Testing × treatment		–	
Selection biases × treatment	–	–	–
Experimental arrangements		?	
Multiple treatments			

Note: + = strength; – = weakness; blank = not relevant; ? = questionable.

Copyright, 1970, by the American Educational Research Association. Adapted by permission of the publisher.

vary as a result of instrumentation problems, and that the loss of participants is not different between the groups (experimental mortality).

Randomized-Groups Design

Note that the randomized-groups design resembles static group comparison except that groups are randomly formed:

$$R \quad T \quad O_1$$
$$R \qquad\ \ O_2$$

If the researcher controls the threats to internal validity that are not controlled by randomization (no easy task), has a sound theoretical basis for the study, and meets the necessary and sufficient rule, then this design allows the conclusion that significant differences between O_1 and O_2 are due to T. An independent t test is used to analyze the difference between O_1 and O_2. This design as earlier depicted represents two levels of one independent variable. It can be extended to any number of levels of an independent variable:

$$R \quad T_1 \quad O_1$$
$$R \quad T_2 \quad O_2$$
$$R \qquad\ \ O_3$$

Here, three levels of the independent variable exist—one is the control, and T_1 and T_2 represent two levels of treatment. This design can be analyzed by simple ANOVA, which contrasts the dependent variable as measured in the three groups (O_1, O_2, O_3). For example, T_1 is training at 70% of $\dot{V}O_2$max, T_2 is training at 40% of $\dot{V}O_2$max, and the control group is not training. The variables O_1, O_2, and O_3 are the measures of cardiorespiratory fitness (12-minute run) in each group taken at the end of the training.

This design can also be extended into a factorial design; that is, more than one independent variable (IV) could be considered. Example 18.1 shows how this works.

Independent variable 1 (IV_1) has three levels (A_1, A_2, A_3), and independent variable 2 (IV_2) has two levels (B_1, B_2). This results in six cells (A_1B_1, A_1B_2, A_2B_1, A_2B_2, A_3B_1, A_3B_2) to which participants are randomly assigned. At the end of the treatments, each cell is tested on the dependent variable (O_1 through O_6). This design is analyzed by a 3×2 factorial ANOVA that tests the effects of IV_1 (F_A), IV_2 (F_B), and their interaction (F_{AB}).

EXAMPLE 18.1

		IV_2	
		B_1	B_2
	A_1	A_1B_1	A_1B_2
IV_1	A_2	A_2B_1	A_2B_2
	A_3	A_3B_1	A_3B_2
	R	A_1B_1	O_1
	R	A_1B_2	O_2
	R	A_2B_1	O_3
	R	A_2B_2	O_4
	R	A_3B_1	O_5
	R	A_3B_2	O_6

Note: Analyzed in a 3×2 factorial ANOVA. F_A = main effect of A; F_B = main effect of B; F_{AB} = interaction of A and B.

This design can also be extended to address even more independent variables and retain all the controls for internal validity previously discussed. Sometimes this design is used with a categorical independent variable. Look again at example 18.1 and suppose that IV_2 (B_1, B_2) represents two age levels. Clearly, the levels of B could not be randomly formed. The design appears as follows:

	R	A_1	O_1
B_1	R	A_2	O_2
	R	A_3	O_3
	R	A_1	O_4
B_2	R	A_2	O_5
	R	A_3	O_6

The levels of A are randomly formed within B, but the levels of B cannot be randomly formed. This no longer qualifies completely as a true experimental design, but it is frequently used in the study of physical activity. This design is analyzed in a 3×2 ANOVA, but the interpretation of results must be done more conservatively.

Any version of randomized-group design may also have more than one dependent variable. Although the essential design remains the same, the statistical analysis becomes multivariate. Where two or more levels of one independent variable exist and several dependent variables are present, discriminant analysis is the appropriate multivariate statistic. In the factorial versions of this design (two or more independent variables), if multiple dependent variables are used, then MANOVA is typically the most appropriate analysis, although practical concerns (e.g., participant numbers) or theoretical matters may dictate alternative statistical analyses.

Pretest–Posttest Randomized-Groups Design

In the pretest–posttest randomized-groups design, the groups are randomly formed, but both groups are given a pretest as well as a posttest:

$$R \quad O_1 \quad T \quad O_2$$
$$R \quad O_3 \qquad O_4$$

The major purpose of this type of design is to determine the amount of change produced by the treatment; that is, does the experimental group change more than the control group? This design threatens the internal validity through testing, but the threat is controlled because the comparison of O_3 to O_4 in the control group as well as the comparison of O_1 to O_2 in the experimental group include the testing effect. Thus, although the testing effect cannot be evaluated in this design, it is controlled.

This design is used frequently in the study of physical activity, but its analysis is complex. There are at least three common ways to do a statistical analysis of this design. First, a factorial repeated-measures ANOVA can be used. One factor (between subjects) is the treatment versus no treatment, whereas the second factor is pretest versus posttest (within subjects or repeated measures). The interest in this design, however, is usually the interaction: Do the groups change at different rates from pretest to posttest? If you choose a repeated-measures ANOVA for this design (in our opinion, the best choice), pay close attention to our chapter 9 discussion of univariate and multivariate issues in repeated measures. Another choice is simple ANCOVA, using the pretest for each group (O_1 and O_3) to adjust the posttest (O_2 and O_4). Recall that some problems are associated with using the pretest as a covariate (see the chapter 9 discussion of ANCOVA). Finally, the experimenter could subtract each participant's pretest value from the posttest value (called a **difference score or gain score**) and perform a simple ANOVA (or, with only two groups, an independent *t* test), using each participant's difference score as the dependent variable. Each of these techniques has strengths and weaknesses, but you will find all three used in the literature.

> **difference score or gain score**—A score that represents the difference (change) from pretest to posttest.

In this design, the important question is: Does one group change more than the other group? Although this issue is frequently called the *analysis of difference scores*, a more appropriate label is the *assessment of change*. Clearly, in a learning study, the change is a gain. But in an exercise physiology study, the change might be decreased performance caused by fatigue. Regardless, the issues are the same. How can this change be assessed appropriately?

The easiest answer is to obtain a difference score by subtracting the pretest from the posttest. Although this method is intuitively attractive, it has some severe problems. First, these difference scores tend to be unreliable. Second, the level of initial values applies: Participants who begin low in performance can improve more easily than those who begin with high scores. Thus, the initial score is negatively correlated with the difference score. How would you like improvement in your tennis performance evaluated if your initial score was high (e.g., 5 successful forehand drives out of 10 trials) in comparison with that of a friend who began with a low score (e.g., 1 successful hit out of 10)? If you improved to 7 out of 10 on the final test and your friend improved to 5 out of 10 (the level of your initial score), your friend has improved twice as much as you have—a gain of four successful hits versus two. Yet your performance is still considerably better, and it was more difficult for you to improve. For these reasons, the use of difference scores is rarely a good method for measuring change.

The issues involved in the proper measurement of change are complex, and we cannot treat these issues here. Much has been written on this topic. We suggest that you read Schmidt, Lee, Winstein, Wulf, and Zelaznik (2019, chapter 9) about this problem in motor learning and performance. Or, for a classic short book on the topic, see Harris (1963).

This design can also be extended into more complex forms. First, more than two (pretest and posttest) repeated measures can be used. This circumstance is common in the areas of exercise physiology, motor behavior, and exercise psychology. Two randomly formed groups of participants in a motor behavior experiment might be measured 30 or more times as they learn a task. The two groups might differ in the information they are given. The design might use as the statistical analysis a 2 (groups) $\times$ 30 (trials) ANOVA with repeated measures on the second factor (trials). Remember from chapter 9 that meeting the assumptions for a repeated-measures ANOVA with many repeated measures is difficult. Thus, in designs like these, trials may be blocked (i.e., several trials averaged, reducing the number of repeated measures) into 3 blocks of 10 trials or 5 blocks of 6 trials.

Sometimes the design is extended in other ways. For example, we could take the design in example 18.1 (a 3 $\times$ 2 factorial) and add a third factor of a pretest and a posttest. This design would result in a three-way factorial with repeated measures on the third factor. All the versions of this design are subject to the first threat to external validity: reactive or interactive effects of testing. The pretest may make the participant more sensitive to the treatment and thus reduce the ability to generalize the findings to a population that was not pretested.

Solomon Four-Group Design

The Solomon four-group design is the only true experimental design to specifically evaluate one of the threats to external validity: reactive or interactive effects of testing. The design is depicted as follows:

$$
\begin{array}{cccc}
R & O_1 & T & O_2 \\
R & O_3 &   & O_4 \\
R &   & T & O_5 \\
R &   &   & O_6 \\
\end{array}
$$

This approach combines the randomized-groups and the pretest–posttest randomized-groups designs. The purpose is explicitly to determine whether the pretest produces an increased sensitivity of the participants to the treatment. This design allows a replication of the treatment effect (is $O_2 > O_4$? and is $O_5 > O_6$?), an assessment of the amount of change due to the treatment (is $O_2 - O_1 > O_4 - O_3$?), an evaluation of the testing effect (is $O_4 > O_6$?), and an assessment of whether the pretest interacts with the treatment (is $O_2 > O_5$?). Thus, this experimental design is extremely powerful. Unfortunately, it is also inefficient because twice as many participants are required. Consequently, this design is in limited use, especially among graduate students doing theses and dissertations. In addition, no good way exists to analyze this design statistically. The best alternative (one that does not use all the data) is a 2 $\times$ 2 ANOVA setup as follows:

	No T	T
Pretested	O_4	O_2
Unpretested	O_6	O_5

Thus, IV_1 has two levels (pretested and not pretested), and IV_2 has two levels (treatment and no treatment). In the ANOVA, the F ratio for IV_1 establishes the effects of pretesting, the F for IV_2 establishes the effects of the treatment, and the F for interaction evaluates the external validity threat of interaction of the pretest with the treatment. Table 18.2 summarizes the control of threats to validity for the true experimental designs.

Quasi-Experimental Designs

Not all research in which an independent variable is manipulated fits clearly into one of the true experimental designs. As researchers attempt to increase external and ecological valid-

TABLE 18.2

True Experimental Designs and Their Control of the Threats to Validity

Validity threat	Randomized groups	Pretest–posttest randomized groups	Solomon four-group
Internal			
History	+	+	+
Maturation	+	+	+
Testing	+	+	+
Instrumentation			
Statistical regression	+	+	+
Selection	+	+	+
Experimental mortality	+	+	+
Selection × maturation	+	+	+
Expectancy	?	?	?
External			
Testing × treatment	+	−	+
Selection biases × treatment	?	?	?
Experimental arrangements	?	?	?
Multiple treatments			

Note: + = strength; − = weakness; blank = not relevant; ? = questionable.

Data from Campbell and Stanley (1963).

ity, the careful and complete control of the true designs becomes increasingly difficult, if not impossible. The purpose of **quasi-experimental designs** is to fit the design to settings more like the real world while still controlling as many of the threats to internal validity as possible. The use of these types of designs in kinesiology, physical education, exercise science, sport science, and other areas (e.g., education, psychology, and sociology) has increased considerably in recent years. The most authoritative text on quasi-experimental designs is that by Shadish et al. (2002).

In quasi-experimental research, the use of randomization to control threats to internal validity is difficult. It makes sense that random assignment cannot be used in many settings. For example, a pedagogy researcher who wants to investigate the effect of a curriculum intervention could not randomly assign children to classes because schools make these decisions based on other criteria that have educational value; no school district would agree to permit a study if it had to change the students' classes. Randomly assigning classes within a school to different treatments would also be difficult because teachers are likely to talk to each other (and even trade ideas from a curriculum that they thought was effective), which would reduce the strength of the treatment intervention. Similarly, if a researcher were studying the effect of an exercise program on the aged in a community setting, random assignment would not work because people select where to enroll in classes based on various factors in their lives (e.g., convenience, membership, transportation needs), not on whether their choice helps a researcher. Asking people to go to another site or to attend class at another time would probably reduce the number of people who agree to participate and increase participant attrition—or, to say it another way, increase experimental mortality.

quasi-experimental design—A research design in which the experimenter tries to fit the design to real-world settings while still controlling as many of the threats to internal validity as possible.

Reversal Design

The reversal design is used increasingly in school and other naturalistic settings and is depicted as follows:

$$O_1 \quad O_2 \quad T_1 \quad O_3 \quad O_4 \quad T_2 \quad O_5 \quad O_6$$

The purpose here (as with the time series) is to determine a baseline measurement (O_1 and O_2), evaluate the treatment (change between O_2 and O_3), evaluate a no-treatment time period (O_3 to O_4), reevaluate the treatment (O_4 to O_5), and evaluate a return to a no-treatment condition (O_5 to O_6). This design is sometimes called *A-B-A-B-A*, or sometimes just *A-B*, where A is the baseline condition and B is the treatment condition.

Lines like A, B, and C in figure 18.1 suggest that the treatment is effective, whereas lines like D, E, and F do not support a treatment effect. Statistical analyses for reversal designs also need to be regression tests of the slopes and intercepts of the lines between various observations.

Nonequivalent-Control-Group Design

A design using a nonequivalent control group is frequently used in real-world settings where groups cannot be randomly formed. The design is as follows:

$$O_1 \quad T \quad O_2$$
$$O_3 \qquad O_4$$

You can recognize this as a pretest–posttest design without randomization. Frequently, researchers compare O_1 and O_3 and declare the groups equivalent if this comparison is not significant. Unfortunately, just because the groups do not differ on the pretest does not mean that they are not different on any number of unmeasured characteristics that could affect the outcome of the research. If the groups differ when O_1 and O_3 are compared, ANCOVA is usually employed to adjust O_2 and O_4 for initial differences. Alternatively, a within (the pretest–posttest comparison, a repeated measure) and between (treatment and control group comparison) two-way ANOVA could be used to analyze whether groups changed from pretest to posttest and whether the change was different for those in the treatment and control groups.

Ex Post Facto Design

In its simplest case, ex post facto design is a static group comparison but with the treatment not under the control of the experimenter. For example, we frequently compare the

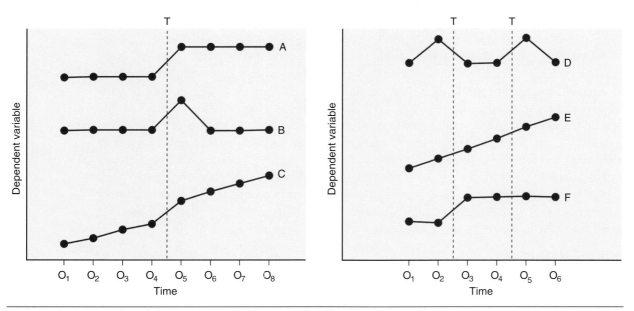

Figure 18.1 Examples of changes across time in reversal designs.
Data from Campbell and Stanley (1963).

characteristics of athletes versus nonathletes, highly fit versus unfit individuals, female versus male performers, and expert versus novice performers. In effect, we are searching for variables that discriminate between these groups. Our interest usually resides in the question: Did these variables influence the way these groups became different? Of course, this design cannot answer this question, but it may provide interesting insights and characteristics for manipulation in other experimental designs. This design is also often called a *causal comparative design.*

The previously mentioned quasi-experimental designs have been frequently reported in research on the study of physical activity. Several additional designs have considerable potential but have seen less use in our research. We hope that the following presentations about two promising designs will increase interest in and use of these designs.

Switched-Replication Design

The switched-replication design (Shadish et al., 2002) can be either a true experiment or quasi experiment depending on whether levels are random or intact groups.

Levels (random or intact groups)	Trials				
	1	2	3	4	5
1	O_1 T	O_2	O_3	O_4	O_5
2	O_6	O_7 T	O_8	O_9	O_{10}
3	O_{11}	O_{12}	O_{13} T	O_{14}	O_{15}
4	O_{16}	O_{17}	O_{18}	O_{19} T	O_{20}

If participants are randomly assigned to levels 1 through 4, the design is a true experiment. If levels 1 through 4 are different intact groups (e.g., tennis players in college, high school, and two age levels of youth leagues), then the design is quasi-experimental. Any number of levels beyond two can be used, but the number of trials must be one greater than the number of levels.

This design has two strong features: The treatment is replicated several times, and long-term treatment effects can be evaluated. There is no standard statistical analysis for this design, but various ANOVAs with repeated measures could be used. Or the design might be analyzed by fitting regression lines to each level and testing how the slopes and intercepts change.

Many opportunities exist to use this design in our field, yet it is seldom used. Since the design is seldom used, fewer scholars are familiar with the design, thus it is used less frequently. Another issue is that the design requires a treatment be repeated, often extending the length of time an experiment must be conducted. Of course, that can lead to higher rates of attrition. This design might be particularly useful in research on sport teams in which either different teams or different players within a team could be assigned to the various levels.

Table 18.3 summarizes the threats to validity of quasi-experimental designs discussed so far.

Time-Series Design

The time-series design has only one group but attempts to show that the change that occurs when the treatment is administered differs from the times when it is not. This design may be depicted as follows:

$$O_1 \quad O_2 \quad O_3 \quad O_4 \quad T \quad O_5 \quad O_6 \quad O_7 \quad O_8$$

The basis for claiming that the treatment causes the effect is two-fold: A constant rate of change can be established from O_1 to O_4 and from O_5 to O_8, but this rate of change varies

TABLE 18.3

Quasi-Experimental Designs and Their Control of the Threats to Validity

Validity threat	Time series	Nonequivalent control	Reversal	Ex post facto	Switched replication
Internal					
History	–	–	–	?	?
Maturation	+	+	+	?	+
Testing	+	+	+		+
Instrumentation					
Statistical regression	+	–	+		+
Selection	+	+	+	–	?
Experimental mortality	+	+	+		?
Selection × maturation	+	–	+	–	
Expectancy	?	?	?	?	?
External					
Testing × treatment	–	–	?	–	?
Selection biases × treatment	?	?	?	?	?
Experimental arrangements	?	?	?	?	+
Multiple treatments			?		+

Note: + = strength; – = weakness; blank = not relevant; ? = questionable.

Data from Campbell and Stanley (1963).

between O_4 and O_5, where T has been administered. For example, in figure 18.2, lines A, B, and C suggest that the treatment (T) results in a visible change between observations, whereas lines D, E, F, and G indicate that the treatment has no reliable effect.

The typical statistical analyses previously discussed do not fit time-series designs very well. For example, a repeated-measures ANOVA with appropriate follow-ups applied to line C in figure 18.2 might indicate that all observations (O_1 to O_8) differ significantly, even though we can visibly see a change in the rate of increase between O_4 and O_5. We do not

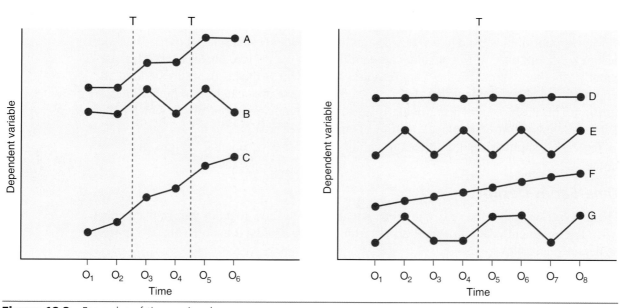

Figure 18.2   Examples of time-series changes.

Data from Campbell and Stanley (1963).

present the details, but regression techniques can be used to test both the slopes and the intercepts in time-series designs.

This type of design appears to control for a number of the threats to internal validity. For example, maturation is constant between observations. Testing effects can also be evaluated, although they could be difficult to separate from maturation. Selection biases also appear to be controlled because the same participants are used at each observation. Of course, history, instrumentation, and mortality are controlled only to the extent that the researcher controls them. See a humorous example of a time-series design in the 7 ± 2: Miller Must Have Been an Assistant Professor sidebar.

A reminder about using intact groups in quasiexperiments: As we noted in the chapter 6 section on sampling, the use of intact groups and, particularly, the timing of when a treatment is delivered to a group, influences the analysis and the number of groups needed for a study. This issue has been discussed at length elsewhere (Silverman, 2004; Silverman & Solmon, 1998), and we will not duplicate that discussion here. We are compelled to state, however, that when groups receive treatments as a group, the appropriate unit is usually the *group*. This circumstance, then, requires a number of groups to have sufficient power to analyze the data. Those planning quasi experiments need to consider this early so that they are not dealing with the issue after all the data have been collected.

Single-Subject Design

A single-subject design is exactly what it sounds like—a researcher is looking for the effect of an intervention on a single subject. These designs also are sometimes called $N = 1$ designs because they often have only one subject. There are many designs within this family, and we could include them as a type of quasi experiment because a researcher using this design is looking for the effect of a treatment without using randomization. We have not done that here because the focus on individual effects instead of group effects makes these designs different. In addition, those who conduct single-subject research look at the changes on graphs and do not analyze the results with statistics.

In our field, single-subject designs are most often used in clinical settings. Examples include observing physical education instruction, pursuing sport psychology work with athletes, studying an outstanding performer (e.g., an Olympic athlete), or studying motor functions of a person with Parkinson's disease. A participant in this type of study is typically measured repeatedly on the task of interest. Many trials are needed for evaluating the influence of the treatment. During some periods, a baseline measurement is obtained, and during other periods, a treatment is administered. The focus is often on participant variability as well as average values. Quasi-experimental time-series, reversal, and switched-replication designs can work as single-subject designs or group designs. When used with single subjects, these designs, as previously mentioned, are A-B or A-B-A-B designs, where A refers to the baseline condition (no treatment) and B refers to when the treatment is administered. Sometimes more than one treatment is administered to the same participant. As in research using a group of participants, counterbalancing the treatment order to separate treatment effects is important. Possible research questions include the following:

- Does the treatment produce the same effect each time?
- Are the effects of the treatment cumulative, or does the participant return to baseline following each treatment period?
- Does the participant's response to the treatment become less variable over multiple treatment periods?
- Is the participant's magnitude of response less sensitive to multiple applications of the treatment?
- Do varying intensities, frequencies, and lengths of treatment produce varying responses?

In figure 18.3, we present a graph with a traditional A-B-A-B design. Note that the A periods are baseline measurements and the B periods are when the intervention occurred. The second A is the reversal, when the treatment was withdrawn.

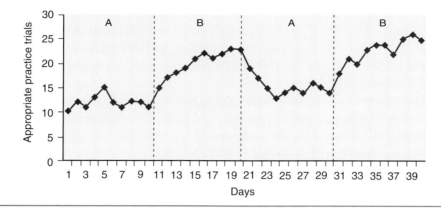

Figure 18.3   Single-subject A-B-A-B design.

---

7 ± 2: Miller Must Have Been an Assistant Professor

Summary—Miller's magical number for memory span, 7 ± 2, is humorously brought to task because of its inability to predict the everyday performance of teenagers, graduate students, and full professors. Explanations are provided for the deficits in memory performance of these subgroups during life span development.

Miller's (1956) classic paper identifying the memory span as 7 ± 2 items must have omitted using full professors as subjects, at least based on an *n* of 1, namely me. I have frequently heard clinical psychologists accused of going into psychology to study their own problems. Maybe that is a more valid explanation than bringing Miller's work to task. For years, my graduate students have said that I study memory for movement because (a) I have little or no memory, and (b) I lose my spatial orientation just walking around the block. However, I prefer to think that Miller's results only apply to a subsample of the population—6- to 10-year-old children, college students, and new assistant professors. Six- to 10-year-old children never forget anything you promised to do (or even things you said you might do . . .). However, as soon as they approach the teen years, their memory spans drop to less than one unit of information.

Teenagers cannot remember to make their beds as they are getting out of them. Evidently, nothing ever happens to teenagers at school, although I prefer to think that they just cannot recall anything that happened. When these teenagers with little memory capacity go to college, an amazing transition in memory occurs. Between the first and fourth years, the young adults' memory capacity increases to at least 7 ± 2 units of information. Loosely defined, this means *they know everything*; conversely, parents know nothing.

After working a few years and coming back to graduate school, memory facility is somewhat reduced. Graduate students have a memory capacity of 3 ± 1 units of information: (a) They know that they are graduate students; (b) they remember to pick up their graduate assistantship paycheck; and (c) they remember to attend their graduate seminars. The ± 1 refers to the fact that they occasionally remember to do the reading for the seminar (+1), but they sometimes forget to come to class (–1).

As soon as graduate students receive their PhDs, Miller's magical number (7 ± 2) is again a good predictor of memory—new assistant professors know everything. However, movement through the academic ranks gradually reduces capacity to remember until the average capacity of full professors (and parents) is reached, 2 ± 1 units of information. Full professors can remember (a)

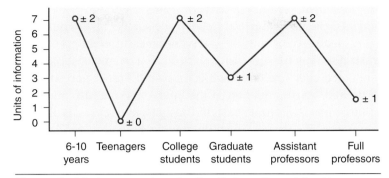

Figure 1   Maximum number of units of information retained in memory across the life span.

that they are full professors and (b) their paychecks come regularly. The ± 1 refers to the fact that full professors sometimes remember that they have graduate students (+1) but occasionally forget to pick up the regular paychecks (–1).

Based on the failure of this model to meet the assumptions of stage theory (i.e., one should never regress to an earlier stage), we must assume that this uneven transition in memory states (see figure 1) is environmentally induced.

But how can the environment cause these wide variations in memory performance? First, I believe we have to assume that Miller is correct about the structural maximum of memory performance because the deficits are not confined to a single point in the life span (i.e., teenagers, graduate students, full professors). The question of interest then becomes *What causes such a serious depression in memory performance for teenagers, graduate students, and full professors?* Given my extensive study of memory as well as personal experience with all the stages, I can deduce an answer.

For teenagers, the drop to about zero memory with no variance is an interaction of carbonated soft drinks, junk food, and other teens. This interaction is not a simple one and has an indirect effect; that is, this interaction causes pimples and the need for braces, both of which are very distracting in and of themselves. Taken in combination with parents saying, "Wash your face!" and "Brush your teeth!", memory capacity in all teenagers is occupied.

The deficit that occurs for graduate students is easy to explain. Three factors are involved. First, the graduate student is expected to work full time (either as a research or teaching assistant) for one-third or less of normal pay—a very distracting and disconcerting circumstance. Second, the graduate student is expected to study and do research day and night. One's major professor assumes graduate students do not need sleep. Finally, the major professor is continually grumbling at the graduate student to "Read this paper!", "Collect these data!", "Write this paper!", and "For crying out loud, work on your dissertation!" In combination, these items reduce memory capacity.

But why does the full professor, who seems to have everything going well, show such poor memory performance? Full professors have a living wage (according to some), cars that run, houses with real furniture, tenure, graduate students to do the work, and time to play golf and tennis. What could possibly explain the deficit in memory performance? Answering that question is the easiest of all:

GRADUATE STUDENTS AND TEENAGERS!

Reference

Miller, G. (1956). The magical number seven, plus or minus two: Some limits on our capacity for processing information. *Psychological Review, 63,* 81-97.

Adapted by permission "7 ± 2: Miller Must Have Been an Assistant Professor," *NASPSPA Newsletter* 12, no. 1 (1987): 10-11.

The single-subject designs described in this section typically have one subject. Other single-subject designs may have more than one subject undergoing the same intervention. In these studies, the subjects start the intervention at different points in time by extending the baseline measurement for successive subjects. For example, subject 1 has baseline measurements conducted over five days, subject 2 over 10 days, and subject 3 over 15 days. This approach permits an examination of the intervention at different points in time. There are many permutations of multiple baseline designs, and reversals also could be added to these designs. In figure 18.4, we present one graphic example.

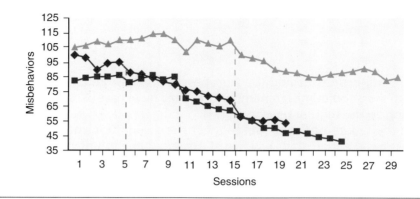

Figure 18.4 Extension of a single-subject design to three subjects over different time spans.

Other quasi-experimental designs exist, but the ones discussed here are the most commonly used. Of course, quasi-experimental designs never quite control internal validity as well as true experimental designs do, but they do allow us to conduct investigations when true experiments cannot be used or when a true experimental design significantly reduces external validity.

Summary

In experimental research, one or more independent variables (the treatment) are manipulated to assess the effects on one or more dependent variables (the response measured). Research studies are concerned with both internal and external validity. Internal validity requires controlling factors so that the results can be attributed to the treatment. Threats to internal validity include history, maturation, testing, instrumentation, statistical regression, selection biases, experimental mortality, selection–maturation interaction, and expectancy.

External validity is the ability to generalize the results to other participants and to other settings. Four threats exist to external validity: reactive or interactive effects of testing, the interaction of selection biases and the experimental treatment, reactive effects of experimental arrangements, and multiple-treatment interference. Having high degrees of both internal and external validity is nearly impossible. The rigid controls needed for internal validity make it difficult to generalize the results to the real world. Conversely, studies with high external validity are usually weak in internal validity. Random selection of participants and random assignment to treatments are the most powerful means of controlling most threats to internal and external validity.

Preexperimental designs are weak because they can control few sources of invalidity. True experimental designs are characterized by random formation of groups, which allows the assumption that the groups were equivalent at the beginning of the study. The randomized-groups, pretest–posttest randomized-groups, and Solomon four-group designs are examples of true experimental designs.

Quasi-experimental designs are often used when it is difficult or impossible to use true experimental designs or when a true experimental design significantly limits external validity. The time-series, reversal, nonequivalent-control-group, and ex post facto designs are the most commonly used; the switched-replication and single-subject designs, though potentially useful, are less frequently used quasi-experimental designs.

☑ Check Your Understanding

Locate two research papers in refereed journals in your area of interest. One paper should have a true experimental design and the other a quasi-experimental design.

1. Describe the type of design. Draw a picture of it using the notation from this chapter.
2. How many independent variables are there? How many levels of each? What are they?
3. How many dependent variables are there? What are they?
4. What type of statistical analysis was used? Explain how it fits the design.
5. Identify the threats to internal validity that are controlled and uncontrolled. Explain each.
6. Identify the threats to external validity that are controlled and uncontrolled. Explain each.

19

Qualitative Research

Experience is a wonderful thing. It enables you to recognize a mistake when you make it again.

—From Lachlan's Laws (Scottish Highland philosopher)

Although qualitative research in physical education, exercise science, and sport science was once considered relatively new, it is now common in many areas of kinesiology. Researchers in education began adapting ethnographic research design to educational settings in the United States in the 1970s (Goetz & LeCompte, 1984), and a great deal of qualitative research in physical education and sport science has been conducted steadily since the 1980s. In 1989, Locke was invited to write an essay and tutorial on qualitative research that was published in *Research Quarterly for Exercise and Sport*, which promoted further qualitative research in the field.

Qualitative research methods have been employed in anthropology, psychology, and sociology for many years. This general form of research has been called by various names, including *ethnographic*, *naturalistic*, *interpretive*, *grounded theory*, *phenomenological*, and *participant observational research*. Although the approaches are all slightly different, each "bears a strong family resemblance to the others" (Erickson, 1986, p. 119).

Those of you who are unfamiliar with the term *qualitative research* may benefit by reviewing your courses or readings in anthropology. Nearly everyone has heard of classic ethnographic studies (even if not by that name), such as the famous cultural research that Margaret Mead conducted when she lived in Samoa. Mead interviewed the Samoans at length about their society, their traditions, and their beliefs. This research is an example of qualitative research. But you do not need to live in a remote area to do it. The qualitative research that we refer to in this chapter is done mainly in everyday settings, such as schools, gymnasiums, sport facilities, fitness centers, and hospitals.

It could be argued that the term *qualitative research* may be too restrictive in that it seems to denote the absence of anything quantitative, but this is certainly not the case. Nonetheless, *qualitative research* seems to be the term used most often in our field. And, as we note in the next chapter, mixed-methods research—using both qualitative and quantitative methods—is being conducted in many fields. Our discussion of qualitative research focuses on the interpretive method as opposed to the so-called thick, rich description that characterized early research in anthropology, psychology, and sociology. This latter approach, involving a long and detailed account of the entity or incident, was espoused by Franz Boas, considered the father of cultural anthropology. In rejecting the armchair speculation that typified late-19th-century work, Boas insisted that researchers not only collect their own data but also report it with as little comment or interpretation as possible (cited in Kirk & Miller, 1986). Now, much of the research produced in kinesiology and related areas involves interpretation. Researchers collect data and make interpretations to address specific questions and to understand what is taking place in the setting(s) or with the person(s) being studied.

We do not intend to present a comprehensive review of qualitative research—we do not have the space to do so. Moreover, we would need to address the considerable disagreement among qualitative researchers concerning their methodologies and theoretical presuppositions. A comprehensive review would be a major undertaking, so throughout this chapter, we will suggest places to find further information. To start, you can gain a solid appreciation of the scope of qualitative research, the approaches to data collection and analysis, and the subtleties of approaches to qualitative research in compact form from Locke, Silverman, and Spirduso (2010) or in a more expansive form from Creswell and Poth (2018). Readable, single-source texts on conducting qualitative research include those of Bogdan and Biklen (2016), Creswell and Creswell (2018), Marshall, Rossman and Blanco (2021), Morse and Richards (2012), and Patton (2014). You will also learn from reading the qualitative study of sport psychology by Bhalla and Weiss (2010).

Procedures in Qualitative Research

Qualitative research is done in a variety of ways. Consequently, the procedures outlined in this section should be viewed simply as an attempt to orient readers who are unfamiliar with qualitative research. As with any type of research, the novice will benefit most from reading books that specifically focus on planning research (e.g., Locke, Spirduso, & Silverman, 2014) or qualitative methods, such as those listed in the previous paragraph, and recently completed qualitative research studies that used particular methods.

Defining the Problem

We do not spend much time on defining the problem in this chapter because we have talked about this step in earlier chapters, and it does not differ appreciably from that used with other research methods. We emphasize that researchers have several methods to choose from for any research problem. Different methods can yield different information about the problem. Thus, when researchers decide to use qualitative research as opposed to some other design, the decision is based on what they want to know about the problem.

Formulating Questions and Theoretical Frameworks

In highlighting some differences between quantitative and qualitative research, we stated that qualitative research usually builds hypotheses and theories in an inductive manner—that is, as a result of observations. Quantitative research often begins with research hypotheses that are subsequently tested. Thus, most qualitative studies do not have hypotheses as a part of the introduction. At one time it was common for those doing qualitative research not to state either hypotheses or questions, but in recent years, it has become more common for qualitative researchers to state questions that will be the study's focus. These questions are not as specific as hypotheses in quantitative research. They do, however, provide the direction that is the research focus. For example, the question, "What is the perspective of children in a competitive youth sport program?" focuses the study and provides a great deal of information about the research direction. Generally, the types of qualitative studies that we address in this chapter start with questions that guide the studies. Although researchers may be able to move in other directions and modify questions based on the ongoing research, they are unlikely to totally abandon the original questions.

Qualitative researchers, like other researchers, develop a case for why it is important to do a study. This justification appears in the introduction of a research paper or as part of the rationale or significance in a thesis' or dissertation's study sections. One approach is to provide a theoretical framework—a theory that guides the research. A theoretical framework frames the study and is used throughout to develop questions, design the method, and analyze data. Although many qualitative studies use a theoretical framework, another possible approach is to provide a rationale for the study by using other published research to show how this research fills a gap in our knowledge (Locke, Silverman, & Spirduso, 2010, 2014; Shulman, 2003). In either case, in most qualitative research, the questions are previously established in research and theory and are crafted with the objective of extending what we already know about a particular area.

Before moving on to other procedures for qualitative research, we should note that determining the theoretical framework that is appropriate for a study may require a great deal of reading and consultation with others. Because the theoretical framework influences all research aspects, you must thoroughly understand it before using it to frame a study, and so you can write about it in a way that others can easily understand. The process of understanding the theory and then applying it to a study often involves time, discussion, and testing the way it will influence future research. If you are venturing into a new theory or beginning to design a qualitative study that will employ theory, we highly recommend Maxwell (2013) and Rossman and Rallis (2017) as the first resources to consult. Then you may want to read more in-depth resources on theory construction such as Jaccard and Jacoby (2020). Having an advisor or specialist in qualitative research with whom to discuss these readings will help you develop the framework.

Collecting the Data

Several components are involved in collecting data for qualitative research, just as in quantitative research. Training and pilot work are still necessary, and so is selecting participants appropriately. In addition, you need to enter the field setting and be as unobtrusive as possible in your data collection.

Training and Pilot Work

In qualitative research, the investigator is the instrument for collecting and analyzing the data. It is imperative, then, that the researcher be adequately prepared. Certainly, coursework, fieldwork reports, and interaction with your advisor are helpful, but ultimately, the only way to become competent is through hands-on experience. You should gain this experience with an advisor who has a background in and experience with qualitative research. At many universities students begin this process in a class in which they learn about and do qualitative research under the direction of an experienced qualitative researcher.

As always, pilot work is essential, and fieldwork experience in a setting similar to that of the proposed study is recommended. Experience in the field setting (e.g., as an instructor, a coach, or a player) is helpful in most respects. Locke (1989), however, made a good point that familiarity tends to spawn "an almost irresistible flood of personal judgments" (p. 7) that, if not recognized and controlled, could become a significant threat to the integrity of the data in a qualitative study.

Selecting Participants

Selecting participants for a qualitative study is basically the same as that described for the case study in chapter 16. Qualitative research studies do not attempt to make inferences from their participants to some larger population. Rather, the participants are selected because they have certain characteristics. Obviously, pragmatic concerns arise about the location and the availability of participants because, in nearly every case, numerous other sites and people with similar characteristics exist.

Probability sampling is not used simply because there is no way to estimate the probability of each person's being selected and no assurance that each person has some chance of being included. Instead, participant selection in qualitative research is purposeful, essentially meaning that researchers select a sample from which they can learn the most. A researcher, for example, may be looking for people with certain levels of expertise or experience. There are several ways to guide participant selection (e.g., see Marshall, Rossman, & Blanco, 2021). In essence, it involves determining who will provide the data that will help answer the research questions while also deciding where the study will take place and how data will be collected. Further, the researcher must determine whether the research focus is on a particular setting, certain individuals, or both (Creswell & Poth, 2018).

> **A purposefully chosen sample is one from which the most can be learned.**

Entering the Setting

The researcher must have access to the field setting to conduct a qualitative study. Moreover, in most cases, the researcher must be able to observe and interview the participants at the appropriate time and location. These details may seem trivial, but nothing is more important than site entry. In any type of research, you must have access to the data, whether it is source material in historical research or participants in an experimental or survey study. However, the problem can be of greater magnitude in qualitative research than in most other methods. The quantitative researcher is simply "borrowing" the participants for a short time for some measurements or is taking a little class time to administer a questionnaire. In qualitative research, the investigator is often at the site for weeks or months. This outsider is listening, watching, coding, and video-recording, as well as imposing on the time of the teacher (or coach or other person) and the students (or players or other participants) for interviews and observations.

I believe all of these participants are small giraffes.

© corbisrffancy/fotolia

We are intentionally belaboring this topic because it is so important. Obviously, diplomacy, personality, and artful persuasion are required to gain site entry. Frankly, some people just cannot do this. Even when entry is achieved, some studies have failed because the investigator rubbed people the wrong way and the participants were not motivated to cooperate fully. The negotiation of gaining access to the participants in their natural setting is important and complex. It starts with the first contact by telephone or letter, extends through data collection, and continues after the researcher has left the site (Rossman & Rallis, 2017).

Before we discuss aspects of data collection, we should elaborate a little more on the topic of participant cooperation. Rapport is everything. Unless the participants believe that they can trust you, they will not give you the information you seek. Obviously, formal informed consent must be obtained, and the stipulations embodied in the whole concept of informed consent must be observed.

Qualitative research involves a number of ethical considerations simply because of the intensive personal contact with the participants. Thus, the participants need to know that provisions will be followed to safeguard their privacy and guarantee anonymity. If, for some reason, it is impossible to keep information confidential, this circumstance must be made clear. The researcher must give a great deal of thought to these matters before collecting data, must be able to explain the purpose and significance of the study effectively, and must convey the importance of cooperation in language participants can understand. One of the largest obstacles in the quest for natural behavior and candor on the part of the participants is their suspicion that the researcher will be evaluating them. The researcher must be convincing in this regard. The most successful studies are those in which the participants feel as though they are a part of the project. In other words, a collaborative relationship should be established.

Methods of Collecting Data

The most common sources of data collection in qualitative research are interviews, observations, and documents (Creswell & Creswell, 2018; Locke et al., 2010; Marshall, Rossman, & Blanco, 2021). The methodology is planned and pilot-tested before the study. Creswell and Creswell (2018) placed the data-collecting procedures into four categories—observations, interviews, documents, and audiovisual materials—and provided a concise table of the four methods, the options within each type, and their advantages and limitations.

We noted previously that the researcher typically has some type of framework that determines and guides data collection. For example, one phase of the research might pertain to the way expert and nonexpert sport performers perceive aspects of a game. This phase could involve having athletes describe their perceptions of what is taking place in a scenario. A second phase of the study might focus on the interactive thought processes and decisions of the two groups of athletes while they are playing. The data for this phase could be obtained from filming them in action and then interviewing them while they are watching their performances on video. Still another aspect of the study could be directed at the knowledge structure of the participants, which could be determined by a researcher-constructed instrument.

Qualitative data collection is a time-intensive process. Simply conducting quick interviews or making short observations will likely prevent you from gaining more knowledge. A well-thought-out plan of being in the environment long enough to understand the nuance of what is occurring is of paramount importance for collecting sound data in qualitative research.

Interviews

The interview is undoubtedly the most common source of data in qualitative studies. The person-to-person format is most prevalent, but occasionally group interviews and focus groups are conducted. Interviews range from the highly structured style, in which questions are determined before the interview, to the open-ended, conversational format. In qualitative research, the highly structured format is used primarily to gather sociodemographic information. For the most part, however, interviews are more open ended and less structured (Merriam, 2007; Rubin & Rubin, 2012). Frequently, the interviewer asks the same questions of all the participants, but the order of the questions, the exact wording, and the type of follow-up questions may vary considerably.

Often, qualitative researchers conduct partially structured interviews in which the main interview questions, developed from theory or the questions guiding the study, are written out along with likely follow-up questions. The major focus of each interview is the same with this format, but follow-up questions can explore other things that come up as the participant answers the main question. This approach allows for great flexibility while keeping the interview focused.

Being a good interviewer requires skill and experience. We emphasized earlier that the researcher must first establish a rapport with the respondents. If the participants do not trust the researcher, they will not open up and describe their true feelings, thoughts, and intentions. Complete rapport is established over time as people get to know and trust one another. An important skill in interviewing is being able to ask questions in such a way that makes respondents feel comfortable enough to believe that they can talk freely.

Kirk and Miller (1986), in a classic paper, described their field research in Peru, where they tried to learn how much urban, lower-middle-class people knew about coca, the organic source of cocaine. Coca is legal and widely available in Peru. In their initial attempts to get the people to tell them about coca, the researchers received the same culturally approved answers from all the respondents. The researchers decided to try a different approach—to change their question style to ease potential concerns that respondents might have about providing possibly self-incriminating answers (e.g., "How did you find out you didn't like coca?"). As a result, the Peruvians not only opened up and elaborated on their knowledge of coca but also, in some instances, admitted to their personal use of coca. Kirk and Miller made a good point about carefully crafting and asking the right questions as well as the value of using various approaches. Indeed, this is a basic argument for the validity of qualitative research.

Skillful interviewing takes practice! Ways to develop this skill include video-recording yourself conducting an interview, observing experienced interviewers, role-playing an interview, and critiquing peers. It is important that the interviewer appear nonjudgmental. This can be difficult in situations in which the interviewee's views are quite different from those of the interviewer. The interviewer must be alert to both verbal and nonverbal messages and be flexible in rephrasing and pursuing certain lines of questioning. The interviewer must use words that are clear and meaningful to the respondent and must be able to ask questions so that the participant understands what is being asked. Above all, the interviewer has to be a good listener. Developing these skills is not easy. When the participants are very young or come from a cultural background different from that of the interviewer, it is even harder. Targeting the questions to the right level and having the sensitivity to both ask questions and listen well so as to formulate follow-up questions require practice in the technique as well as with the sample being studied.

The use of a digital recorder is undoubtedly the most common method of recording interview data because it has the obvious advantage of preserving the entire verbal part of the interview for later analysis. Although some respondents may be nervous to talk while being recorded, this uneasiness usually disappears in a short time. The main drawback with recording is the possibility of equipment malfunction. This problem is vexing and frustrating when it happens during the interview, but it is devastating when it occurs afterward, when you are trying to replay and analyze the interview. Certainly, you should have fresh batteries or have recently charged the recorder, and you should make sure it is working properly early in the interview. You should also stop and play back some of the interview to see whether the person is speaking into the microphone loudly and clearly enough and whether you are getting the data. Some participants (especially children) love to hear themselves speak, so playing back the recording for them can also serve as motivation. Remember, however, that machines can malfunction at any time.

Video-recording may seem to be the best method because you preserve not only what people say but also their nonverbal behavior. The drawback to using video, however, is that it can be awkward and intrusive. Therefore, it is used infrequently. Taking notes during the interview is another common method and often is used in addition to recording, primarily when the interviewer wishes to note certain points of emphasis or make additional notations. Taking notes without recording prevents the interviewer from recording all that is said. Also, because taking notes interferes with the interviewer's thoughts and observations while the respondent is talking, researchers rarely take extensive notes. In highly structured interviews and when using some types of formal instrument, the interviewer can more easily take notes by checking off items and writing short responses.

In addition to formal interviews, many researchers employ informal interviews to follow up on things they observed or about which they have questions. For example, a researcher who observes a physical education teacher quickly changing to another task during a class can conduct an informal interview to learn why this occurred. These interviews often are very short (e.g., taking place between physical education classes or on a walk from an exercise class to the locker room), and the researcher makes notes afterward. Although taking notes after a formal interview would be disastrous, doing so is common with informal interviews.

Focus Groups

Another type of qualitative research technique employs interviews on a specific topic with a small group of people, or **focus group**. This technique can be efficient because the researcher can gather information from several people in one session. The group is usually homogeneous, such as a group of students, an athletic team, or a group of teachers.

focus group—A small group of people interviewed about a specific topic as a method of qualitative research.

Patton (2014) argued that focus group interviews might provide quality controls because participants tend to provide checks and balances on one another that can curb false or extreme views. Focus group interviews are usually enjoyable for participants, who may be less fearful of being evaluated by the interviewer because of the group setting. The group members get to hear what others in the group have to say, which may stimulate them to rethink their own views.

In the focus group interview, the researcher is not trying to persuade the group to reach consensus. It is an interview. Taking notes can be difficult, but an audio recorder will solve that problem. Certain group dynamics, such as power struggles and a reluctance to state views publicly, are limitations of the focus group interview. The number of questions that can be asked in one session is limited. As with most qualitative data collection techniques, the focus group should be used in combination with other data-gathering techniques.

Developing the skills to collect focus group data requires practice and training similar to that required to develop interview skills. The work of Krueger and Casey (2015) is an excellent place to begin learning about focus groups. Be sure to practice, as discussed in the preceding section on interviews, to develop the ability to ask follow-up questions and keep the group engaged.

Observation

Observation in qualitative research generally involves spending a prolonged amount of time in the setting. Field notes taken throughout the observations focus on what is seen. Many researchers also record notes to help them determine what the observed events might mean and to help them answer the research questions during subsequent data analysis (Bogdan

& Biklen, 2017; Pitney, 2020). Although some researchers use cameras to record what is occurring at the research site, most use field notes.

Most researchers develop a note-taking technique adapted to the setting. For example, a researcher who is studying participants' perspectives about a group exercise class finds a place with an unobstructed view from which to observe and take notes without being obtrusive. The researcher takes notes on what occurred and the participants' reactions and also makes additional comments to help focus future observations or provide reminders about how the observations tie back to the study's theoretical framework. All of this is done by using different pages or different colored pens for each type of note to make everything easily identifiable at a later time.

One major drawback to observation is obtrusiveness: a stranger with a notebook and pen, a computer, or a camera to record behaviors. The keyword here is *stranger*. The task of qualitative researchers is to make sure the participants become accustomed to having them (and, if appropriate, their recording devices) around. They may do this by visiting the site for at least a couple of days before data collection begins.

In an artificial setting, researchers can use one-way mirrors and observation rooms. In a natural setting, the limitations that stem from the presence of an observer can never be ignored. Locke (1989) observed that most naturalistic field studies are reports of what goes on when a visitor is present. Here, we run into an important question: How important and limiting is this? Locke suggested ways to suppress reactivity, such as the visitors being in the setting long enough that they are no longer a novelty and being as unobtrusive as possible in everything from dress to choice of location in a room.

Visual Methods

Researchers' use of visual methods for data collection has significantly increased over the past few years. Visual images create the basis for more in-depth data collection; a variety of techniques can be employed. Examples include using pictures during an interview to elicit more information from participants, having participants use a camera to take photos and then describe why the photos have meaning for them, having participants create a scrapbook about their life that focuses on a specific variable (e.g., physical activity or body image), and others. An example of using visual methodology in a physical activity setting is Safron (2020). In her study, she focused on how teenagers in an after-school physical activity program interact with health and fitness methods. She employed participant-driven diaries and scrapbooks as a major aspect of data collection, which enhanced the depth of the data. For more information on visual methods, we recommend Banks (2018).

Other Data-Gathering Methods

Among the many sources of data in qualitative research are self-reports of knowledge and attitude. The researcher can also develop scenarios, in the form of descriptions of situations or videos, that are acted out for participants to observe. The participants then give their interpretations of what was going on in the scenarios. The participants' responses reveal their perceptions, interpretations, and awareness of the situation and of the interplay of the actors in the scenarios.

Other sources of data include documents used in the setting (e.g., employee handbooks, procedure manuals, handouts and emails, and fliers), photos of the setting, participants' diaries, Internet chat transcripts, and websites. In most studies, researchers use multiple sources of data to reach conclusions during data analysis.

Data Analysis

Data analysis in qualitative research differs considerably from that in quantitative (normal) research. For example, data analysis occurs not only after but also during data collection in qualitative studies (in quantitative research, it only happens afterward). During data collection, researchers sort and organize the data and speculate and develop tentative hypotheses to guide them to other sources and types of data. Qualitative research is somewhat similar to multiple-experiment research, in which discoveries made during the study shape each successive phase of the study. Simultaneous data collection and analysis allows researchers to work more effectively. Analysis then becomes more intensive after the data have been collected (Merriam, 2007). Another major difference is the way in which data are presented. Qualitative data are generally presented in words, descriptions, and images, whereas quantitative data are typically presented in numbers.

> **Subsequent phases of qualitative research are shaped by discoveries made in previous phases.**

Data analysis in a qualitative study can take various forms, depending on the nature of the investigation and the defined purposes. Consequently, we cannot go into great depth in discussing analysis without tying it to a specific study. For this reason, we have summarized the general phases of analysis synthesized from descriptions in several qualitative research texts. The general phases are (a) sorting and analyzing during data collection, (b) analyzing and categorizing, and (c) interpreting and constructing theory. First, however, comes the process of data reduction.

Data Reduction

Prior to doing actual data analysis, all of the data must be reduced so it can be analyzed by the researcher using a computer (Locke et al., 2014). This involves a number of steps and often is very labor intensive. Formal interviews and focus group interviews must be transcribed word for word. Often, researchers add comments (e.g., "laughing," "angry," "crying") to denote emotion so that what was said is not confused during analysis. Informal interviews also are transcribed from the researcher's notes. After a formal interview is transcribed, the researcher should read the printout and listen to the interview to make certain it is accurate, adding or adjusting comments as necessary. Researchers also must check informal interviews to make sure the notes agree with what was entered into the computer.

Researchers enter field notes into the computer while taking care to clearly delineate what they observed and what they wrote down to inform their analysis or theory. They enter field notes many different ways, and the process is often as idiosyncratic as the note-taking itself. Documents are also entered into the computer by scanning (e.g., a flier found on the wall at the research site), by importing from other electronic sources (e.g., digital pictures), or by cutting and pasting from other documents (e.g., an Internet chat transcript). The accuracy and completeness of the data must be verified before analysis can begin.

Sorting, Analyzing, and Categorizing Data

The simultaneous collection and analysis of data is an important feature of qualitative research. As noted, researchers add comments about their analyses while taking notes. This occurs at other times as well: when reducing data and then beginning analysis while still collecting other data. This approach enables researchers to focus on certain questions and, in turn, to better direct data collection. Although researchers have specific questions in mind when data collection begins, they may shift their focus as the data unfold.

Researchers need to keep in close touch with the data. Waiting until after the data are collected to analyze them is foolish. Decisions must be made concerning scope and direction—otherwise, researchers could end up with unfocused or repetitious data or an overwhelming volume of material to process (Merriam, 2007). Also, gaps may occur due to the failure of collecting vital evidence. Researchers want data that answer their questions. Close examination of data during data collection reveals areas of data saturation (overabundance) and areas less complete and in need of a greater focus.

Researchers typically write many comments to stimulate critical thinking while taking notes or entering data into the computer. They should not be merely human recording machines; rather, they should try out new ideas and consider how certain data relate to large theoretical, methodological, and substantive issues (Bogdan & Biklen, 2017). It is quite common for researchers to reexamine their original questions to make sure their analysis is on track or to decide that the data lead to additional or different questions (Miles, Huberman, & Saldaña, 2014).

Analysis is the process of making sense of the data. Once the data are entered into the computer and the researcher has confirmed that they are accurate, analysis begins. Initial analytical notes, categories from theory, and questions inform the analysis. This is the beginning of the stages of organizing, abstracting, integrating, and synthesizing, which ultimately permit the researcher to report what they have seen, heard, and read. They may develop an outline to search for patterns that can be transformed into categories.

The qualitative researcher faces a formidable task in sorting the data for content analysis. Obviously, many categories can be devised for any set of data, depending on the problem being studied. For example, a researcher could categorize observations of a physical education class in terms of the teacher's management style, the social interactions among students, sex differences in behavior or treatment, or verbal and nonverbal instructional behaviors. Categories can range in complexity from relatively simple units of behavior types to conceptual typologies or theories (Merriam, 2007; Miles et al., 2014).

Researchers use a variety of techniques for sorting data. Index cards and file folders were once used but are not common today. Computer software is now used to store, sort, and retrieve data. Most universities at which qualitative researchers are active have one or two of the most popular qualitative software programs (e.g., NVivo or Atlas.ti) available through their computer centers. In addition, many of these universities offer classes (both credit and noncredit) in how to use the software. If you are considering a qualitative study, these classes offer an excellent opportunity to see the characteristics of the software programs and evaluate their usefulness. Of course, if you have the opportunity to take a class in which you collect pilot data and then use the software, you can improve your eventual efficiency. And, if you have a faculty member with whom you can consult, you can get help and input throughout the analysis process.

Data categorization is a key facet of true qualitative research. Instead of using mere description, the researcher may use descriptive data as examples of the concepts that are being advanced. The data need to be studied and categorized so that the researcher can retrieve and analyze information across categories as part of the inductive process. Qualitative computer analysis programs permit great flexibility in designing the categories and marking data to examine possible conclusions.

Interpreting the Data

When the data have been organized and sorted, the researcher then attempts to merge them into a holistic portrayal of the phenomenon. An acknowledged goal of qualitative research is to produce a vivid reconstruction of what happened during data collection. This goal is accomplished through the **analytical narrative**, or analytical memoing (Miles et al., 2014). This can consist of a descriptive narrative organized chronologically or topically. In our concept of qualitative research, however, we support the position of Flick that "interpretation is the core activity of qualitative data analysis" (Flick, 2014, p. 375). As Peshkin observed, "Pure, straight description is a chimera; accounts that attempt such a standard are sterile and boring" (Peshkin, 1993, p. 24). Goetz and LeCompte (1984) and many others (e.g., Creswell & Poth, 2018; Rossman & Rallis, 2017) maintained that by leaving readers to their own conclusions, the researcher risks misinterpretation and perhaps trivialization of the data by readers who are unable to make the implied connections. They further suggested that the researcher who can find no implications beyond the data should never have undertaken the study in the first place.

analytical narrative— A short, interpretive description of an event or situation used in qualitative research; also called *analytical memoing.*

The analytical narrative (or memo) is the foundation of qualitative research. Researchers (especially novices) are often reluctant to take the bold action needed to assign meaning to the data. Erickson (1986) suggested that to stimulate analysis early in the process, researchers should force themselves to make an assertion, choose an excerpt from the field notes that substantiates that assertion, and then write a **narrative vignette** that portrays the validity of the assertion. In the process of making decisions concerning the event to report and the descriptive terms to use, the researcher becomes more explicitly aware of the perspectives emerging from the data. This awareness stimulates and facilitates further critical reflection.

The narrative vignette is one of the fundamental components of qualitative research. As opposed to the typical analysis sections in quantitative research studies (which are about as interesting as watching paint dry), the vignette captures the reader's attention, thereby helping the researcher make their point. The vignette gives the reader a sense of being there. A well-written description of a situation can convey a sense of holistic meaning that is definitely advantageous in providing evidence for the researcher's assertions. In characterizing qualitative research, Locke stated that the researcher can describe the physical education scene so vividly that "you can smell the lockers and hear the thud of running feet" (Locke, 1989, p. 4). Griffin and Templin (1989, p. 399) provided the following example of a vignette:

> **The second period physical education class at Big City Middle School is playing soccer. The teacher has placed two piles of sweatshirts at each end of a large open field to serve as goals. There are no field markings. Four boys run up and down the field following the ball. Several other students stand silently in their assigned positions until the ball comes near, then they move tentatively toward the ball to kick it away. Three girls stand talking in a tight circle near the far end of the field. They are startled when the ball rolls into their group, and two boys yell at them to get out of the way. They do and then regroup after the ball and the boys go to the other side of the field. Two boys, who have not touched the ball during the class, engage in a playful wrestling match near one goal. The teacher stands in the center of the field with a whistle in his mouth. He hasn't said anything since he divided students into teams at the beginning of class. He has blown the whistle twice to call fouls. The students play around him as if he were not there. A bell rings, and all the students drop their pinnies where they are and start toward the school building. Belatedly, the teacher blows his whistle to end the game and begins to move around the field picking up pinnies.**
>
> **After class, as we walk back to the building, the teacher says, "These kids are wild. If you can just run off some of their energy, they don't get into so much trouble in school. This group especially, not too many smarts (taps his temple), don't get into much game strategy." (He sees a boy and girl from the class standing near the door of the girls' locker room talking.) He yells, "Johnson, get your butt to the shower and stop bothering the ladies." He smiles at me. "You've got to be on them all the time." He looks up and sighs, "Well, two [classes] down, three to go."**

We caution you, however, that Locke and other scholars do not believe that richness of detail alone is what makes a narrative vignette valid. Siedentop (1989) warned that whether data are to be trusted should not be based on the narrative skills of the researcher. According to Erickson, a valid account is not simply a description but an analysis: "A story can be an accurate report of a series of events, yet not portray the meaning of the actions from the perspectives taken by the actors in the event. . . . It is the combination of richness and interpretive perspective that makes the account valid" (Erickson, 1986, p. 150). Vignettes should not be left to stand by themselves. The researcher should make interpretive connections between narrative vignettes and other forms of description, such as direct quotations and quantitative materials (Miles et al., 2014).

narrative vignette—A component of a qualitative research report that gives detailed descriptions of an event, including what people say, do, think, and feel in that setting.

Direct quotations from interviews with the participants taken from field notes and audio or video recordings are another form of vignette that enriches the analysis and furnishes documentation for the researcher's point of view. They can demonstrate agreement (or disagreement) about a phenomenon. Direct quotations from the same people on different occasions can provide evidence that certain events are typical or demonstrate a pattern or trend in perceptions over time. For example, the quotations Woods and Lynn (2014) used throughout their results provide some of the original data that helped them arrive at their conclusions. Without these quotations, the material would be more difficult for most readers to grasp and would not represent the vitality of the data collected.

Qualitative researchers should communicate their perspectives clearly to their readers. The function of the narrative is to present the researcher's interpretive point in a clear and meaningful manner.

Theory Construction

theorizing—The cognitive process of discovering abstract categories and the interrelationships of those categories.

theory—An explanation of some aspect of practice that permits the researcher to draw inferences about future happenings.

Our use of the terms *theory* and *theorizing* should not unduly alarm you or discourage you from undertaking qualitative research. We are not talking about developing a model on the scale of the theory of relativity here. **Theorizing**, according to Goetz and LeCompte (1984), is a cognitive process of discovering abstract categories and the interrelationships of those categories. A **theory** is an explanation of some aspect of practice that permits the researcher to draw inferences about future events. This process is a fundamental tool for developing or confirming explanations. You process information, compare the findings based on your experience and sets of values, and then make decisions. If the decisions are not correct, you will need to revise the theory or model.

In many but not all qualitative research traditions, data analysis depends on theorizing. The tasks of theorizing are "perceiving; comparing, contrasting, aggregating, and ordering; establishing linkages and relationships; and speculating" (Goetz & LeCompte, 1984, p. 167). Perceiving involves the consideration of all sources of data and all aspects of the phenomenon being studied during as well as after data collection. The perceptual process of determining which factors to analyze guides the collection of data.

Comparing, contrasting, aggregating, and ordering are primary functions in qualitative research. The researcher decides which units are similar or dissimilar and what is important about the differences and similarities. Analytical description cannot occur until the researcher builds the categories of like and unlike properties and carries out a systematic content analysis of the data. Establishing linkages and relationships constitutes a kind of detective work that qualitative researchers do in the theorizing process. They use both inductive and deductive methods of establishing relationships "while developing a theory or hypothesis that is grounded in the data" (Goetz & LeCompte, 1984, p. 172). As Corbin and Strauss (2015) noted, multiple interpretations of the same data are possible, but all interpretations should be grounded in data.

grounded theory—A theory that is based on and evolves from data.

A **grounded theory** is based on data from which it also evolves (Glaser & Strauss, 2017). In applied research, grounded theories can be used for explaining observed phenomena, understanding relationships, and drawing inferences about future activities.

Trustworthiness in Qualitative Research

Qualitative researchers do not attempt to provide numerical evidence that their data are reliable and valid. In fact, the terms *reliability* and *validity* are rarely used in qualitative research. However, that does not mean that the qualitative researcher is not concerned with getting good data and reaching conclusions in which readers can have faith. Both issues are extremely important, and without attention to them, the quality of the research may be suspect. Just as for quantitative data, those reporting qualitative data make a case for the quality of the data *and* the conclusions that were derived from the data analysis.

Several terms are used to describe quality in qualitative research. The terms have changed and still are evolving. Lincoln and Guba (1985) used **trustworthiness**, a term most often used, to describe the overall quality of the results from the study, and this term is used most often. Rossman and Rallis (2017) broke trustworthiness down into two questions: Is the study competently conducted? and Is it ethically conducted?

trustworthiness—A quality achieved in a study when the data collected generally are applicable, consistent, and neutral.

Conducting an Ethical Qualitative Study

We discuss the issue of ethics first. Ethics often plays a prominent role in judging qualitative research. We can think of ethics in many ways, but it boils down to two major areas. First, the ethical treatment of study participants (Locke et al., 2014) is important in qualitative research. Qualitative researchers often spend a great deal of time with participants. From their initial approach to have them be a part of the study through the reporting of data, researchers should treat participants with dignity. If researchers have promised anonymity, they should maintain it. This idea does not suggest that those doing quantitative research should not treat participants ethically. The relationships and the type of data are different in qualitative research, however, and often require a more sustained commitment to ethics. Second, Rossman and Rallis (2017) provided an in-depth discussion of the ethics of qualitative research. These authors and others (Creswell & Poth, 2018; Locke et al., 2014) suggested confronting ethical issues while designing the study. Without forethought, you can get into an ethical quagmire that can make it impossible to complete or publish the study.

This discussion brings us to a question: What constitutes a competently conducted qualitative study? Again, we can think of this subject in a variety of ways. Lincoln and Guba (1985) described four useful concepts—**credibility**, *transferability*, *dependability*, and *confirmability*—that researchers still use today for thinking about quality when designing studies and completing data analyses. The manner in which the researcher understands context, participants, and settings informs the interpretation of the results of qualitative research. This, paired with the description presented in a paper, are critical to judging other parts of the study and its overall credibility. If it is unclear who the participants are and where the study is being completed, readers will have difficulty evaluating the conclusions and the study's credibility will suffer.

credibility—A quality achieved when the participants and setting of a study are accurately described.

transferability—In qualitative research, a quality achieved when the results have the potential to be transferred to other settings.

Transferability addresses whether the results would be useful to researchers in either similar settings or other settings. You may be thinking that this concept is like generalizability, and in some ways it is. All researchers would like their research to help others, yet both quantitative and qualitative research fail to benefit from random selection from a large population. Transferability is a question of argument and perception. The researcher may present reasons a given study might apply in other settings (e.g., many schools operate in a similar fashion, or the participants in an exercise class are similar to those found in many community settings), but ultimately, the readers or users of the research must determine whether the qualitative study applies to their work environments or future research. Again, this often is the case in quantitative research—qualitative researchers just start with that concept up front.

In many qualitative studies, researchers make changes in the phenomenon being studied or in the methods used based on previous data collection. For a relatively straightforward example, a researcher may use a structured interview to ask questions but then change the way to ask follow-up questions depending on the answers. A second set of interviews might be based on the answers provided during the first set of interviews. If the researcher asked questions in lockstep without making adjustments, the quality of the data would suffer. In this example, the researcher could not possibly plan for all possible contingencies. How the researcher deals with the changes determines the **dependability** of the data.

dependability—A characteristic that addresses the quality of the data in a qualitative study, including how well the researcher deals with change.

Finally, **confirmability** deals with the issue of researcher bias. As we discuss later in this section, qualitative research differs from quantitative research in that researchers do not calculate reliability statistics. Rather, they use approaches to confirm their findings so that readers are assured a study has produced reliable results.

confirmability—A characteristic of qualitative research that addresses whether readers are to believe a study's results as reliable.

Providing Evidence of Trustworthiness

Researchers can provide evidence of the trustworthiness of their studies (i.e., whether they were conducted well) in a variety of ways based on the preceding categories. We will review the common techniques that qualitative researchers use to gather good data and reach objective conclusions. These techniques are used during both data collection and data analysis. As noted earlier, analysis often occurs during data collection, so separating these tasks into temporal categories is difficult. Not all researchers use all techniques in every study, but researchers commonly use multiple methods to increase the trustworthiness of their data and conclusions. The following techniques are commonly used:

- *Prolonged engagement.* The collection of qualitative data requires that the researcher spend sufficient time in a setting to collect good data and develop an in-depth understanding of them rather than reaching superficial conclusions.

- *Audit trail.* The method and focus in qualitative research change during a study. An audit trail describes the changes that occurred and how they influenced the study. These changes are often reported in the methods section, where the researcher explains how the changes improved the study.

- *Rich, thick description.* As noted earlier, describing the setting and context is important for credibility. A thorough description in the report is needed so that readers can both understand the study and assess whether the setting and results will transfer to their settings or future research.

- *Clarification of researcher bias.* All researchers come into a study with biases. The management of these biases is particularly important in qualitative research in which the researcher is the instrument of data collection. Evidence that researchers acknowledged their biases and dealt with them is essential. In a move that we believe enhances the quality of research, many qualitative researchers address this directly in the methods section by "coming clean"—presenting their biases and explaining how they worked to control them.

- *Triangulation.* This technique is often used in qualitative data analysis. It requires the use of three (hence the prefix *tri-*) independent sources of data to support a conclusion. For example, a researcher may use student interview data, teacher interview data, and observation to support a result. Note that this is not simple vote counting (i.e., three people said this in an interview), but the use of independent types of data to support a conclusion.

- *Negative case check.* Researchers use this technique to look at instances in which what they expected to happen did not. For example, in an exercise class, the researcher notes that as time goes on, many participants seem more upbeat and are integrating the class into their lives. To check, the researcher must ask several questions: Does this happen for everyone? For those for whom it does not happen, is the phenomenon not as pervasive? Did I selectively look at the data to reach a conclusion? or Is something different happening for some participants? Performing a negative case check helps address bias and enables the researcher to investigate the phenomenon further.

- *Member check.* A member check occurs when the researcher shares the conclusions with participants to ascertain whether the participants agree with them. Sometimes an intermediate step occurs in which the researcher asks the participants to review interview transcripts, clarify statements, and add anything they believe is missing. That by itself is not a true member check. Presenting the participants with the conclusions goes much further and provides a confirmation of the analysis. In some cases, if participants believe that the researcher may have gotten it wrong, the conclusions may be modified to reflect the participants' views. In other cases, the researcher may believe that bias (e.g., when a report does not reflect positively on a person) influenced a participant's evaluation and would then include in the report an analysis of that person's evaluation.

• *Peer debriefing.* As in much of what we do, a new set of eyes can bring new light to a data set and our conclusions. A peer debriefer is another person who examines the data and the conclusions and serves as a devil's advocate, questioning the researcher to see whether the findings hold up. Often, the person performing this has a background in the phenomenon being investigated and expertise in qualitative methods and data analysis. This step can improve the conclusions and how they are presented in the research report.

Writing the Report

There is no standard (or correct) format for a qualitative research report, just as there are no formats that are rigorously followed for any other type of research. Here we simply mention some main components of a qualitative research report and their placement in the report. Your department or university may have an order of components that you must follow.

The components of a qualitative study are similar to those of other conventional research reports. The first part introduces the problem and provides background and related literature. A description of method is an integral part of the report. Although this section is not as extensive in a journal article as it is in a thesis, it is usually much more extensive than it is in other forms of research. The reasons are obvious, as we explained. The methodology is integrally related to the analysis and is also important in terms of trustworthiness and credibility.

The results and discussion section forms most of the report. If used, charts, tables, and figures are contained in this section and must be integrated into the narrative. A major contributor to the bulk of this section is the description contained in a qualitative study. As we said earlier, narrative vignettes and direct quotations are basic to this type of research. The qualitative study strives to provide enough detail to show the reader that the author's conclusions make sense (Creswell & Creswell, 2018; Walcott, 2009). The author is faced with the delicate problem of finding a balance between the rich descriptive materials and the analysis and interpretation.

Curious Incidents During the Collection of Qualitative Data

1. On two occasions, researcher reported to gymnasium (carrying camera and notes) to find door locked. Had to walk all the way around the building—once in the rain.
2. Reported to school to collect data only to find classes wouldn't be held because of
   a. a doughnut sale,
   b. teacher being ill (twice), and
   c. holiday music rehearsal.
3. Interview with teacher interrupted by principal due to crisis over a parent and a charity clothing sale. Interview postponed.
4. Audio recorder became inoperable during interview.
5. Audio recorder became inoperable while transcribing. Had to repeat interviews.
6. Group recording session had to be rescheduled because of complete loss of control after a third-grade student belched into the microphone.
7. Had to buy own recorder because department's digital recorder was inoperable.
8. Broke own expensive new recorder. Repair took two weeks. Interviews rescheduled.
9. While making backup of audio files and original recordings were stolen. Police nabbed culprit, and data were recovered . . . but the researcher was comatose.

Evaluating the Quality of Qualitative Research

Linda L. Bain

California State University, Northridge
AAHPERD, Indianapolis

Definition of the Problem

_____ Has a clearly stated purpose

_____ Focuses on a significant issue

_____ Seeks to understand meaning of experiences for the participants

_____ Provides holistic view of the setting

Data Collection

_____ Researcher has training in methods used

_____ Pilot work done in similar setting using similar methods

_____ Rationale provided for selection of the sample

_____ Researcher has trusting, collaborative relationship with participants

_____ Methods for data collection are unobtrusive, where appropriate

_____ Data collection procedures provide thorough description of events

_____ Prolonged engagement in field

Data Analysis

_____ Analysis done during and after data collection

_____ Triangulation of data sources and search for convergence

_____ Search for negative cases

_____ Provides interpretation and theory as well as description of events

_____ Provides opportunity for participants to corroborate interpretation (member check)

_____ Arranges for peer evaluation of procedures and interpretation

Preparation of Report

_____ Complete description of setting

_____ Complete description of procedures

_____ Includes description of researcher's values, assumptions, and bias and how each was addressed

_____ Uses vignettes and quotations to support conclusions and interpretation

General Assessment

_____ Internal validity: How much confidence do you have in the quality of the description and interpretation of events in the particular research setting?

_____ External validity: What is your assessment of the extent to which the results of this study apply to a different setting with which you are familiar?

Reprinted by permission from L.L. Bain, *Evaluating the Quality of Qualitative Research* 1992, 350.

All academic writing requires rewriting and editing, and this is especially true of qualitative research. Writers should provide the theoretical framework for the study, describe the method, provide data to support the results (which often requires more words than in most quantitative studies), and discuss the contributions of the results to our understanding and to theory. Because this task requires some balancing, writers should take the time to decide which evidence will best elucidate the results and then integrate it with the other parts of the manuscript.

It is doubtful that any graduate student is under the false impression that doing qualitative research is a quick-and-easy task. To emphasize this point, we call your attention to the Curious Incidents During the Collection of Qualitative Data sidebar, which lists some of the trials and tribulations that befell one doctoral student while she was attempting to collect data.

Concluding Remarks

At one time, qualitative research was thought to be confined primarily to the area of pedagogy. Getting that impression is easy because of the ever-increasing volume of literature on qualitative research in educational journals and textbooks. Qualitative research, however, is now common in many areas of research, including sport and exercise psychology, sport sociology, athletic training, and exercise prescription and adherence. If you read and use research in your professional work or actively conduct research, you will likely come across many qualitative studies.

This chapter has focused on interpretive qualitative research. Another approach, **critical theory**, is emerging in the literature. The main difference between the two approaches is found in the research goals (Locke et al., 2010). Interpretive research is largely free of value judgments, whereas critical theory research is based on value judgments. In other words, critical theory aims to provide the research participants the insight necessary to making choices that will improve their lives or empower them. Often, the research is grounded in a theory that seeks to ameliorate the negative aspects of the human condition (e.g., feminism, neo-Marxism, the empowering pedagogy of Paulo Freire, the writings of Michel Foucault). Such theoretical perspectives challenge the status quo and strive for greater equality. You can begin exploring critical forms of qualitative research by reading Marshall, Rossman, and Blanco (2021) and Locke et al. (2010, 2014).

> **critical theory**—Qualitative research based on value judgments.

The Evaluating the Quality of Qualitative Research sidebar provides an assessment that can be used to evaluate the quality of qualitative research studies. Originally developed by Linda Bain, it was presented at the 1992 national convention of the American Alliance for Health, Physical Education, Recreation and Dance (AAHPERD). We modified it for this edition of this book and also incorporated detailed information from Locke et al. (2010) for critiquing both qualitative and quantitative research.

Summary

Qualitative research methods include field observations, interviews, and material collection. The researcher gathers data in a natural setting such as a gymnasium, classroom, fitness center, or sport facility.

Qualitative research does not have the preconceived hypotheses that characterize quantitative research. Inductive reasoning is stressed, whereby the researcher seeks to develop hypotheses from observations. The focus is on the essence of the phenomena. The researcher should exhibit sensitivity and perception when collecting and analyzing the data.

We stressed the importance of gaining access to data in the field setting. Establishing rapport and gaining the participants' trust are essential. The most common methods of collecting data are interviews and observations. Data should be analyzed during and after collection. The researcher must input the data into a computer to sort and organize it and develop tentative conclusions from the results that lead to other sources and types of data.

Data analysis involves organizing, abstracting, integrating, and synthesizing. The analytical narrative is the foundation of qualitative research. The narrative vignette gives the reader a sense of being present for the observation; it conveys a sense of holistic meaning to the situation. It is not unusual for a qualitative study to include quantitative analysis.

The qualitative researcher often attempts to construct a theory through the inductive process to explain relationships between categories of data. A grounded theory evolves from data. In the written report, the qualitative researcher must achieve a balance between the rich description and the analysis and interpretation.

Trustworthiness is used to determine whether the study was competent. Ethical issues are important in determining trustworthiness, as are issues related to how the study was conducted (i.e., credibility, transferability, dependability, confirmability). Many techniques are used during data collection and analysis to enhance the quality of the data and the conclusions. Among these are prolonged engagement, an audit trail, a substantial and rich description, a clarification on researcher bias, triangulation, negative case and member checks, and peer debriefing. Not all of these can be used in one study. However, incorporating as many techniques as possible is a common practice that not only enhances the study's integrity but also strengthens the reader's trust in its conclusions and transferability to other settings.

Qualitative research is a viable approach to solving problems in our field. It applies to pedagogy in physical education, to exercise science, and to sport science. Answers to the question *What is happening here?* can best be obtained in natural settings through the systematic observation and interactional methodology of qualitative research.

✓ Check Your Understanding

1. Locate a qualitative study and write an abstract of approximately 300 words on the methods used in gathering and analyzing the data (observations, interviews, triangulation, member checks, and so on) and in presenting the results (narrative vignettes, quotations, tables, and so on).

2. Locate a qualitative study and write an abstract of approximately 300 words that describes why the study was conducted and what theoretical basis supported the study. Note whether the results answer the questions and add to the theoretical understanding that the authors presented in the introduction.

20

Mixed-Methods Research

While they were saying among themselves it cannot be done, it was done.

—Helen Keller

Mixed-methods research in kinesiology is relatively new. For a long time, scholars in many areas of research believed that combining quantitative and qualitative methods into a single study or research program was impractical. Qualitative and quantitative researchers bickered over the superiority of their methods (Denzin, 2008; Gelo, Braakmann, & Benetka, 2008; Hatch, 2006; Reichardt & Rallis, 1994; Shavelson & Towne, 2002). Gage (1989) characterized these debates as the paradigm wars and suggested that they would not advance research. Others, however, continued to suggest that one method was superior to another.

The quotation from Helen Keller captures what happened during this debate. Researchers began combining methods—and asking questions that neither method could independently answer. Books were published that discussed mixed-methods research (e.g., see Green & Caracelli, 1997, for one of the first). Researchers found that the use of mixed methods helped answer complementary questions on a similar topic. For example, one of the authors of this book has done research on student attitude in physical education using primarily quantitative methods. While working with a graduate student, it became clear that the quantitative data allow for addressing certain questions (e.g., understanding the underlying structure of attitude and getting student attitude scores for various subgroups of students). It also became apparent that other methods were needed to address questions about what experiences influenced the attitudes of students. This discovery resulted in a series of interrelated studies (Subramaniam & Silverman, 2000, 2002, 2007) that used both quantitative and qualitative methods to approach questions about student attitude. As we discussed earlier in this book, different questions require different methods—even when the questions are closely related.

Teddlie and Tashakkori (2009) noted that mixed-methods research is largely a pragmatic enterprise. People do it because it can help answer questions and advance their research programs. That is not to suggest, however, that mixing methods is easy. Researchers often must work with others who have expertise in another methodological approach or start from scratch and learn new research skills. As with any research enterprise, you must become skilled in the method so that you can plan the study, collect data, analyze the results, and communicate with others about the research (Plano-Clark, Huddleston-Casas, Churchill, Green, & Garrett, 2008).

In addition to learning about other research methods, you will gain a greater understanding about combining methods when engaging in mixed-methods research. This chapter provides an overview of that process and allows you to begin planning a mixed-methods study. We recommend other resources that can assist you in the process: Both Flick (2014) and Robson (2002) provide solid, pragmatic introductions to the topic of combining methods. Tashakkori, Johnson, and Teddlie (2021), Creswell and Plano-Clark (2007), and Thomas (2003) provide more detail in easy-to-read books. If further information is needed, the edited book by Tashakkori and Teddlie (2010) provides an in-depth discussion of mixed-methods research.

Combining Quantitative and Qualitative Methods

There is no one way to do mixed-methods research. As we noted earlier, mixed-methods research is pragmatic, and the questions influence the method selected. A study may be primarily quantitative with a part that is qualitative. Conversely, it may be primarily qualitative with a quantitative component. Of course, it may be anything in between (see Morse, 2003, for an example of the many possible weightings of qualitative and quantitative research within one study). Figure 20.1 presents a continuum of mixed-methods research. The balance between quantitative and qualitative components will depend on the study, as will the methods used within each part of the study.

A researcher undertaking a mixed-methods study might collect descriptive quantitative measures and qualitative interview data based on an aspect of the quantitative data. An experimental intervention could be combined with a qualitative study evaluating participants' perceptions of the intervention. Interviews and focus groups could be supplemented by descriptive data. The permutations of mixed-methods designs are infinite. Any combination of questions that suggest mixed methods can be accommodated on condition that the researcher or research team has the skills to conduct all parts

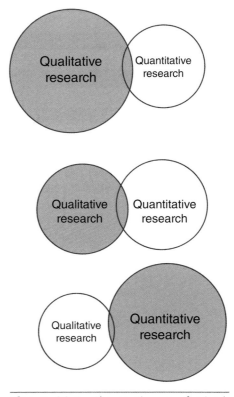

Figure 20.1 The continuum of mixed-methods research from highly qualitative to highly quantitative.

of the study. Lopez-Aymes et al. (2021) formed a team with the necessary skills to assess parental perceptions of physical activity and quality of life during the COVID-19 pandemic. The study was conducted in four countries using quantitative (questionnaires) and qualitative (four open-ended questions) techniques. The participants were recruited through contacts with professional organizations in each country using the snowball technique. Questionnaire data was analyzed comparing age and gender. Based on the quantitative data of parental response, participants were placed in one of two groups; those with sufficient physical activity and those with insufficient activity. The open-ended questions, for example "Explain why you have sufficient or insufficient physical activity," were analyzed and the two groups compared. This is an example of a parallel design that will be discussed in the next section.

Although many mixed-method designs are possible (e.g., see Creswell, 2009, or Leech & Onwuegbuzie, 2009, for discussions of other ways to describe designs), the way a study is conducted often is characterized in one of two ways (Creswell & Creswell, 2018; Tashakkori & Teddlie, 2010; Tashakkori, Johnson, & Teddlie 2009). The first, and most common, type of mixed-methods design—**sequential mixed methods**—uses a staggered arrangement of quantitative and qualitative components, where one part (regardless of order) comes first and the other part follows. Often, the results of the first part of the study influence what occurs in the study's second part. For example, a researcher might collect children's fitness test scores to provide a description of the fitness levels in some group. Based on the BMI scores above 30, participants would be recruited for semistructured interviews focusing on how they perceived that their body composition and fitness levels affected their interactions with their peers.

In **parallel**, or concurrent, **mixed-methods** research, the quantitative and qualitative components occur at the same time or are independent. From the beginning of the study, both methodological approaches are used, and the results of one part do not dictate the participants or methods for the second part. For example, a research team may be examining the influence of physical activity and diet on health indicators such as blood pressure, heart rate, fitness, and blood lipids by having participants engage in physical activity of various intensities and diet programs that range from easy implementation to those which require large changes in eating behavior. The researchers want to know not only the effectiveness of various intervention combinations but also the perceptions of those who were successful and unsuccessful in maintaining the diet and physical activity interventions to which they were assigned. The qualitative portion of the study would use observation and interviews to address this question. The results could indicate that one combination of diet and physical activity was the most effective in improving the health indicators. They would also indicate, however, that another program was nearly as effective and perceived as more enjoyable by the participants.

Lindsay-Smith et al. (2019) conducted a sequential mixed-methods study in Australia to assess the impact of community involvement on physical activity in 35 older adults (average age 67 years). A questionnaire was administered at three time periods (baseline 6 and 12 months) quantifying physical activity, group, and quality of life. Two focus groups were formed at the end of the study based on the group: social or physical activity. The physical activity group maintained their level of physical activity and physical quality of life over the study year, while the social group declined on both. Focus group data suggested that the physical activity group was motivated by the group nature of the program to attend and extend their physical activity beyond the program. The social aspects of both groups provided mental health benefits for both groups based on the focus group interviews.

Designing Mixed-Methods Research

Everything we noted in the previous chapters applies to planning mixed-methods research. The methods are adapted to the study, and the researcher must plan the study so that all parts of the methods are strong and directly address the questions or hypotheses.

Like all research, a mixed-methods study offers no shortcut to good data. Beginners using mixed-methods research often design a study in which one part is well designed and the other is not. This occurs because researchers are adding a component they are unfamiliar with. Research training and pilot studies are necessary to complete a quality study. Or, as often occurs, a research team works together to plan and conduct a study. Members design and collect data for the parts of the study that draw on their strengths. This cooperative approach helps ensure that all of the final study's parts are linked.

sequential mixed methods—A study design in which either the quantitative or qualitative component comes first and then is followed by the other component.

parallel mixed methods—A study design in which the quantitative and qualitative components occur at the same time or independently; also called *concurrent mixed methods*.

Issues in Mixed-Methods Research

Although the parts of a mixed-methods study are combined to make a whole, a researcher or research team must consider several issues—from planning the study through data analysis and writing the research paper. Some of these decisions are made intuitively. Others require forethought and cooperation among the research team. The following issues should be addressed in a mixed-methods study:

• *Questions and hypotheses.* If complementary quantitative and qualitative questions arise as researchers develop questions and hypotheses, it is important to develop each component and make certain it has a strong supporting rationale. It is not enough to tack on a methodological component because it is simple or because the researcher has strength in that methodological approach. All parts of the study should be interrelated and add to the literature.

• *Design selection.* After finalizing the questions and hypotheses, a design must be developed. The design should be based on the specific demands of the study, and the researchers should be pragmatic in addressing the qualitative and quantitative parts: Is one larger than the other? Will data collection occur sequentially or in parallel?

• *Sampling.* In many mixed-methods studies, the samples of one part of the study are based on the samples of the other part. The study may use a larger sample and a subsample. The quantitative component may have a larger sample based on random selection or on a convenience sample of volunteers, and the qualitative portion may then select from those participants to get a purposeful sample to answer the qualitative questions. Conversely, a qualitative purposeful sample may be supplemented by another sample to obtain descriptive data to answer the quantitative questions. Whether the samples are the same or different, the researcher must plan so that the participants for each part of the study are appropriate.

• *Data collection.* All parts of data collection should be planned and pilot-tested. If the data are to be collected concurrently, the research team should be ready to start at the same time. Researchers should carefully design participants' interactions in advance of data collection. The interactions should be a good use of their time so that participants are not turned off by multiple requests that appear to waste time.

• *Obtaining good data.* As data are collected and analyzed, attention must be paid to whether there is evidence of reliability and validity or trustworthiness and credibility. These decisions must be based on the traditions and demands for each type of data. This step requires researchers to individualize data collection and analysis so that each part relies on good data. All the attention to these issues for which separate studies mandate are also required of a mixed-methods study.

• *Presenting the study.* Many options are available for presenting the study. In a sequential study, the introduction and discussion are typically combined, and the methods and results are separate and based on the order of the parts. In a parallel study, the methods or results may also be separate, or they can be intermixed to suit the presentation. Generally, in a parallel study, more integration is present throughout the discussion section (Creswell & Creswell, 2018). In some sequential studies, in which both the qualitative and quantitative components are themselves major studies, separate papers may be published on each part. If a large study is presented in two papers, the researcher should note this in both papers and inform the editors when submitting them for publication (American Psychological Association, 2020).

Examples of Mixed-Methods Research

A growing number of mixed-methods research papers are appearing in fields related to physical activity. As researchers see the value in combining methods, more research using this

Sometimes mixing methods is good, sometimes not.

© SDI Production/E+/Getty Images

approach is likely. Here we provide two examples that are typical of the types of methods and designs used in mixed-methods research. You will benefit by reading each of the papers as you think about the material in this chapter.

Thøgerson-Ntoumani and Fox (2005) completed a mixed-methods sequential study that addressed well-being typologies in corporate employees. They administered a quantitative questionnaire on the Internet ($N = 312$) and used hierarchical cluster analysis to develop clusters for physical activity and well-being. They employed a variety of statistical techniques to produce evidence that the data provided a valid model for these participants. They followed this with semistructured interviews from participants ($n = 10$) who represented the clusters that were developed in the quantitative phase. This sequential study clearly reported which data were collected and the order in which they were collected. In addition, the authors provided information (e.g., trustworthiness) that showed they had paid attention to getting good data. The results were presented with the quantitative data followed by the qualitative data, and the discussion combined the two parts of the study in examining the results for each cluster.

Harvey and colleagues (2009) provided an example of a parallel, or concurrent, mixed-methods study in which the qualitative portion was a larger part of the study than the quantitative portion. Twelve boys from a purposeful sample, six with and six without attention-deficit hyperactivity disorder (ADHD), were tested for motor skills and interviewed using structured and semistructured interview questions to explore their knowledge of and preferences for physical activity. The testing and interviews occurred simultaneously. Comparisons were made between groups on the skills assessment and through thematic analysis of the interviews. The results were presented first for the skill assessment and then for the qualitative results. The results of both types of research were integrated in the discussion, but as would be expected, when the qualitative part of the study is much larger, it focuses more on the qualitative results. The authors provided a good overview of the steps they took to obtain good data and were candid about how a pilot study influenced the final method.

Summary

Mixed-methods research is a pragmatic method of addressing interrelated questions that can best be answered by combining quantitative and qualitative approaches. Generally, these methods are used sequentially or concurrently, but as Creswell (2009) noted, the ways mixed-methods designs are described are varied and likely to change over time. Researchers who do mixed-methods research need to be skilled in both approaches, so two or more researchers often collaborate to address the parts of the study.

✓ Check Your Understanding

1. Locate a mixed-methods study and write an abstract of approximately 300 words about the methods used to gather and analyze the data and present the results. Include information about which methods are qualitative and which are quantitative. Identify the relative emphasis on quantitative and qualitative methods and whether the design is sequential or parallel.

2. Locate a mixed-methods study and write a summary of how the researchers worked to obtain good data. Focus on issues of reliability, validity, trustworthiness, and credibility. Note any differences in the data and discuss whether additional information would have been helpful.

IV

Writing the Research Report

> **Brevity in writing is the best insurance for its perusal.**
>
> —Rudolf Virchow

Part I discusses the research proposal, its purpose, and the structure of the various parts. Parts II and III provide the details needed for understanding and conducting research, including statistics, measurement, and types of research. This part completes the research process with instructions on preparing the research report. You may also want to refer to chapter 2, which discusses some rules and recommendations for writing the review of literature, because that is an important part of the research report.

Following are the purposes of a thesis or dissertation:

1. Establishing the expertise of the graduate student
2. Warehousing the artifacts that demonstrate scientific competence and chronicling the detailed processes used in research
3. Expanding the knowledge base of the field of kinesiology

Chapter 21 examines all the parts of the research proposal that have already been discussed. In addition, we offer some of our thoughts about the nature of the meeting to review the research proposal. Up until now, we have tried to explain how to understand other research and how to plan your own research. Here, we help you organize and write the results and discussion sections (or chapters) about research you have conducted. We also explain how to prepare tables, figures, and illustrations and where to place them in the research report.

Finally, in chapter 22, we suggest ways of using journal and traditional styles to organize and write theses and dissertations. We also present a brief section on writing for scientific journals and a short discourse on preparing and giving oral and poster presentations.

21

Completing the Research Process

> Ginger Rogers did everything Fred Astaire did only backwards and in high heels.
>
> —1982 *Frank and Ernest* cartoon

Completing the research process involves writing your proposal, getting it approved, carrying out the research, and writing it up in the results and discussion section of your thesis. In the following sections, we provide guidelines to help you complete this process. After you have read these guidelines, you can find greater detail in Locke, Spirduso, and Silverman (2014).

Research Proposal

The research proposal contains the definition, scope, and significance of the problem and the methods that will be used to solve it. If the journal format (advocated in this book and reviewed in detail in chapter 22) is used for the thesis or dissertation, the proposal consists of the introduction and methods sections and appropriate tables, figures, and appendixes (e.g., score sheets, cover letters, questionnaires, informed-consent forms, and pilot-study data). In a four-section thesis or dissertation (introduction, methods, results, and discussion—IMRD), the proposal consists of the first two sections. In studies using a five-section format, in which the introduction and the review of literature are separated into two sections or chapters, the proposal encompasses the first three sections.

One of the goals of this book is to help prepare you to develop a research proposal. We have already discussed the contents of the proposal. Chapters 2, 3, and 4 pertain specifically to the body of the research proposal. Other chapters relate to facets of planning a study: the hypotheses, measurements, designs, and statistical analyses. In this chapter, we attempt to bring the proposal together. We also discuss the proposal meeting and committee actions. Finally, we touch on basic considerations involved in grant proposals—specifically, how they differ from thesis and dissertation proposals.

Thesis and Dissertation Proposals

Most departments require a proposal (sometimes called a *prospectus*) for a thesis or dissertation prior to completing the research and writing the final copy. However, graduate schools do not typically require the formal filing of a proposal; it is usually a department document. The proposal is formatted according to one of the categories listed previously for the journal style of a thesis or dissertation. It may include previous work completed (published, in press, and submitted), and it presents the plan for the upcoming research. The upcoming research plan is also in journal style including an introduction (with a short literature review) and the methods. If you have not completed previous research on this topic (more likely with master's students than with doctoral students), you should include pilot data in the proposal. The purpose of pilot data is to show that the research instrumentation (or materials) and methods actually work under the conditions of the research—that you can administer them and participants can complete them, and that they yield valid and reliable data.

The research proposal should also contain references as well as figures or tables. There may be several appendixes that include a more detailed literature review, the pilot data, other methods information, and so on. If the proposal is done correctly, it should serve as the basis for the final research paper(s) and thesis or dissertation.

Advisor and Dissertation Committee

You should work closely with your advisor in developing your thesis or dissertation proposal. The planned research should not come as a surprise to anyone associated with your work. This allows the proposal meeting to be about how to best do the research rather than addressing the question *Is this a good idea?* Once the proposal is approved by your advisor and the committee, it constitutes an agreement that if the research is completed as planned and written effectively, it meets the requirement for a thesis or dissertation. If adjustments are needed during the research, clear this with your advisor and the dissertation committee.

Special consideration must be given to the role of the dissertation committee when previously published and in-press papers are included in the dissertation. The committee should be involved in the planning and execution of all work. Once work is published, the committee has no avenue for revision; thus, its approval during the process is important. The committee will approve the entire dissertation or thesis, not just the unpublished studies. Therefore, it should be involved as the studies are conducted, completed, and prepared for publication. You should be first author on all studies in your thesis or dissertation; the committee members and your advisor may or may not be coauthors depending on their contributions to the work. It is a good practice to discuss authorship during the process from proposal to submission of articles for review.

The Good Scholar Must Research and Write

In today's world, the good scholar must have the knowledge and skill to carry out quality research that answers important questions. However, this is not enough. The good scholar must also publish that research in quality journals. "Thus, the scientist must not only 'do' science but must 'write' science" (Gastel & Day, 2016, p. xvi). Of course, changing the format and style of reporting research does not improve the research. Bad science cannot be improved with good formatting or writing, but good science can certainly be negatively affected by bad writing and formatting. Use of the journal style allows good science to be reported in a standard outlet, the scholarly journal, while meeting all requirements of the graduate school for the thesis or dissertation.

Scientific Writing

A writer of scientific papers is not a writer in the literary sense (Gastel & Day, 2016). The writer of science should use simple and straightforward language to report research findings.

Do not swallow a thesaurus; keep your words simple and understandable, and use jargon minimally. We note the following sentence that appeared in an abstract:

> **Amalgamating the decision maker's inputs is a new and unique decision model that can be classified as a fuzzily parameterized, multistage, forward- and backward-chained, displaced ideal, two-dimensional attribute model of business-level strategic objectives and the functional-level strategies that realize those objectives.**

Although the words are clear, the meaning is not.

Nearly all fields of research have jargon that is used by the scientific writer. However, Gastel and Day (2016, p. 208-09) explained the use of inappropriate jargon in the following paragraph:

> **The most common type of verbosity that afflicts authors is jargon. This syndrome is characterized, in extreme cases, by the total omission of one-syllable words. Writers with this affliction never *use* anything—they *utilize*. They never *do*—they *perform*. They never *start*—they *initiate*. They never *end*—they *finalize* (or *terminate*). They never *make*—they *fabricate*. They use *initial* for *first*, *ultimate* for *last*, *prior* for *before*, *subsequent to* for *after*, *militate against* for *prohibit*, *sufficient* for *enough*, and *a plethora* for *too much*. An occasional author will slip and use the word *drug*, but most will salivate like Pavlov's dogs in anticipation of using *chemotherapeutic agent*. (We do hope that the name Pavlov rings a bell.) Who would use the three-letter word *now* instead of the elegant expression *at this point in time*?**

Another reason for straightforward scientific writing is that English is generally accepted as the language of science. Yet English is a second language to many scientists around the world. Not only do these scientists need to read and understand papers in English, but also, in most cases, they need to write their scientific papers in English.

First Things Are Sometimes Best Done Last

Research writing is often best done by not using the IMRD model in a linear way. For example, the final versions of the paper's title and abstract are typically better done last than first. Greater detail on any of the following topics can be found in Gastel and Day (2016).

Title

Although your thesis and dissertation (and the journal-style papers that are included) will need a working title, often writing the final version of the title is best done last. In today's electronic world, people will locate your paper by the title; thus, it needs to be informative about the research. However, we have often seen titles that consisted of the statement of purpose and the methods. Such a title is likely too long, whereas a three- or four-word title is likely too short. Gastel and Day (2016) defined a good title as "the fewest possible words that adequately describe the contents of the paper" (p. 41). In other words, your title should be concise but informative.

Here is an excellent title from the June 2012 issue of *Research Quarterly for Exercise and Sport*: "Quiet Eye Duration Is Responsive to Variability of Practice and to the Axis of Target Changes." This title provides the outcome of the research in 16 words. For fun (probably only for a professor), look at the thesis and dissertation titles from a recent graduation program at your institution. The diversity of research foci can be quite amazing. Here are several examples of thesis or dissertation titles that effectively communicate the research focus of each study:

- "Effect of Age and Fitness on Vascular Function and Oxidative Stress During Acute Inflammation"—The title effectively highlights independent (age, fitness) and dependent (vascular function, oxidative stress) variables under a particular circumstance (acute inflammation).

- "Immunomodulation of Influenza Infection by Echinacea and Obesity"—Even though this is a relatively short title, it clearly draws attention to how immune system response to flu virus is affected by two factors (echinacea, obesity).

- "Working Baby Boomers' Knowledge of Retiree Health Benefits and Costs"—This is a good title that conveys the purpose of the study and is understandable by an informed reader.

Always check your titles for waste phrases such as *the effects of, the influence of, a study of, an experiment on,* and *an observation of.* Typically, these phrases can be eliminated from the title without detrimental effects.

Figures and Tables

The journal papers in your thesis or dissertation will typically have tables and figures. Tables are useful for storing data, whereas figures are better for showing changes and trends, particularly as groups or subgroups change differently. Put only important tables and figures in your journal paper(s); relegate the rest to the appendix.

Tables and figures often contain average and mean values. These are useless without estimates of variability. Therefore, wherever you give a mean value, provide either the standard deviation or a confidence interval. We personally prefer standard deviations in tables and confidence intervals in figures, but that is just our preference.

References

The first rule for references is: Check your references carefully. Make sure every citation in the paper has a reference at the end of the paper. If other references are needed (e.g., for the extended literature review), put them in the appendix. Then make sure all your references are accurate. Journal citation error rates have been reported for many fields (e.g., Kristof, 1997; Rivkin, 2020; Siebers & Holt, 2000; Stull, Christina, & Quinn, 1991; Zasa, 2015). For example, Siebers and Holt reported reference error rates ranging from 4.1 to 40.3% for five well-regarded medical journals. In the physical activity field, a 47% error rate was found for the 1988 and 1989 volumes of *Research Quarterly for Exercise and Sport* (Stull, Christina, & Quinn, 1991). Zasa reported an average error rate of 12% for five well-known exercise science and sports medicine journals with error rates ranging from 2.5% to 22.5%. The problem of journal citation errors is widespread and continues to be a serious problem. It can be highly frustrating when you are unable to locate a relevant publication from a journal article reference list because the citation has not been reported accurately. Careful scholars check every reference against the text of the paper and against the original source.

Developing a Good Introduction

Your most important task is to convince the committee (whether the proposal committee, a journal reviewer, or a reviewing committee for a granting agency) that the problem is important and worth investigating. The first section of the proposal should do this, and it should also attract the reader's interest to the problem. The review of literature provides background information and a critique of the previous research done on the topic, pointing out weaknesses, conflicts, and areas needing study. A concise statement of the problem informs the reader of your exact purpose—that is, what you intend to do. Hypotheses or questions are advanced on the basis of previous research and perhaps a theoretical model.

The first part of the introduction explains the constructs that define the study and *briefly* determines the relationships between them. Generally, this section, just a few pages in length, goes from broader to narrower; when the reader gets to the problem or purpose statement, it becomes clear which variables are being studied. For example, in a study of student attitude in physical education, the introduction might first address the valuable health consequences of physical activity. Subsequent paragraphs might discuss how physical education provides and promotes physical activity, how attitude influences whether students participate in physical activity, how elementary school physical education attitudes may influence future attitudes, and how attitudes change with time. The constructs of physical education, attitude, and elementary school children are defined and their relationships are described before the discussion moves on to other sections of the introduction.

For many universities, the introduction includes a section on the significance of or rationale for the study; this is where you make the case that the study will contribute to the field. This section may address theoretical and practical reasons for the study. Many students find it relatively easy to address practical reasons, but describing how the study contributes to theory is crucial in determining whether a paper from the study can be published. Making a case for theoretical contributions requires going beyond saying that there is not a lot of literature on the topic. You must specifically state how the study will fit with and extend the current literature and theory.

Throughout the proposal, you need to provide operational definitions to explain exactly how you are using certain terms; see the Operational Definition of *Cricket* (as Explained to Us by an Australian Friend) sidebar for an example. Some departments have the tradition of a section devoted to operational definitions, but it has become more common to define terms operationally as they are used. This approach is similar to the way research papers are written and helps the reader by defining terms when they are encountered. Limitations and possible shortcomings of the study also are acknowledged by the researcher and are generally the result of the delimitations that the researcher imposes. This discussion may be completed in a separate section of the introduction. Alternatively, a sentence acknowledging limitations and possible shortcomings may be included in the methods section when samples and techniques are discussed. As a third possibility, limitations may be highlighted in the section in which results are discussed. This approach is particularly common in journal articles.

Chapter 2 of this book concerns the literature review and includes a discussion on the inductive and deductive reasoning processes used in developing the problem and formulating hypotheses. Chapter 3 covers the other parts typically required in the introduction section or chapter of a proposal.

Innumerable hours are required to create the first section (introduction) of the proposal, especially in preparing the literature review and formulating the significance of or rationale

---

Operational Definition of *Cricket* (as Explained to Us by an Australian Friend)

You have two sides: one side in the field and one out.

Each player that's on the side that's in goes out, and when out comes in, the next player goes in until out.

When they are all out, the side that's out comes in and the side that has been in goes out and tries to get those coming in out.

Sometimes you get players still in and not out.

When both sides have been in and out including the not out, that ends the game!

for the problem. You will most likely depend heavily on your advisor and on completed studies for examples of formatting and descriptions. Before we continue, however, we need to comment on the appropriate tense to use in the proposal. Since the research has not yet been done, a proposal is often presented in future tense (e.g., that so many participants will be selected and that certain procedures will be carried out). Theoretically, if the proposal is carefully planned and well written, you need only change from future to past tense to have the first two sections or chapters of the thesis or dissertation. Presenting methods in past tense in the proposal, however, would eliminate the need for this conversion. It is worth discussing with your advisor whether future or past tense is preferred for the proposal. Of course, pilot studies that have been completed should be in the past tense. To reiterate, the importance of the study and its contribution to the physical activity field are the main focus of the first section of the proposal, and this constitutes the basis for approval or disapproval.

Describing the Methods

The methods section of the proposal frequently draws the most questions from committee members in the proposal meeting. In the methods section or chapter, you must clearly describe how the data will be collected to solve the problem set forth in the first section. You must specify who the participants will be and how they will be chosen, how many participants are planned, whether any special characteristics of importance are present, how the participants' rights and privacy will be protected, and how informed consent will be obtained. Methods of obtaining measurements are detailed, and the validity and reliability of these measures are documented. Next, the procedures are described, typically in chronological order as employed in the study. If, for example, the study is a survey, discuss the steps in developing the instrument and cover letter, mailing the questionnaires, and following up. If the study is experimental, describe the treatments (or experimental programs) explicitly, the control procedures that will be exercised, and the data collection and analysis methods. Finally, explain the experimental design and planned statistical analysis of the data.

We have previously emphasized the importance of conducting pilot studies before gathering data. If pilot work has been done, it should be described, and the results reported. Often, the committee members have major concerns about whether the treatments can produce meaningful changes, whether the measurements are accurate and can reliably discriminate among participants, and whether the investigator can satisfactorily perform the measurements and administer the treatments. The pilot study should provide answers to these questions.

If the proposed study is qualitative, then it is important to address the method and tailor its presentation to the study. This is similar to what we suggested in the previous paragraph. For example, most qualitative proposal methods sections include the sample, data collection techniques and pilot-testing procedures (e.g., development, pilot studies, and revisions for semistructured interview questions), data analysis, and the ways trustworthiness and credibility (or reliability and validity) are addressed in both data collection and analysis. We present advice for preparing the qualitative proposal later in this chapter.

We recommended in chapter 4 that you use the literature to determine your methods. Answers to questions about whether certain treatment conditions are sufficiently long, intense, and frequent to produce anticipated changes can be defended with the results of previous studies.

The Proposal Process

We reiterate the contents of the proposal: the introduction (including the review of literature) and the methods to be used. The proposed purpose—in conjunction with pertinent background information, plausible hypotheses or questions, operational definitions, and delimitations—determines whether the study is worthwhile. Consequently, the first section (or chapter) is instrumental in stimulating interest in the problem and establishing the rationale for and significance of the study. The committee's decision to approve or disapprove rests primarily with the persuasiveness exhibited in the first section.

Actually, the basic decision about the merit of the topic should already have been made before the proposal meeting. You should consult with your advisor and most—if not all—of the committee members to reach a consensus about the worth of the study before the proposal meeting is scheduled. If you cannot convince the majority of the committee that your study is worthwhile, do not convene a formal proposal meeting. We strongly recommend working closely with your major professor and scheduling the proposal meeting only when they are certain that the proposal is ready and that the meeting will result in a successful conclusion. You may have a problem if your proposal is returned with a checklist similar to the one in the Interim Thesis or Dissertation Evaluation sidebar. Of course, we're kidding in suggesting such a checklist has been used.

> **To avoid problems, convince the majority of the committee of your study's worth before convening the proposal meeting.**

What to Expect of the Proposal Committee

Let us digress a moment to discuss the composition of the proposal committee, which varies from one institution to another. We can safely say that most thesis committees consist of at least three members, and most dissertation committees have at least five. The major and minor professors are included at institutions where that occurs, although the master's student is often not required to have a minor. Other members should be chosen on the basis of their knowledge about the subject or their expertise in other aspects of the research, such as design and statistical analysis. Sometimes, the institution or department specifies how many members of the committee must be from inside or outside the department. Usually, no limit is placed on the number of committee members.

Interim Thesis or Dissertation Evaluation

Dear _____ :

Greetings! I regret that my busy schedule prohibits me from rendering a detailed written evaluation of your [thesis or dissertation]. However, I have checked the appropriate actions or comments that apply to your proposal.

_____ If at first you don't succeed, try, try again.

_____ If I agreed with you, we would both be wrong.

_____ Don't sell your research methods textbook (Thomas, Silverman, Martin, and Etnier, of course); you'll need to take the course again.

_____ You were not required to write your paper in a foreign language (Burmese, or whatever it was).

_____ I couldn't read beyond the third page; one does not have to eat a whole pie to know it is bad.

_____ I hear they are hiring at Sam's Diner.

_____ You are entitled to your own opinions but not your own facts.

_____ Have you paid your tuition and fees yet? If not. . . .

_____ Please send me an email to arrange an appointment with me. Consider taking a tranquilizer before you arrive.

_____ I have been serving on [thesis or dissertation] committees for over 15 years and can now honestly say that I've seen it all.

You should work closely with your advisor in developing your thesis or dissertation proposal. The planned research should not come as a surprise to anyone associated with your work. This allows the proposal meeting to focus on how to best do the research rather than addressing whether it is a good idea. Once the proposal is approved by your advisor and the committee, it constitutes an agreement that if the research is completed as planned and written effectively, it meets the requirement for a thesis or dissertation. If adjustments are needed during the research, clear this with your advisor and the dissertation committee.

Special consideration must be given to the role of the dissertation committee when previously published and papers that are in press are included in the dissertation. The committee should be involved in the planning and execution of all work. Once the work is published, the committee has no avenue for revision; thus, their approval during the process is important. The committee will approve the entire dissertation or thesis, not just the unpublished studies. Therefore, they should be involved as the studies are conducted, completed, and prepared for publication. The student should be first author on all studies in the thesis or dissertation; the committee members and advisor may or may not be coauthors depending upon their individual contributions to the work. It is a good practice to discuss authorship during the process from proposal to submission of articles for review.

How to Prepare the Formal Proposal

The formal proposal should be carefully prepared. Typically, the proposal undergoes multiple rounds of editing, review by the major professor, and revisions before it is ready for a proposal meeting. If the proposal contains errors of grammar, spelling, or format, committee members may conclude that you lack the interest, motivation, or competence to do the proposed research. With the availability of computers and word processing software, no reason exists to present a poorly prepared proposal. Remember, however, that a spelling and grammar checker may not always identify the misuse of a correctly spelled word, as the opening lines of the poem entitled "Candidate for a Pullet Surprise" (Journal of Irreproducible Results, 39, January-February 1994, p.13) by Jerrold H. Zar illustrate:

I have a spelling checker.

It came with my PC.

It plane lee marks four my review

Miss steaks aye can knot see.

Depending on your committee's preference, you may distribute your proposal either on paper copy or in electronic format. If they require paper copies, make sure the paper and printer you use are high quality. Check that all pages are included—look at every individual page in each copy—copy machines often do not work perfectly. If you distribute your proposal through email, now the more preferred method, be certain that you've attached all materials before clicking "Send." Check that the subject line indicates what is included and that the body of your email contains all pertinent information, such as any expected time lines. This will ensure that committee members recognize what they are receiving, what is attached, and when subsequent events will occur.

Committee members should not ignore errors in proposals with the idea that the student will correct these later. Doing so may lead the student to assume that carelessness is acceptable in data collection or in the final thesis or dissertation. Copies of the proposal should be given to the committee members well in advance of the meeting. The department or university usually specifies the number of days or weeks.

What Happens at the Proposal Meeting

In the typical proposal meeting, the committee asks the student to summarize the rationale for the study, its significance, and the methodology. Good visual aids enhance this part of

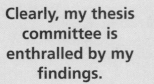

Clearly, my thesis committee is enthralled by my findings.

the presentation. The remainder of the session consists of questioning by the committee members. Practicing the presentation and answering questions in front of your major professor and other graduate students will improve your presentation and may alert you to areas that require more preparation. If the topic is acceptable, the questions will concern mainly your methods and your competence to conduct the study. Exhibit tactful confidence in presenting the proposal. A common mistake that students make is to be so humble and pliable that they agree to every suggestion made, even those that radically change the study. Your advisor should help ward off these so-called helpful suggestions, but you must also be able to defend the scope of your study and your methodology. If adequate planning has gone into the proposal, you should be able to recognize (with the assistance of your major professor) useful suggestions and defend against those that seem to offer minimal aid.

Projected schedules of procedures are sometimes required for thesis or dissertation proposals. Regardless of whether a time frame is formally required, you should establish one anyway. A common mistake is underestimating how long each phase of the study will take. Don't assume that everyone involved will drop the entirety of what they are doing to accommodate the study or that all its phases will proceed without a hitch. Instead, take a proactive approach in dealing with aspects of the study that will be more complicated and time-consuming than you'd originally anticipated. Much grief can be avoided through careful planning and realistic projections for how long the steps in your research process will take. In other words, always build in extra time to account for unforeseen discontinuities.

After the proposal is approved, most institutions treat it as a contract: The committee expects the study to be done in the manner specified in the proposal. Moreover, you can expect that if your study is conducted and analyzed as planned and is well written, it will be approved. If any unforeseen changes are required during the course of the study, your advisor must approve them. Substantial changes usually must be reviewed by some or all of the committee members.

> Avoid grief and headaches by adding time to careful and realistic projections of the expected duration of each phase.

Preparing and Presenting Qualitative Research Proposals

The content, procedures, and expectations for the proposal that we have discussed so far largely pertain to quantitative theses and dissertations. Although many similarities exist in the kinds of information presented in quantitative and qualitative research proposals, some salient differences are worth mentioning.

Preparing the proposal is much easier if all committee members are familiar with qualitative research. Today, many university faculty members have a basic understanding of qualitative research and its methodological differences with quantitative research. If proposed members of the committee are not familiar with qualitative research, problems may arise. If you are contemplating doing a qualitative study, we strongly urge you to read Locke, Spirduso, and Silverman's *Proposals That Work* (2014) or Marshall, Rossman, and Blanco's *Designing Qualitative Research* (2021).

One of the problems the qualitative researcher may face concerns potential shifts in focus and methods during the study. Earlier, we likened the proposal to a contract, whereby the committee members expect you to carry out the study as specified in the proposal. Unlike quantitative research, however, some qualitative studies undergo changes typically in the focus of the research question or the sources of data, the methodology, and the analysis of data. This aspect of qualitative research demands an open contract (Locke et al., 2014), and the methods for handling changes should be included in the proposal.

Another difference between the quantitative and qualitative proposal may be found in the literature review. In some rare cases in qualitative research, closely related studies may be purposefully omitted when planning the study because the researcher does not want to be influenced by the views and perceptions of others. Still another difference is the need for qualitative researchers to address their own values, biases, and perceptions so that they can more clearly understand the context of the research setting. This discussion often is included as a paragraph in the methods section where data collection and analysis are addressed.

All committee members, whether or not familiar with qualitative research, are interested in methodological concerns: Where will the study take place? Who will be the participants? How will the researcher gain access to the site? What data sources will be used? Does the researcher possess the needed research skills? What pilot work has been done? How will ethical concerns be handled? What strategies will be employed in collecting, sorting, categorizing, and analyzing the data?

It should be apparent by now, depending on the study, that there may be differences in the quantitative and qualitative proposal format and in the nature of the proposal process. These differences must be recognized and accepted. As with any type of research, your major professor (and, in some cases, the committee members) plays a vital role in preparing you, in showing support, and in helping you get out of tight spots. Most of all, with the support of your major professor, you must be able to assure the committee that you will carry out the study in a competent and scholarly manner.

Writing Proposals for Granting Agencies

All sources of grants, whether government agencies or private foundations, require research proposals so that they can decide which projects to fund and how much to award. Granting agencies nearly always publish guidelines for applicants to follow in preparing proposals.

A well-written proposal is everything. Researchers rarely get a chance to explain or defend the purpose or procedures. Thus, the decision is based entirely on the written proposal. The basic format for a grant proposal is similar to that of a thesis or dissertation proposal. However, some additional information is required, and certain procedural deviations may exist.

Researchers must follow the guidelines of the granting agency. Granting agencies tolerate few (if any) deviations from their guidelines. All sections of the application must be addressed, and deadlines must be strictly followed. Frequently, a statement of intent to submit a proposal is required a month or so before the proposal is filed.

Grant proposals include the following:

- Abstract of the proposed project
- Statement of the problem and its relevance to one of the granting agency's priorities
- Methods to be followed
- Time frame
- Budget
- Curricula vitae of the investigators

A limited review of literature is often required to demonstrate familiarity with previous research. Occasionally, the funding agency imposes some restrictions on design and methods. For example, an agency may prohibit the use of control groups if the treatment is hypothesized to be effective. In other words, the agency may not want anyone to be denied treatment. This limitation can pose problems for the researcher in scientifically evaluating the project's outcomes.

Because every granting agency has stipulations regarding the types of programs it can or cannot fund, a detailed budget is required, as is a justification for each budget item. A time frame is usually expected so that the reviewers can see how the project will be conducted and when the phases will be accomplished. This time frame provides information about the depth and scope of the study, justification for the length of time that support is requested, contributions of the personnel, and when the granting agency can expect progress reports.

The competence of the researcher has to be documented. Each researcher must include a curriculum vitae and often a written statement describing appropriate preparation, experience, and accomplishments. The adequacy of facilities and sources of support must also be addressed. Letters of support are occasionally encouraged. These are included in the appendix. Many agencies now require that the proposal be submitted electronically. This necessitates double-checking that all parts are prepared in accordance with the request for proposals, because after the proposal is uploaded, the funder's site will determine, for example, whether all parts have been addressed and whether the word count exceeds the number allowed. A panel of reviewers evaluates the proposal in accordance with criteria regarding the contribution to knowledge, relevance, significance, and soundness of design and methodology.

The preparation of a grant proposal is a time-consuming and exacting process. Several types of information are required, and time is needed for gathering the information and stating it in the manner prescribed by the guidelines. Applicants are advised to begin preparing the proposal as soon as the guidelines are available.

Finally, it is usually wise to contact the granting agency before preparing and submitting a proposal. Seldom are proposals funded that are submitted without some prior contact with the granting agency. Ascertaining the agency's interests and needs is a time-saving venture. A visit to, or an extended phone conversation with, the research officer is often advisable. Looking at the agency's previously funded proposals and seeking guidance from researchers who have been funded can be helpful.

Submitting Internal Proposals

Many colleges and universities (particularly larger research universities) offer internal funding for graduate student research, although these grants usually are not large. These grants typically require a two- to five-page proposal that has the support of the major professor and often the department chair. The contents usually include an abstract, a budget, and a short narrative focusing on the proposed methodology and why the research is important. Notices that internal funding is available are routinely posted or advertised around the college or university. A good place to find internal funding sources is at the office of your graduate school or vice president for research.

Completing Your Thesis or Dissertation

After your proposal meeting, you collect the data to evaluate the hypotheses that you proposed. Of course, you follow the methods that you specified in your proposal carefully and consult with your major professor if problems arise or changes need to be made. After the data are collected, you complete the agreed-upon analysis and discuss the outcomes with your major professor (and possibly some committee members, particularly if a member has statistical expertise). Then you are ready to write the results and discussion to complete your research.

Results and Discussion

The final two sections of a thesis or dissertation are the results and discussion. Results are what you have found, and the discussion explains what the results mean. More often than not, sections are separate, although they are sometimes combined (particularly in multiple-experiment papers).

How to Write the Results Section

The results section is the most important part of the research report. The introduction and literature review indicate why you conducted the research, the methods section explains how you did it, and the results section presents your contribution to knowledge—in other words, your findings. The results should be concise and effectively organized and include appropriate tables and figures.

> The results section is the most important section because it presents the study's findings.

There are several ways to organize the results section. The best approach may be to address each of the tested hypotheses; on other occasions, organizing the results around the independent or dependent variables of interest may present itself as a better option. Occasionally, you may need to document that participants adhered to specified procedures. For example, in a field-based exercise intervention study, you need data that documents adherence to the exercise prescription before addressing other outcomes. In addition, before discussing other findings, you may want to show the previously replicated standard and expected effects. For example, in developmental studies of motor performance tasks, older children typically perform better than younger ones. You may want to report the replication of this effect before discussing the other results. When looking at the effects of training on several dependent variables, you may first want to establish that a standard dependent variable known to respond to training did, in fact, respond. For example, before looking at the effects of cardiorespiratory training as potentially reducing cognitive stress, you need to show that a change occurred in cardiorespiratory response as a result of the training.

Some items should always be reported in the results. The means and standard deviations for all dependent variables under the important conditions should be included. These are basic descriptive data that allow other researchers to evaluate your findings. Sometimes only the means and standard deviations of important findings are included in the results, but all the remaining means and standard deviations should be included in the appendix.

The results section should also feature tables and figures that display appropriate findings. Accompanying each table and figure should be a meaningful title or caption that doesn't simply note what data are being presented, but rather, highlights important outcomes. Figures are particularly useful for percentage data, interactions, time-based variables, and summaries of related findings. When using figures representing group effects, always present an estimate of variability with the mean data—either standard deviations or confidence intervals are appropriate. Only the important tables and figures should be included in the results; those remaining should be placed in the appendix, if they are included at all.

Statistical information should be summarized in the text where possible. Statistics from ANOVA and MANOVA should always be summarized in the text, and complete tables should be relegated to the appendix. Make sure, however, that you include the appropriate statistical information in the text. For example, when giving the F ratio, report the degrees of freedom, the probability, and an estimate of the effect size: $F(1, 36) = 6.23$, $p < .02$, ES = 0.65. Above all, the statistics reported should be meaningful. Gastel and Day (2016, p. 73) reported a classic case that read "33 1/3% of the mice used in this experiment were cured by the test drug; 33 1/3% of the test population were unaffected by the drug and remained in a moribund condition; the third mouse got away."

Sometimes a better way to present this information is in a table. If the perfect scientific paper is ever written, it will read, "Results are in table 1." However, this suggestion does not mean that the results section should consist mostly of tables and figures. Having to thumb through eight tables and figures placed between two pages of text is disconcerting. But even worse is having to turn 50 pages to the appendix to find a necessary table or figure. Read what you have written. Are all the important facts there? Have you provided more information than the reader can absorb? Is some of the information peripheral to the questions or hypotheses guiding the study? If the answer to any of these questions is "yes," you should revise the results section.

Do not be redundant and repetitive. A common error is to include a table or figure in the results and then repeat it in the text, or to repeat data from a table in a figure. Describing tables and figures in a general way or pointing out particularly important outcomes is appropriate, but do not repeat every finding. Also, be sure that you do not call tables figures, and vice versa. As Gastel and Day (2016, p. 74) reported, some writers are so concerned with reducing verbiage that they lose track of antecedents, particularly for the pronoun *it:*

> **"The left leg became numb at times and she walked it off. . . . On her second day, the knee was better, and on the third day it had completely disappeared." The antecedent for both *it*s was presumably "the numbness," but I rather think that the wording in both instances was a result of dumbness.**

Reporting Statistical Data

A consistent dilemma among researchers, statisticians, and journal editors concerns the appropriate reporting of statistical information for published research papers (e.g., Fritz & Morris, 2012; Nuzzo, 2014; Wasserman & Lazar, 2016). In recent years, some progress has been made that involves two issues in particular: (a) reporting some estimate of the size and meaningfulness of the finding and (b) the reliability or significance of the finding. Two organizations of importance to our field—the American Physiological Society and the American Psychological Association (2020)—have now published guidelines regarding these issues. Following are summaries of general guidelines taken from these two sources:

- Information on how sample size was determined is always important. Indicate the information (e.g., effect sizes) used in the power analysis to estimate sample size. When the study is analyzed, confidence intervals are best used to describe the findings.

- Always report any complications that have occurred in the research, including missing data, attrition, and nonresponse, as well as how these problems were handled in data analysis. Before you compute *any* statistics, *look at your data*. Always screen your data (this is not tampering with data) to be sure the measurements make sense.

- Select minimally sufficient analyses; using complicated methods of quantitative analyses may appropriately fit data and lead to useful conclusions, but many designs fit basic and simpler techniques. When they do, these should be the statistics of choice. Your job is not to impress your reader with your statistical knowledge and expertise but to analyze the research appropriately and present it so that a reasonably well-informed person can understand it.

- Report actual *p* values; confidence intervals are even better. Always report an estimate of the magnitude of the effect. If the measurement units (e.g., maximal oxygen consumption) have real meaning, then reporting them in an unstandardized way such as mean difference is useful. Otherwise, standardized reporting such as effect size or r^2 is useful. In addition, placing these findings in practical and theoretical context adds much to the report.

- Control multiple comparisons through techniques such as Bonferroni.

- Always report variability using the standard deviation. Standard error characterizes the uncertainty associated with a population and is most useful in determining confidence intervals.

- Report your data at the level (e.g., how many decimal places) that is appropriate for scientific relevance.

Reporting Qualitative Data

Most of our general suggestions for preparing the results section hold for reporting qualitative data. In a qualitative study, you will report the results from the data analysis as themes or conclusions. Each of these themes or subthemes must be supported with data—from interviews, observations, or material collection. Present data to make the case that the results come from multiple sources and are clearly supported. For example, if you said that triangulation among student interviews, teacher interviews, and observations was used during data analysis, all three types of data should be included. Writing a qualitative results section may require multiple revisions so that the results seem plausible and are clear to the committee.

What to Include in the Discussion Section

Although the results are the most important part of the research report, the discussion is the most difficult to write. There are no cute tricks or clear-cut ways to organize the discussion, but the following rules define what to include:

- Discuss your results—not what you wish they were, but what they are.
- Relate your results back to the introduction, previous literature, and hypotheses.
- Explain how your results fit within theory.
- Interpret your findings.
- Recommend or suggest applications of your findings.
- Discuss any significant limitations of the methods and outcomes.
- Summarize and state your conclusions with appropriate supporting evidence.

Your discussion should point out both where data support or fail to support the hypotheses and important findings. But do not confuse significance with meaningfulness in your discussion. In fact, be especially careful to point out where they may not coincide. In particular, the discussion should point out factual relationships between variables and situations, thus leading to a presentation of the significance of the research. Of course, this is an essential place not to confuse cause and effect with correlation. For example, do not say that a characteristic had an effect or influence on a variable when you mean that they were related to each other.

The discussion should end on a positive note, possibly a summarizing statement of the most important finding and its meaning. Never end your discussion with a variation of the old standby of graduate students: *More research is needed.* Who would have thought otherwise?

The discussion should also point out any methodological problems that occurred in the research. But using a methodological cop-out to explain the results is unacceptable. *If you did not find predicted outcomes and you resort to methodological failure as an explanation, you did not do sufficient pilot work.*

Graduate students sometimes want their results to sound wonderful and to solve all the problems of the world. Thus, in their discussions, they often make claims well beyond what their data indicate. Your major professor and committee are likely to know a lot about your topic and therefore are unlikely to be fooled by these claims. They can see the data and read the results. They know what you have found and the claims that can be made. A much better strategy is to make your points effectively in your discussion and not try to generalize these points into grandiose ideas that solve humanity's major problems. Write so that your limited contribution to knowledge is highlighted. If you make broader claims, knowledgeable readers are likely to discount the importance of your legitimate findings. Your discussion should not sound like the *Calvin and Hobbes* cartoon (by Bill Watterson) in which Calvin said, "I used to hate writing assignments, but now I enjoy them. I realized that the purpose of writing is to inflate weak ideas, obscure poor reasoning, and inhibit clarity."

Another point about writing your discussion is to write so that reasonably informed and intelligent people can understand what you have found. We strongly recommend having others proofread your work and identify any passages with which they had difficulty. Do not use a thesaurus to replace your normal vocabulary with multisyllabic words and complex sentences. Your writing should not look like the examples in the In Other Words sidebar. (By translating, you can probably recognize these sentences as some well-known sayings.)

Your discussion can generally be guided by the following questions taken from the *Publication Manual of the American Psychological Association* (American Psychological Association, 2020):

- What have I contributed here?
- How has my study helped to resolve the original problem?
- What conclusions and theoretical implications can I draw from my study?

Writing the Discussion

Problem statement: Why did the chicken cross the road?

Method: One chicken observed by several individuals.

Results: Said chicken crossed the road.

Discussion: Following are the explanations given for the chicken crossing the road.

Dr. Seuss—Did the chicken cross the road? Did he cross it with a toad? Yes, the chicken crossed the road. But why it crossed I've not been told.

Sigmund Freud—I dream of a world where chickens can cross the road without having their motives questioned.

Captain Kirk—To boldly go where no chicken has gone before.

Colonel Sanders—Did I miss one?

Graduate student—Is that regular or extra crispy?

In Other Words

1. As a case in point, other authorities have proposed that slumbering canines are best left in a recumbent position.
2. It has been posited that a high degree of curiosity proved lethal to a feline.
3. There is a large body of experimental evidence that clearly indicates that smaller members of the genus *Mus* tend to engage in recreational activity while the feline is remote from the locale.
4. From time immemorial, it has been known that the ingestion of an "apple" (i.e., the pome fruit of any tree of the genus *Malus*, said fruit being usually round in shape and red, yellow, or greenish in color) on a diurnal basis will, with absolute certainty, keep a primary member of the health care establishment from one's local environment.
5. Even with the most sophisticated experimental protocol, it is exceedingly unlikely that you can instill in a superannuated canine the capacity to perform novel feats of legerdemain.
6. A sedimentary conglomerate in motion down a declivity gains no addition of mossy material.
7. The resultant experimental data indicate that there is no utility in belaboring a deceased equine.

The responses to these questions are the core of your contribution, and readers have a right to clear and direct answers. If the reader asks, "So what?" after reading your discussion, then you have failed in your research reporting.

Handling Multiple Experiments in a Single Report

Graduate students are conducting more research that involves multiple experiments. These experiments may ask several related questions about a particular problem or may build on one another, with the outcomes of the first leading to questions for the second. This trend is positive, but it sometimes leads to problems within the traditional (chapter structure) thesis or dissertation format. Chapter 22 discusses the journal and traditional formats for organizing theses and dissertations.

Multiple experiments in journal format typically involve a general introduction and literature review. If the experiments use a common methodology, then a general methods section might follow. This is followed by a presentation of each experiment with its own short introduction and the citation of a few critical studies, methods (specific to this experiment), results, and a discussion (remember, sometimes the results and discussion are combined). The report concludes with a general discussion of the series of experiments and their related findings.

Within the traditional framework, multiple experiments are probably best handled by separate chapters or major sections within the results chapter. The first chapter includes the introduction, theoretical framework, literature review, statement of the research problem, and related definitions and delimitations. Subsequent chapters describe each experiment. Each of these chapters includes a brief introduction; a discussion of the specific problem and hypotheses; and the methods, results, and discussion sections. The final chapter is a general discussion in which the experiments are tied together. It contains the features of the discussion previously presented.

Using Tables and Figures

Preparing tables and figures is a difficult task. Howard Wainer (1992) wrote one of the best papers on this topic. We begin with a quotation that he used.

> Drawing graphs, like motor-car driving and love-making, is one of those activities which almost every educator thinks can be done well without instruction. The results are of course usually abominable. (paraphrased, with my [Wainer's] apologies, from Margerison, 1965, p. 53)

Wainer suggested that tables and figures should allow the reader to answer questions at three levels:

- Basic: Extraction of data
- Intermediate: Trends in parts of the data
- Advanced: Overall questions involving the deep structure of the data (seeing trends and comparing groupings)

These levels can be thought of as an ordered effect:

1. Variables by themselves (data)
2. One variable in relation to another
3. Overall comparisons and relationships in the data

Preparing Tables

Getting information from a table is like extracting sunlight from a cucumber, to paraphrase Farquhar and Farquhar (1891). Remember, tables are for communicating to the reader, not storing data. First, consider whether you need a table. In settling on an answer, two characteristics are important: Is the material more easily understood in a table? and Does the table interfere with reading the results? If you decide that you do need a table (not all numbers require tables), then follow these basic rules:

- Like characteristics should read vertically in the table.
- Headings of tables should be clear.
- The reader should be able to understand the table without referring to the text.

Examples of Poor Tables and Good Tables

Table 21.1 is an example of a useless table; the data could be more easily presented in the text. This table can be handled in one sentence: The experimental group ($M = 17.3$, $s = 4.7$) was significantly better than the control group ($M = 12.1$, $s = 3.9$), $t(28) = 3.31$, $p < .05$. Table 21.2 is also unnecessary. Of the 10 comparisons among group means, only 1 was significant. The values in the table are the equivalent of t tests. This table can also be presented in one sentence: The Scheffé test was used to make comparisons of the age-group means, and the only significant difference was between the youngest (7-year-olds) and oldest (15-year-olds) groups, $t = 8.63$, $p < .05$.

We have borrowed an example of a useful table (table 21.3) from Safrit and Wood (1983). As you can see, like characteristics appear vertically. Also, an extensive amount of text would be required to present these results, yet they are easy to understand in this brief table.

TABLE 21.1   USELESS TABLE 1

Means, Standard Deviations, and *t* Test for Distance Cartwheeled While Blindfolded

Groups	N	M	s	t
Experimental	15	17.3 m	4.7 m	3.31*
Control	15	12.1 m	3.9 m	

*$p < .05$.

TABLE 21.2 USELESS TABLE 2
Scheffé's Test for Differences Between Age Levels in Ability to Wiggle Their Ears

Age	7	9	11	13	15
7	–	1.20	1.08	1.79	8.63*
9		–	1.32	1.42	1.57
11			–	1.58	1.01
13				–	0.61
15					–

$*p < .05.$

TABLE 21.3 USEFUL TABLE
Characteristics of Participants

	Urban	Suburban	Rural	All participants
Sex				
Female	280 (55.8%)	219 (50.7%)	101 (47.2%)	600 (52.3%)
Male	222 (44.2%)	213 (49.3%)	113 (52.8%)	548 (47.7%)
Age				
10	36 (7.2%)	14 (4.0%)	6 (2.8%)	59 (5.1%)
11	103 (20.5%)	96 (22.2%)	42 (19.6%)	241 (21.0%)
12	183 (36.5%)	127 (29.4%)	69 (32.2%)	379 (33.0%)
13	162 (32.3%)	158 (36.6%)	82 (38.3%)	402 (35.0%)
14	18 (3.6%)	34 (7.9%)	15 (7.0%)	67 (5.8%)
BMI				
Under 20	57 (11.4%)	32 (7.4%)	8 (3.7%)	97 (8.4%)
20.0-25.0	193 (38.4%)	162 (37.5%)	117 (54.7%)	472 (41.1%)
25.1-30.0	207 (41.2%)	171 (39.6%)	61 (28.5%)	439 (38.2%)
30.1 and over	45 (9.0%)	67 (15.5%)	28 (13.1%)	149 (12.2%)
Self-report as physically active				
Yes	209 (41.6%)	178 (41.2%)	109 (50.9%)	496 (43.2%)
No	293 (58.4%)	254 (58.8%)	105 (49.1%)	652 (56.8%)

Total $N = 1,148$.

Improving Tables

An important topic is how to improve tables so that they are more useful, more informative, and easier to interpret. Wainer (1992) offers the following three good rules for developing tables:

1. The columns and rows should be ordered so that they make sense. For example, the row elements are often placed in alphabetical order according to the label for the row (e.g., names, places). This approach is seldom useful. Order the rows naturally, for instance, by time (e.g., from the past to the future) or by size (e.g., put the biggest or smallest value of mean or frequency first).

2. When values go to multiple decimal places, round them off. Two digits are about the most that people can understand, that can be measured with precision, or that anyone cares about. For example, what does a $\dot{V}O_2max$ value mean when it is carried to four decimal places? We do not understand it, we cannot measure it that precisely, and no one cares. Sometimes attempts at precision become humorous, bringing to mind the report that the average U.S. family has 2.4 children. (We thought that children came only in whole units!) A whole child is the smallest (most discrete) unit of measurement available.

3. Use and pay attention to the summaries of rows and columns. The summary data, often provided as the last row or column, are important because these values (sometimes sums, means, or medians) provide a standard of comparison (or usualness). Often, setting these values apart in some way (e.g., bold type) is valuable.

Let us try these improvements on an actual table. In our research methods class, we often give students the assignment to find a table in *Research Quarterly for Exercise and Sport* and improve it by applying Wainer's suggestions. Our graduate students have not been hesitant about finding tables from our scholarly work to improve. (Professors: It is not only our work that can be improved. Offer your students one of your published figures or tables. They will improve it, too.) A good example from a paper by Thomas, Salazar, and Landers (1991) that appeared in *Research Quarterly for Exercise and Sport* was provided by James D. George when he was a doctoral student at Arizona State University (our thanks to Jim for allowing us to use his work). The example in A Fine Table Made Better—The Original Table sidebar shows the data as presented by Thomas et al. (1991). The other example in A Fine Table Made Better—The Better Way sidebar is George's rearrangement. Observing the improvements in data presentation and understanding is easy. First, the rows have been reordered by the column ES info, and all the *yes* responses are followed by all the *no* responses (it might be just as well to remove the *no* responses and list the authors' names at the bottom of the table). Then, they have been reordered from smallest to largest by the next column, *N* (sample size). Finally, an additional label (study's most important effects) was inserted under Primary ES to clarify the meaning of those three columns.

A Fine Table Made Better—The Original Table

TABLE 2

Data on Articles in Volume 59, 1988

First author	ES info	N^c	Primary ES*		
Doody	no				
Kamen[b]	yes	9	0.64	0.72	0.14
Alexander	yes	26-48	0.33	0.73*	0.39
Era	yes	5-6	0.50*	0.10	1.42*
Kokhonen	yes	9-12	−1.97*	−2.64*	−1.78*
Farrell	yes	45-368	0.77*	−0.51*	0.37*
Heinert	no				
Kamen	no[a]	10	1.14	0.81	0.90
Ober	no				
Simard	yes	7	−1.59*	0.52	−2.71*
Berger	no				
Stewart	no				
Abernethy	no				
Etnyre	no				
Nelson[b]	yes	13	0.73	1.76	0.85
Wesson	no				
Housh	yes	20	−0.53*	−2.11*	0.25*
Hutcheson	yes	34	−0.06	0.63*	−0.30*

*Comparison of Ms forming ES was significant, $p < .05$.

[a]No significant main effects.

[b]The main effect is significant, but no information is provided regarding the significance of the post hoc comparison.

[c]Per comparison group.

How do these changes relate to Wainer's three levels of questions? The revised table makes clear the sample size and ES information (basic level) for each study (but so did the original table). At the intermediate and advanced levels, however (trends, relationships, and overall structure), the revised table is a considerable improvement. For example, you can more easily observe that sample size and ES are unrelated; that is, neither studies with large samples nor those with small samples are more likely to produce larger or smaller treatment effects (as estimated by ES).

Another example of the gratuitous use of numbers is often the reporting of statistical values. Just because computer printouts carry the statistics (e.g., F, r) and probabilities (p) to five or more places beyond the decimal does not mean that the numbers should be reported to that level. Two or (at most) three places are adequate. But applying this guideline can result in rather odd probabilities: $t(22) = 14.73$, $p < .000$. Now, $p < .000$ means no chance of error; this cannot occur because if the chance of error is zero, it cannot be a probability. What happened is that the exact probability was something like $p < .00023$ and the researcher rounded it to $p < .000$. **You must not do this.** As indicated earlier, we believe that it is more appropriate to report the exact probability (e.g., $p = .025$) and whether this probability exceeded the alpha set for the experiment (e.g., $p < .05$). But the last digit in the probability must always be 1 or higher. The previous example, $p < .00023$, if reported to three decimals, should read $p < .001$.

Other numbers are frequently used without careful thought as well. In reviewing for a research journal, one of us encountered a study in which children were given a 12-week

A Fine Table Made Better—The Better Way

TABLE 2

Data on Articles in Volume 59, 1988

First author	ES info	N^c	Primary ES* (study's most important effects)		
Era	yes	5-6	0.10	0.50*	1.42*
Simard	yes	7	−2.71*	−1.59*	0.52
Kamen[a]	yes	9	0.14	0.64	0.72
Kamen	yes[b]	10	0.81	0.90	1.14
Kokhonen	yes	9-12	−2.64*	−1.97*	−1.78*
Nelson[b]	yes	13	0.73	0.85	1.76
Housh	yes	20	−2.11*	−0.53*	0.25*
Hutcheson	yes	34	−0.30*	−0.06	0.63*
Alexander	yes	26-48	0.33	0.39	0.73*
Farrell	yes	45-368	−0.51*	0.37*	0.77*
Abernethy	no				
Berger	no				
Doody	no				
Etnyre	no				
Heinert	no				
Ober	no				
Stewart	no				
Wesson	no				

[a]Per comparison group.

[b]Comparison of Ms forming ES.

[c]The main effect is significant, but no information is provided regarding the significance of the post hoc comparison.

*ES was significant, $p < .05$.

treatment. The author reported the mean age and standard deviation of the children before and after the 12-week treatment. Not surprisingly, the children had all aged 12 weeks. The author also calculated a *t* test between the pre- and the posttreatment means for age that was, of course, significant. That is, the fact that the children had aged 12 weeks during the 12-week period was a reliable finding.

Preparing Figures and Illustrations

Many suggestions about table construction also apply to figures and illustrations. A figure is often another way to present a table. Before using a table or a figure, ask: Does the reader need the actual numbers, or is a picture of the results more useful? A more important question pertains to whether you need either. Can the data be presented more concisely and easily in the text? Figures and tables do not add scientific validity to your research report. In fact, they may only clutter the results. Gastel and Day (2016, p. 103) suggested a reasonable means for deciding whether to use a table or a figure: "If the data show pronounced trends making an interesting picture, use a graph. If the numbers just sit there, with no exciting trend in evidence, a table should be satisfactory."

Several other considerations are important in preparing figures. Selection of the type of figure is somewhat arbitrary, but some distinctions make the choice of one type of figure more appropriate than another (see table 21.4).

> A table or figure should be used only if the data is not better presented in the text.

TABLE 21.4
Charts and Diagrams

	Bar and column charts—Bars (horizontal) are best for comparing amounts; arrange by size, small to large or large to small. Columns (vertical) are good for comparing amounts over time, especially if trends are evident. Shading may be used to distinguish or stack bars or columns.
	Curve graph—Best for showing change over time; time is horizontal, and quantity is vertical. Allows more than one curve to be compared. Sometimes the area between curves can be shaded to show the amount of change; shading, broken lines, symbols, or colored lines can be used to distinguish lines.
	Dot graph—Shows patterns of individual scores; each dot represents a score on both the vertical and horizontal axes. Different dots or dot symbols can be used to distinguish groups.
	Flow chart—Shows relationships in process; often useful for demonstrating steps in a process when more than one option exists (e.g., if *yes*, then *this*; if *no*, then *this*).
	Pie chart—Circle for pie equals 100%; use a maximum of six segments. Best for showing proportions of segments. Order segments from large to small, starting at 12 o'clock; highlight segments with shadings, making the smallest segment the darkest.
	Schematics—Show relationships between variables or concepts (e.g., overlap in two correlated variables).

Developed from J.V. White, *Using Charts and Graphs: 1000 Ideas for Visual Presentation* (New Providence, NJ: R.R. Bowker Co. 1984).

To evaluate whether you have used a figure appropriately, make sure that it

- does not duplicate text,
- contains important information,
- does not have visual distractions,
- is easy to read and understand,
- is consistent with other figures in the text, and
- contains a way to evaluate the variability of the data (e.g., standard deviation bars or confidence intervals).

Figures are useful for presenting interactions and data points that change over time (or across multiple trials). The dependent variable is placed on the *y*-axis, and some independent or categorical variable is placed on the *x*-axis. If you have more than one independent variable, how do you decide which to put on the *x*-axis? We have already partially answered that question. If time or multiple trials are used, put them on the *x*-axis. For example, if a study found an interaction between the dependent variable by age level (7-, 9-, 11-, 13-, and 15-year-olds) and the treatment (experimental versus control), age with five levels is usually the more appropriate choice for the *x*-axis. Note that this is a general rule; specific circumstances may dictate otherwise. A good example of the use of a figure to present an interaction is shown in figure 21.1. Note that both age and time are independent variables, so time is placed on the *x*-axis.

Figure 21.2 is a good example of a useless figure. The results can be summarized in two sentences: "During acquisition, all three groups reduced their frequency of errors but did not differ significantly from one another. At retention, experimental group 2 further reduced its number of errors, whereas

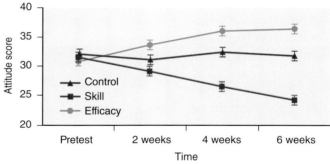

Figure 21.1   Appropriate use of a figure to depict an interaction.

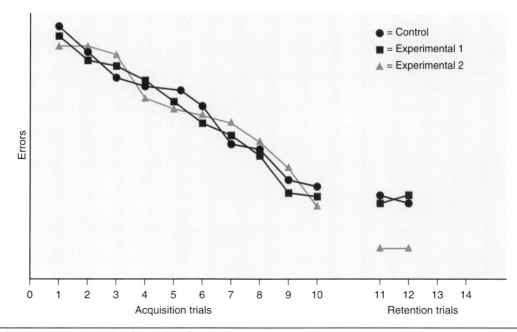

Figure 21.2   Example of a useless figure.

experimental group 1 and the controls remained at the same level." When results of different groups follow a similar pattern, a figure frequently appears cluttered. If figures appear cluttered because standard deviation bars or confidence intervals overlap, try putting the bars from one group going upward on the graph and the bars from a group in close proximity downward.

Figures should include error bars representing the variability of each mean data point that is shown. The error bars can be either standard deviations for each mean value (e.g., figure 21.1) or 95% confidence intervals (CI) for each mean value. An advantage of using the 95% CI is that when the 95% CIs for mean values do not overlap on the *x*-axis at the same point, the means are significantly different. This circumstance allows the effective interpretation of significant interactions when they appear as a figure. Standard errors should not be used as the representation of variability because they represent the variability of sample means about the true population mean, not variability in an individual sample. The most appropriate use of standard error is to calculate the 95% CI.

One final consideration is the construction of the *y*-axis. In general, use between 8 and 12 intervals to encompass the range of values. To avoid wasting space, do not extend the *y*-axis outside the range of values. Again, consider whether you need a figure at all. Sometimes, theses and dissertations include examples such as that in figure 21.3. Looking at that figure, you immediately see a strong and significant interaction between knowledge of results (KR) and goal setting. Now look at the *y*-axis, on which the dependent variable is shown. Note that the scores are given to the nearest hundredth of a second. There is a difference of less than 500 ms among the four groups on a task in which average performance is about 18 s (18,000 ms). In fact, this interaction is not significant and clearly accounts for little variance. The

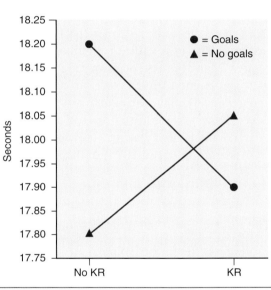

Figure 21.3 A nonsignificant interaction made to appear significant by the scale of the dependent variable.

result should merely be reported as nonsignificant with no figure included. The researcher made the interaction appear important by the scale used on the *y*-axis.

Sometimes multiple figures reporting data across several independent variables are used. For example, a comparison of running speed of Chinese, African, and U.S. boys and girls in five age groups might be reported. A single figure would be cluttered with all this data, so several figures (one for each continent of origin, each age, or each sex) might be used. The *y*-axis on each figure should have the same 8 to 12 points for running speed. Otherwise, visual comparisons of the data are difficult.

With current computer technology and software, producing figures from data takes a matter of seconds. Try different formats for figures to see which looks the best and displays the data most effectively. Put data in a bar graph and then in a line graph and compare the two. Be creative; you can really improve your presentation of results by using tables and figures effectively.

Illustrations (photographs and line drawings) are also used in research reporting. Most frequently, illustrations are of experimental arrangements and equipment. They should not be used when the equipment is of a standard design or make; a brief description suffices. Any unusual arrangement or novel equipment should be described and either a picture or line drawing included. If specifications and relationships are important to include within the illustration rather than in the text, a line drawing is preferable because it can be more easily labeled.

Remember that tables, figures, and illustrations are used to present results, and therefore are appropriate for the results chapter but not for the discussion. An exception to this rule is a report of multiple experiments, in which the general discussion section could contain a table or figure to display common findings or a summary of several experiments.

To review, when determining whether tables, figures, and illustrations should appear in the text or in the appendix, consider the following recommendations:

- Put important tables, figures, and illustrations in the text and all others in the appendix.
- Try not to clutter the results with too many tables, figures, and illustrations.
- Do not put summary tables for ANOVA and MANOVA in the results. Place the important statistics from these tables in the text and put the tables in the appendix.

Remember that all journals have prescribed formats and styles for articles submitted to them (e.g., those of the American Psychological Association or American Physiological Society). The instructions for authors usually include directions for preparing tables and figures. Many of these decisions are arbitrary. Read and look at what you have written and then use common sense. Select the tables, figures, and illustrations that are needed for reading and understanding the results. Everything else goes in the appendix. The appropriate use of tables, figures, and illustrations can add to readers' interest, understanding, and motivation. We recommend that you get a copy of the *Publication Manual of the American Psychological Association* (APA, 2020) or the style manual used in your department.

Summary

The research proposal describes the definition, scope, and significance of a problem and the methodology that will be used to study it. The proposal is essentially the plan for the study. The introduction provides the background and literature on the problem, the problem statement, hypotheses, definitions, assumptions and limitations, and the significance of the study. The methods section describes the participants, instrumentation, procedures, and design and analysis.

These two sections are presented to the research committee as the plan for inquiry. The committee determines the worth of the study, suggests needed alterations, and ultimately must agree that the study should be done. The proposal must be carefully prepared with appropriate pilot work so that the committee is convinced that you can complete the research plan.

Proposals to granting agencies are similar but are typically required to have specific lengths and formats. Talk with outside agencies and more experienced grant writers before preparing and submitting proposals. Many colleges and universities offer internal grants for which graduate students can apply to support thesis and dissertation research.

The results and discussion sections are written after the data have been collected and analyzed. The results tell what you found; the discussion explains what the results mean. The results are the most important part of the research. They represent the unique findings of your study and your contribution to knowledge. The discussion ties the findings back to the literature review, theory, and empirical findings from other studies. Your interpretation of findings should be in the discussion. In journal format, the results and discussion are sections in the body of the paper (more on this in chapter 22). In the traditional format, the results are chapter 3 and the discussion is chapter 4. In multiple-experiment studies, however, each experiment may best be reported in a separate section (or chapter) with its own brief introduction, methods, results, and discussion. A section (or chapter) of general discussion of the multiple experiments often follows.

Tables should be used in the results to summarize and present data when they are more effective than text presentations. Figures and illustrations are also used in the results, most frequently to demonstrate more dramatic findings. Careful construction of tables and figures is important in ensuring that information is well communicated.

✓ Check Your Understanding

1. Learn the steps for writing a thesis or dissertation proposal at your school. List them in chronological order (e.g., select a major professor and committee, prepare a proposal, and get the proposal approved). Explain the process at each step.

2. Select a research report of interest from a refereed journal in your area of interest. Read the paper, but concentrate on the results and discussion. Answer the following questions in a brief report.

   a. Results:

      How is the results section organized?

      Compare the order of reported findings with the introduction, literature review, and statement of the problem. What, if any, relationships do you see?

      How else might the results have been organized? Would a different scheme have been better or worse? Why?

   b. Tables, figures, and illustrations:

      Are there any? How many of each?

      Why are they used?

      Could the data have been reported more easily in the text?

      When either a table or figure is used, would the other have been as good? Better? Why?

      Identify ways to restructure one of the tables or figures to improve it.

   c. Discussion:

      How is the discussion organized?

      Are all results discussed?

      Is the discussion accurate in terms of the results?

      Are previous findings and theory woven into the discussion?

      Are all conclusions and supporting evidence clearly presented?

      Did the authors use any methodological cop-outs?

Ways of Reporting Research

If I agreed with you, we'd both be wrong.

—Onelinerz.net

We feel that the following quotation from our favorite authors, Gastel and Day (2016, p. xv), provides an appropriate introduction to this chapter on ways of reporting research:

> A scientific experiment, no matter how spectacular the results, is not completed until the results are published. In fact, the cornerstone of the philosophy of science is based on the fundamental assumption that original research *must* be published; only thus can new scientific knowledge be authenticated and then added to the existing database that we call scientific knowledge.

In chapter 21, we covered the research proposal: how to write the introduction, the literature review, the problem statement, and the methodology for the thesis or dissertation. Then we explained how to organize and write the results and discussion sections. Effectively coordinating all this information into a thesis or a dissertation is this chapter's focus. We present both the journal style of organization, which we advocate (Thomas, Nelson, & Magill, 1986), and the traditional chapter style. In addition, we include information about writing for publication in research journals, preparing abstracts, and presenting papers orally (including in a poster format).

Basic Writing Guidelines

A thesis or dissertation demonstrates that you can write in an acceptable scientific style. However, because it is useful to determine the expectations of your advisor, department, and university, you should do the following:

• Collect all the documents that outline university, graduate school, and department policy for theses and dissertations. Then actually read these documents because someone at some level will eventually check to see whether you have followed them. Pay particular attention to graduate school and department deadlines that impact the scheduling of your defense, submission of your final document to the graduate school, and ultimately the term in which you officially graduate.

- Review the theses and dissertations of past graduate students whose work is well regarded by your institution. Identify common elements in their work and pattern your work after theirs.

- Allow twice as much time as you think you will need. Remember Murphy's law, "Whatever can go wrong will go wrong," and its special corollary, "When several things can go wrong, the one that will go wrong is the one that will cause the greatest harm."

The thesis and dissertation have basically the same parts found in any scientific paper: an introduction and a literature review and methods, results, and discussion sections. In the traditional format, each part becomes a chapter. Sometimes the introduction and the literature review are separate chapters, leading to five, rather than four, chapters. However, the model of introduction and literature review, methods, results, and discussion (IMRD) is pretty standard for scientific theses and dissertations. Qualitative, historical, and philosophy papers often vary considerably from this format. We advocate for the use of journal style in theses and dissertations because it results in a paper ready for journal submission. Nearly all universities allow the use of this style, but you should check with your graduate school office.

A Brief Word About Acknowledgments

acknowledgments— The section of a scholarly paper that credits people important to the development of the work.

A section seldom mentioned in articles and books on preparing theses and dissertations is the **acknowledgments**. You should acknowledge the people without whom the research would have been impossible. Acknowledgments prepared by graduate students can take some odd turns. We saw one in which a student acknowledged her ex-husband. She said that if he had not been so difficult to live with, she would never have gone back to the university and done graduate work. She found an area in which she had a tremendous interest, and her life changed significantly. That is about the only positive statement that we have ever heard about an ex-spouse. Some other amusing acknowledgments follow:

- My parents, husband, and children provided inspiration and support throughout, but I was able to complete my thesis anyway.

- My major professor, Dr. I.M. Published, coordinated the work and made an occasional contribution.

- Professor B.A. Snobb wants everyone to know he had nothing to do with this thesis.

- Finally, we would like to thank our proofreader, I.D. Best, without whose kelp; thes document wld net be pausible.

More seriously, you can acknowledge appropriate people, but keep the list short and to the point and do not be overly sentimental. As in other areas, use proper English. In their acknowledgments, graduate students often use *wish* when they mean *want*, as in *I wish to acknowledge I.B.A. Fink*. Does that mean they might have acknowledged him if his contribution had not been so lousy? The graduate student really wants to acknowledge him. But more appropriately, why not just say it directly: *I acknowledge I.B.A. Fink*.

Thesis and Dissertation Format: Traditional Versus Journal

We devote considerable space here to the journal format for thesis and dissertation writing. Our material is partially reproduced from an article two of us wrote with a colleague (Thomas, Nelson, & Magill, 1986). We acknowledge the contribution of Dr. Richard Magill from Teachers College, Columbia University, and we appreciate *Quest* (published by Taylor & Francis) for allowing us to adapt the article for this book.

Graduate education in the United States, especially at the doctoral level, was modeled after graduate education in Germany. The German university's spirit of the search for knowledge and the concomitant emphasis on productive research were transplanted in large measure in America (Rudy, 1962). Although some changes have occurred over the years in the requirements for the doctoral degree, the basic aims and expectations have in essence remained unaltered. In most cases, the doctoral degree is conferred in recognition of a candidate's scholarship and research accomplishments in a specific field of learning through an original contribution of significant knowledge and ideas (Boyer, 1973). Some universities now have alternatives to the traditional dissertation in which the production of new research is the goal for those pursuing the doctorate for professional practice (e.g., an evaluation study with suggestions to improve practice based on research). These types of dissertations are appropriate for those who will not do research after graduating with their degrees. Our suggestions here, however, are focused on dissertations and theses in which the reporting of research is the focus.

The process of becoming an independent scholar, a primary goal of doctoral training, requires extensive engagement in the research process. This usually begins with the student being integrated into the research program and projects of their mentor and is culminated with the dissertation. Stated simply, research is the foundation of the doctoral program, and the dissertation is the most distinctive aspect of the doctoral degree. It has been reported that the dissertation of U.S. graduate students occupies an average of 39% of the time devoted to obtaining a degree in the fields of biochemistry, electrical engineering, psychology, physics, sociology, and zoology (Porter, Chubin, Rossini, Boeckmann, & Connally, 1982). While the goals and needs of doctoral students have become more diverse over the past 50 years, gaining research experience and developing independent scholars remain important emphases of doctoral training (Gardner, 2008).

A Case for the Journal Format

The major purpose of the thesis and the dissertation is to contribute new knowledge with scientific merit (American Psychological Association, 2020; Berelson, 1960; Porter & Wolfe, 1975; Robinson & Dracup, 2008). A purpose typically cited in university bulletins is that the thesis or dissertation provides evidence of competency in planning, conducting, and reporting research. In terms of program objectives, the study is a valuable learning experience in that theses and dissertations are functional exercises in executing the steps in the scientific method of problem-solving. Even ABDs—the "all but done" people who have completed all work for the PhD except the dissertation—acknowledge the contribution that dissertations make to science and the scientific method (Jacks, Chubin, Porter, & Connally, 1983).

The importance of disseminating results as part of the research process is well established. It follows that one of the purposes of the thesis or dissertation is to serve as a vehicle to carry the results of independent investigation undertaken by the graduate student. Thus, the dissertation or thesis becomes part of the dissemination process.

Despite the potential contribution that the thesis and dissertation can make to a field of study, the fact remains that no more than about one-quarter to one-half of the dissertations (and even fewer theses), depending on the field of study, become available to the profession through publications (Anwar, 2004; Evans et al., 2018; Lee & Kamler, 2008; Porter et al., 1982; Silverman & Manson, 2003). Several reasons may account for why a thesis or dissertation is never published. For one, despite the emphasis placed on research by the institutions, many students do not consider research an important aim, especially those who enter graduate programs focused on professional practice. A.L. Porter and colleagues (1982) reported that 24% of the doctoral recipients surveyed expressed this feeling. Arlin (1977) went so far as to claim that most educators never do another piece of published research after they complete the master's thesis or doctoral dissertation. In addition, job placements fix varying degrees of importance on research and publication. Another inescapable fact is that not all theses and dissertations are worthy of publication.

Conceivably, another contributing factor in the low rate of publication is the traditional style and format often used in the thesis or dissertation. Granted, the highly motivated new PhD recipient will spend the time and effort to rewrite the dissertation into the proper format (IMRD) for journal review, but the less-motivated PhD recipient may not. Why should dissertations and theses be written in a format that requires rewriting before publication if a vital part of the research process is publication of the research study?

We contend that the traditional format of dissertations and theses is archaic. Doctoral students (especially those working with productive major professors) may have published in several scholarly refereed publications by the time the PhD is awarded. Should such people, who have shown that their previous work has scientific merit, be required to go through the ritual of writing a dissertation involving separate chapters that spell out every detail of the research process? It seems more logical to have the body of the dissertation prepared in the appropriate format and style for submission to a journal, which is the acceptable model for communicating results of research and scholarly works in many of the arts, sciences, and professions. We explain the contents of a thesis or dissertation prepared using a journal format in this chapter. Professionals in other fields have also been advocating for journal style (e.g., Duke & Beck, 1999; Smaldone et al., 2019).

Limitations of Chapter Style

Conventional theses and dissertations typically contain four or five chapters. Traditionally, the chapters are intended to reflect the scientific method for solving problems: developing the problem and formulating the hypotheses, gathering the data, and analyzing and interpreting the results. These steps are usually embodied in the introduction (which sometimes contains the literature review; at other times the literature review is a separate chapter), methods, results, and discussion chapters.

The thesis or dissertation also has several introductory pages as prescribed by the institution, usually consisting of the title page, acknowledgments, abstract, table of contents, and lists of figures and tables. The references and one or more appendixes that contain items such as participant consent forms, tabular materials not presented in the text, more detailed descriptions of procedures, instructions to participants, and raw data are at the end of the study. A brief biographical sketch (the curriculum vitae) is usually the last entry in the conventional thesis or dissertation.

The chapter format is, of course, steeped in scholastic tradition. In defense of this tradition, developing the discipline required to accomplish the steps involved in the scientific method is usually viewed as an educationally beneficial experience. Moreover, for the master's student, the thesis is normally the first research effort, and there may be merit in formally addressing such steps as operationally defining terms, delimiting the study, stating the basic assumptions, and justifying the study's significance.

A more serious limitation of the chapter style relates to the dissemination of the results of the study; that is, publishing the manuscript in a research journal. Considerable rewriting is usually required to publish a thesis or dissertation because the journal format differs from the chapter format. Granted, the information for the journal article is provided in the thesis and the dissertation, but a considerable amount of deleting, reorganizing, and consolidating is necessary to transform the study into a journal article.

Students usually want to publish this product of months (or years) of time and painstaking effort. But in terms of expediency, the chapter format is decidedly counterproductive. The rewriting required may be made more difficult by the fact that the typical new PhD recipient immediately begins a new job that demands considerable time and energy. Unless the person is truly motivated, the transformation of the conventional dissertation into journal format may be delayed—sometimes indefinitely. A.L. Porter and colleagues (1982) claimed that new PhD recipients who fail to publish within two years of the awarding of the degree are unlikely to publish later.

The master's thesis is even more unlikely to be published. One reason is that the master's student generally does not consider publishing unless the major professor suggests it. Furthermore, the master's student may not be as well prepared to write for publication as the doctoral student is. Thus, the burden for publication is on the major professor, who understandably is often unwilling to spend the additional time necessary to supervise (or do) the rewriting. Thus, most theses are not submitted for publication. Ironically, the time-honored scholarly chapter style of the thesis and dissertation impedes an integral part of the research process: the dissemination of results.

Structure of the Journal Format

The journal format overcomes the limitations of the traditional chapter format while maintaining the contents of a complete research report. The journal format has three major parts: preliminary materials, main body, and appendixes. Preliminary materials include the title page, table of contents, acknowledgments, and abstract. The body of the thesis and dissertation is a complete manuscript prepared in journal form. Included are the standard parts of a research report: the introduction, methods, results, discussion, references, figures, and tables. The appendixes often include a more thorough literature review, additional detail about methods, and additional results not placed in the body of the manuscript.

Journal style (IMRD) in theses and dissertations is appropriate when the goal is publication as an article in a journal, which should be the goal for all scientific studies. There are exceptions, however. The traditional chapter format may be appropriate for studies of the history or philosophy of sport because the likely publication would be in book form. Some modification of the IMRD format may be best for areas such as the sociology of sport as well.

Following are the parts for quantitative studies using the journal format:

1.0 Preliminary Materials

1.1	Title page
1.2	Acknowledgments
1.3	Abstract
1.4	Table of contents
1.5	List of tables
1.6	List of figures

2.0 Body of the Thesis or Dissertation (IMRD)

2.1	Introduction
2.2	Methods
2.3	Results
2.4	Discussion
2.5	References
2.6	Tables
2.7	Figures

3.0 Appendixes

3.1	Extended literature review
3.2	Additional methodology

4.0 One-Page Curriculum Vitae

How can this format overcome the limitations of the chapter style? For both master's and doctoral students, a manuscript is developed that is ready for journal submission. All that remains is to add the title page and abstract, and the paper can be sent to a suitable journal.

The advantage of the journal format for doctoral students should be apparent. Because PhD recipients who fail to publish their dissertations within two years are unlikely to publish afterward, a more functional format encourages publication. Especially when we consider that dissertations appear to make important contributions to knowledge, the evaluation and subsequent publication of that knowledge through refereed journals is an important step. Although master's theses are not as likely as dissertations to be published, any format that encourages the publication of quality thesis work is desirable.

We want to make one final point before proceeding to the structure of the journal format: Your graduate school probably requires that the thesis or dissertation follow a standard style manual (or at least the style of a journal). The three most common are the *Publication Manual of the American Psychological Association* (American Psychological Association, 2020), American Physiological Society style (Curran-Everett & Benos, 2004), and *The Chicago Manual of Style* (University of Chicago Press, 2017). The journal format adapts nicely to any of these styles. University regulations usually do not specify a particular style, but frequently, an academic department may adopt one or two styles. If the journal format is to be used, a department might want to allow more than one style. For example, many journals reporting exercise physiology and biomechanical studies use the American Physiological Society style. Journals publishing articles in motor behavior, sport psychology and sociology, and professional preparation frequently use the APA manual. Journals that publish articles on the history and philosophy of sport frequently use the CMS manual. Graduate students benefit considerably by having the flexibility to choose the style recommended by the journal to which the paper will be submitted.

Preliminary Materials

Most of the information in the preliminary materials is required by the institution and usually appears at the beginning of the thesis or dissertation. The exact nature of this material is often specified by an institution's graduate school, so make sure you have the necessary information. Journals (and the style manuals mentioned previously) typically require abstracts between 100 and 150 words. More specifically, we recommend that you consider which journal you will target for publication of your research and use that journal's requirement for abstract length. (We provide further discussion on abstracts later in this chapter.)

Body

The length of the body should be kept within the bounds set by the journal or the institution. Journals typically use word count for setting maximum length of a submission. You and your major professor must be rigorous in keeping the paper within a length that a journal will consider. In our experience, maintaining an acceptable length is highly challenging and requires careful and concise scientific writing. If the dissertation has multiple parts and will be published as more than one paper, you would include two or more reports in this section.

You should consult several sources in preparing the body. One is the journal to which submission is anticipated. Guidelines to authors and instructions regarding the appropriate style manual are typically published annually in journals. Read a number of similar papers in the selected journal to see how specific issues are handled (tables, figures, unusual citations, multiple experiments).

Typical Scientific Studies Format Reported in Journal Style

As we have suggested, good scientific style is often labeled IMRD—introduction (including the literature review), methods, results, and discussion. Here is the format for journal style:

- Introduction and brief literature review—where you catch the reader's attention and say what the study is about, why it is important, what closely related research tells you, and what your hypotheses are
- Methods—what you did and how you did it
- Results—your contribution to knowledge and the details of what you found
- Discussion—what it means and why it is important
- References—only those cited in the paper
- Tables—important data; rest goes in the appendix
- Figures—important trends and changes; rest goes in the appendix
- Appendixes—typically include a comprehensive literature review, additional methodological procedures, other tables and figures, and any other material the major professor and committee require

Multiple Studies Format

Many graduate students, particularly doctoral students, have a continuous line of research with their advisor, which is a desired result in good and productive graduate student–advisor relationships. The dissertation should not be the first research the graduate student has completed; instead it should be one step along the way of multiple research studies and publications. In fact, the dissertation may include multiple studies and therefore articles for publications.

Some graduate schools have specific formats for multiple studies. Most require that students be first authors on any papers included in the thesis or dissertation. Following are the parts of the thesis or dissertation:

- Standard graduate school materials
- Chapter or section I—introduction to the series of studies
- Multiple chapters or sections—one for each study using the single-study format (IMRD) previously given
- Final chapter or section—overview and summary of the findings from all the studies
- Appendixes—extended literature review, additional methods, additional results, other items required by the major professor or committee

In this case, each chapter or section for a study is a stand-alone paper that includes all the standard parts previously listed. Each paper is ready to submit to a journal.

Series of Related Studies

If the paper reports a series of related studies (e.g., experiment 1, experiment 2), they may be included as parts of a single paper that is ready to go to a single journal. Typically, no introduction or final summary is required as a separate chapter or section because this is already part of the paper itself.

Series of Studies: Some Published, Some Not Published

Another situation that may occur is the use of published and in-press papers in the dissertation. This is expected when the graduate student has been doing research and writing papers over several years of work with the advisor, especially when the studies are related. In this instance, the

(continued)

(continued)

dissertation might include one or more published papers, one or more in-press papers, and one or more studies just completed. Most graduate schools allow this model as long as the student is first author on all the papers. However, a specific format may be required, such as the following:

- Chapter or section I—introduction to the series of papers
- Chapter or section II—previously published work
- Chapter or section III—in-press work
- Chapter or section IV—recently completed studies
- Chapter or section V—summary of studies
- Appendixes—same as previously discussed

Adapted from J.R. Thomas, J.K. Nelson, and R.A. Magill, "A Case for an Alternative Format for the Thesis/Dissertation," *Quest* 38, no. 2 (1986): 116-124.

Appendixes

In the traditional format, the appendixes serve primarily as a depository for nonessential information. To a degree, this notion still characterizes what should be put in the appendixes when the journal format is used. However, some additional features give appendixes unique worth. Usually, the number and content of the appendixes are determined by the agreement of the student, the advisor, and the supervisory committee. In the journal format, we suggest four types of appendixes. Other appendixes could be included, but we see these four as the most common: comprehensive literature review, additional methods information, additional results information, and other materials.

Each appendix should begin with a description of what is in it and how that information relates to the body of the thesis or dissertation. This description enables the reader to get the most use of the information in the appendix.

COMPREHENSIVE LITERATURE REVIEW   An important and useful appendix is a comprehensive literature review. The introduction in the body of the thesis or dissertation includes a discussion of related research but presents only minimal information to establish an appropriate background for the one or more studies that follow. One purpose of the thesis or dissertation is to provide an opportunity to demonstrate knowledge of the research literature related to the topic. In the body of the journal format, limited opportunity is available for this demonstration because journals typically have concise introductions. A comprehensive literature review should be included as the first appendix to provide an appropriate mechanism for demonstrating knowledge of the relevant literature. We recommend that this review also be written in journal style.

At least two additional purposes are served by including this literature review as an appendix. First, the information is made available to committee members, some of whom may not be sufficiently familiar with the literature related to the thesis or dissertation. Second, if properly developed, the literature review can go directly to a journal for publication.

It can take several forms, the most popular of which is probably the comprehensive narrative that synthesizes and evaluates research. This review links studies and establishes a strong foundation on which to build the research. It also reveals how the research represented by the thesis or dissertation extends the existing body of knowledge.

A second form is the meta-analysis: a quantitative literature review that synthesizes previous research by analyzing the results of many research studies using specified statistical methods (see chapter 14). An example of a meta-analysis related to the study of physical activity can be seen in an article by Thomas and French (1985).

ADDITIONAL METHODS INFORMATION   A second important appendix presents additional methods information not included in the body of the thesis or dissertation. Journals encourage authors to provide methods information that is brief yet sufficient to describe essential details related to the participants, apparatuses, and procedures used in the research. Information suitable for this section includes more detailed participant characteristics; more comprehensive experimental design information; fuller descriptions (and perhaps photographs) of tests, testing apparatuses, or interview protocols; copies of tests, inventories, or questionnaires; and instructions given to participants.

ADDITIONAL RESULTS INFORMATION   This appendix should include information that is not essential for the results section. Journal editors typically want only summary statements about the analyses and a minimal number of figures and tables. Thus, considerable material that is not included in the results section can be placed here, such as means and standard deviations, ANOVA tables, multiple-correlation tables, validity and reliability information, and additional tables and figures.

This information serves several purposes. First, it provides evidence to the committee and your advisor that you accurately described the data and properly performed the analyses. Second, the committee members have an opportunity to evaluate the statistical analyses and interpretations from the body of the thesis or dissertation. Third, other researchers have access to more detailed data and statistical information should they want it. For example, you may want to include this thesis or dissertation in a meta-analysis. Because the additional results presented in the appendix may have many future uses, their importance to the total thesis or dissertation cannot be overemphasized.

OTHER MATERIALS   This appendix includes information such as the approval form from the human research committee, participants' informed consent forms, sample data-recording forms, and perhaps the raw data from each participant. Also, detailed descriptions of any pilot work done before the study could be included in this or a separate appendix.

One-Page Curriculum Vitae

Many colleges and universities request that the last page of the thesis or dissertation be a one-page curriculum vitae about the student. This document should be a professional version of a curriculum vitae and include items such as education, previous work experience, and indicators of research productivity (e.g., presentations, publications).

Example Theses and Dissertations Using Journal Format

Numerous theses, such as Farren, 2014; and Glidden, 2010, and dissertations, such as DeShaw, 2019; Scrabis-Fletcher, 2007; and Zaman, 2020, use journal format. These theses and dissertations provide excellent examples of the format discussed here.

Helpful Hints for Successful Journal Writing

We highly recommend the eighth edition of a work to which we have frequently referred: *How to Write and Publish a Scientific Paper* (Gastel & Day, 2016). In our opinion, this book is the researcher's best resource for preparing a paper for submission to a research journal. The book is short, informative, and readable, with doses of humor speckled throughout. Although the small section here cannot replace Gastel and Day's more thorough treatment, we do offer a few suggestions.

First, decide to which journal you will submit your research paper. Carefully read its guidelines and follow the recommended procedures. Be sure to select a recent issue because the guidelines may have changed from older issues. Guidelines usually explain the publication style of the journal; procedures for preparing tables, figures, and illustrations; the process for submitting manuscripts; acceptable lengths; and sometimes estimated review time. Nearly all journals require that manuscripts not be submitted elsewhere simultaneously. To do so is unethical (see chapter 5).

Journals follow a standard procedure for papers submitted. Most now use electronic submissions. One of us (Thomas) was editor-in-chief of *Research Quarterly for Exercise and Sport* for six years (1983-1989); and another (Silverman) for three years (2002-2005). Here is what happens to a manuscript between submission to the editor and return to the author (the average time from submission to the first post-review editorial decision for *Research Quarterly for Exercise and Sport* is 65 days):

1. The editor checks to see whether the paper is within the scope of the journal, is the appropriate length, and uses correct style. Failure to meet these characteristics may result in the editor rejecting your submission.

2. The editor looks through the paper to determine that all appropriate materials are included (e.g., abstract, tables, and figures).

3. The editor reads the abstract, looks at the keywords, and evaluates the reference list to identify potential reviewers.

4. Depending on the size of the journal, there may be numerous sections overseen by associate editors. For example, *Research Quarterly for Exercise and Sport* has five sections, each overseen by multiple associate editors. The editor places the paper in the appropriate section or with an appropriate associate editor.

5. The editor, sometimes in consultation with section or associate editors, assigns usually two or three reviewers.

6. The reviewers and section editor (or associate editor) receive an email informing them that the paper is posted electronically on a particular site. They have a specific date by which they should complete the review.

7. The reviewers submit reviews on the journal's website to the section editor (or associate editor), who evaluates the paper and the reviews and makes a recommendation to the editor.

8. The editor reads the paper, the reviews, and the evaluations and writes the author concerning the status of the paper. Usually, the editor outlines the major reasons for the decision.

Smaller, more narrowly focused journals may not have section or associate editors, so reviews go directly to the editor. Publication decisions usually fall into three or more general categories (e.g., see table 22.1, which shows the number of decision categories of numerous kinesiology journals), such as the following:

- Acceptable (sometimes with varying degrees of revision)
- Unacceptable (sometimes called *rejected without bias*), which means that the editor will have the paper evaluated again if it is revised
- Rejected, which usually means that the journal will no longer consider the paper

Review criteria are fairly common and are similar to those presented in chapter 2 for a review of published literature. Ratios of published papers to rejected papers vary by the quality of the journal and the area of the paper and are frequently available in an issue of the journal. *Research Quarterly for Exercise and Sport* accepts about 12% of the papers submitted.

If this is your first paper, seek some advice from a more experienced author, such as your major professor or another faculty member. Papers often are rejected because they fail to provide important information. More experienced authors pick up on this immediately.

TABLE 22.1

Peer-Review Procedures in Kinesiology Journals

Journal	Blinding	Criteria*	Decision categories
British Journal of Sports Medicine	Author	3 (e.g., ethics in research and publication)	4
European Journal of Sport Science	Double	6 (e.g., impact, originality, rank)	4
Journal of Applied Biomechanics	Author	9 (e.g., clear question, new insights)	5
Journal of Motor Behavior	Double	6 (e.g., varying levels of analysis)	4
Journal of Orthopaedic & Sports Physical Therapy	Double	5 (e.g., importance or interest)	6
Journal of Physical Activity and Health	Double	4 (e.g., study design and interpretation)	3
Journal of Sport & Exercise Psychology	Double	4 (e.g., experimental design sound)	4
Journal of Sports Sciences	Double	3 (e.g., quality of data and methods)	5
Journal of Strength & Conditioning Research	Double	4 (e.g., practical applications impact)	6
Measurement in Physical Education and Exercise Science	Double	None provided	4
Medicine & Science in Sports & Exercise	Author	4 (e.g., percentile rank)	4
Pediatric Exercise Science	Double	11 (e.g., relevance, originality)	4
Research Quarterly for Exercise and Sport	Double	4 (e.g., impact to whole field)	4
Sports Biomechanics	Author	10 (e.g., writing and organization)	5

*Note: Because of space limitations, only sample evaluation criteria for each journal are listed.

Reprinted by permission from D.V. Knudson, J R. Morrow, and J. R. Thomas, "Advancing Kinesiology Through Improved Peer Review," *Research Quarterly for Exercise and Sport* Vol. 85 (2014): 127-35.

Journal editors and reviewers do not have time to teach you good scientific reporting. Not everyone is fond of journal editors. One jokester was reported to have said, "Editors have only one good characteristic—if they can understand a scientific paper, anyone can!" Another said, "Editors are, in my opinion, a low form of life—inferior to the viruses and only slightly above academic deans" (Day, 1983, p. 80). Teaching good reporting is the responsibility of your major professor, research methods courses, and books such as this one, but it is your responsibility to acquire the skill.

The peer-review process is widely used to judge scientific papers. However, there is disagreement about its value and success. Knudson, Morrow, and Thomas (2014) made several recommendations to improve the process; for example, publishing clear evaluation standards, establishing collaborative evaluation procedures and editorial team roles, using online submission data to improve reviewer comments, and creating author-appeal procedures.

Revising Research Papers

Do not be discouraged if your paper is rejected. Every researcher has had papers rejected. Carefully evaluate the reviews and determine whether the paper can be salvaged. If it can, then rewrite it, taking into account the reviewers' and editors' criticisms, and submit it to another journal. Do not send the paper out to another journal without evaluating the reviews and making appropriate revisions. Often, another journal uses the same reviewers, who will not be happy if you did not take their original advice. If your research and paper cannot be salvaged, learn from your mistakes.

Reviewers and editors will not always agree in what they say about your paper. This is frustrating because sometimes they tell you conflicting things. Be aware that editors select reviewers for different reasons. For example, in a sport psychology study that develops a questionnaire, the editor may decide that one or more content reviewers and a statistical reviewer are needed to evaluate the work. You would not expect the content and statistical reviewers to discuss the same issues. The research literature contains numerous assessments and commentaries on strengths and weaknesses of peer review, interrater reliability of reviews, and suggestions for improving the process, including several in kinesiology (e.g., Fischman, 2014; Knudson, Morrow, & Thomas, 2014; Morrow, Bray, Fulton, & Thomas, 1992). Upon receiving reviews of a paper you have submitted, our advice to you as a graduate student is to view reviewer comments as objectively as possible and to seek help from your advisor and other faculty members in making revisions to your manuscript.

If you are fortunate enough to be asked to submit a revision, you should start by reading the reviewers' and editors' comments. Often, these comments seem overwhelming at first, so it may be a good idea to read the comments thoroughly and then put them away for a few days, but no longer. This interval allows your ego to recover from the onslaught of criticism before you prepare a revision. We recommend the following steps for preparing a revision:

1. Reread all the comments from the reviewers and editors. Make notes as you are doing this so that you can prepare the revision and provide the rationale for the changes you made. In some cases, the requested changes (e.g., additional data from participants whom you can no longer contact) may be impossible, and you may decide to revise the paper for another journal. This is recommended if the editor makes it clear that something is required that you cannot provide in a revision.

2. Put all the comments from the reviews in a table with three columns: (a) comment number (or original page number if the review was organized that way); (b) the comment from the reviewer or editor; and (c) how you addressed the comment. Have a section for each editor and reviewer, and use the table to make clear which comment is which (i.e., use spaces or lines so that it will be clear to you—and to those reading it with the submitted revision).

3. Look at each comment again and determine whether any themes are running across comments and reviewers. If so, determine whether major changes will address all these comments. For example, reviewers may have commented that they did not understand how the results apply to other settings and that you did not provide enough information about the participants. Providing additional information about the participants and the setting context may address all these comments.

4. Look at individual comments and determine what changes are appropriate for the revised manuscript. You do not need to make all changes, but you should address most of the comments in a revision—particularly those that (a) are straightforward, (b) improve the manuscript, and (c) are identified by the reviewers and editors as important. Our advice is to address as many of the comments as you can if the revisions do not change the paper focus or add issues that may be problematic in a subsequent review. You cannot address every comment, but you should have a research or theoretical reason for not addressing a comment in the revision.

5. As you are making revisions, record how you addressed each comment in the third column of the table you prepared in step 2. Be specific and provide detail (e.g., *We have revised the section on participants and added demographic information, as well as a description of the study setting. Please see page XX, lines XX-XX.*). If the suggestions are grammatical, you may just indicate that the change was made. If you did not make a change, provide a rationale for why you did not (i.e., you thought about making a change but doing so was not appropriate for the following reasons, or you made other changes that eliminated the need to make the first change). This kind of explanation may necessitate a slightly longer response than that required for the changes you did make. If you do not provide a rationale, when the revision

is returned to the reviewers and editors, they will think you did not take the time to consider their suggestions and will be less likely to feel positive about your work.

6. After you have made all the changes, including updating the references, proofread the paper to ensure that it is grammatically correct and reads well. If the manuscript now exceeds the journal page limits, consider ways of editing it down to the acceptable limits. Include pertinent information in the responses to the comments.

7. Add any general comments to the file with the comments and revisions table to address major changes. Proofread the table of comments and your responses to ensure that page and line numbers are correct, that everything is clear and well written, and that your comments are presented objectively and are well justified. Being combative with the reviewers and editors will not help your case that the revision is an improvement and should now be accepted. You may find it helpful to have others read the table of comments to ensure that the tone and writing are appropriate.

8. Submit the revision by the deadline stated in the letter from the editor.

Occasionally, you may feel that you have received an unfair review. If so, write back to the editor, point out the biases, and ask for another assessment by a different reviewer. Editors are generally open to this type of correspondence if it is handled in a professional way. Opening with a statement such as "Look what you and the stupid reviewers said" is unlikely to achieve success (for a humorous example, see the A Letter We've All Wanted to Write sidebar). Recognize that appealing to the editor is less likely to be successful if two or more reviewers have agreed on the important criticisms. Editors cannot be experts in every area, and they must rely on reviewers. If you find yourself resorting to this tactic often, the problem is likely in your work. You do not want to end up like Snoopy in *Peanuts* (by Charles Schulz), who received this note from the journal to which he had submitted a manuscript: "Dear Contributor, We are returning your manuscript. It does not suit our present needs. P.S. We note that you sent your story by first-class mail. Junk mail may be sent third-class."

Writing Abstracts

Whether you are writing for a journal, using the chapter or journal format for preparing a thesis or dissertation, or readying a conference paper, you need to include an abstract. Abstracts for each purpose require slightly different orientations, but nearly all have constraints on length and form.

Thesis and Dissertation Abstracts

The abstract for your thesis or dissertation probably has several constraints, including length, form, style, and location. First, consult your university or graduate school regulations and then follow their policies carefully. For dissertations, the graduate school regulations nearly always include the form for submission to *Proquest Dissertations & Theses Global*. The exact headings, length, and margins are provided in a handout available from your graduate school office. In writing the abstract, consider who will read it. Computer searches will locate it by the title and keywords. (We discussed the importance of these in chapter 2.) Write the abstract so that anyone reading it can decide whether to look at the total thesis or dissertation. Clearly identify the theoretical framework, the problem, the participants, the measurements, and your findings. Do not use all the space writing about your sophisticated statistical analyses or minor methodological problems. Keep jargon to a minimum. People in related areas read the abstract to see whether your work is relevant to theirs.

A Letter We've All Wanted to Write

Dear Sir, Madam, or Other:

Enclosed is our latest version of MS #85-02-22-RRRR; that is, the re-re-re-revised revision of our paper. Choke on it. We have again rewritten the entire manuscript from start to finish. We even changed the #@*! running head! Hopefully we have suffered enough by now to satisfy even you and your bloodthirsty reviewers.

I shall skip the usual point-by-point description of every single change we made in response to the critiques. After all, it is fairly clear that your reviewers are less interested in details of scientific procedure than in working out their personality problems and frustrations by seeking some kind of demented glee in the sadistic and arbitrary exercise of tyrannical power over hapless authors like ourselves who happen to fall into their clutches. We do understand that, in view of the misanthropic psychopaths you have on your editorial board, you need to keep sending them papers, for if they weren't reviewing manuscripts, they'd probably be out committing heinous crimes.

We couldn't do anything about some of the reviewers' comments. For example, if (as reviewer C suggested) several of my recent ancestors were indeed drawn from other species, it is too late to change that. Other suggestions were implemented, however, and the paper has improved and benefited. Thus, you suggested that we shorten the manuscript by 5 pages, and we were able to accomplish this very effectively by altering the margins and printing the paper in a different font with a smaller typeface. We agree with you that the paper is much better this way.

Our perplexing problem was dealing with suggestions 13 through 28 by reviewer B. As you may recall (that is, if you even bother reading the reviews before writing your decision letter), that reviewer listed 16 works that we should cite in this paper. These were on a variety of different topics, none of which has any relevance to our work that we can see. Indeed, one was an essay on the Spanish-American War from a high school literary magazine. The only common thread was that all 16 were by the same author, presumably someone whom reviewer B greatly admires and feels should be more widely cited. To handle this, we have modified the introduction and added, after the review of relevant literature, a subsection titled *Review of Irrelevant Literature* that discusses these articles and also duly addresses some of the more asinine suggestions in the other reviews.

We hope that you will be pleased with this revision and will finally recognize how deserving of publication this work is. If not, then you all are unscrupulous, depraved monsters with no shred of human decency. If you do accept it, however, we wish to thank you for your patience and wisdom throughout this process and to express our appreciation of your scholarly insights. To repay you, we would be happy to review some manuscripts for you; please send us the next manuscript that any of these reviewers submits to your journal.

Assuming you accept this paper, we would also like to add a footnote acknowledging your help with this manuscript and pointing out that we liked the paper much better the way we originally wrote it, but that you held the editorial shotgun to our heads and forced us to chop, reshuffle, restate, hedge, expand, shorten, and, in general, convert a meaty paper into stir-fried vegetables.

We couldn't or wouldn't have done it without your input.

Sincerely,
Trying to Publish but Perishing Anyway

Reprinted by permission from Didier Delignieres, "A Letter We've All Wanted to Write," *Bulletin of the Canadian Society for Psychomotor Learning and Sport Psychology,* 1991.

Abstracts for Published Papers

When looking for literature related to a topic, a researcher first looks at the title. If that gets the researcher's attention, the next thing read is the abstract. Nearly all electronic search engines provide the title and abstract of the paper. Thus, a well-written abstract is essential to the research paper. The abstract should have all the components of the paper—introduction and problem, methods, results, and discussion. Too often, graduate students become so enamored with their methods and sophisticated analyses that they spend a disproportionate amount of their abstracts on them. The abstract should tell the purpose and value of the study in the opening sentence or two; then who the participants were and what the main treatments and measurements were, what the most important outcomes were, and why the study was important. Use synonyms for words in the title because this increases the hit rate on electronic searches.

There will be two constraints on your abstract's length—one by your graduate school (usually following the length of *Proquest Dissertations & Theses Global*) and the other by the journal to which you plan to submit. The beginning of the thesis or dissertation should be the length specified by your graduate school. You must have an abstract appropriate in length in front of the article(s) you are preparing for the journal(s). To see what length is allowed, go to the journal sites. Be certain that your abstract provides only information given in the article. Finally, the most useless sentence in any abstract is *Results were discussed*. Who would have thought otherwise?

Conference Abstracts

Abstracts for conferences are slightly different from other types of abstracts. Usually, you are allowed a little more space—from 400 to 2,000 words, depending on the conference—because the reviewers must be persuaded to accept your paper for presentation. In these abstracts, follow these procedures:

1. Write a short introduction to set up the problem statement.
2. State the problem.
3. Describe the methodology briefly, including the following:
   a. Participants
   b. Instrumentation
   c. Procedures
   d. Design and analysis
4. Summarize the results.
5. Explain why the results are important.

If the submission guidelines ask for other formats or information, you must follow those guidelines (for tips on writing conference papers and abstracts, see Jalongo, 2012). If a word limit is in place, use the word count function of your word processing software to confirm you do not exceed the limit. Most conference websites allow only the stated maximum number of words and truncate abstracts when that number is reached.

The results and their importance are a critical part of a conference abstract. If you do this in a nondescript way, the reviewer may conclude that you have not completed the study. This judgment generally leads to rejection. The conference planners cannot turn down completed research when it is possible that yours will not be finished.

Finally, most conferences require that the paper be presented before publication. Therefore, if you have a paper in review at a journal, submitting it to a conference that is 8 to 12 months away could be hazardous. Also, many conferences require that the paper not have been previously presented. Be aware of this and follow the regulations. Violating them will never enhance your professional status, and other scholars will quickly become aware of it.

Making Oral and Poster Presentations

After your conference paper is accepted, you are faced with presenting it. The presentation is conducted either orally or in a poster session.

How to Give Oral Presentations

oral presentation—A method of presenting a paper in which the author speaks before a group of colleagues at a conference following this format: introduction, statement of the problem, methods, results, discussion, questions.

Oral presentations usually cause panic among graduate students and new faculty members. The only way to get over this is to do several oral presentations. However, you can help assuage the apprehension. Usually, the time allowed for oral research reports is 10 to 20 minutes, depending on the conference. You will be notified of the time limit when your paper is accepted. Because you must stay within the time limit and cannot possibly present a complete report within this period, what do you do? We suggest that you present the essential features of the report using the following divisions for a 15-minute presentation. Visual aids, such as slides, computer presentation software, and overheads, are helpful.

- Introduction that cites a few important studies: 3 minutes
- Statement of the problem: 1 minute
- Methods: 3 minutes
- Results (Present tables and figures; figures are usually better because tables are hard to read.): 3 minutes
- Discussion (main points): 2 minutes
- Questions and discussion: 3 minutes
- Total time: 15 minutes

The most frequent errors in oral presentations are spending too much time on methods and presenting results poorly (see the Do Not Use These Quotations From British Sport Commentators in Your Presentation sidebar). Proper use of visual aids is the key to an effec-

I'm really due for an equipment upgrade.

© Artur Bogacki/Fotolia

tive presentation, particularly in the results section. Regarding the number of slides to use, a simple rule of thumb is one minute per slide. Thus, 12 slides ($\pm$2 slides) for a 12-minute presentation (excluding time for questions and answers) is about right. Place a brief statement of the problem on a slide and show it while you talk. A slide of the experimental arrangements reduces much of the verbiage in the methods discussion. Always use slides to illustrate the results. Pictures of the results (particularly figures and graphs) are much more effective than either tables or a verbal explanation. Keep the figures and graphs simple and concise. Tables and figures prepared for a written paper are seldom effective when converted for visual presentations. They typically contain too much information and font sizes are often too small to be readable by people in the audience. Have a pointer available to indicate significant features. Finally, remember Thomas, Silverman, Martin, and Etnier's Four Laws of Oral Presentations as outlined in the sidebar. Practicing your presentation is one solution to these and many other problems. At our universities, we gather graduate students and faculty who are presenting papers at upcoming conferences and conduct practice sessions. Everyone presents their paper and has it timed. Then the audience asks questions and offers suggestions to clarify the presentation and visual aids. Practice sessions improve the quality of presentations and strengthen presenters' confidence.

> **Prepare different tables and figures for a visual presentation. Paper versions often contain too much information and are too small.**

Using a Poster Presentation to Best Advantage

The poster session is another way to present a conference paper. For many professional meetings (e.g., annual meeting of the American College of Sports Medicine), poster presentations far outnumber oral presentations on the conference program. Poster sessions take place in a large room in which presenters place summaries of their research on the wall or on poster stands. The session is scheduled for a specific period, during which presenters stand by their work while those interested walk around, read the material, and discuss items of interest with the presenters.

We prefer this format to oral presentations. The audience can look at the papers in which they are interested and have detailed discussions with the authors. Within 75 minutes, 15 to 40 poster presentations can be made available in a large room, whereas only five 15-minute oral presentations can take place in the same length of time. In addition, in a session of oral presentations, the audience must sit through several papers. Audience members may lose interest or create a disturbance by arriving or leaving.

Do Not Use These Quotations From British Sport Commentators in Your Presentation

"Moses Kiptanul—the 19-year-old Kenyan who turned 20 a few weeks ago."

"We now have exactly the same situation we had at the start of the race, only exactly the opposite."

"He's never had major knee surgery on any other part of his body."

"She's not Usain Bolt—but then who is?"

"I owe a lot to my parents, especially my mother and father."

"The Port Elizabeth ground is more of a circle than an oval. It's long and square."

"The racecourse is as level as a billiard ball."

"Watch the time—it gives you an indication of how fast they are running."

"That's inches away from being a millimeter perfect."

"If history repeats itself, I think we can expect the same thing again."

Thomas, Martin, Etnier, and Silverman's Four Laws of Oral Presentations

1. Something always goes wrong with the projection system. More specifically:
   a. Unless you check it beforehand, it will not work.
   b. The projected image resolution is altered, and all of your slides are distorted and unreadable.
   c. If the projector has worked perfectly for the three previous presenters, it will overheat and automatically shut down during your presentation.
2. The screen will be too small for the room.
3. Your paper will be the last one scheduled during the conference. Only the moderator, you, and the previous presenter will be there, the latter of whom will leave on completing their paper. Or, conversely, your paper will be scheduled at 7:30 a.m., at which time everyone except the session moderator and other speakers will still be in bed.
4. At your first presentation, the most prestigious scholar in your field will show up, be misquoted by you, and ask you a question.

Poster sessions allow audience members to pace the way they look at material. We have often heard oral presentations about which we wanted more information. We may have wished that a table or figure presented had been left up longer. Those concerns are not an issue in poster sessions. Participants are more likely to ask questions or comment on points in a poster session when they have sufficient time to consider the material. Longer and more in-depth conversations that benefit both the presenter and the spectator often ensue.

Good poster design maximizes all the benefits that poster sessions offer. The poster should be arranged to highlight the important points and eliminate extraneous information. Posters, which are typically produced using a single PowerPoint slide with custom dimensions, are mounted on walls or portable stands. Usually, the conference will notify you about how much space you have, but a 3-foot-by-4-foot (0.9 by 1.2 m) space is common. The poster should be arranged in five or six panels or sections that represent the parts of the poster presentation—abstract, introduction and problem statement, methods, results (including figures and tables), discussion, and references. Include clear, easy-to-read headings for each section to help readers follow poster content. In figure 22.1, you can see a logical way to arrange the panels using the vertical flow of information (horizontal flow can also be used, but vertical flow is usually preferred). Often, an effective approach is to place tables and figures at the center of the poster and arrange panels around them. The tables and figures represent what you have found and should be the central feature of the poster.

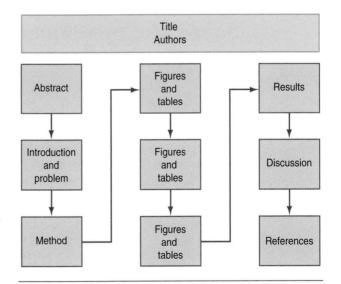

Figure 22.1   Poster with vertical flow.

Posters have text, but this text should have specific features to increase how easily it can be read and understood:

- Simply structured short sentences
- To-the-point text (no filler information)
- Common words (limited jargon)
- Little abstraction
- Bulleted lists

Use design elements that improve posters and make them more understandable:

- Select a plain and clear typeface (common fonts such as Arial and Times New Roman).
- Use font sizes of 24 to 30 points for text, 36 to 44 points for headings, and 72 to 80 points for titles. A well-designed poster should be readable from 2 to 3 meters.
- Most current evidence suggests that using both upper- and lowercase increases readability.
- Major headings on each panel with section headings inside the panel guide the eye in reading the poster.
- Within sections, use left justification and ragged right rather than full justification.
- If there is a reasonable choice between a figure and a table, use the figure; if you must use a table, put in minimal information, use large numbers, and use labeling (e.g., shading, colors) to guide the reader to important information; *a table or figure prepared for publication is seldom acceptable for a poster (even when enlarged).*
- Use color judiciously; it can help the poster if it sets off points, unifies sections, and guides the reader's eye. If you are having trouble deciding how much color to use, you are probably using too much; tone it down.

If you want your poster to be exceptional and attract interested readers, follow these suggestions, prepare it well in advance, and then have your friends and faculty advisors look at it to make suggestions and comments.

Summary

The journal format for theses and dissertations has the advantage of being the format for journal publication, one of the main ways to disseminate scholarly work. Yet it retains the essential characteristics of the complete reporting that is valued in the thesis or dissertation. This format comprises preliminary materials (e.g., title page, abstract), the body (e.g., introduction, methods, results, discussion), and a set of appendixes (e.g., comprehensive literature review, additional results). The goal of this reporting style is to promote rapid publication of high-quality research.

Abstracts are frequently used as a form for submitting papers to scholarly meetings so that they can be evaluated for possible presentation. Usually, the abstract is limited in length and format by the scholarly group to which it is submitted. If the abstract paper is accepted for presentation, the form may be oral or by poster. Oral presentations usually last 10 to 20 minutes, and poster presentations are typically limited to a display space.

✓ **Check Your Understanding**

1. Writing: Select a study from a journal and write a 150-word abstract in APA style (or whatever style your department uses).

2. Oral presentations: To emphasize the importance of time limits on oral presentations, we suggest that each student give the following presentations in class:

   a. Select a published research study and present a 2-minute summary.

   b. Select another study but give a 5-minute summary.

3. Poster presentation: Prepare a poster presentation of a research study from a journal. Put your poster on the classroom wall, along with those of the other students in your class. Critique each other's posters (or some percentage of posters for larger classes).

APPENDIX

Statistical Tables

TABLE 1 The Standard Normal Curve

Z	One-tailed p value	Remainder	Two-tailed p value	Remainder	Z	One-tailed p value	Remainder	Two-tailed p value	Remainder
1.75	0.040059	0.959941	0.080118	0.919882	2.88	0.001988	0.998012	0.003977	0.996023
1.76	0.039204	0.960796	0.078408	0.921592	2.89	0.001926	0.998074	0.003852	0.996148
1.77	0.038364	0.961636	0.076727	0.923273	2.90	0.001866	0.998134	0.003732	0.996268
1.78	0.037538	0.962462	0.075076	0.924924	2.91	0.001807	0.998193	0.003614	0.996386
1.79	0.036727	0.963273	0.073454	0.926546	2.92	0.001750	0.998250	0.003500	0.996500
1.80	0.035930	0.964070	0.071861	0.928139	2.93	0.001695	0.998305	0.003390	0.996610
1.81	0.035148	0.964852	0.070296	0.929704	2.94	0.001641	0.998359	0.003282	0.996718
1.82	0.034380	0.965620	0.068759	0.931241	2.95	0.001589	0.998411	0.003178	0.996822
1.83	0.033624	0.966376	0.067250	0.932750	2.96	0.001538	0.998462	0.003076	0.996924
1.84	0.032884	0.967116	0.065768	0.934232	2.97	0.001489	0.998511	0.002978	0.997022
1.85	0.032157	0.967843	0.064314	0.935686	2.98	0.001441	0.998559	0.002882	0.997118
1.86	0.031443	0.968557	0.062886	0.937114	2.99	0.001395	0.998605	0.002790	0.997210
1.87	0.030742	0.969258	0.061484	0.938516	3.00	0.001350	0.998650	0.002700	0.997300
1.88	0.030054	0.969946	0.060108	0.939892	3.01	0.000968	0.999032	0.002612	0.997388
1.89	0.029379	0.970621	0.057858	0.942142	3.02	0.000935	0.999065	0.002528	0.997472
1.90	0.028717	0.971283	0.057433	0.942567	3.03	0.001223	0.998777	0.002446	0.997554
1.91	0.028067	0.971933	0.056133	0.943867	3.04	0.001183	0.998817	0.002366	0.997634
1.92	0.027429	0.972571	0.054858	0.945142	3.05	0.001144	0.998856	0.002288	0.997712
1.93	0.026803	0.973197	0.053607	0.946393	3.06	0.001107	0.998893	0.002213	0.997787
1.94	0.02619	0.973810	0.05238	0.94762	3.07	0.001070	0.998930	0.002141	0.997859
1.95	0.025588	0.974412	0.051176	0.948824	3.08	0.001035	0.998965	0.002070	0.997930
1.96	0.024998	0.975002	0.049996	0.950004	3.09	0.001001	0.998999	0.002002	0.997998
1.97	0.024419	0.975581	0.048838	0.951162	3.10	0.009680	0.990320	0.001935	0.998065
1.98	0.023852	0.976148	0.047704	0.952296	3.11	0.009350	0.990650	0.001871	0.998129
1.99	0.023295	0.976705	0.046591	0.953409	3.12	0.000904	0.999096	0.001809	0.998191
2.00	0.022750	0.977250	0.045500	0.954500	3.13	0.000874	0.999126	0.001748	0.998252
2.01	0.022216	0.977784	0.044431	0.955569	3.14	0.000845	0.999155	0.001689	0.998311
2.02	0.021692	0.978308	0.043383	0.956617	3.15	0.000816	0.999184	0.001633	0.998367
2.03	0.021178	0.978822	0.042357	0.957643	3.16	0.000789	0.999211	0.001578	0.998422
2.04	0.020675	0.979325	0.041350	0.958650	3.17	0.000762	0.999238	0.001524	0.998476
2.05	0.020182	0.979818	0.040364	0.959636	3.18	0.000736	0.999264	0.001473	0.998527
2.06	0.019699	0.980301	0.039399	0.960601	3.19	0.000711	0.999289	0.001423	0.998577
2.07	0.019226	0.980774	0.038452	0.961548	3.20	0.000687	0.999313	0.001374	0.998626
2.08	0.018763	0.981237	0.037526	0.962474	3.21	0.000664	0.999336	0.001327	0.998673
2.09	0.018309	0.981691	0.036618	0.963382	3.22	0.000641	0.999359	0.001282	0.998718
2.10	0.017864	0.982136	0.035729	0.964271	3.23	0.000619	0.999381	0.001238	0.998762
2.11	0.017429	0.982571	0.034858	0.965142	3.24	0.000598	0.999402	0.001195	0.998805
2.12	0.017003	0.982997	0.034006	0.965994	3.25	0.000577	0.999423	0.001154	0.998846
2.13	0.016586	0.983414	0.033172	0.966828	3.26	0.000557	0.999443	0.001114	0.998886
2.14	0.016177	0.983823	0.032355	0.967645	3.27	0.000538	0.999462	0.001075	0.998925
2.15	0.015778	0.984222	0.031555	0.968445	3.28	0.000519	0.999481	0.001038	0.998962
2.16	0.015386	0.984614	0.030773	0.969227	3.29	0.000501	0.999499	0.001002	0.998998
2.17	0.015003	0.984997	0.030007	0.969993	3.30	0.000483	0.999517	0.000967	0.999033
2.18	0.014629	0.985371	0.029257	0.970743	3.31	0.000466	0.999534	0.000933	0.999067

	One-tailed		Two-tailed			One-tailed		Two-tailed	
Z	p value	Remainder	p value	Remainder	Z	p value	Remainder	p value	Remainder
2.19	0.014262	0.985738	0.028524	0.971476	3.32	0.000450	0.999550	0.000900	0.999100
2.20	0.013903	0.986097	0.0278707	0.9721293	3.33	0.000434	0.999566	0.000868	0.999132
2.21	0.013553	0.986447	0.027105	0.972895	3.34	0.000419	0.999581	0.000838	0.999162
2.22	0.013209	0.986791	0.026419	0.973581	3.35	0.000404	0.999596	0.000808	0.999192
2.23	0.012874	0.987126	0.025747	0.974253	3.36	0.000390	0.999610	0.000779	0.999221
2.24	0.012545	0.987455	0.025091	0.974909	3.37	0.000376	0.999624	0.000752	0.999248
2.25	0.012224	0.987776	0.024449	0.975551	3.38	0.000362	0.999638	0.000725	0.999275
2.26	0.011911	0.988089	0.023821	0.976179	3.39	0.000349	0.999651	0.000699	0.999301
2.27	0.011604	0.988396	0.023208	0.976792	3.40	0.000337	0.999663	0.000674	0.999326
2.28	0.011304	0.988696	0.022608	0.977392	3.41	0.000325	0.999675	0.000650	0.999350
2.29	0.011011	0.988989	0.022021	0.977979	3.42	0.000313	0.999687	0.000626	0.999374
2.30	0.010724	0.989276	0.021448	0.978552	3.43	0.000302	0.999698	0.000604	0.999396
2.31	0.010444	0.989556	0.020888	0.979112	3.44	0.000291	0.999709	0.000582	0.999418
2.32	0.010170	0.989830	0.020341	0.979659	3.45	0.000280	0.999720	0.000561	0.999439
2.33	0.009903	0.990097	0.019806	0.980194	3.46	0.000270	0.999730	0.000540	0.999460
2.34	0.009642	0.990358	0.019284	0.980716	3.47	0.000260	0.999740	0.000520	0.999480
2.35	0.009387	0.990613	0.018773	0.981227	3.48	0.000251	0.999749	0.000501	0.999499
2.36	0.009137	0.99086	0.018275	0.981725	3.49	0.000242	0.999758	0.000483	0.999517
2.37	0.008894	0.991106	0.017788	0.982212	3.50	0.000233	0.999767	0.000465	0.999535
2.38	0.008656	0.991344	0.017313	0.982687	3.51	0.000224	0.999776	0.000448	0.999552
2.39	0.008424	0.991576	0.016848	0.983152	3.52	0.000216	0.999784	0.000432	0.999568
2.40	0.008198	0.991802	0.016395	0.983605	3.53	0.000208	0.999792	0.000416	0.999584
2.41	0.007976	0.992024	0.015953	0.984047	3.54	0.000200	0.999800	0.000400	0.999600
2.42	0.007760	0.992240	0.015521	0.984479	3.55	0.000193	0.999807	0.000385	0.999615
2.43	0.007549	0.992451	0.015099	0.984901	3.56	0.000185	0.999815	0.000371	0.999629
2.44	0.007344	0.992656	0.014687	0.985313	3.57	0.000178	0.999822	0.000357	0.999643
2.45	0.007143	0.992857	0.014286	0.985714	3.58	0.000172	0.999828	0.000344	0.999656
2.46	0.006947	0.993053	0.013894	0.986106	3.59	0.000165	0.999835	0.000331	0.999669
2.47	0.006756	0.993244	0.013511	0.986489	3.60	0.000159	0.999841	0.000318	0.999682
2.48	0.006569	0.993431	0.013138	0.986862	3.61	0.000153	0.999847	0.000306	0.999694
2.49	0.006387	0.993613	0.012774	0.987226	3.62	0.000147	0.999853	0.000295	0.999705
2.50	0.006210	0.993790	0.012419	0.987581	3.63	0.000142	0.999858	0.000283	0.999717
2.51	0.006037	0.993963	0.012073	0.987927	3.64	0.000136	0.999864	0.000273	0.999727
2.52	0.005868	0.994132	0.011735	0.988265	3.65	0.000131	0.999869	0.000262	0.999738
2.53	0.005703	0.994297	0.011406	0.988594	3.66	0.000126	0.999874	0.000252	0.999748
2.54	0.005543	0.994457	0.011085	0.988915	3.67	0.000121	0.999879	0.000243	0.999757
2.55	0.005386	0.994614	0.010772	0.989228	3.68	0.000117	0.999883	0.000233	0.999767
2.56	0.005234	0.994766	0.010467	0.989533	3.69	0.000112	0.999888	0.000224	0.999776
2.57	0.005085	0.994915	0.010170	0.989830	3.70	0.000108	0.999892	0.000216	0.999784
2.58	0.004940	0.995060	0.009880	0.990120	3.71	0.000104	0.999896	0.000207	0.999793
2.59	0.004799	0.995201	0.009598	0.990402	3.72	0.000100	0.999900	0.000199	0.999801
2.60	0.004661	0.995339	0.009322	0.990678	3.73	0.000096	0.999904	0.000191	0.999809
2.61	0.004527	0.995473	0.009054	0.990946	3.74	0.000092	0.999908	0.000184	0.999816
2.62	0.004396	0.995604	0.008793	0.991207	3.75	0.000088	0.999912	0.000177	0.999823

(continued)

Table 1 *(continued)*

| | One-tailed | | Two-tailed | | | One-tailed | | Two-tailed | |
Z	p value	Remainder	p value	Remainder	Z	p value	Remainder	p value	Remainder
2.63	0.004269	0.995731	0.008538	0.991462	3.76	0.000085	0.999915	0.000170	0.999830
2.64	0.004145	0.995855	0.008291	0.991709	3.77	0.000082	0.999918	0.000163	0.999837
2.65	0.004025	0.995975	0.008049	0.991951	3.78	0.000078	0.999922	0.000157	0.999843
2.66	0.003907	0.996093	0.007814	0.992186	3.79	0.000075	0.999925	0.000151	0.999849
2.67	0.003793	0.996207	0.007585	0.992415	3.80	0.000072	0.999928	0.000145	0.999855
2.68	0.003681	0.996319	0.007362	0.992638	3.81	0.000069	0.999931	0.000139	0.999861
2.69	0.003573	0.996427	0.007145	0.992855	3.82	0.00067	0.999933	0.000133	0.999867
2.70	0.003467	0.996533	0.006934	0.993066	3.83	0.000064	0.999936	0.000128	0.999872
2.71	0.003364	0.996636	0.006728	0.993272	3.84	0.000062	0.999938	0.000123	0.999877
2.72	0.003264	0.996736	0.006528	0.993472	3.85	0.000059	0.999941	0.000118	0.999882
2.73	0.003167	0.996833	0.006333	0.993667	3.86	0.000057	0.999943	0.000113	0.999887
2.74	0.003072	0.996928	0.006144	0.993856	3.87	0.000054	0.999946	0.000109	0.999891
2.75	0.002980	0.997020	0.005960	0.994040	3.88	0.000052	0.999948	0.000104	0.999896
2.76	0.002890	0.997110	0.005780	0.994220	3.89	0.000050	0.999950	0.000100	0.999900
2.77	0.002803	0.997197	0.005606	0.994394	3.90	0.000048	0.999952	0.000097	0.999904
2.78	0.002718	0.997282	0.005436	0.994564	3.91	0.000046	0.999954	0.000093	0.999908
2.79	0.002635	0.997365	0.005271	0.994729	3.92	0.000044	0.999956	0.000089	0.999911
2.80	0.002555	0.997445	0.00511	0.994890	3.93	0.000042	0.999958	0.000085	0.999915
2.81	0.002477	0.997523	0.004954	0.995046	3.94	0.000041	0.999959	0.000081	0.999919
2.82	0.002401	0.997599	0.004802	0.995198	3.95	0.000039	0.999961	0.000078	0.999922
2.83	0.002327	0.997673	0.004655	0.995345	3.96	0.000037	0.999963	0.000075	0.999925
2.84	0.002256	0.997744	0.004511	0.995489	3.97	0.000036	0.999964	0.000072	0.999928
2.85	0.002186	0.997814	0.004372	0.995628	3.98	0.000034	0.999966	0.000069	0.999931
2.86	0.002118	0.997882	0.004236	0.995764	3.99	0.000033	0.999967	0.000066	0.999934
2.87	0.002052	0.997948	0.004105	0.995895	4.00	0.000032	0.999968	0.000063	0.999937

TABLE 2   Critical Values of Correlation Coefficients

	One-tailed					
	.25	.10	.05	.025	.01	.005
	Two-tailed					
	.50	.20	.10	.05	.02	.01
df = 1	.7071	.9511	.9877	.9969	.9995	.9999
2	.5000	.8000	.9000	.9500	.9800	.9900
3	.4040	.6870	.8054	.8783	.9343	.9587
4	.3473	.6084	.7293	.8114	.8822	.9172
5	.3091	.5509	.6694	.7545	.8329	.8745
6	.2811	.5067	.6215	.7067	.7887	.8343
7	.2596	.4716	.5822	.6664	.7498	.7977
8	.2423	.4428	.5494	.6319	.7155	.7646
9	.2281	.4187	.5214	.6021	.6851	.7348
10	.2161	.3981	.4973	.5760	.6581	.7079
11	.2058	.3802	.4762	.5529	.6339	.6835
12	.1968	.3646	.4575	.5324	.6120	.6614
13	.1890	.3507	.4409	.5140	.5923	.6411
14	.1820	.3383	.4259	.4973	.5742	.6226
15	.1757	.3271	.4124	.4821	.5577	.6055
16	.1700	.3170	.4000	.4683	.5425	.5897
17	.1649	.3077	.3887	.4555	.5285	.5751
18	.1602	.2992	.3783	.4438	.5155	.5614
19	.1558	.2914	.3687	.4329	.5034	.5487
20	.1518	.2841	.3598	.4227	.4921	.5368
21	.1481	.2774	.3515	.4132	.4815	.5256
22	.1447	.2711	.3438	.4044	.4716	.5151
23	.1415	.2653	.3365	.3961	.4622	.5052
24	.1384	.2598	.3297	.3882	.4534	.4869
25	.1356	.2546	.3233	.3809	.4451	.4869
26	.1330	.2497	.3172	.3739	.4372	.4785
27	.1305	.2451	.3115	.3673	.4297	.4705
28	.1280	.2407	.3061	.3610	.4226	.4629
29	.1258	.2366	.3009	.3550	.4158	.4556
30	.1237	.2327	.2960	.3494	.4093	.4487
31	.1217	.2289	.2914	.3440	.4032	.4421
32	.1197	.2254	.2869	.3388	.3916	.4296
33	.1179	.2220	.2826	.3338	.3916	.4296
34	.1161	.2187	.2785	.3291	.3862	.4238
35	.1144	.2156	.2746	.3246	.3810	.4182
36	.1128	.2126	.2709	.3202	.3760	.4128
37	.1113	.2097	.2673	.3160	.3712	.4076
38	.1098	.2070	.2638	.3120	.3665	.4026
39	.1084	.2043	.2605	.3081	.3621	.3978
40	.1070	.2018	.2573	.3044	.3578	.3932

(continued)

Table 2 *(continued)*

	One-tailed					
	.25	.10	.05	.025	.01	.005
	Two-tailed					
	.50	.20	.10	.05	.02	.01
41	.1057	.1993	.2542	.3088	.3536	.3887
42	.1044	.1970	.2512	.2973	.3496	.3843
43	.1032	.1947	.2483	.2940	.3457	.3801
44	.1020	.1925	.2455	.2907	.3420	.3761
45	.1008	.1903	.2429	.2876	.3374	.3721
46	.0997	.1883	.2403	.2845	.3348	.3683
47	.0987	.1863	.2377	.2816	.3314	.3646
48	.0976	.1843	.2343	.2787	.2381	.3610
49	.0966	.1825	.2329	.2759	.3249	.3575
50	.0956	.1806	.2306	.2732	.3218	.3542
51	.0947	.1789	.2284	.2706	.3188	.3509
52	.0938	.1772	.2262	.2681	.3158	.3477
53	.0929	.01755	.2241	.2656	.3129	.3445
54	.0912	.1723	.2201	.2609	.3074	.3385
55	.0912	.1723	.2201	.2609	.3074	.3385
56	.0904	.1708	.2181	.2586	.3048	.3357
57	.0896	.1693	.2162	.2564	.3022	.3328
58	.0888	.1678	.2144	.2542	.2997	.3301
59	.0880	.1664	.2126	.2521	.2972	.3274
60	.0873	.1650	.2108	.2500	.2948	.3248
61	.0866	.1636	.2097	.2480	.2925	.3223
62	.0858	.1623	.2075	.2461	.2902	.3198
63	.0852	.1610	.2058	.2441	.2880	.3173
64	.0845	.1598	.2042	.2423	.2858	.3150
65	.0828	.1586	.2027	.2606	.2837	.3126
66	.0832	.1574	.2012	.2387	.2816	.3104
67	.0826	.1562	.1997	.2367	.2796	.3081
68	.0820	.1550	.1982	.2352	.2776	.3060
69	.0814	.1539	.1968	.2335	.2756	.3038
70	.0808	.1528	.1954	.2319	.2737	.3017
71	.0802	.1517	.1940	.2303	.2718	.2997
72	.0796	.1507	.1927	.2287	.2700	.2977
73	.0791	.1497	.1914	.2272	.2682	.2957
74	.0786	.1486	.1901	.2257	.2664	.2938
75	.0780	.177	.1888	.2242	.2647	.2919
76	.0775	.1467	.1876	.2227	.2630	.2900
77	.0770	.1457	.1864	.2213	.2613	.2882
78	.0765	.1448	.1852	.2199	.2597	.2864
79	.0760	.1439	.1841	.2186	.2581	.2847
80	.0755	.1430	.1829	.2172	.2565	.2830
81	.0751	.1421	.1818	.2159	.2550	.2813

	One-tailed					
	.25	.10	.05	.025	.01	.005
	Two-tailed					
	.50	.20	.10	.05	.02	.01
82	.0746	.1412	.1807	.2146	.2535	.2796
83	.0742	.1404	.1796	.2133	.2520	.2780
84	.0737	.1396	.1786	.2120	.2505	.2764
85	.0733	.1387	.1775	.2108	.2491	.2748
86	.0728	.1379	.1756	.2096	.2477	.2732
87	.0724	.1371	.1755	.2084	.2463	.2717
88	.0720	.1364	.1745	.2072	.2449	.2790
89	.0716	.1356	.1735	.2061	.2435	.2687
90	.0712	.1348	.1726	.2050	.2422	.2673
91	.0708	.1341	.1716	.2039	.2409	.2659
92	.0704	.1334	.1707	.2028	.2396	.2645
93	.0700	.1327	.1698	.2017	.2384	.2631
94	.0697	.1320	.1689	.2006	.2371	.2617
95	.0693	.1313	.1680	.1996	.2359	.2604
96	.0689	.1306	.1671	.1986	.2347	.2591
97	.0686	.1299	.1663	.1975	.2335	.2578
98	.0682	.1292	.1654	.1966	.2324	.2565
99	.0679	.1286	.1646	.1956	.2312	.2552
100	.0675	.1279	.1638	.1946	.2301	.2540

TABLE 3   Transformation of r to Zr

r	Zr	r	Zr	r	Zr
0.00	0.0000	0.33	0.3428	0.67	0.8107
0.01	0.0100	0.34	0.3541	0.68	0.8291
0.02	0.0200	0.35	0.3654	0.69	0.8480
0.03	0.0300	0.36	0.3769	0.70	0.8673
0.04	0.0400	0.37	0.3884	0.71	0.8872
0.05	0.0500	0.38	0.4001	0.72	0.9076
0.06	0.0601	0.39	0.4118	0.73	0.9287
0.07	0.0701	0.40	0.4236	0.74	0.9505
0.08	0.0802	0.41	0.4356	0.75	0.9730
0.09	0.0902	0.42	0.4477	0.76	0.9962
0.10	0.1003	0.43	0.4599	0.77	1.0203
0.11	0.1104	0.44	0.4722	0.78	1.0454
0.12	0.1206	0.45	0.4847	0.79	1.0714
0.13	0.1307	0.46	0.4973	0.80	1.0986
0.14	0.1409	0.47	0.5101	0.81	1.1270
0.15	0.1511	0.48	0.5230	0.82	1.1568
0.16	0.1614	0.49	0.5361	0.83	1.1881
0.17	0.1717	0.50	0.5493	0.84	1.2212
0.18	0.1820	0.51	0.5627	0.85	1.2562
0.19	0.1923	0.52	0.5763	0.86	1.2933
0.20	0.2027	0.53	0.5901	0.87	1.3331
0.21	0.2132	0.54	0.6042	0.88	1.3758
0.22	0.2237	0.55	0.6184	0.89	1.4219
0.23	0.2342	0.56	0.6328	0.90	1.4722
0.24	0.2448	0.57	0.6475	0.91	1.5275
0.25	0.2554	0.58	0.6625	0.92	1.5890
0.26	0.2661	0.59	0.6777	0.93	1.6584
0.27	0.2769	0.60	0.6931	0.94	1.7380
0.28	0.2877	0.61	0.7089	0.95	1.8318
0.29	0.2986	0.62	0.7250	0.96	1.9459
0.30	0.3095	0.63	0.7414	0.97	2.0923
0.31	0.3205	0.64	0.7582	0.98	2.2976
0.32	0.3316	0.65	0.7753		

TABLE 4   Critical Values of *t*

	One-tailed					
	0.25	0.1	0.05	0.025	0.01	0.005
	Two-tailed					
	0.5	0.2	0.1	0.05	0.02	0.01
df = 1	1.000	3.078	6.314	12.706	31.821	63.657
2	0.816	1.886	2.920	4.303	6.965	9.925
3	0.765	1.638	2.353	3.182	4.541	5.841
4	0.741	1.533	2.132	2.776	3.747	4.604
5	0.727	1.476	2.015	2.571	3.365	4.032
6	0.718	1.440	1.943	2.447	3.143	3.707
7	0.711	1.415	1.895	2.365	2.998	3.499
8	0.706	1.397	1.860	2.306	2.896	3.355
9	0.703	1.383	1.833	2.262	2.821	3.250
10	0.700	1.372	1.812	2.228	2.764	3.169
11	0.697	1.363	1.796	2.201	2.718	3.106
12	0.695	1.356	1.782	2.179	2.681	3.055
13	0.694	1.350	1.771	2.160	2.650	3.012
14	0.692	1.345	1.761	2.145	2.624	2.977
15	0.691	1.341	1.753	2.131	2.602	2.947
16	0.690	1.337	1.746	2.120	2.583	2.921
17	0.689	1.333	1.740	2.110	2.567	2.898
18	0.688	1.330	1.734	2.101	2.552	2.878
19	0.688	1.328	1.729	2.093	2.539	2.861
20	0.687	1.325	1.725	2.086	2.528	2.845
21	0.686	1.323	1.721	2.080	2.518	2.831
22	0.686	1.321	1.717	2.074	2.508	2.819
23	0.685	1.319	1.714	2.069	2.500	2.807
24	0.685	1.318	1.711	2.064	2.492	2.797
25	0.684	1.316	1.708	2.060	2.485	2.787
26	0.684	1.315	1.706	2.056	2.479	2.779
27	0.684	1.314	1.703	2.052	2.473	2.771
28	0.683	1.313	1.701	2.048	2.467	2.763
29	0.683	1.311	1.699	2.045	2.462	2.756
30	0.683	1.310	1.697	2.042	2.457	2.750
40	0.681	1.303	1.684	2.021	2.423	2.704
50	0.679	1.299	1.676	2.009	2.403	2.678
60	0.679	1.296	1.671	2.000	2.390	2.660
70	0.678	1.294	1.667	1.994	2.381	2.648
80	0.678	1.292	1.664	1.990	2.374	2.639
90	0.677	1.291	1.662	1.987	2.368	2.632
100	0.677	1.290	1.660	1.984	2.364	2.626
z	0.674	1.282	1.645	1.960	2.326	2.576

TABLE 5   Critical Values of F

df	1	2	3	4	5	6	7	8	9	10	15	20	30	40	60	120	∞
1	161.4476	199.5000	215.7073	224.5832	230.1619	233.9860	236.7684	238.8827	240.5433	241.8817	245.9499	248.0131	250.0951	251.1432	252.1957	253.2529	254.3144
	4,052.1807	4,999.5000	5,403.3520	5,624.5833	5,763.6496	5,858.9861	5,928.3557	5,981.0703	6,022.4732	6,055.8467	6,157.2846	6,208.7302	6,260.6486	6,286.7821	6,313.0301	6,339.1275	6,365.5460
2	18.5128	19.0000	19.1643	19.2468	19.2964	19.3295	19.3532	19.371	19.3848	19.3959	19.4291	19.4458	19.4624	19.4707	19.4791	19.4874	19.4957
	98.5025	99.0000	99.1662	99.2494	99.2993	99.3326	99.3564	99.3742	99.3881	99.3992	99.4325	99.4492	99.4658	99.4742	99.4825	99.4908	99.4991
3	10.1280	9.5521	9.2766	9.1172	9.0135	8.9406	8.8867	8.8452	8.8123	8.7855	8.7029	8.6602	8.6166	8.5944	8.5720	8.5494	8.5264
	34.1162	30.8165	29.4567	28.7099	28.2371	27.9107	27.6717	27.4892	27.3452	27.2287	26.8722	26.6880	26.5045	26.4108	26.3165	26.2211	26.1263
4	7.7086	6.9443	6.5914	6.3882	6.2561	6.1631	6.0942	6.0410	5.9988	5.9644	5.8578	5.8025	5.7459	5.7170	5.6877	5.6581	5.6281
	21.1977	18.0000	16.6944	15.9770	15.5219	15.2069	14.9758	14.7989	14.6591	14.5459	14.1982	14.0196	13.8377	13.7454	13.6522	13.5581	13.4642
5	6.6079	5.7861	5.4095	5.1922	5.0503	4.9503	4.8759	4.8183	4.7725	4.7351	4.6188	4.5581	4.4957	4.4638	4.4314	4.3985	4.3650
	16.2582	13.2739	12.0600	11.3919	10.9670	10.6723	10.4555	10.2893	10.1578	10.0510	9.7222	9.5526	9.3793	9.2912	9.2.20	9.1118	9.0215
6	5.9874	5.1433	4.7571	4.5337	4.3874	4.2839	4.2067	4.1468	4.0990	4.0600	3.9381	3.8742	3.8082	3.7743	3.7398	3.7047	3.6689
	13.7450	10.9248	9.7795	9.1483	8.7459	8.4661	8.2600	8.1017	7.9761	7.8741	7.5590	7.3958	7.2285	7.1432	7.0567	6.9690	6.8811
7	5.5914	4.7374	4.3468	4.1203	3.9715	3.8660	3.7870	3.7257	3.6767	3.6365	3.5107	3.4445	3.3758	3.3404	3.3043	3.2674	3.2298
	12.2464	9.5466	8.4513	7.8466	7.4504	7.1914	6.9928	6.8400	6.7188	6.6201	6.3143	6.1554	5.9920	5.9084	5.8236	5.7373	5.6506
8	5.3177	4.4590	4.0662	3.8379	3.6875	3.5806	3.5005	3.4381	3.3881	3.3472	3.2184	3.1503	3.0794	3.0428	3.0053	2.9669	2.9276
	11.2586	8.6491	7.5910	7.0061	6.632	6.3707	6.1776	6.0289	5.9106	5.8143	5.5151	5.3591	5.1981	5.1156	5.03162	4.9461	4.8599
9	5.1174	4.2565	3.8625	3.6331	3.4817	3.3738	3.2927	3.2296	3.1789	3.1373	3.0061	2.9365	2.8637	2.8259	2.7872	2.7475	2.7067
	10.5614	8.0215	6.9919	6.4221	6.0569	5.8018	5.6129	5.4671	5.3511	5.2565	4.9621	4.8080	4.6486	4.5666	4.4831	4.3978	4.3116
10	4.9646	4.1028	3.7083	3.4780	3.3258	3.2172	3.1355	3.0717	3.0204	2.9782	2.8450	2.7740	2.6996	2.6609	2.6211	2.5801	2.5379
	10.0443	7.5594	6.5523	5.9943	5.6363	5.3858	5.2001	5.0567	4.9424	4.8491	4.5581	4.4054	4.2469	4.1653	4.0819	3.9965	3.9100
11	4.8443	3.9823	3.5874	3.3567	3.2039	3.0946	3.0123	2.948	2.8962	2.8536	2.7186	2.6464	2.5705	2.5309	2.4901	2.4480	2.4045
	9.6460	7.2057	6.2167	5.6683	5.3160	5.0692	4.8861	4.7445	4.6315	4.5393	4.2509	4.0990	3.9311	3.8596	3.7761	3.6904	3.6035
12	4.7472	3.8853	3.4903	3.2592	3.1059	2.9961	2.9134	2.8486	2.7964	2.7534	2.6169	2.5436	2.4663	2.4259	2.3842	2.3410	2.2962
	9.3302	6.9266	5.9525	5.4120	5.0643	4.8206	4.6395	4.4994	4.3875	4.2961	4.0096	3.8584	3.7008	3.6192	3.5355	3.4494	3.3619
13	4.6672	3.8056	3.4105	3.1791	3.0254	2.9153	2.8321	2.7669	2.7144	2.6710	2.5331	2.4589	2.3803	2.3392	2.2966	2.2524	2.2064
	9.0738	6.7010	5.7394	5.2053	4.8616	4.6204	4.4410	4.3021	4.1911	4.1003	3.8154	3.6646	3.5070	3.4253	3.3413	3.2548	3.1665
14	4.6001	3.7389	3.3439	3.1122	2.9582	2.8477	2.7642	2.6987	2.6458	2.6022	2.4630	2.3879	2.3082	2.2664	2.2229	2.1778	2.1307
	8.8616	6.5149	5.5639	5.0354	4.6950	4.4558	4.2779	4.1399	4.0297	3.9394	3.6557	3.5052	3.3476	3.2656	3.1813	3.0942	3.0051
15	4.5431	3.6823	3.2874	3.0556	2.9013	2.7905	2.7066	2.6408	2.5876	2.5437	2.4034	2.3275	2.2468	2.2043	2.1601	2.1141	2.0658
	8.6831	6.3589	5.4170	4.8932	4.5556	4.3183	4.1415	4.0045	3.8948	3.8049	3.5222	3.3719	3.2141	3.1319	3.0471	2.9595	2.8695
16	4.4940	3.6337	3.2389	3.0069	2.8524	2.7413	2.6572	2.5911	2.5377	2.4935	2.3522	2.2756	2.1938	2.1507	2.1058	2.0589	2.0096
	8.5310	6.2262	5.2922	4.7726	4.4374	4.2016	4.0259	3.8896	3.7804	3.6909	3.4089	3.2587	3.1007	3.0182	2.9330	2.8447	2.7540

df	1	2	3	4	5	6	7	8	9	10	15	20	30	40	60	120	∞
17	4.4513	3.5915	3.1968	2.9647	2.8100	2.6987	2.6143	2.5480	2.4943	2.4499	2.3077	2.2304	2.1477	2.1040	2.0584	2.0107	1.9604
	8.3997	6.1121	5.1850	4.6670	4.3360	4.1015	3.9267	3.7910	3.6822	3.5931	3.3117	3.1615	3.0032	2.9205	2.8348	2.7459	2.6432
18	4.4139	3.5546	3.1599	2.9277	2.7729	2.6613	2.5767	2.5102	2.4563	2.4117	2.2686	2.1906	2.1071	2.0629	2.0166	1.9681	1.9168
	8.2854	6.0129	5.0919	4.5790	4.2479	4.0146	3.8406	3.7054	3.5971	3.5082	3.2273	3.0771	2.9185	2.8354	2.7493	2.6597	2.5671
19	4.3807	3.5219	3.1274	2.8951	2.7401	2.6283	2.5435	2.4768	2.4227	2.3779	2.2341	2.1555	2.0712	2.0264	1.9795	1.9302	1.8780
	8.1849	5.9259	5.0103	4.5003	4.1708	3.9286	3.7653	3.6305	3.5225	3.4338	3.1533	3.0031	2.8442	2.7608	2.6742	2.5839	2.4904
20	4.3512	3.4928	3.0984	2.8661	2.7109	2.5990	2.5140	2.4471	2.3928	2.3479	2.2033	2.1242	2.0391	1.9938	1.9464	1.8963	1.8432
	8.0960	5.8489	4.9382	4.4307	4.1027	3.8714	3.6987	3.5644	3.4567	3.3682	3.0880	2.9377	2.7785	2.6847	2.6077	2.5168	2.4224
21	4.3248	3.4668	3.0725	2.8401	2.6848	2.5727	2.4876	2.4205	2.3660	2.3210	2.1757	2.0960	2.0102	1.9645	1.9165	1.8657	1.8117
	8.0166	5.7804	4.8740	4.3688	4.0421	3.8117	3.6396	3.5056	3.3981	3.3098	3.0300	2.8796	2.7200	2.6359	2.5484	2.4568	2.3615
22	4.3009	3.4434	3.0491	2.8167	2.6613	2.5491	2.4638	2.3965	2.3419	2.2967	2.1508	2.0707	1.9842	1.9380	1.8894	1.8380	1.7831
	7.9454	5.7190	4.8166	4.3134	3.9880	3.7583	3.5867	3.4530	3.3458	3.2576	2.9779	2.8274	2.6675	2.5831	2.4951	2.4029	2.3067
23	4.2793	3.4221	3.0280	2.7955	2.6400	2.5277	2.4422	2.3748	2.3201	2.2747	2.1282	2.0476	1.9605	1.9139	1.8648	1.8128	1.7570
	7.8811	5.6637	4.7649	4.2636	3.9392	3.7102	3.5390	3.4057	3.2986	3.2106	2.9311	2.7805	2.6202	2.5355	2.4471	2.3542	2.2571
24	4.2597	3.4028	3.0088	2.7763	2.6207	2.5082	2.4226	2.3551	2.3002	2.2547	2.1077	2.0267	1.9390	1.8920	1.8424	1.7896	1.7330
	7.8229	5.6136	4.7181	4.2184	3.8951	3.6667	3.4959	3.363	3.2560	3.1681	2.8887	2.7380	2.5773	2.4923	2.4035	2.3100	2.2119
25	4.2417	3.3852	2.9912	2.7587	2.6030	2.4904	2.4047	2.3371	2.2821	2.2365	2.0889	2.0075	1.9192	1.8718	1.8217	1.7684	1.7110
	7.7698	5.5680	4.6755	4.1774	3.8550	3.6272	3.4568	3.3239	3.2172	3.1294	2.8502	2.6993	2.5383	2.4530	2.3637	2.2696	2.1706
26	4.2252	3.3690	2.9752	2.7426	2.5868	2.4741	2.3883	2.3205	2.2655	2.2197	2.0716	1.9898	1.9010	1.8533	1.8027	1.7488	1.6906
	7.7213	5.5263	4.6366	4.1400	3.8183	3.5911	3.4210	3.2884	3.1818	3.0941	2.8150	2.6640	2.5026	2.4170	2.3273	2.2325	2.1327
27	4.2100	3.3541	2.9604	2.7278	2.5719	2.4591	2.3732	2.3053	2.2501	2.2043	2.0558	1.9736	1.8842	1.8361	1.7851	1.7306	1.6717
	7.6767	5.4881	4.6009	4.1056	3.7848	3.5580	3.3882	3.2558	3.1494	3.0618	2.7827	2.6316	2.4699	2.3840	2.2938	2.1985	2.0978
28	4.1960	3.3404	2.9467	2.7141	2.5581	2.4453	2.3593	2.2913	2.2360	2.1900	2.0411	1.9586	1.8687	1.8203	1.7689	1.7138	1.6541
	7.6356	5.4529	4.5681	4.0740	3.7539	3.5276	3.3581	3.2259	3.1196	3.0320	2.7530	2.6017	2.4397	2.3535	2.2629	2.1670	2.0655
29	4.1830	3.3277	2.9340	2.7014	2.5454	2.4324	2.3463	2.2783	2.2229	2.1768	2.0275	1.9446	1.8543	1.8055	1.7537	1.6981	1.6376
	7.5977	5.4204	4.5378	4.0449	3.7254	3.4995	3.3303	3.1982	3.0920	3.0045	2.7256	2.574	2.4118	2.3253	2.2344	2.1379	2.0355
30	4.1709	3.3158	2.9223	2.6896	2.5336	2.4205	2.3343	2.2662	2.2107	2.1646	2.0148	1.9317	1.8409	1.7918	1.7396	1.6835	1.6223
	7.5625	5.3903	4.5097	4.0179	3.6990	3.4735	3.3045	3.1726	3.0665	2.9791	2.7002	2.5487	2.3860	2.2992	2.2079	2.1108	2.0075
40	4.0847	3.2317	2.8387	2.6060	2.4495	2.3359	2.2490	2.1802	2.1240	2.0772	1.9245	1.8389	1.7444	1.6928	1.6373	1.5766	1.5089
	7.3141	5.1785	4.3126	3.8283	3.5138	3.2910	3.1238	2.9930	2.8876	2.8005	2.5216	2.3689	2.2034	2.1142	2.0194	1.9172	1.8061
60	4.0012	3.1504	2.7581	2.5252	2.3683	2.2541	2.1665	2.0970	2.0401	1.9926	1.8364	1.7480	1.6491	1.5943	1.5343	1.4673	1.3893
	7.0771	4.9774	4.1239	3.6490	3.3388	3.1187	2.9530	2.8233	2.7185	2.6318	2.3523	2.1978	2.0285	1.9360	1.8363	1.7263	1.6023
120	3.9201	3.0718	2.6802	2.4472	2.2899	2.1750	2.0868	2.0164	1.9588	1.9105	1.7505	1.6587	1.5543	1.4952	1.4290	1.3519	1.2539
	6.8509	4.7865	3.9491	3.4795	3.1735	2.9559	2.7918	2.6629	2.5586	2.4721	2.1915	2.0346	1.8600	1.7628	1.6557	1.5330	1.3827
∞	3.8415	2.9957	2.6049	2.3719	2.2141	2.0986	2.0096	1.9384	1.8799	1.8307	1.6664	1.5705	1.4591	1.3940	1.3180	1.2214	1.0000
	6.637	4.6073	3.7836	3.3210	3.0191	2.8038	2.6411	2.5130	2.4091	2.3227	2.0403	1.8801	1.6983	1.5943	1.4752	1.3273	1.0476

p = .05 level; p = .01 level

TABLE 6 Critical Values of Chi Square

Probability under H^0 that $X^2 \geq$ chi square

df	.99	.95	.90	.85	.80	.75	.50	.25	.15	.10	.05	.02	.01	.001
1	0.00016	0.00393	0.0158	0.0358	0.06418	0.1015	0.4549	1.3233	2.0722	2.7055	3.841	5.412	6.6349	10.8276
2	0.02010	0.10259	0.2107	0.3250	0.4463	0.5754	1.386	2.7726	3.7942	4.6052	5.9915	7.8240	9.2103	13.8155
3	0.1148	0.3518	0.5844	0.7978	1.0052	1.2125	2.3660	4.1083	5.3170	6.2514	7.8147	9.8274	11.345	16.266
4	0.2971	0.7107	1.0636	1.3665	1.6488	1.9226	3.3567	5.3853	6.7449	7.7794	9.4877	11.6678	13.2767	18.4668
5	0.5543	1.1455	1.61031	1.9938	2.3425	2.6746	4.3514	6.6257	8.1152	9.2364	11.0705	13.3882	15.08627	20.5150
6	0.8721	1.6354	2.2041	2.6613	3.0701	3.4546	5.3481	7.8408	9.4461	10.6446	12.5916	15.0332	16.8119	22.4577
7	1.2390	2.1673	2.8331	3.3583	3.8223	4.2549	6.3458	9.0371	10.7479	12.0170	14.0671	16.6224	18.4753	24.3219
8	1.6465	2.7326	3.4895	4.0782	4.5936	5.0706	7.3441	10.2189	12.0271	13.3616	15.5073	18.1682	20.0902	26.1245
9	2.0880	3.3251	4.1682	4.8165	5.3801	5.8988	8.3428	11.3888	13.2880	14.6837	16.9190	19.6790	21.6660	27.8772
10	2.5582	3.9403	4.8652	5.5701	6.1791	6.7372	9.3418	12.5489	14.5339	15.9872	18.3070	21.6077	23.2093	29.5883
11	3.0535	4.5748	5.5778	6.3364	6.9887	7.5841	10.3410	13.7007	15.7671	17.2750	19.6751	22.6180	24.7250	31.2641
12	3.5706	5.2260	6.3038	7.1138	7.8073	8.4384	11.3403	14.8454	16.9893	18.5493	21.0261	24.0540	26.2170	32.9095
13	4.1069	5.8919	7.0415	7.9008	8.6339	9.2991	12.3398	15.9839	18.2020	19.8119	22.3620	25.4715	27.6882	34.5282
14	4.6604	6.5706	7.7895	8.6963	9.4673	10.1653	13.3393	17.1169	19.4062	21.0641	23.6848	26.7828	29.1412	36.1233
15	5.2293	7.2610	8.5468	9.4993	10.3070	11.0365	14.3389	18.2451	20.6030	22.3071	24.9958	28.2595	30.5779	37.6973
16	5.8122	7.9616	9.3122	10.3090	11.1521	11.9122	15.3385	19.3589	21.7931	23.5418	26.2962	29.6332	31.9999	39.2524
17	6.4078	8.6718	10.0852	11.1249	12.0023	12.7919	16.3382	20.4887	22.9770	24.7690	27.5871	30.9950	33.4087	40.7902
18	7.0149	9.3905	10.8649	11.946	12.8570	13.6753	17.3379	21.6049	24.1555	25.9894	28.8693	32.3462	34.8053	42.3124
19	7.6327	10.1170	11.6509	12.7727	13.7158	14.5620	18.3377	22.7178	25.3289	27.2036	30.1435	33.6874	36.1909	43.8202
20	8.2604	10.8508	12.4426	13.6039	14.5784	15.4518	19.3374	23.8277	26.4976	28.4120	31.4104	35.0196	37.5662	45.3147
21	8.8972	11.5913	13.2396	14.4393	15.4446	16.3444	20.3372	24.9348	27.6620	29.6151	32.6706	36.3434	38.9322	46.7970
22	9.5425	12.3380	14.0415	15.2788	16.3140	17.2396	21.3370	26.0393	28.8225	30.8133	33.9244	37.6595	40.2894	48.2679
23	10.1957	13.0905	14.8480	16.1219	17.1865	18.1373	22.3369	27.1413	29.9792	32.0069	35.1725	38.9683	41.6384	49.7282
24	10.8564	13.8484	15.6587	16.9686	18.0618	19.0373	23.3367	28.2412	31.1325	33.1962	36.4150	40.2703	42.9798	51.1786
25	11.5240	14.6114	16.4734	17.8184	18.9398	19.9393	24.3366	29.3389	32.2825	34.3816	37.6525	41.5661	44.3141	52.6197
26	12.1981	15.3792	17.2919	18.6714	19.8202	20.8434	25.3365	30.4346	33.4295	35.5632	38.8851	42.8558	45.6417	54.0520
27	12.8785	16.1514	18.1139	19.5272	20.7030	21.7494	26.3363	31.5284	34.5736	36.7412	40.1133	44.1400	46.9629	55.4760
28	13.5647	16.9279	18.9392	20.3857	21.5880	22.6572	27.3362	32.6205	35.7150	37.9160	41.3371	45.4188	48.2782	56.8923
29	14.2565	17.7084	19.7677	21.2468	22.4751	23.5666	28.3361	33.7109	36.8538	39.0875	42.5570	46.6927	49.5879	58.3012
30	14.9525	18.4927	20.5992	22.1103	23.3641	24.4776	29.3360	34.7997	37.9903	40.2560	43.7730	47.9618	50.8921	59.7031
40	22.1643	26.5093	29.0505	30.8563	32.3450	33.6603	39.3353	45.6160	49.2439	51.8051	55.7585	60.4361	63.6907	73.4020
50	29.7067	34.7643	37.6886	39.7539	41.4492	42.9421	49.3349	56.3336	60.3460	63.1671	67.5048	72.6133	76.1539	86.6608
60	37.4849	43.1880	46.4589	48.7587	50.6406	52.2938	59.3347	66.9815	71.3411	74.3970	79.0819	84.5799	88.3794	99.6072
70	45.4417	51.7393	55.3289	57.8443	59.8978	61.6983	69.3345	77.5767	82.2554	85.5270	90.5312	96.3875	100.4252	112.3169
80	53.5401	60.3915	64.2778	66.9938	69.2069	71.1445	79.3343	88.1303	93.1058	96.5782	101.8794	108.0693	112.3288	124.8392
90	61.7541	69.1260	73.2911	76.1954	78.5584	80.6247	89.3342	98.6499	103.9041	107.5650	113.1453	119.5485	124.1163	137.2084
100	70.6489	77.9295	82.3581	85.4406	87.9453	90.1332	99.3341	109.1412	114.6588	118.4980	124.3421	131.1417	135.8067	149.4493

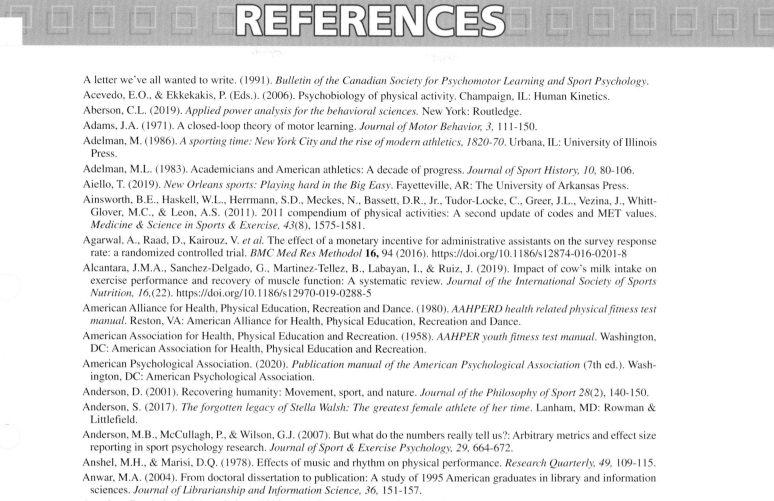

A letter we've all wanted to write. (1991). *Bulletin of the Canadian Society for Psychomotor Learning and Sport Psychology.*

Acevedo, E.O., & Ekkekakis, P. (Eds.). (2006). Psychobiology of physical activity. Champaign, IL: Human Kinetics.

Aberson, C.L. (2019). *Applied power analysis for the behavioral sciences.* New York: Routledge.

Adams, J.A. (1971). A closed-loop theory of motor learning. *Journal of Motor Behavior, 3,* 111-150.

Adelman, M. (1986). *A sporting time: New York City and the rise of modern athletics, 1820-70.* Urbana, IL: University of Illinois Press.

Adelman, M.L. (1983). Academicians and American athletics: A decade of progress. *Journal of Sport History, 10,* 80-106.

Aiello, T. (2019). *New Orleans sports: Playing hard in the Big Easy.* Fayetteville, AR: The University of Arkansas Press.

Ainsworth, B.E., Haskell, W.L., Herrmann, S.D., Meckes, N., Bassett, D.R., Jr., Tudor-Locke, C., Greer, J.L., Vezina, J., Whitt-Glover, M.C., & Leon, A.S. (2011). 2011 compendium of physical activities: A second update of codes and MET values. *Medicine & Science in Sports & Exercise, 43*(8), 1575-1581.

Agarwal, A., Raad, D., Kairouz, V. *et al.* The effect of a monetary incentive for administrative assistants on the survey response rate: a randomized controlled trial. *BMC Med Res Methodol* **16,** 94 (2016). https://doi.org/10.1186/s12874-016-0201-8

Alcantara, J.M.A., Sanchez-Delgado, G., Martinez-Tellez, B., Labayan, I., & Ruiz, J. (2019). Impact of cow's milk intake on exercise performance and recovery of muscle function: A systematic review. *Journal of the International Society of Sports Nutrition, 16,*(22). https://doi.org/10.1186/s12970-019-0288-5

American Alliance for Health, Physical Education, Recreation and Dance. (1980). *AAHPERD health related physical fitness test manual.* Reston, VA: American Alliance for Health, Physical Education, Recreation and Dance.

American Association for Health, Physical Education and Recreation. (1958). *AAHPER youth fitness test manual.* Washington, DC: American Association for Health, Physical Education and Recreation.

American Psychological Association. (2020). *Publication manual of the American Psychological Association* (7th ed.). Washington, DC: American Psychological Association.

Anderson, D. (2001). Recovering humanity: Movement, sport, and nature. *Journal of the Philosophy of Sport 28*(2), 140-150.

Anderson, S. (2017). *The forgotten legacy of Stella Walsh: The greatest female athlete of her time.* Lanham, MD: Rowman & Littlefield.

Anderson, M.B., McCullagh, P., & Wilson, G.J. (2007). But what do the numbers really tell us?: Arbitrary metrics and effect size reporting in sport psychology research. *Journal of Sport & Exercise Psychology, 29,* 664-672.

Anshel, M.H., & Marisi, D.Q. (1978). Effects of music and rhythm on physical performance. *Research Quarterly, 49,* 109-115.

Anwar, M.A. (2004). From doctoral dissertation to publication: A study of 1995 American graduates in library and information sciences. *Journal of Librarianship and Information Science, 36,* 151-157.

American Psychological Association. (1983). APA Statement on Authorship of Research Papers. *Chronicle of Higher Education, 27*(7).

Arlin, M. (1977). One-study publishing typifies educational inquiry. *Educational Researcher*, *6*(9), 11-15.

Armstrong, R.A. (2017). Recommendations for analysis of repeated-measures designs: Testing and correcting for sphericity and use of manova and mixed model analysis. *Ophthalmic and Physiological Optics*, *37*(5), 585-593.

Atencio, M., Beal, B., Wright, M., McClain, Z. (2018). *Moving Borders: Skateboarding and the changing landscape of urban youth sport.* Fayetteville: The University of Arkansas Press.

Atkinson, R.F. (1978). *Knowledge and explanation in history.* Ithaca, NY: Cornell University Press.

Atkinson, F., Short, S.E., & Martin, J. (2018). College soccer players' perceptions of coach and team efficacy. *The Sport Psychologist, 32*(3), 237-243.

Austin, B. (2015). *Democratic sports: Men's and women's college athletics during the Great Depression.* Fayetteville, AR: The University of Arkansas Press.

Bachynski, K. (2019). *No Game for Boys to Play: The History of Youth Football and the Origins of a Public Health Crisis.* Chapel Hill: University of North Carolina Press.

Bain, L.L. (1989). Interpretive and critical research in sport and physical education. *Research Quarterly for Exercise and Sport, 60,* 21-24.

Bain, L.L.(1992) *Evaluating the quality of qualitative research, a paper presented at* American Alliance for Health, Physical Education, Recreation and Dance**,** Indianapolis, Indiana, 350.

Baker, W.J. (1983). The state of British sport history. *Journal of Sport History, 10,* 53-66.

Baker, W.J. (1994). Press games: Sportswriters and the making of American football. *Reviews in American History, 22,* 530-537.

Banks, M. (2018). Using Visual Data in Qualitative Research (2nd ed.). London: Sage.

Barnett, V., & Lewis, T. (1978). *Outliers in statistical data.* New York: Wiley.

Barney, R.K. & Seagrave, J.O. (2014). From vision to reality: The pre-history of NASSH and the fermentation of an idea. *Journal of Sport History, 41,* 381-400.

Barney, R.K., Wenn, S.R., & Martyn, S.G. (2002). *Selling the five rings: The International Olympic Committee and the rise of Olympic commercialism.* Salt Lake City, UT: University of Utah Press.

Baron, R.M., & Kenny, D.A. (1986). The moderator-mediator variable distinction in social psychological research: Conceptual, strategic, and statistical considerations. *Journal of Personality and Social Psychology, 5,* 1173-1182.

Barton, M., Yeatts, P.E., Henson, R.K., & Martin, S.B. (2016). Moving beyond univariate post-hoc testing in exercise science: A primer on descriptive discriminate analysis. *Research Quarterly for Exercise and Sport, 87*(4), 365-375.

Bass, A. (2002). *Not the triumph but the struggle: The 1968 Olympics and the making of the black athlete.* Minneapolis: University of Minnesota Press.

Baumgartner, T.A. (1989). Norm-referenced measurement: Reliability. In M.J. Safrit & T.M. Wood (Eds.), *Measurement concepts in physical education and exercise science* (pp. 45-72). Champaign, IL: Human Kinetics.

Baumgartner, T.A., Jackson, A.S., Mahar, M.T., & Rowe, D.A. (2016). *Measurement for evaluation in kinesiology* (9th ed.). Burlington, MA: Jones & Bartlett.

Becker R, Möser S, Glauser D. Cash vs. vouchers vs. gifts in web surveys of a mature panel study--Main effects in a long-term incentives experiment across three panel waves. Soc Sci Res. 2019 Jul;81:221-234. doi: 10.1016/j.ssresearch.2019.02.008. Epub 2019 Feb 23. PMID: 31130198.

Bellows LL, Davies PL, Courtney JB, Gavin WJ, Johnson SL, Boles RE. (2017). Motor skill development in low-income, at-risk preschoolers: A community-based longitudinal intervention study. Journal Science and Medicine in Sport :20(11):997-1002. doi: 10.1016/j.jsams.2017.04.003. Epub 2017 Apr 26. PMID: 28506451.

Bentham, J. (1970). *Introduction to the principles of morals and legislation.* London: Athalone Press.

Berelson, B. (1960). *Graduate education in the United States.* New York: McGraw-Hill.

Berg, B.L. (2001). *Qualitative research methods for the social sciences* (4th ed.). Boston: Allyn & Bacon.

Berryman, J.W. (1973). Sport history as social history? *Quest, 20,* 65-73.

Berryman, J.W. (1982). Sport, health, and the rural–urban conflict: Baltimore and John Stuart Skinner's American farmer 1819-1820. *Conspectus of History, 1,* 43-61.

Berryman, J.W., & Park, R.J. (Eds.). (1992). *Sport and exercise science: Essays in the history of sports medicine.* Urbana, IL: University of Illinois Press.

Betts, J.R. (1951). *Organized sport in industrial America.* Doctoral presentation, Columbia University, New York.

Betts, J.R. (1974). *America's sporting heritage, 1850-1950.* Reading, MA: Addison Wesley.

Betz, N.E. (1987). Use of discriminant analysis in counseling psychology research. *Journal of Counseling Psychology, 34,* 393-403.

Bhalla, J.A., & Weiss, M.R. (2010). A cross-cultural perspective of parental influence on female adolescents' achievement beliefs and behaviors in sport and school domains. *Research Quarterly for Exercise and Sport, 81,* 494-505.

Biddle, S.J.H., Markland, D., Gilbourne, D., Chatzisarants, N.L.D., & Sparkes, A.C. (2001). Research methods in sport and exercise psychology: Quantitative and qualitative issues. *Journal of Sport Sciences, 19,* 777-809.

Bird, S.J. (2001). Mentors, advisors, and supervisors: their role in teaching responsible research conduct. *Science and Engineering Ethics, 7,* 455-68.

Blackwell, D.L., & Clarke, T.C. (2018). State variation in meeting the 2008 federal guidelines for both aerobic and muscle-strengthening physical activity among adults aged 18-64: United States 2010-2015. *National Health Statistics Reports, 112.* Hyattsville, MD: National Center for Health Statistics.

Blair, S.N. (1993). Physical activity, physical fitness, and health (1993 C. H. McCloy Research Lecture). *Research Quarterly for Exercise and Sport, 64,* 365-376.

Bloom, J. (2000). *To show what an Indian can do: Sports at Native American boarding schools.* Minneapolis, MN: University of Minnesota Press.

Bogdan, R.C., & Biklen, S.K. (2016). *Qualitative research for education: An introduction to theories and method* (5th ed.). Boston: Allyn & Bacon.

Booth, D. (1997). Sports history: What can be done? *Sport, Education and Society, 2,* 191-204.

Booth, D. (1999). Sport history. In D. Booth & A. Jutel (Eds.), *Sporting Traditions, 16,* 5-15.

Booth, D. (2001). From ritual to record: Allen Guttmann's insights into modernization and modernity. *Sport History Review, 32,* 19-27.

Booth, D. (2003). Theory: Distorting or enriching sport history? *Sport History Review, 34,* 1-32.

Booth, D. (2005). *The field: Truth and fiction in sport history.* New York: Routledge.

Booth, D., & Loy, J. (1999). Sport, status, and style. *Sport History Review, 30,* 1-26.

Borg, G.A. (1962). *Physical performance and perceived exertion.* Lund, Sweden: Gleerup.

Borg, W.R., & Gall, M.D. (1989). *Educational research* (5th ed.). New York: Longman.

Borish, L.J. (1996). Do not neglect exercise nor recreation: Rural New Englanders, sport and health concerns. *Colby Quarterly, 32,* 25-35.

Bouchard, C., Blair, S.N., , &Haskell, W.L. (Eds.). (2012). *Physical activityand health (2nd ed.).* Champaign, IL: Human Kinetics.

Bourque, L.B., & Fielder, E.P. (2003). *How to conduct in-person interviews for surveys* (2nd ed.). Thousand Oaks, CA: Sage.

Boyer, C.J. (1973). *The doctoral dissertation as an informational source: A study of scientific information flow.* Metuchen, NJ: Scarecrow Press.

Briefing (2013). Trouble at the Lab. *The Economist.* October 19, p. 2. https://www.economist.com/briefing/2013/10/18/trouble-at-the-lab

Briesach, E. (1994). *Historiography: Ancient, medieval, & modern* (2nd ed.). Chicago: University of Chicago Press.

Brown, J.A.C. (1954). *The social psychology of industry.* Middlesex, England: Penguin.

Brownstein, N.C., Louis, T.A., O'Hagan, A., & Pendergast, J. (2019). The role of expert judgment in statistical inference and evidence-based decision-making. *The American Statistician, 73*(supplement 1), 56-68.

Brownell, S. (2001). The problems with ritual and modernization theory, and why we need Marx: A commentary on From Ritual to Record. *Sport History Review, 32,* 28-41.

Brunnström, K. and Barkowsky, M. (2018). Statistical quality of experience analysis: on planning the sample size and statistical significance testing, *Journal of Electronic Imaging*, 27, 053013. https://doi.org/10.1117/1.JEI.27.5.053013

Buckner, S.L., Jessee, M.B., Mouser, J.G., Dankel, S.J., Mattocks, K.T., Bell, Z.W., Abe, T., & Loenneke, J.P. (2020). The basics of training for muscle size and strength: a brief review on the theory. *Medicine and Science in Sports and Exercise*, *52*, 645-53.

Burke, P. (1992). *History and social theory*. Ithaca, NY: Cornell University Press.

Cahn, S. (1994). *Coming on strong: Gender and sexuality in twentieth-century women's sport*. New York: The Free Press.

Campbell, D.T., & Stanley, J.C. (1963). *Experimental and quasi-experimental designs for research*. Chicago: Rand McNally.

Cardinal, B.J., & Thomas, J.R. (2005). The 75th anniversary of *Research Quarterly for Exercise and Sport*: An analysis of status and contributions. *Research Quarterly for Exercise and Sport*, *76*(supplement 2), S122-S134.

Carlberg, C.C., Johnson, D.W., Johnson, R., Maruyama, G., Kavale, K., Kulik, C., Kulik, J.A., Lysakowski, R.S., Pflaum, S.W., & Walberg, H. (1984). Meta-analysis in education: A reply to Slavin. *Educational Researcher, 13*(4), 16-23.

Carron, A.V., Widmeyer, W.N., & Brawley, L.R. (1985). The development of an instrument to assess cohesion in sport teams: The group environment questionnaire. *Journal of Sport Psychology, 7,* 244-266.

Caspersen, C.J. (1989). Physical activity epidemiology: Concepts, methods, and applications to exercise science. *Exercise and Sport Sciences Reviews, 17,* 423-474.

Cavello, D. (1981). *Muscles and morals: Organized playgrounds and urban reform*. Philadelphia: University of Pennsylvania Press.

Chan, Z.Y.S., Zhang, J.H., Au, I.P.H., An, W.W., Shum, G.L.K., Ng, G.Y.F., & Cheung, R.T.H. (2018). Gait retraining for reduction of injury occurrence in novice distance runners: 1-year follow-up of a randomized control trial. *American Journal of Sports Medicine*, 46, 388-95.

Chang, A.H., Song, J., Lee, J., Chang, R.W., Semanik, P.A., Dunlop, D.D. (2020). Proportion and associated factors of meeting the 2018 Physical Activity Guidelines for Americans in adults with or at risk for knee osteoarthritis. *Osteoarthritis Cartilage, 28*(6), 774-781.

Cheffers, J.T.F. (1973). The validation of an instrument design to expand the Flanders' system of interaction analysis to describe nonverbal interaction, different varieties of teacher behavior and pupil responses (Doctoral dissertation, Temple University, Philadelphia, 1972). *Dissertation Abstracts International, 34,* 1674A.

Chien, I. (1981). Appendix: An introduction to sampling. In L. H. Kidder (Ed.), *Selltiz, Wrightsman, and Cook's research methods in social relations* (4th ed.). New York: Holt, Rinehart & Winston.

Christina, R.W. (1989). Whatever happened to applied research in motor learning? In J. S. Skinner et al. (Eds.), *Future directions in exercise and sport science research* (pp. 411-422). Champaign, IL: Human Kinetics.

Clarke, H.H. (Ed.). (1968, December). *Physical Fitness Newsletter, 14*(4).

Clarke, H.H., & Clarke, D.H. (1970). *Research processes in physical education, recreation, and health*. Englewood Cliffs, NJ: Prentice Hall.

Cohen, D.J., & Rosenzweig, R. (2006). *Digital history: A guide to gathering, preserving, and presenting the past on the web*. Philadelphia: University of Pennsylvania Press.

Cohen, J. (1988). *Statistical power analysis for the behavioral sciences* (2nd ed.). New York: Academic Press.

Cohen, J. (1990). Things I have learned (so far). *American Psychologist, 45,* 1304-1312.

Cohen, J. (1994). The earth is round ($p < .05$). *American Psychologist, 49,* 997-1003.

Cohen, J., Cohen, P, West, S.G., & Aiken, L.S. (2003). *Applied multiple regression / correlation analysis for the behavioral sciences*. Mahwah, NJ: Lawrence Erlbaum Associates.

Cooper, H., & Hedges, L.V. (Eds.). (1994). *The handbook of research synthesis*. New York: Sage Foundation.

Coorough, C., & Nelson, J.K. (1997). The dissertation in education from 1950 to 1990. *Educational Research Quarterly, 20*(4), 3-14.

Corbin, J., & Strauss, A. (2015). *Basics of qualitative research* (4rd ed.). Los Angeles: Sage.

Cornett, A.C., Duski, A., Wagner, S., Wright, B.V., & Stager, J.M. (2017). Maturational timing and swim performance in collegiate female swimmers. *Journal of Swimming Research, 25*(1), 11-19.

Costill, D.L. (1985). Practical problems in exercise physiology research. *Research Quarterly for Exercise and Sport, 56,* 378-384.

Cox, R.W., & Salter, M.A. (1998). The IT Revolution and the practice of sport history: An overview and reflection on Internet research and teaching resources. *Journal of Sport History, 25,* 283-302.

Cozens, F., & Stumpf, F. (1953). *Sports in American life*. Chicago: University of Chicago Press.

Crase, D., & Rosato, F.D. (1992). Single versus multiple authorship in professional journals. *Journal of Physical Education, Recreation and Dance, 63*(7), 28-31.

Creswell, J.W. (2009). Mapping the field of mixed methods research. *Journal of Mixed Methods Research, 3,* 95-108.

Creswell, J.W. & Creswell, J.D. (2018). Research Design; Qualitative, Quantitative and Mixed Methods Approaches(5th ed.). Thousand Oaks, CA: Sage.

Creswell, J.W. & Poth, C.N. (2018). *Qualitative inquiry: Choosing among five approaches* (4th ed.). Thousand Oaks, CA: Sage.

Creswell, J.W. (2014). *Research design: Qualitative, quantitative and mixed methods approaches* (4th ed.). Los Angeles: Sage.

Creswell, J.W., & Plano-Clark, V.L. (2006). *Designing and conducting mixed methods research*. Thousand Oaks, CA: Sage.

Cronbach, L. (1951). Coefficient alpha and the internal structure of tests. *Psychometrika, 16,* 297-334.

Curran-Everett, D., & Benos, D.J. (2004). Guidelines for reporting statistics in journals published by the American Physiological Society. *Journal of Applied Physiology, 97,* 457-459.

Danion, F., & Latask, M.L. (Eds.). (2011). *Motor control: theories, experiments, and applications*. New York, NY: Oxford University Press.

Davis, D. (2015). *Waterman: The life and times of Duke Kahanamoku*. Lincoln, NE: University of Nebraska Press.

Davis, R.O. (2014). *The main event: Boxing in Nevada from the mining camps to the Las Vegas Strip*. Las Vegas, NV: University of Nevada Press.

Day, R.A. (1983). *How to write and publish a scientific paper* (2nd ed.). Philadelphia: Institute for Scientific Information.

Delaney, M., Warren, M., Kinslow, B., de Heer, H., Ganley, K. (2019). Association and dose–response relationship of self-reported physical activity and disability among adults ≤50 years: National Health and Nutrition Examination Survey, 2011-2016. *Journal of Aging and Physical Activity, 28,* 434-441.

Demas, L. (2017). *Game of privilege: An African American history of golf.* Chapel Hill, NC: The University of North Carolina Press.

Dennett, D. (1991). *Consciousness explained.* Boston: Little, Brown.

Denton, J.J., & Tsai, C. (1991). Two investigations into the influence of incentives and subject characteristics on mail survey responses in teacher education. *Journal of Experimental Education, 59,* 352-366.

Denzin, N.K. (2008). The new paradigm dialog and qualitative inquiry. *International Journal of Qualitative Studies in Education, 21,* 315-325.

Diekfuss, J.A., Rhea, C.K., Schmitz, R.J., Grooms, D.R., Wilkins, R.W., Slutsky, A. B., & Raisbeck, L.D. (2019). The influence of attentional focus on balance control over seven days of training. *Journal of motor behavior, 51*(3), 281-292.

DeShaw, K.J. (2019). *Methods and evaluation of a health coach training practicum experience for healthy lifestyle behavior change.* (Publication No. 13904797) [Doctoral dissertation, Iowa State University]. ProQuest Dissertations & Theses Global.

Dinces, S. (2018). *Bulls markets: Chicago's basketball business and the new inequality.* Chicago: University of Chicago Press.

Drowatzky, J.N. (1993). Ethics, codes, and behavior. *Quest, 45,* 22-31.

Drowatzky, J.N. (1996). *Ethical decision making in physical activity research.* Champaign, IL: Human Kinetics.

Duke, N.K. & Beck, S.W. (April 1999). Education should consider alternative formats for the dissertation. *Educational Researcher, 28*(3), 31-36.

Dulles, F.R. (1965). *A history of recreation: America learns to play.* New York: Appleton-Century-Crofts.

Dunn, A.L., Garcia, M. E., Marcus, B. H., Kampert, J. B., Kohl, H. W., III, & Blair, S. N. (1998). Six-month physical activity and fitness changes in Project Active, a randomized trial. *Medicine & Science in Sports & Exercise, 30,* 1076-1083.

Dyreson, M. (1998). *Making the American team: Sport, culture and the Olympic experience.* Urbana: University of Illinois Press.

Dyreson & Schultz (2015). American National Pastimes-A History. London: Routledge.

Edwards, A.L. (1957). *Techniques of attitude and scale construction.* New York: Appleton-Century-Crofts.

Eisen, G., & Wiggins, D.K. (1994). *Ethnicity and sport in North American history and culture.* Westport, CT: Greenwood Press.

Elcombe, T.L (2018). Sport philosophy inquiry in 3D: A response to the (sport) philosophy paradox. *Sport, Ethics and Philosophy, 12*(3), 317-333.

Ellingson, L.D., Meyer, J.D., Shook, R.P., Dixon, P.M., Hand, G.A., Wirth, M.D., Paluch, A.E., Burgess, S., Hebert, J.R., & Blair, S.N. (2018). Changes in sedentary time are associated with changes in mental wellbeing over 1 year in young adults. *Preventive Medicine Reports, 11,* 274-81.

Erickson, F. (1986). Qualitative methods in research on teaching. In M.C. Wittrock (Ed.), *Handbook of research on teaching* (3rd ed., pp. 119-161). New York: Macmillan.

Evans, L., & Rees, S. (2012). An interpretation of digital humanities. In D.M. Barry (Ed.), *Understanding digital humanities.* London: Palgrave MacMillan.

Evans, S.C., Amaro, C.M., Herbert, R., Blossom, J.B., & Roberts, M.C. (2018). "Are you gonna publish that?" Peer-reviewed publication outcomes of doctoral dissertations in psychology. *PLoS One,* 13(2): e0192219.

Eye, A., & Mun, E. (2012). *Log-linear modeling: Concepts, interpretation, and application.* Hoboken, NJ: John Wiley & Sons.

Eyler, M.H. (1974). Objectivity and selectivity in historical inquiry. *Journal of Sport History, 1,* 63-76.

Fahlberg, L.L., & Fahlberg, L.A. (1994). A human science for the study of movement: An integration of multiple ways of knowing. *Research Quarterly for Exercise and Sport, 65,* 100-109.

Fair, J.D. (1999). *Muscletown USA: Bob Hoffman and the manly culture of York barbell.* University Park: Pennsylvania State University Press.

Farquhar, A.B., & Farquhar, H. (1891). *Economic and industrial delusions: A discourse of the case for protection.* New York: Putnam.

Farren, G.L. (2014). *Factors related to meeting physical activity guidelines in college students: a social cognitive perspective.* (Publication No. 1691230) [Master's thesis, University of North Texas]. ProQuest Dissertations & Theses Global.

Festle, M.J. (1996). *Playing nice: Politics and apologies in women's sports.* New York: Columbia University Press.

Fields, S.K. (2005). *Female gladiators: Gender, law, and contact sport in America.* Urbana, IL: University of Illinois Press.

Fine, M.A., & Kurdek, L.A. (1993). Reflections on determining authorship credit and authorship order on faculty–student collaborations. *American Psychologist, 48,* 1141-1147.

Fink, A (2016). How to Conduct Surveys: A Step-by-Step Guide. (6th ed.). Thousand Oaks, CA: Sage.

Fink, A. (2003). *How to sample in surveys* (2nd ed.). Thousand Oaks, CA: Sage.

Fink, A. (2005). *How to sample in surveys: A step-by-step guide* (3rd ed.). Thousand Oaks, CA: Sage.

Fischman, M.G. (2014). The importance of peer review: thoughts on Knudson, Morrow, and Thomas (2014). *Research Quarterly for Exercise and Sport, 85,* 449-50.

Fischman, M.G., Christina, R.W., & Anson, J.G. (2008). Memory drum theory's C movement: Revelations from Franklin Henry. *Research Quarterly for Exercise and Sport, 79,* 312-318.

Flanders, N.A. (1970). *Analyzing teaching behavior.* Reading, MA: Addison-Wesley.

Flick, U. (2014). *An introduction to qualitative research* (5th ed.). Los Angeles: Sage.

Forsyth, J., & Giles, A.R., (2013). Aboriginal peoples and sport in Canada: Historical foundations and contemporary issues. Vancouver: University of British Columbia Press.

Fowler, F.J., Jr. (2009). *Survey research methods* (4th ed.). Thousand Oaks, CA: Sage.

Freedson, P.S., Melanson, E., Sirard, J. (1998). Calibration of the Computer Science and Applications, Inc. accelerometer. *Medicine & Science in Sport & Exercise, 30*(5), 777-81.

French, K.E., & Thomas, J.R. (1987). The relation of knowledge development to children's basketball performance. *Journal of Sport Psychology, 9,* 15-32.

Fritz, C.O., & Morris, P.E. (2012). Effect size estimates: current use, calculations, and interpretation. *Journal of Experimental Psychology: General, 141,* 2-18.

Gage, N.L. (1989). The paradigm wars and their aftermath: A "historical" sketch of research on teaching since 1989. *Educational Researcher, 18*(7), 4-10.

Gall, M.D., Borg, W.R., & Gall, J.P. (1996). *Educational research: An introduction* (6th ed.). White Plains, NY: Longman Publishing.

Gall, M.D., Gall, J.P., & Borg, W.R. (2006). *Educational research: An introduction* (8th ed.). Boston: Allyn & Bacon.

Gardner, S.K. (2008). "What's too much and what's too little?": the process of becoming an independent researcher in doctoral education. *The Journal of Higher Education, 79:3,* 326-50. 10.1080/00221546.2008.11772101.

Gastel, B. & Day, R.A. (2016). *How to write and publish a scientific paper* (8th ed.). Westport, CT: Greenwood Press.

Gelo, O., Braakmann, D., & Benetka, G. (2008). Quantitative and qualitative research: Beyond the debate. *Integrative Psychological and Behavioral Science, 42,* 266-290.

Gems, G.R. (2013). *Sport and the shaping of Italian-American identity.* Syracuse, NY: Syracuse University Press.

Gems, G.R. (1997). *Windy City wars: Labor, leisure, and sport in the making of Chicago.* Lanham, MD: Scarecrow Press.

Gill, D.L., & Deeter, T.E. (1988). Development of the sport orientation questionnaire. *Research Quarterly for Exercise and Sport, 59,* 191-202.

Glaser, B.G., & Strauss, A.L. (2017). *The discovery of grounded theory.* Abingdon, Oxon: Routledge.

Glass, G.V. (1976). Primary, secondary, and meta-analysis. *Educational Researcher, 5,* 3-8.

Glass, G.V. (1977). Integrating findings: The meta-analysis of research. *Review of Research in Education, 5,* 351-379.

Glass, G.V., McGaw, B., & Smith, M. (1981). *Meta-analysis in social research.* Beverly Hills, CA: Sage.

Glass, G.V., & Smith, M.L. (1979). Meta-analysis of research on the relationship of class-size and achievement. *Evaluation and Policy Analysis, 1,* 2-16.

Glassford, R.G. (1987). Methodological reconsideration: The shifting paradigms. *Quest, 39*(3), 295-312.

Glidden, A.R. (2010). *The blunted insulin release after exercise and the relationship with gastric inhibitory polypeptide and glucagon-like peptide-1.* (Publication No. 1487975) [Master's thesis, Iowa State University]. ProQuest Dissertations & Theses Global.

Goetz, J.P., & LeCompte, M.D. (1984). *Ethnography and qualitative design in educational research.* Orlando, FL: Academic Press.

Goldstein, W. (1989). *Playing for keeps: A history of early baseball.* Ithaca, NY: Cornell University Press.

Gordon, E.S., Tucker, P., Burke, S.M., & Carron, A.V. (2013). Effectiveness of physical activity interventions for preschoolers: A meta-analysis. *Research Quarterly for Exercise and Sport, 84,* 287-294.

Gorn, E.J. (1986). *The manly art: Bare-knuckle prize fighting in America.* New York: Cornell University Press.

Gorn, E.J., & Oriard, M. (1995, March 24). Taking sports seriously. *Chronicle of Higher Education,* p. A52.

Gottschalk, L. (1969). *Understanding history: A primer of historical method.* New York: Knopf.

Gravetter, F.J., & Wallnau, L.B. (2017). Statistics for the behavioral sciences (10th ed.). Belmont, CA: Wadsworth.

Green, H. (1986). *Fit for America: Health, fitness and sport in American society.* Baltimore: Johns Hopkins University Press.

Green, J.C., & Caracelli, V.J. (Eds.). (1997). *New directions for evaluation, number 74: Advances in mixed-method evaluation, the challenges and benefits of integrating diverse paradigms.* San Francisco: Jossey-Bass.

Green, K.E. (1991). Reluctant respondents: Differences between early, late, and nonresponders to a mail survey. *Journal of Experimental Education, 59,* 268-276.

Griffin, P., & Templin, T.J. (1989). An overview of qualitative research. In P.W. Darst, D.B. Zakrajsek, & V.H. Mancini (Eds.), *Analyzing physical education and sport instruction* (2nd ed., pp. 399-410). Champaign, IL: Human Kinetics.

Gromeier, M., Koester, D., & Schack, T. (2017). Gender differences in motor skills of the overarm throw. *Frontiers in psychology, 8*(212), 1-12.

Grundy, P. & Shackelford, S. (2007). *Shattering the glass: The remarkable history of women's basketball.* Chapel Hill, NC: University of North Carolina Press.

Grossman, G.D., & DeVries, D.R. (2019). Authorship decisions in ecology, evolution, organismal biology and natural resource management: who, why, and how. *Animal Biodiversity and Conservation, 42,* 337-46.

Gruneau, R. (1983). *Class, sports, and social development.* Amherst: University of Massachusetts Press.

Guba, E.G., & Lincoln, Y.S. (1981). *Effective evaluation.* San Francisco: Jossey-Bass.

Gunnell, K.E., Gareau, A., & Gaudreau, P. (2016). In Ntounamis, N., & Myers, N.D. (Eds.), *An introduction to intermediate and advance statistical analyses for sport and exercise scientists.* Wiley & Sons.

Gunnell, K.E., Poitras, V.J., & Tod, D. (2020) Questions and answers about conducting systematic reviews in sport and exercise psychology, *International Review of Sport and Exercise Psychology,* https://doi.org/10.1080/1750984X.2019.1695141

Guttmann, A. (1978). *From ritual to record: The nature of modern sport.* New York: Columbia University Press.

Guttmann, A. (1983). Recent work in European sport history, *Journal of Sport History, 10,* 35-52.

Guttmann, A. (1984). *The games must go on: Avery Brundage and the Olympic movement.* New York: Columbia University Press.

Guttmann, A. (1991). *Women's sport: A history.* New York: Columbia University Press.

Guttmann, A. (1992). *The Olympics: A history of the modern games.* Urbana: University of Illinois Press.

Guttmann, A. (2001). From ritual to record: A retrospective critique. *Sport History Review, 32,* 2-11.

Haase, R.F., & Ellis, M.V. (1987). Multivariate analysis of variance. *Journal of Counseling Psychology, 34,* 404-413.

Hagen, R.L. (1997). In praise of the null hypothesis statistical test. *American Psychologist, 52,* 15-24.

Halverson, L.E., Roberton, M.A., & Langendorfer, S. (1982). Development of the overarm throw: Movement and ball velocity changes by seventh grade. *Research Quarterly for Exercise and Sport, 53,* 198-205.

Hamlyn, D.W. (1990). *In and out of the black box.* Oxford, England: Basil Blackwell.

Hardy, S. (1981). The city and the rise of American sport, 1820-1920. *Exercise and Sports Sciences Reviews, 9,* 183-229.

Hardy, S. (1982). *How Boston played: Sport, recreation and community, 1865-1915.* Boston: Northeastern University Press.

Hardy, S. (1990). Adopted by all the leading clubs: Sporting goods and the shaping of leisure. In R. Butsch (Ed.), *For fun and profit: The transformation of leisure into consumption* (pp. 71-92). Philadelphia: Temple University Press.

Harlow, L.L., Mulaik, S.A., & Steigher, J.H. (2016). *What if there were no significance tests?* New York: Routledge.

Harris, C. (1963). *Problems in measuring change.* Madison: University of Wisconsin Press.

Hartmann, D. (2003). *Race, culture, and the revolt of the black athlete: The 1968 Olympic protests and their aftermath.* Chicago: University of Chicago Press.

Harvey, W.J., Reid, G., Bloom, G.A., Staples, K., Grizenko, N., Mbebou, V., Ter-Stepanian, M., & Joober, R. (2009). Physical activity experiences of boys with and without ADHA. *Adapted Physical Activity Quarterly, 26,* 131-150.

Harwell, M.R. (1990). A general approach to hypothesis testing for nonparametric tests. *Journal of Experimental Education, 58,* 143-156.

Hatch, A. (2006). Research: Musings of a former *QSE* editor. *International Journal of Qualitative Studies in Education, 19,* 403-409.

Haverkamp, N., & Beauducel, A. (2017). Violation of the sphericity assumption and its effect on type-I error rates in repeated measures ANOVA and multi-level linear models (MLM). *Frontiers in psychology, 8,* 1841.

Hayes, A.F. (2009). Beyond Baron and Kenny: Statistical mediation analysis in the new millennium. *Communication monographs, 76*(4), 408-420.

Hedges, L.V. (1981). Distribution theory for Glass's estimator of effect size and related estimators. *Journal of Educational Statistics, 6,* 107-128.

Hedges, L.V. (1982a). Estimation of effect size from a series of independent experiments. *Psychological Bulletin, 92,* 490-499.

Hedges, L.V. (1982b). Fitting categorical models to effect sizes from a series of experiments. *Journal of Educational Statistics, 7,* 119-137.

Hedges, L.V. (1984). Estimation of effect size under nonrandom sampling: The effects of censoring studies yielding statistically insignificant mean differences. *Journal of Educational Statistics, 9,* 61-85.

Hedges, L.V., & Olkin, I. (1980). Vote counting methods in research synthesis. *Psychological Bulletin, 88,* 359-369.

Hedges, L.V., & Olkin, I. (1983). Regression models in research synthesis. *American Statistician, 37,* 137-140.

Hedges, L.V., & Olkin, I. (1985). *Statistical methods for meta-analysis.* New York: Academic Press.

Helmstadter, G.C. (1970). *Research concepts in human behavior.* New York: Appleton-Century-Crofts.

Henderson, K.A., Bialeschki, M.D., Shaw, S.M., & Freysinger, V.J. (1996). *Both gains and gaps: Feminist perspectives on women's leisure.* State College, PA: Venture.

Henry, F.M., & Rogers, D.E. (1960). Increased response latency for complicated movements and a "memory drum" theory of neuromotor reaction. *Research Quarterly, 31,* 448-458.

Herkowitz, J. (1984). Developmentally engineered equipment and playgrounds. In J.R. Thomas (Ed.), *Motor development during childhood and adolescence.* Minneapolis, MN: Burgess.

Higgins, J.P.T., Thomas, J., Chandler, J., Cumpston, M., Li, T., Page, M., & Welch, V. (Eds.) (2021). *Cochrane handbook for systematic reviews of interventions version 6.2.* Retrieved from www.training.cochrane.org/handbook

Hill, A.B. (1965). The environment and disease: Association or causation? *Journal of the Royal Society of Medicine, 58,* 295-300.

Hill, J. (1996). British sports history: A post-modern future. *Journal of Sport History, 23,* 1-19.

Hoberman, J. (1997). *Darwin's athletes: How sport has damaged black America and preserved the myth of race.* Boston: Houghton Mifflin.

Holliman, J. (1931). *American sports (1785-1835).* Durham, NC: Seeman Press.

Hollingsworth, M.A., & Fassinger, R.E. (2002). The role of faculty mentors in the research training of counseling psychology doctoral students. *Journal of Counseling Psychology, 49*(3), 324-330.

Holt, R. (1998). Sport and history: British and European traditions. In L. Allison (Ed.), *Taking sport seriously* (pp. 7-30). Aachen, Germany: Meyer and Meyer.

Howell, C. (1998). On Metcalfe, Marx, and materialism: Reflections on the writing of sport history in the postmodern age. *Sport History Review, 29,* 96-102.

Howell, C.D. (2001). Of remembering and forgetting: *From Ritual to Record* and beyond. *Sport History Review, 32,* 12-18.

Hunter, J.E., & Schmidt, F.L. (2004). *Methods of meta-analysis: Correcting error and bias in research findings* (2nd ed.). Thousand Oaks, CA: Sage.

Hutchinson, J.C., Jones, L., Vitti, S.N., Moore, A., Dalton, P.C., & O'Neil, B.J. (2018). The influence of self-selected music on affect-regulated exercise intensity and remembered pleasure during treadmill running. *Sport, Exercise, and Performance Psychology, 7*(1), 80-92.

Hyde, J.S. (1981). How large are cognitive gender differences? A meta-analysis using v^2 and *d. American Psychologist, 36*(8), 892-901.

Iacovetta, F. & Mitchinson, W. (Eds.). (1998). *On the case: Explorations in social history.* Toronto: University of Toronto Press.

Iacovetta, F., & Mitchinson, W. (1998). Social history and case files research. In F. Iacovetta and W. Mitchinson (Eds.), *On the case: Explorations in social history.* Toronto: University of Toronto Press.

ICSSPE Bulletin, 27, Fall 1999.

Iggers, G.G. (1997). *Historiography in the twentieth century: From scientific objectivity to postmodern challenge.* Middletown, CT: Wesleyan University Press.

Ingrassia, B.M. (2012). *The rise of Gridiron University: Higher education's uneasy alliance with big-time football.* Lawrence, KS: University Press of Kansas.

Jaccard, J, & Jacoby, J. (2020). *Theory construction and model-building skills (2nd ed.).* New York: Guilford.

Jacks, P., Chubin, D.E., Porter, A.L., & Connally, T. (1983). The ABCs of ABDs: An interview study of incomplete doctorates. *Improving College and University Teaching, 31,* 74-81.

Jacobs, D., Anderson, J., & Blackburn, H. (1979). Diet and serum cholesterol: Do zero correlations negate the relationship? *American Journal of Epidemiology, 110,* 77-87.

Jalongo, M.R. (2012). Getting on the conference program and writing a practical article: Templates for success. *Early Childhood Education Journal.* https://doi.org/10.1007/s10643-012-0533-x

Jewczyn, N. (2013). Theory guidance in social science and finance application. *International Journal of Social Science, 1,* 72-82.

Jiang, Y., Huang, G., & Fisher, B. (2019). Air quality, human behavior and urban park visit; A case study in Beijing. *Journal of Cleaner Production.* 240. https://doi.org/10.1016/j.jclepro.2019.118000

Johal, G.S., Hulbert, S.H., & Briggs, S.P. (1995). Disease lesion mimics of maize: a model for cell death in plants. *BioEssays, 17,* 685-92.

Johnson, R.L. (1979). *The effects of various levels of fatigue on the speed and accuracy of visual recognition.* Unpublished doctoral dissertation, Louisiana State University, Baton Rouge.

Johnson, E.B., Ziegler, G., Penny, W., Tabrizi, S.J., Scahill, R.I., Gregory, S. (2021). Dynamics of cortical degeneration over a decade in Huntington's disease. Biological Psychiatry, 89: 807-816. DOI:https://doi.org/10.1016/j.biopsych.2020.11.009

John, E.M, Koo, J., and Horn-Ross, P.L. (2010.) "Lifetime Physical Activity and Risk of Endometrial Cancer," *Cancer Epidemiology Biomarkers and Prevention* 19(5): 1276-83. DOI: 10.1158/1055-9965.EPI-09-1316.

Joint Committee on Quantitative Assessment of Research. (2008). *Citation statistics.* Berlin, Germany: International Mathematical Union.

Jones, E.R. (1988, Winter). Philosophical tension in a scientific discipline: So what else is new? *NASP–SPA Newsletter, 14*(1), 10-16.

Kalinowski, P., & Fidler, F. (2010). Interpreting significance: the differences between statistical significance, effect size, and practical importance. *Newborn and Infant Nursing Reviews, 10*(1), 50-54.

Kaliss, G.J. (2012). *Men's college athletics and the politics of racial equality: Five pioneer stories of black manliness, white citizenship, and American democracy.* Philadelphia, PA: Temple University Press.

Kang, N., Summers, J.J., & Cauraugh, J.H. (2015). Transcranial direct current stimulation facilitates motor learning post-stroke: a systematic review and meta-analysis. *Journal of Neurology, Neurosurgery and Psychiatry, 87,* 1-11.

Kalichman, M., Sweet, M., & Plemmons, D. (2014). Standards of scientific conduct: are there any? *Science and Engineering Ethics, 20,* 885, 96.

Keating, X.D., & Silverman, S. (2004). Physical education teacher attitudes toward fitness test scale: Development and validation. *Journal of Teaching in Physical Education, 23,* 143-161.

Kendall, M.G. (1959). Hiawatha designs an experiment. *American Statistician, 13,* 23-24.

Kennedy, M.M. (1979). Generalizing from single case studies. *Evaluation Quarterly, 3,* 661-679.

Keppel, G. & Wickens, T.D. (2004). *Design and analysis: A researcher's handbook.* Upper Saddle River, NJ: Pearson Prentice Hall.

Kim, H.Y. (2013). Statistical notes for clinical researchers: assessing normal distribution (2) using skewness and kurtosis. *Restorative Dentistry & Endodontics, 38*(1), 52-54.

Kirk, J., & Miller, M.L. (1986). *Reliability and validity in qualitative research.* Newbury Park, CA: Sage.

Kirk, R.E. (1985). *Experimental design: Procedures for the behavioral sciences* (3rd ed.). Pacific Grove, CA: Brooks.

Kirk, R.E. (2013). Experimental design (4th ed.). Los Angeles: Sage.

Kirsch, G.B. (1989). *The creation of American team sports: Baseball and cricket, 1838-72.* Urbana, IL: University of Illinois Press.

Knudson, D.V., Morrow, J.R., & Thomas, J.R. (2014). Advancing kinesiology through improved peer review. *Research Quarterly for Exercise and Sport, 85* (2), 127-135.

Kobayashi, L.C., Janssen, I., Richardson, H., Lai, A.S., Spinelli, J.J., & Aronson, K.J. (2013). Moderate-to-vigorous intensity physical activity across the life course and risk of pre- and post-menopausal breast cancer. *Breast Cancer Research and Treatment, 139,* 851-861.

Kohl, H.W., III, Dunn, A.L., Marcus, B.H., & Blair, S.N. (1998). A randomized trial of physical activity interventions: Design and baseline data from Project Active. *Medicine & Science in Sports & Exercise, 30,* 275-283.

Kołodziejczyk, M., Chmura, P., Milanovic, L., Konefał, M., Chmura, J., Rokita, A., & Andrzejewski, M. (2021). How did three consecutive matches with extra time affect physical performance? A case study of the 2018 football Men's World Cup. *Biology of sport, 38*(1), 65–70.

Krathwohl, D.R. (1993). *Methods of educational and social science research.* New York: Longman.

Kraus, H., & Hirschland, R.P. (1954). Minimum muscular fitness tests in school children. *Research Quarterly, 25,* 177-188.

Kretchmar, R.S. (1997). Philosophy of sport. In J.D. Massengale & R.A. Swanson (Eds.), *The history of exercise and sport science* (pp. 181-202). Champaign, IL: Human Kinetics.

Kretchmar, R.S. (2005). *Practical philosophy of sport and physical activity* (2nd ed.). Champaign, IL: Human Kinetics.

Kretchmar, R.S. (November, 2007). What to do with meaning? A research conundrum. *Quest, 59,* 373-383.

Kretchmar, R.S. (2015). Pluralistic internalism. *Journal of the Philosophy of Sport, 42*(1), 83-100.

Kretchmar, S. (2005). Jigsaw puzzles and riverbanks: Two ways of picturing our future. *Quest, 57*(1), 171-177.

Kretchmar, S. (2008). The utility of silos and bunkers in the evolution of kinesiology. *Quest, 60*(1), 3-12.

Kretchmar, S., & Elcombe, T. (2007). In defense of competition and winning. In W. Morgan (Ed.), *Ethics in sport* (2nd ed., pp. 181-194). Champaign, IL: Human Kinetics.

Kristof, C. (1997). Accuracy of reference citations in five entomology journals. *American Entomologist, 43,* 246-51.

Kroll, W.P. (1971). *Perspectives in physical education.* New York: Academic Press.

Krout, J.A. (1929). *Annals of American sport.* New Haven, CT: Yale University Press.

Krueger, R.A., & Casey, M.A. (2015). *Focus groups: A practical guide for applied research* (5th ed.). Los Angeles: Sage.

Kruger, A. (1990). Puzzle solving: German sport historiography of the eighties. *Journal of Sport History, 17,* 261-277.

Kruskal, W., & Mosteller, F. (1979). Representative sampling, III: The current statistical literature. *International Statistical Review, 47,* 245-265.

Kuhn, T.S. (1970). *The structure of scientific revolutions.* Chicago: University of Chicago Press.

Kulinna, P.H., Scrabis-Fletcher, K., Kodish, S. Phillips, S., & Silverman, S. (2009). A decade of research literature in physical education pedagogy. *Journal of Teaching in Physical Education, 28,* 119-140.

Kulinna, P.H., & Silverman, S. (1999). The development and validation of scores on a measure of teachers' attitudes toward teaching physical activity and fitness. *Educational and Psychological Measurement, 59,* 507-517.

Labban, J.D., & Etnier, J.L. (2011). Effects of acute exercise on long-term memory. *Research Quarterly for Exercise and Sport, 82*(4), 712-721.

Laderman, S. (2014). Empire in waves: A political history of surfing. Berkeley: University of California Press.

Lakens, D. (2013). Calculating and reporting effect sizes to facilitate cumulative science: A practical primer for t-tests and ANOVAs. *Frontiers in psychology, 4,* 863.

Lane, K.R. (1983). *Comparison of skinfold profiles of black and white boys and girls ages 11-13.* Unpublished master's thesis, Louisiana State University, Baton Rouge.

Lariviere, V. & Gingras, Y. (2009). The impact factor's Matthew Effect: A natural experiment in bibliometrics. *Journal of the Association for Information Science and Technology, 61,* 424-427.

Last, J.M. (1988). *A dictionary of epidemiology.* New York: Oxford University Press.

Lau E.Y., Dowda M., McIver K.L., Pate R.R. (2017). Changes in Physical Activity in the School, Afterschool, and Evening Periods During the Transition From Elementary to Middle School. *J Sch Health. Jul;87*(7):531-537.

Lawrence, P.A. (2003). The politics of publication. *Nature, 422,* 259-261.

Lawson, H.A. (1993). After the regulated life. *Quest, 45* (4), 523-545.

Lee, A., & Kamler, B. (2008). Bringing pedagogy to doctoral publishing. *Teaching in Higher Education, 13,* 511-523.

Lee, D.C., Artero, E.G., Sui, X., Blair, S.N. (2010). Mortality trends in the general population: the importance of cardiorespiratory fitness. *Journal of Psychopharmacology, 24,* 27-35.

Lee, D.C., Pate, R.R., Lavie, C.J., Sui, X., Church, T.S., Blair, S.N. (2014). Leisure-time running reduces all-cause and cardio-vascular mortality risk. *Journal of the American College of Cardiology, 64*(5), 472-81.

Lee, C.D., Folsom, A.R., & Blair, S.N. (2003). Physical activity and stroke risk: a meta-analysis. *Stroke, 34,* 2475-81.

Lee, I.M., & Paffenbarger, R.S., Jr. (2000). Associations of light, moderate, and vigorous intensity physical activity with longevity: The Harvard Alumni Health Study. *American Journal of Epidemiology, 151,* 293-299.

Leech, N.L., & Onwuegbuzie, A.J. (2009). *Quality and Quantity, 43,* 265-275.

Lester, R (1995). *Stagg's University: The rise, decline, and fall of big-time football* at *Chicago.* Urbana, IL: University of Illinois Press.

Levine, P. (1992). *Ellis Island to Ebbetts Field: Sport and the American Jewish experience.* New York: Oxford University Press.

Levine, R.V. (1990, September-October). The pace of life. *American Scientist, 78,* 450-459.

Light, R.J., Singer, J.D., & Willett, J.B. (1994). The visual presentation and interpretation of meta-analysis. In H. Cooper and L.V. Hedges (Eds.), *The handbook of research synthesis* (pp. 439-453). New York: Russell Sage Foundation.

Lincoln, Y.S., & Guba, E.G. (1985). *Naturalistic inquiry.* Newbury Park, CA: Sage.

Lindsay-Smith, G., Eime, R., O'Sullivan, G., O'Sullivan, G., Harvey, J. & van Uffelen, J.G.Z. (2019). A mixed-methods case study exploring the impact of participation in community activity groups for older adults on physical activity, health and wellbeing. *BMC Geriatrics, 19:* 243. https://doi.org/10.1186/s12877-019-1245-5

Lippe, G. von der. (2001). Sportification processes: Whose logic? Whose rationality? *Sport History Review, 32,* 42-55.

Llewellyn, M.P., Gleaves, J. &Wilson, W. (2015). *The 1984 Los Angeles Olympic Games: Assessing the 30-year legacy.* London: Routledge.

Liberti, R. & Smith, M.M. (2017). *San Francisco Bay Area Sports: Golden Gate athletics, recreation, and community.* Fayette-ville, AR: The University of Arkansas Press.

Locke, L.F. (1989). Qualitative research as a form of scientific inquiry in sport and physical education. *Research Quarterly for Exercise and Sport, 60,* 1-20.

Locke, L.F., Silverman, S.J., & Spirduso, W.W. (2010). *Reading and understanding research* (3rd ed.). Thousand Oaks, CA: Sage.

Locke, L.F., Spirduso, W.W., & Silverman, S.J. (2014). *Proposals that work: A guide for planning dissertations and grant proposals* (6th ed.). Los Angeles: Sage.

Löfström, E., & Pyhältö, K. (2019). What are ethics in doctoral supervision, and how do they matter? Doctoral students' perspective. *Scandinavian Journal of Educational Research, 64,* 535-50.

Logan, S., Kipling Webster, E., Getchell, N., Pfeiffer, K., Robinson, L. (2015). Relationship between fundamental motor skill competence and physical activity during childhood and adolescence: A systematic review. *Kinesiology Review, 4,* 416-426. https://doi.org/10.1123/kr.2013-0012

Loland, S. (2020). Caster Semenya, athlete classification, and fair equality of opportunity in sport. *Journal of Medical Ethics, 46,* 584-590.

Looney, M.A., Feltz, C.S., & VanVleet, C.N. (1994). The reporting and analysis of research findings for within subjects designs: Methodological issues for meta-analysis. *Research Quarterly for Exercise and Sport, 65,* 363-366.

Lopez-Aymes, G., de los Dolores Valdez, M., Rodrigues-Naveiras, E., Dastellanos-Simons, D., Aguirre, T., & Borges, A. (2021). A mixed method research study of parental preception of physical activity and quality of life of children under home lock down in the COVID-19 pandemic. Frontiers in Psychology. https://doi.org/10.3389/fpsyg.2021.649481

Lord, F.M. (1969). Statistical adjustments when comparing preexisting groups. *Psychological Bulletin, 72,* 336-337.

Love, R., Adams, J., van Sluijs, E., Foster, C., & Humphreys, D. (2018). A cumulative meta-analysis of the effects of individual physical activity interventions targeting healthy adults. *Obesity reviews, 19,* 1164-1172. https://doi.org/10.1111/obr.12690

Mabley, J. (1963, January 22). Mabley's report. *Chicago American,* p. 62.

MacAloon, J. (1981). *This great symbol: Pierre de Coubertin and the origins of the modern Olympic Games.* Chicago: University of Chicago Press.

Manchester, H. (1931). *Four centuries of sport in America, 1490-1890.* New York: The Derrydale Press.

Mangan, J.A., & Park, R. (Eds.). (1987). *From "fair sex" to feminism: Sport and the socialization of women in the Industrial and Post-Industrial Eras.* London: Frank Cass.

Mann, D., Williams, A., Ward, P. & Janelle, C. (2007). Perceptual-cognitive expertise in sport: A meta-analysis. *Journal of Sport & Exercise Psychology, 29,* 457-78. https://doi.org/10.1123/jsep.29.4.457

Margerison, T. (1965, January 3). Review of *Writing Technical Reports* (by B.M. Copper). *Sunday Times.* In R.L. Weber (Compiler) & E. Mendoza (Ed.), *A random walk in science* (p. 49). New York: Crane, Russak.

Marsh, H.W., Marco, I.T., & Aþçý, F.H. (2002). Cross-cultural validity of the physical self-description questionnaire: Comparison of factor structures in Australia, Spain, and Turkey. *Research Quarterly for Exercise and Sport, 73,* 257-270.

Marshall, C., Rossman, G.B. & Blanco, G. (2021). *Designing qualitative research* (7th ed.). Los Angeles: Sage.

Martens, R. (1973, June). People errors in people experiments. *Quest, 20,* 16-20.

Martens, R. (1977). *Sport competition anxiety test.* Champaign, IL: Human Kinetics.

Martens, R. (1979). About smocks and jocks. *Journal of Sport Psychology, 1,* 94-99.

Martens, R. (1987). Science, knowledge, and sport psychology. *Sport Psychologist, 1,* 29-55.

Matt, K.S. (1993). Ethical issues in animal research. *Quest, 45*(1), 45-51.

Matthews, P.R. (1979). The frequency with which the mentally retarded participate in recreation activities. *Research Quarterly, 50,* 71-79.

Maurissen, J.P., & Vidmar, T.J. (2017). Repeated-measure analyses: Which one? A survey of statistical models and recommendations for reporting. *Neurotoxicology and Teratology, 59,* 78-84.

Maxwell, J.A. (2013). *Qualitative research design* (3rd ed.). Los Angeles: Sage.

McBain, K., Shrier, I., Schultz, R., Meeuwisse, W.H., Klügl, M., Garza, D., and Matheson, G.O. (2012). Prevention of sports injury I: A systematic review of applied biomechanics and physiology outcomes research. *British Journal of Sports Medicine, 46,* 169-173.

McCaughtry, N., & Rovegno, I. (2003). Development of pedagogical content knowledge: Moving from blaming students to predicting skillfulness, recognizing motor development, and understanding emotion. *Journal of Teaching in Physical Education, 22,* 355-368.

McCloy, C.H. (1930). Professional progress through research. *Research Quarterly, 1,* 63-73.

McCullagh, P., & Meyer, K.N. (1997). Learning versus correct models: Influence of model type on the learning of a free-weight squat lift. *Research Quarterly for Exercise and Sport, 68,* 56-61.

McLeroy, K.R., Bibeau, D., Steckler, A., & Glanz, K. (1988). An ecological perspective on health promotion programs. *Health Education Quarterly, 15,* 351-377.

McNamara, J.F. (1994). *Surveys and experiments in education research.* Lancaster, PA: Nechnomic.

McNeil, D. (2001). The spectacle of protest and punishment: Newspaper coverage of the Melksham weavers' riot of 1738. *Media History, 7,* 71-86.

Merriam, S.B. (1988). *Case study research in education: A qualitative apporach.* San Francisco: Jossey-Bass.

Merriam, S.B. (2007). *Qualitative research and case study approaches in education (2nd ed.).* San Francisco: Jossey-Bass.

Merritt, B. (2015). *State physical activity policies and racial/ethnic disparities in adult physical activity: 2006-2013* (Publication No. 3708631) [Doctoral dissertation, University of Oklahoma Health Sciences Center]. ProQuest Dissertations & Theses Global.

Micceri, T. (1989). The unicorn, the normal curve, and other improbable creatures. *Psychological Bulletin, 105,* 156-166.

Miles, M.B., Huberman, A.M., & Saldaña, J. (2014). *Qualitative data analysis: A methods sourcebook* (3rd ed.). Los Angeles: Sage.

Miller, D.K. (2006). *Measurement by the physical educator: Why & how.* New York: McGraw-Hill.

Miller, G. (1956). The magical number seven, plus or minus two: Some limits on our capacity for processing information. *Psychological Review, 63,* 81-97

Moher, D., Liberati, A., Tetzlaff, J., & Altman, D.G. The PRISMA Group. (2009). Preferred reporting items for systematic reviews and meta-analyses: the PRISMA statement. *PLoS Medicine,* 6(7): e1000097. doi: 10.1371/journal.pmed.1000097. Epub 2009 Jul 21. PMID: 19621072; PMCID: PMC2707599.

Möhring, W., Klupp, S., Zumbrunnen, R., Segerer, R., Schaefer, S., & Grob, A. (2021). Age-related changes in children's cognitive–motor dual tasking: Evidence from a large cross-sectional sample. *Journal of Experimental Child Psychology, 206,* Article 105103.

Montoye, H., Kemper, H., Saris, W., & Washburn, R. (1996). *Measuring physical activity and energy expenditure.* Champaign, IL: Human Kinetics.

Mooney, K.C. (2014). *Race horse men: How slavery and freedom were made at the race track.* Cambridge, MA: Harvard University Press.

Moore, L. (2017). *We will win the day: The Civil Rights Movement, the black athlete, and racial equality.* Santa Barbara, CA: Praeger.

Morgan, D.L. (1996). *Focus groups as qualitative research* (2nd ed.). Thousand Oaks, CA: Sage.

Morgan, W. (Ed.). (2007). *Ethics in sport* (2nd ed.). Champaign, IL: Human Kinetics.

Morgenstern, N.L. (1983). Cogito ergo sum: Murphy's refutation of Descartes. In G.H. Scherr (Ed.), *The best of* The Journal of Irreproducible Results (p. 112). New York: Workman Press.

Morland, R.B. (1958). A philosophical interpretation of the educational views held by leaders in American physical education. *Health, Physical Education and Recreation Microform Publications, 1,* (October 1949-March 1965), PE394.

Morris, J.N., Heady, J.A., Rattle, P.A.B., Roberts, C.G., & Parks, J.W. (1953). Coronary heart disease and physical activity of work. *Lancet, 2,* 1053-1057.

Morrow, D. (1983). Canadian sport history: A critical essay. *Journal of Sport History, 10,* 67-79.

Morrow, J.R., Jr., Bray, M.S., Fulton, J.E., & Thomas, J.R. (1992). Interrater reliability of 1987-1991. *Research Quarterly for Exercise and Sport* reviews. *Research Quarterly for Exercise and Sport, 63,* 200-204.

Morse, J.M. (2003). Principles of mixed methods and multimethod research design. In A. Tashakkori & C. Teddlie (Eds.), *Handbook of mixed methods in social and behavioral research* (pp. 189-208). Thousand Oaks, CA: Sage.

Morse, J.M., & Richards, L. (2002). *Read me first for a user's guide to qualitative methods.* Thousand Oaks, CA: Sage.

Muir, B.C., Haddad, J.M., van Emmerick, R.E.A., & Rietdyk, S. (2019). Changes in the control of obstacle crossing in middle age become evident as gait task increases. *Gait & Posture, 70,* 254-59.

Nathan, D.A. (2016). *Baltimore sports: Stories from Charm City.* Fayetteville, AR: The University of Arkansas Press.

Nathan, D.A. (2003). *Saying it's so: A cultural history of the Black Sox scandal.* Champaign, IL: University of Illinois Press.

National Center for Health Statistics. National Health Interview Survey, 2015-2018.

National Children and Youth Fitness Study. (1985). *Journal of Physical Education, Recreation and Dance, 56*(1), 44-90.

National Children and Youth Fitness Study II. (1987). *Journal of Physical Education, Recreation and Dance, 56*(9), 147-167.

National Physical Activity Plan Alliance (2016). *U.S. National Physical Activity Plan.* Retrieved from http://physicalactivityplan.org/docs/2016NPAP_Finalforwebsite.pdf

Nauright, J. (1999). The end of sports history? From sports history to sport studies. In D. Booth & A. Jutel (Eds.), *Sporting Traditions, 16,* 5-15.

Nelson, J.K. (1988, March). Some thoughts on research, measurement, and other obscure topics. LAHPERD Scholar Lecture presented at the LAHPERD Convention, New Orleans, LA.

Nelson, J.K. (1989). Measurement methodology for affective tests. In M.J. Safrit & T.M. Wood (Eds.), *Measurement concepts in physical education and exercise science* (pp. 229-248). Champaign, IL: Human Kinetics.

Nelson, J.K., Thomas, J.R., Nelson, K.R., & Abraham, P.C. (1986). Gender differences in children's throwing performance: Biology and environment. *Research Quarterly for Exercise and Sport, 57,* 280-287.

Nelson, J.K., Yoon, S.H., & Nelson, K.R. (1991). A field test for upper body strength and endurance. *Research Quarterly for Exercise and Sport, 62,* 436-447.

Nesselroade, K.P., Jr., & Grimm, L.G. (2019). *Statistical applications for the behavioral and social sciences* (2nd ed.) Hoboken, NJ: Wiley & Sons.

Newell, K.M. (1987). On masters and apprentices in physical education. *Quest, 39*(2), 88-96.

Newell, K.M., & Hancock, P.A. (1984). Forgotten moments: A note on skewness and kurtosis as influential factors in inferences extrapolated from response distributions. *Journal of Motor Behavior, 16,* 320-335.

Nieman, D.C. (1994). Exercise, upper respiratory tract infection, and the immune system. *Medicine & Science in Sports & Exercise, 26B,* 128-139.

Norton, K., Norton, L., Sadgrove, D. (2010). Position statement on physical activity and exercise intensity terminology. *Journal of Science and Medicine in Sport, 13*(5), 496-502.

Norwood, S.H. (2018). *New York sports: Glamour and grit in the Empire City.* Fayetteville, AR: The University of Arkansas Press.

Nunnaly, J.C. (1978). *Psychometric theory* (2nd ed.). New York: McGraw-Hill.

Nuzzo, R. (2014). Statistical errors. *Nature, 506,* 150-52.

O'Connor, D., Larkin, P., & Mark Williams, A. (2016). Talent identification and selection in elite youth football: An Australian context. *European Journal of Sport Science, 16*(7), 837-844.

Øiestad, B.E., Eitzen, C.B., & Thorlund, J.B. (2015). *Knee extensor muscle weakness increases the risk of knee osteoarthritis. a systematic review and meta-analysis.* Osteoarthritis and Cartilage, 23, *171-177.*

Oishi, S.M. (2003). *How to conduct in-person interviews for surveys* (2nd ed.). Thousand Oaks, CA: Sage.

O'Keefe, D.J. (2007). Brief report: post hoc power, observed power, a priori power, retrospective power, prospective power, achieved power: sorting out appropriate uses of statistical power analyses. *Communication Methods and Measures, 1*(4), 291-299.

O'Malley, M., & Rosenzweig, R. (1997). Brand new world or blind alley? American history on the World Wide Web. *Journal of American History, 84,* 132-155.

Oriard, M. (1993). *Reading football: How the popular press created an American spectacle.* Chapel Hill: University of North Carolina Press.

Oriard, M. (2001). *King Football: Sport & spectacle in the golden age of radio & newsreels, movies & magazines, the weekly & the daily press.* Chapel Hill: The University of North Carolina Press.

Oriard, M. (2007a). *Brand NFL: Making and selling America's favorite sport.* Chapel Hill: The University of North Carolina Press.

Oriard, M. (2007b). *Bowled over: Big-time college football from the sixties to the BCS era.* Chapel Hill: The University of North Carolina Press.

Osmond, G. & Phillips, M.G. (2015). *Sport History in the Digital Era.* Urbana, IL: University of Illinois Press.

Osmond, G. (2019). Decolonizing dialogues: Sport, resistance, and Australian Aboriginal settlements. *Journal of Sport History, 46,* 288-301.

Paffenbarger, R.S., Jr., Wing, A.L., & Hyde, R.T. (1978). Physical activity as an index of heart attack risk in college alumni. *American Journal of Epidemiology, 108,* 161-175.

Park, R. (1983). Research and scholarship in the history of physical education and sport: The current state of affairs. *Research Quarterly for Exercise and Sport, 54,* 93-103.

Parratt, C.M. (1998). About turns: Reflecting on sport history in the 1990s. *Sport History Review, 29,* 4-17.

Pati, D., & Lorusso, L.N. (2018). How to Write a Systematic Review of the Literature. *HERD: Health Environments Research & Design Journal, 11,* 15-30. https://doi.org/10.1177/1937586717747384

Patton, M.Q. (2014). *Qualitative research and evaluation methods* (4th ed.). Thousand Oaks, CA: Sage.

Paxson, F.L. (1917). The rise of sport. *Mississippi Valley Historical Review, 4,* 143-168.

Pedhazur, E.J. (1982). *Multiple regression in behavioral research: Explanation and prediction* (2nd ed.). New York: Holt, Rinehart & Winston.

Peshkin, A. (1993). The goodness of qualitative research. *Educational Researcher, 22* (2), 23-29.

Pek, J., & Flora, D.B. (2018). Reporting effect sizes in original psychological research: A discussion and tutorial. *Psychological Methods, 23*(2), 208.

Peters, E.M., & Bateman, E.D. (1983). Ultramarathon running and upper respiratory tract infections: An epidemiological survey. *South African Medical Journal, 64,* 582-584.

Pett, M.A., Lackey, N.R., & Sullivan, J.J. (2003). *Making sense of factor analysis: The use of factor analysis for instrument development in health care research.* Thousand Oaks, CA: Sage.

Phillips, M. G. (1999). Navigating uncharted waters: The death of sports history? In D. Booth & A. Jutel (Eds.), *Sporting Traditions, 16,* 5-15.

Phillips, M.G. (2006). *Deconstructing sport history: A postmodern analysis.* Albany: State University of New York Press.

Phillips, M.G. (2001). Deconstructing sport history: The postmodern challenge. *Journal of Sport History, 28,* 327-343.

Phillips, M.G. & Osmond, G. (2018). Australian Indigenous sport historiography: A review. *Kinesiology Review, 7,* 193-198.

Pitney, W.A., & Parker, J. (2020). *Qualitative research in the health professions*Thorofare, NJ: SLACK Inc.

Plano-Clark, V.L., Huddleston-Casas, C.A., Churchill, S., Green, D.O., & Garrett, A.L. (2008). Mixed methods approaches in family science research. *Journal of Family Studies, 29,* 1543-1566.

Plaven-Sigray, P., Matheson, G.J., Schiffler, B.C., & Thompson, W.H. (2017). The readability of scientific texts is decreasing over time. *eLife* 2017;6:e27725

Polanyi, M. (1958). *Person knowledge: Toward a post-critical philosophy.* Chicago: University of Chicago Press.

Pope, S.W. (1997a). *Patriotic games: Sporting traditions in the American imagination, 1876-1926.* London: Oxford University Press.

Pope, S.W. (1997b). *The new American sport history: Recent approaches and perspectives.* Urbana, IL: University of Illinois Press.

Pope, S.W. (1998). Sport history: Into the 21st century. *Journal of Sport History, 25,* i-x.

Porter, A.C., & Raudenbush, S.W. (1987). Analysis of covariance: Its model and use in psychological research. *Journal of Counseling Psychology, 34,* 383-392.

Porter, A.L., Chubin, D.E., Rossini, F.A., Boeckmann, M.E., & Connally, T. (1982, September-October). Views: The role of the dissertation in scientific careers: The first decade of professional life offers insights into the effectiveness of the doctoral dissertation in scientific training. *American Scientist, 70*(5), 475-481.

Porter, A.L., & Wolfe, D. (1975). Utility of the doctoral dissertation. *American Psychologist, 30,* 1054-1061.

Porter, D.H. (1981). *The emergence of the past: A theory of historical explanation.* Chicago: University of Chicago Press.

Punch, K.F. (2003). *Survey research: The basics.* Thousand Oaks, CA: Sage.

Punch, M. (1986). *The politics and ethics of fieldwork.* Beverly Hills, CA: Sage.

Puri, M.L., & Sen, P.K. (1969). A class of rank order tests for a general linear hypothesis. *Annals of Mathematical Statistics, 40,* 1325-1343.

Puri, M.L., & Sen, P.K. (1985). *Nonparametric methods in general linear models.* New York: Wiley.

Rail, G. (Ed.). (1998). *Sport and postmodern times.* Albany: State University of New York Press.

Rawdon, T. Sharp, R.L., Shelley, M., & Thomas, J.R. (2012). Meta-analysis of the placebo effect in nutritional supplement studies of muscular performance. *Kinesiology Reviews, 1,* 137-148.

Reah, D. (1998). *The language of newspapers.* London: Routledge.

Regalado, S. (1998). *Viva baseball! Latin major leaguers and their special hunger.* Urbana, IL: University of Illinois Press.

Reichardt, C.S., & Rallis, S.F. (Eds.). (1994). *The qualitative-quantitative debate: New perspectives.* San Francisco: Jossey-Bass.

Rhodes, R. E., Hunt Matheson, D., & Mark, R. (2010). Evaluation of social cognitive scaling response options in the physical activity domain. *Measurement in Physical Education and Exercise Science, 14*(3), 137-150.

Richards, L., & Morse, J.H. (2012). README FIRST for a User's Guide to Qualitative Methods (3rd ed.). Thousand Oaks, CA: Sage.

Ries, L.A.G., Kosary, C.L., Hankey, B.F., Miller, B.A., Clegg, L., & Edwards, B.K. (Eds.). (1999). *SEER cancer statistics review, 1973-1996.* Bethesda, MD: National Cancer Institute.

Riess, S.A. (1980). *Touching base: Professional baseball and American culture in the progressive era.* Westport, CT: Greenwood Press.

Riess, S.A. (1989). *City games: The evolution of American urban society and the rise of sports.* Urbana, IL: University of Illinois Press.

Riess, S.A. (Ed.). (1998). *Sports and the American Jew.* Syracuse, NY: Syracuse University Press.

Riess, S.A. (2011). *The sport of kings and the kings of crime: Horse racing, politics, and organized crime in New York, 1865-1913.* Syracuse, NY: Syracuse University Press.

Rink, J., & Werner, P. (1989). Qualitative measures of teaching performance scale. In P. Darst, D. Zakrajsak, & P. Mancini (Eds.), *Analyzing physical education and sports instruction* (2nd ed., pp. 269-276). Champaign, IL: Human Kinetics.

Rivkin, A. (2020). Manuscript referencing errors and their impact on shaping current evidence. *American Journal of Pharmaceutical Education, 84*(7), Article 7846, 877-80.

Roberts, G.C. (1993). Ethics in professional advising and academic counseling of graduate students. *Quest, 45,* 78-87.

Roberts, G.C., Kavussanu, M., & Sprague, R.L. (2001). *Science and Engineering Ethics, 7,* 525-37.

Roberts, R. (Ed.). (2000). *Pittsburgh sports: Stories from the Steel City.* Pittsburgh: University of Pittsburgh Press.

Robertson, M.A., & Konczak, J. (2001). Predicting children's overarm throw ball velocities from their developmental levels in throwing. *Research Quarterly for Exercise and Sport, 72,* 91-103.

Robinson, S., & Dracup, K. (2008). Innovative options for the doctoral dissertation in nursing. *Nursing Outlook, 56,* 174-178.

Robinson, P., & Lowe, J. (2015). Literature reviews vs systematic reviews. *Australian and New Zealand Journal of Public Health, 39,* 103-103.

Robson, C. (2002). *Real world research: A resource for social scientists and practitioner-researchers* (2nd ed.). Malden, MA: Wiley-Blackwell.

Rosenthal, R. (1966). Sport, art, and particularity: The best equivocation. *Journal of the Philosophy of Sport, 13,* 49-63.

Rosenthal, R. (1991). Cumulating psychology: An appreciation of Donald T. Campbell. *Psychological Science, 2,* 213, 217-221.

Rosenthal, R. (1994). Parametric measures of effect size. In H. Cooper & L.V. Hedges (Eds.), *The handbook of research synthesis* (pp. 231-244). New York: Russell Sage Foundation.

Rosenzweig, R. (1983). *Eight hours for what we will: Workers and leisure in an industrial city, 1870-1920.* Cambridge, England: Cambridge University Press.

Rosenzweig, R. (2003). Scarcity of abundance? Preserving the past in a digital era. *The American Historical Review, 108,* 735-762.

Rosenzweig, R., & Blackmar, E. (1992). *The park and the people: A history of Central Park.* Ithaca, NY: Cornell University Press.

Rosnow, R.L., & Rosenthal, R. (1989). Statistical procedures and the justification of knowledge in psychological science. *American Psychologist, 44,* 1276-1284.

Rossman, G.B., & Rallis, S.F. (2017). *Learning in the field: An introduction to qualitative research* (4rd ed.). Los Angeles: Sage.

Rothman, K.J. (1986). *Modern epidemiology.* Boston: Little Brown.

Rubin, H.J., & Rubin, I.S. (2020). *Qualitative interviewing: The art of hearing data* (4th ed.). Los Angeles: Sage.

Ruck, R. (2011). *Raceball: How the Major Leagues Colonized the Black and Latin Game.* Boston: Beacon Press.

Ruck, R. (1987). *Sandlot seasons: Sport in black Pittsburgh.* Urbana, IL: University of Illinois Press.

Ruck, R. (2018). *Tropic of football: The long and perilous journey of Samoans to the NFL.* New York: The New Press.

Rudy, W. (1962). Higher education in the United States, 1862-1962. In W.W. Brickman & S. Lehrer, (Eds.), *A century of higher education: Classical citadel to collegiate colossus* (pp. 20-21). New York: Society for the Advancement of Education.

Ryan, E.D. (1970). The cathartic effect of vigorous motor activity on aggressive behavior. *Research Quarterly, 41,* 542-551.

Safrit, M.J. (Ed.). (1976). *Reliability theory.* Washington, DC: American Alliance for Health, Physical Education and Recreation.

Safrit, M.J., Cohen, A.S., & Costa, M.G. (1989). Item response theory and the measurement of motor behavior. *Research Quarterly for Exercise and Sport, 60*(4), 325-335.

Safrit, M.J., & Wood, T.M. (1983). The health-related fitness test opinionnaire: A pilot survey. *Research Quarterly for Exercise and Sport, 54,* 204-207.

Safron, C. (2020). Experimenting with affective bodies: Young people, health and fitness in an urban after-school program [Doctoral Dissertation]. Teachers College, Columbia University.

Sage, G., Dyreson, M., & Kretchmar, S. (June, 2005). Sociology, history, and philosophy in the *Research Quarterly. Research Quarterly Supplement, 76*(2), 88-107.

Salkind, N.J. (2007). *Encyclopedia of measurement and statistics (Vols. 1-0).* Thousand Oaks, CA: Sage Publications.

Salzinger, K. (2001, February 16). Scientists should look for basic causes, not just effects. *Chronicle of Higher Education, 157*(23), B14.

Sammarco, P.W. (2008). Journal visibility, self-citation, and reference limits: Influences in impact factor and author performance review. *Ethics in Science and Environmental Politics, 8,* 121-125.

Sammons, J.T. (1988). *Beyond the ring: The role of boxing in American society.* Urbana, IL: University of Illinois Press.

Scheffé, H. (1953). A method for judging all contrasts in analysis of variance. *Biometrika, 40,* 87-104.

Schein, E.H. (1987). *The clinical perspective in fieldwork.* Newbury Park, CA: Sage.

Scherr, G.H. (Ed.). (1983). *The best of* The Journal of Irreproducible Results (p. 152). New York: Workman Press.

Schieffer, T.M., & Thomas, K.T. (2012). Fifteen years of promise in school-based physical activity interventions: A meta-analysis. *Kinesiology Reviews, 1,* 155-169.

Schmidt, F.L., & Hunter, J.E. (2014). *Methods of meta-analysis: Correcting error and bias in research findings* (3rd ed.). Los Angeles: Sage.

Schmidt, R.A. (1975). A schema theory of discrete motor skill learning. *Psychological Review, 82,* 225-260.

Schmidt, R.A., Lee, T.D., Winstein, C. Wulf, G & Zelaznik, H. (2019). *Motor control and learning* (6th ed.). Champaign, IL: Human Kinetics.

Schultz, J. (2014). *Qualifying times: Points of change in U.S. women's sports.* Urbana, IL: University of Illinois Press.

Schultz, J. (2016). *Moments of impact: Injury, racialized memory, and reconciliation in college football.* Lincoln, NE: University of Nebraska Press.

Schultz, J. (2017). New directions and future considerations in American sport history. In L.J. Borish, D.K. Wiggins, & G.R. Gems (Eds.), *The Routledge history of American sport* (pp.17-29). London: Routledge.

Schultz, J. (2018). *Women's sports: What everyone needs to know*. New York: Oxford University Press.

Schumacker, R.E., & Lomax, R.G. (2004). *A beginner's guide to structural equation modeling* (2nd ed.). Mahwah, NJ: Erlbaum.

Schutz, R.W. (1989). Qualitative research: Comments and controversies. *Research Quarterly for Exercise and Sport, 60,* 30-35.

Schutz, R.W., & Gessaroli, M.E. (1987). The analysis of repeated measures designs involving multiple dependent variables. *Research Quarterly for Exercise and Sport, 58,* 132-149.

Scrabis-Fletcher, K. (2007). *Student attitude, perception of competence, and practice in middle school physical education.* Available from ProQuest Dissertations and Theses database. (UMI No. 3285162)

Seidman, I. (2019). *Interviewing as qualitative research: A guide for researchers in education and the social sciences* (5rd ed.). New York: Teachers College Press.

Semega, J., Koolar, M., Shrider, E.A. and Creamer, J. (2021) Income and Poverty in the United States: 2019. U.S. Census Bureau. Report number P60-270.

Senn, A.E. (1999). *Power, politics, and the Olympic Games: A history of the power brokers, events, and controversies that shaped the Games.* Champaign, IL: Human Kinetics.

Serlin, R.C. (1987). Hypothesis testing, theory building, and the philosophy of science. *Journal of Counseling Psychology, 34,* 365-371.

Shadish, W.R., Cook, T.D., & Campbell, D.T. (2002). *Experimental and quasi-experimental designs for generalized causal inference.* Boston: Houghton Mifflin.

Shafer, R. (1980). *A guide to historical method* (3rd ed.). Homewood, IL: Dorsey Press.

Shavelson, R.J., & Towne, L. (Eds.). (2002). *Scientific research in education.* Washington, DC: National Academy Press.

Sheets-Johnstone, M. (1999). *The primacy of movement.* Amsterdam and Philadelphia: John Benjamins.

Shore, E.G. (1991, February). Analysis of a multi-institutional series of completed cases. Paper presented at Scientific Integrity Symposium. Harvard Medical School, Boston.

Shropshire, K.L. (1996). *In black and white: Race and sports in America.* New York: New York University Press.

Shulman, L.S. (2003, April). Educational research and a scholarship of education. Charles DeGarmo Lecture presented at the annual meeting of the American Educational Research Association, Chicago.

Sibley, B.A., & Etnier, J.L. (2003). The relationship between physical activity and cognition in children: a meta-analysis. *Pediatric Exercise Science, 15,* 243-56.

Siebers, R. & Holt, S. (2000). Accuracy of references in five leading medical journals. *The Lancet, 356,* October 21, 2000, 1445.

Siedentop, D. (1980). Two cheers for Rainer. *Journal of Sport Psychology, 2,* 2-4.

Siedentop, D. (1989). Do the lockers really smell? *Research Quarterly for Exercise and Sport, 60,* 36-41.

Siedentop, D., Birdwell, D., & Metzler, M. (1979, March). A process approach to measuring teaching effectiveness in physical education. Paper presented at the American Alliance for Health, Physical Education, Recreation and Dance national convention, New Orleans, LA.

Siedentop, D., Trousignant, M., & Parker, M. (1982). *Academic learning time—Physical education: 1982 coding manual.* Columbus: Ohio State University, School of Health, Physical Education, and Recreation.

Sikes, M. (2019). Enduring legacies and convergent identities: The male-dominated origins of the Kenyan running explosion. *Journal of Sport History, 46,* 273-287.

Silverman, S. (2004). Analyzing data from field research: The unit of analysis issue. *Research Quarterly for Exercise and Sport, 74,* iii-iv.

Silverman, S. (2017). Attitude research in physical education: A review. *Journal of Teaching in Physical Education, 36*(3), 303-312.

Silverman, S. (Ed.). (2005). *Research Quarterly for Exercise and Sport: 75th Anniversary Issue, 76*(supplement).

Silverman, S., & Keating, X.D. (2002). A descriptive analysis of research methods classes in departments of kinesiology and physical education in the United States. *Research Quarterly for Exercise and Sport, 73,* 1-9.

Silverman, S., Kulinna, P.H., & Phillips, S.R. (2014). Physical education pedagogy faculty perceptions of journal quality. *Journal of Teaching in Physical Education, 33,* 134-154.

Silverman, S., & Manson, M. (2003). Research on teaching in doctoral programs: A detailed investigation of focus, method, and analysis. *Journal of Teaching in Physical Education, 22,* 280-297.

Silverman, S., & Solmon, M. (1998). The unit of analysis in field research: Issues and approaches to design and data analysis. *Journal of Teaching in Physical Education, 17,* 270-284.

Silverman, S., & Subramaniam, P.R. (1999). Student attitude toward physical education and physical activity: A review of measurement issues and outcomes. *Journal of Teaching in Physical Education, 19,* 97-125.

Slavin, R.E. (1984a). Meta-analysis in education: How it has been used. *Educational Researcher, 13*(4), 6-15.

Slavin, R.E. (1984b). A rejoinder to Carlberg et al. *Educational Researcher, 13*(4), 24-27.

Smaldone, A., Heitkemper, E., Jackman, K., Woo, K.J., & Kelson, J. (2019). Dissemination of PhD dissertation research by dissertation format: a retrospective cohort study. *Journal of Nursing Scholarship, 51,* 599-607.

Smith, M.L. (1980). Sex bias in counseling and psychotherapy. *Psychological Bulletin, 87,* 392-407.

Smith, R.A. (2010). *Pay for play: A history of big-time college athletic reform.* Urbana, IL: University of Illinois Press.

Smith, R.A. (1988). *Sports and freedom: The rise of big-time college athletics.* New York: Oxford University Press.

Snyder, C.W., Jr., & Abernethy, B. (Eds.). (1992). *The creative side of experimentation.* Champaign, IL: Human Kinetics.

Somers, D.A. (1972). *The rise of sports in New Orleans, 1850-1900.* Baton Rouge: Louisiana State University Press.

Spray, J.A. (1987). Recent developments in measurement and possible applications to the measurement of psychomotor behavior. *Research Quarterly for Exercise and Sport, 58,* 203-209.

Spray, J.A. (1989). New approaches to solving measurement problems. In M.J. Safrit & T.M. Wood (Eds.), *Measurement concepts in physical education and exercise science* (pp. 229-248). Champaign, IL: Human Kinetics.

Stathokostas, L., Little, R., Vandervoort, A.A. & Paterson, D. (2012). Flexibility training and functional ability in older adults: A systematic review. *Journal of Aging Research.* https://doi.org/10.1155/2012/306818

Steinhauser, G. (2009). The nature of navel fluff. *Medical Hypotheses, 10,* 1016.

Ste-Marie, D. M., Carter, M. J., Law, B., Vertes, K., & Smith, V. (2016). Self-controlled learning benefits: exploring contributions of self-efficacy and intrinsic motivation via path analysis. *Journal of Sports Sciences, 34*(17), 1650-1656.

Steneck, N.H. (2007). *ORI Introduction to the Responsible Conduct of Research* (revised edition). Washington, D.C., U.S. Department of Health & Human Services Office of Research Integrity.

Stock, W.A. (1994). Systematic coding for research synthesis. In H. Cooper & L.V. Hedges (Eds.), *The handbook of research synthesis.* New York: Sage Foundation.

Stodden, D., Langendorfer, S., & Roberton, M.A. (2009). The association between motor skill competences and physical fitness in young adults. *Research Quarterly for Exercise and Sport, 80,* 223-229.

Struna, N.L. (1985). In glorious disarray: The literature of American sport history. *Research Quarterly for Exercise and Sport, 56,* 151-160.

Struna, N.L. (1997). Sport history. In J.D. Massengale & R.A. Swanson (Eds.), *The history of exercise and sport science* (pp. 143-179). Champaign, IL: Human Kinetics.

Struna, N.L. (2000). Social history in sport. In J. Coakley & E. Dunning (Eds.), *Handbook of sport studies* (pp. 187-203). London: Sage.

Stull, G.A., Christina, R.W., & Quinn, S.A. (1991). Accuracy of references in the *Research Quarterly for Exercise and Sport. Research Quarterly for Exercise and Sport, 62,* 245-248.

Subramaniam, P.R., & Silverman, S. (2000). The development and validation of an instrument to assess student attitude toward physical education. *Measurement in Physical Education and Exercise Science, 4,* 29-43.

Subramaniam, P.R., & Silverman, S. (2002). Using complimentary data: An investigation of student attitude in physical education. *Journal of Sport Pedagogy, 8,* 74-91.

Subramaniam, P.R., & Silverman, S. (2007). Middle school students' attitudes toward physical education. *Teaching and Teacher Education, 22,* 602-611.

Suminski, R. R., Petosa, R. L., & Stevens, E. (2006). A method for observing physical activity on residential sidewalks and streets. Journal of urban health : bulletin of the New York Academy of Medicine, 83(3), 434–443. https://doi.org/10.1007/s11524-005-9017-2

Survey Monkey (2020). https://www.surveymonkey.com/mp/sample-size-calculator/

Sydnor, S. (1998). A history of synchronized swimming. *Journal of Sport History, 25,* 252-267.

Sylvester, B.D., Curran, T., Standage, M., Sabiston, C.M., & Beauchamp, M.R. (2018). Predicting exercise motivation and exercise behavior: A moderated mediation model testing the interaction between perceived exercise variety and basic psychological needs satisfaction. *Psychology of Sport and Exercise, 36,* 50-56.

Tamburrini, C.M. (2000). What's wrong with doping? In T. Tansjo & C. Tamburrini (Eds.), *Values in sport: Elitism, nationalism, and gender equality and the scientific manufacture of winners* (pp. 200-216). London: Spon Press.

Tarnas, R. (1991). *The passion of the Western mind: Understanding the ideas that have shaped our world view.* New York: Ballantine Books.

Tashakkori, A., & Teddlie, C. (1998). *Mixed methodology: Combining qualitative and quantitative approaches.* Thousand Oaks, CA: Sage.

Tashakkori, A.B., & Teddlie, C.B . (Eds.) (2010). *Handbook of mixed methods in social and behavioral research (2nd ed.).* Thousand Oaks, CA: Sage.

Tashakkori, A., Johnson, R.B. & Teddlie, C. (2020). *Foundations of mixed methods research: Integrating quantitative and qualitative approaches in the social and behavioral sciences (2nd ed.).* Thousand Oaks, CA: Sage.

Teramoto, M., & Golding, L.A. (2009). Regular exercise and plasma lipid levels associated with the risk of coronary health disease: A 20-year longitudinal study. *Research Quarterly for Exercise and Sport, 80,* 138-145.

Tew, J., & Wood, M. (1980). *Proposed model for predicting probable success in football players.* Houston, TX: Rice University Press.

Thøgersen-Ntoumani, C., & Fox, K.R. (2005). Physical activity and mental well-being typologies in corporate employees: A mixed methods approach. *Work & Stress, 19,* 50-67.

Thomas, D.L. (2012). *Globetrotting: African American athletes and Cold War politics.* Urbana, IL: University of Illinois Press.

Thomas, J.R. (1977). A note concerning analysis of error scores from motor-memory research. *Journal of Motor Behavior, 9,* 251-253.

Thomas, J.R. (1980). Half a cheer for Rainer and Daryl. *Journal of Sport Psychology, 2,* 266-267.

Thomas, J.R. (Ed.). (1983). Publication guidelines. *Research Quarterly for Exercise and Sport, 54,* 219-221.

Thomas, J.R. (Ed.). (1984). *Motor development during childhood and adolescence.* Minneapolis, MN: Burgess.

Thomas, J.R. (Ed.). (1986). Editor's viewpoint: Research notes. *Research Quarterly for Exercise and Sport, 57,* iv-v.

Thomas, J.R. (1989). An abstract for all seasons. *NASPSPA Newsletter, 14*(2), 4-5.

Thomas, J.R. (2014) Commentary—Improved data reporting in *RQES:* From volumes 45, 59, to 84. *Research Quarterly for Exercise and Sport, 85,* 446-448.

Thomas, J.R. & French, K.E. (1985). Gender differences across age in motor performance: a meta-analysis. *Psychological Bulletin,* 98, 260-82.

Thomas, J.R., & French, K.E. (1986). The use of meta-analysis in exercise and sport: A tutorial. *Research Quarterly for Exercise and Sport, 57,* 196-204.

Thomas, J.R., French, K.E., & Humphries, C.A. (1986). Knowledge development and sport skill performance: Directions for motor behavior research. *Journal of Sport Psychology, 8,* 259-272.

Thomas, J.R., & Gill, D.L. (Eds.). (1993). The academy papers: Ethics in the study of physical activity [special issue]. *Quest, 45*(1).

Thomas, J.R., Lochbaum, M.R., Landers, D.M., & He, C. (1997). Planning significant and meaningful research in exercise science: Estimating sample size. *Research Quarterly for Exercise and Sport, 68,* 33-43.

Thomas, J.R., Nelson, J.K., & Magill, R.A. (1986, Spring). A case for an alternative format for the thesis/dissertation. *Quest, 38,* 116-124.

Thomas, J.R., Nelson, J.K., & Thomas, K.T. (1999). A generalized rank-order method for non-parametric analysis of data from exercise science: A tutorial. *Research Quarterly for Exercise and Sport, 70,* 11-23.

Thomas, J.R., Salazar, W., & Landers, D.M. (1991). What is missing in *p* < .05? Effect size. *Research Quarterly for Exercise and Sport, 62,* 344-348.

Thomas, J.R., Thomas, K.T., Lee, A.M., Testerman, E., & Ashy, M. (1983). Age differences in use of strategy for recall of movement in a large scale environment. *Research Quarterly for Exercise and Sport, 54,* 264-272.

Thomas, K.T., Keller, C.S., & Holbert, K. (1997). Ethnic and age trends for body composition in women residing in the U.S. Southwest: II. Total fat. *Medicine & Science in Sport & Exercise, 29*(1), 90-98.

Thomas, K.T., & Thomas, J.R. (1999). What squirrels in the trees predict about expert athletes. *International Journal of Sport Psychology, 30,* 221-234.

Thomas, R.M. (2003). *Blending qualitative and quantitative research methods in theses and dissertations.* Thousand Oaks, CA: Corwin Press.

Thomson Reuters. (2014). *Journal citation reports.* Retrieved from http://admin-apps.webofknowledge.com/JCR/JCR?wsid=2 FIwwEcIWwo5G6DWD9I&ssid=&SID=2FIwwEcIWwo5G6DWD9I

Todd, J. (1998). *Physical culture and the body beautiful: Purposive exercise in the lives of American women, 1800-1870.* Macon, GA: Mercer University Press.

Tuckman, B.W. (1978). *Conducting educational research* (2nd ed.). New York: Harcourt Brace Jovanovich.

Tudor-Locke, C., Williams, J.E., Reis, J.P., & Pluto, D. (2004). Utility of pedometers for assessing physical activity construct validity. *Sports Medicine, 34* (5), 281-291.

Uminowicz, G. (1984). Sport in a middle-class utopia: Asbury Park, New Jersey, 1871-1895. *Journal of Sport History, 11,* 51-73.

University of Chicago Press. (2017). *The Chicago manual of style* (17th ed.). Chicago: Author.

U.S. Department of Health and Human Services. (1996). *Physical activity and health: A report of the Surgeon General.* Washington, DC: U.S. Department of Health and Human Services.

U.S. Department of Health and Human Services. Office of Disease Prevention and Health Promotion. (2020). *Healthy people 2030.* Washington, DC. Retrieved from https://health.gov/healthypeople

U.S. Department of Health and Human Services. (2018). *Physical activity guidelines for Americans, 2nd edition.* Washington, DC: U.S. Department of Health and Human Services.

U.S. Department of Health and Human Services. Physical Activity Guidelines Advisory Committee. (2008). *Physical activity guidelines advisory committee report, 2008.* Washington, DC: U.S. Department of Health and Human Services.

U.S. Department of Health and Human Services, Centers for Disease Control and Prevention and National Cancer Institute; U.S. Cancer Statistics Working Group. (June 2020). *U.S. cancer statistics data visualizations tool, based on 2019 submission data (1999-2017).* Retrieved from www.cdc.gov/cancer/dataviz

van Munster, M. A., Lieberman, L. J., & Grenier, M. A. (2019). Universal design for learning and differentiated instruction in physical education. *Adapted Physical Activity Quarterly, 36*(3), 359-377.

Vazou, S., Pesce, C., Lakes, K., & Smiley-Oyen, A. (2019). More than one road leads to Rome: a narrative review and meta-analysis of physical activity intervention effects on cognition in youth. *International Journal of Sport and Exercise Psychology, 17,* 153-78.

Vealey, R.S. (1986). The conceptualization of sport-confidence and competitive orientation: Preliminary investigation and instrument development. *Journal of Sport Psychology, 8,* 221-246.

Venkatraman, V. (2010, April 16). *Conventions of scientific authorship.* Science.https://www.sciencemag.org/careers/2010/04/ conventions-scientific-authorship

Verbrugge, M.H. (1988). *Able-bodied womanhood: Personal health and social change in nineteenth-century Boston.* New York: Oxford University Press.

Verducci, F.M. (1980). *Measurement concepts in physical education.* St. Louis, MO: Mosby.

Veri, M.J. & Liberti, R. (2019). *Gridiron gourmet: Gender and food at the football tailgate.* Fayetteville, AR: The University of Arkansas Press.

Verma, J.P. (2016). *Repeated measures design for empirical researchers.* Hoboken, NJ: Wiley & Sons.

Vermilio, R.E., Thomas, K., Thomas, J.R., & Morrow, J.R. (2013). Good to great: Consistent PGA tour money winners—A 6-year longitudinal study—2005-2010. [Unpublished manuscript]. Department of Health, Kinesiology and Recreation, University of North Texas.

Vertinsky, P. (2021). Searching for Balance: A Historian's View of the Fractured World of Kinesiology, *Kinesiology Review, 10 (2),* 126-132. https://doi.org/10.1123/kr.2020-0061.

Vertinsky, P. (1990). *The eternally wounded woman: Women, exercise, and doctors in the late nineteenth century.* Manchester, England: Manchester University Press.

Vertinsky, P. & McKay, S. (2004). *Disciplining bodies in the gymnasium: Memory, movement, modernism.* London: Routledge.

Vines, T.H., Albert, A.Y.K., Andrew, R.L., et al. (2014). The availability of research data declines rapidly with article age. *Current Biology, 24,* 94-7.

Virani, S.S., Alonso, A., Benjamin, E.J., et al. (2020). Heart disease and stroke statistics-2020 update: A report from the American Heart Association. *Circulation, 141*(9), e139-e596.

Vockell, E.L. (1983). *Educational research.* New York: Macmillan.

Wainer, H. (1992). Understanding graphs and tables. *Educational Researcher, 2* (1), 14-23.

Walcott, H.F. (2009). *Writing up qualitative research* (3rd ed.). Los Angeles: Sage.

Wallace, B.A. (2000). *The taboo of subjectivity: Toward a new science of consciousness.* New York: Oxford University Press.

Walvin, J. (1984). Sport, social history and the historian. *The British Journal of Sports History, 1,* 5-13.

Ward, P. (2013). Last man picked: Do mainstream historians need to play with sport historians? *International Journal of the History of Sport, 30,* 6-13.

Ware, S. (2011). *Title IX: A brief history with documents.* Long Grove, IL: Waveland Press, Inc.

Wasserman, R.L., & Lazar, N.A. (2016). The ASA statement on p-values: context, process, and purpose. *The American Statistician, 70,* 129-33.

Weaver, D., Reis, M.H., Albanese, C., Costantini, F., Baltimore, D., & Imanishi-Kari, T. (1986). Altered repertoire of endogenous immunoglobulin gene expression in transgenic mice containing rearranged Mu heavy chain gene. *Cell,* 247-59.

Webb, E.J., Campbell, D.T., Schwartz, R.D., & Sechrest, L. (1966). *Unobtrusive measures: Nonreactive research in the social sciences.* Chicago: Rand McNally.

Weiss, M.R., Bredemeier, B.J., & Shewchuk, R.M. (1985). An intrinsic/extrinsic motivation scale for the youth sport setting: A confirmatory factor analysis. *Journal of Sport Psychology, 7,* 75-91.

Weiss, M.R., McCullagh, P., Smith, A.L., & Berlant, A.R. (1998). Observational learning and the fearful child: Influence of peer models on swimming skill performance and psychological responses. *Research Quarterly for Exercise and Sport, 69,* 380-394.

White, P.A. (1990). Ideas about causation in philosophy and psychology. *Psychological Bulletin, 108,* 3-18.

Whorton, J.C. (1982). *Crusaders for fitness: The history of American health reformers.* Princeton, NJ: Princeton University Press.

Wiggins, D.K. (2018). *More than a game: A history of the African American experience in sport.* Lanham, MD: Rowman & Littlefield.

Wiggins, D.K. & Swanson, R.A. (2016). *Separate games: African American sport behind the walls of segregation.* Fayetteville, AR: The University of Arkansas Press.

Wiggins, D.K. (1983a). Sport and popular pastimes: Shadow of the slavequarter. *Canadian Journal of the History of Sport and Physical Education, X,* 61-88.

Wiggins, D.K. (1983b). The play of slave children in the plantation communities of the old south, 1820-1860. *Journal of Sport History, 7,* 21-39.

Wiggins, D.K. (1986). From plantation to playing field: Historical writing on the black athlete in American sport. *Research Quarterly for Exercise and Sport, 57,* 101-116.

Wiggins, D.K. (1995). Victory for Allah: Muhammad Ali, the Nation of Islam, and American society. In E.J. Gorn (Ed.), *Muhammad Ali, the people's champ* (pp. 88-116). Urbana, IL: University of Illinois Press.

Wiggins, D.K. (1997). *Glory bound: Black athletes in a white America.* Syracuse, NY: Syracuse University Press.

Wiggins, D.K. (2000). The African American athlete experience. In A.E. Strickland & R.E. Weems, Jr., *The African American experience: An historiographical and bibliographical guide* (pp. 255-277). Westport, CT: Greenwood Press.

Wiggins, D.K., & Miller, P.B. (2003). *The unlevel playing field: A documentary history of the African American experience in sport.* Urbana, IL: University of Illinois Press.

Willett, W. (1990). *Nutritional epidemiology.* New York: Oxford University Press.

Wilson, W. & Wiggins, D.K. (2018). *LA Sports: Play, games, and community in the City of Angels.* Fayetteville, AR: The University of Arkansas Press.

Winkleby, M.A., Taylor, C.B., Jatulis, D., & Fortmann, S.P. (1996). The long-term effects of a cardiovascular disease prevention trial: The Stanford Five-City Project. *American Journal of Public Health, 86,* 1773-1779.

Witherspoon, K.B. (2008). *Before the eyes of the world: Mexico and the 1068 Olympics.* DeKalb, IL: Northern Illinois University Press.

Woods, A.M., & Lynn, S.K. (2014). One physical educator's career cycle: Strong start, great run, approaching finish. *Research Quarterly for Exercise and Sport, 85,* 68-80.

Woods, R.A. (2017). *Spotlight on statistics sports and exercise.* U.S. Bureau of Labor Statistics.

Yan, Z., & Cardinal, B.J. (2013). Perception of physical activity participation of Chinese female graduate students: A case study. *Research Quarterly for Exercise and Sport, 84,* 384-396.

Yin, R.K. (2017). *Case study research: Design and methods* (6th ed.). Thousand Oaks, CA: Sage.

Young, D.C. (1996). *The modern Olympics: A struggle for revival.* Baltimore: The Johns Hopkins University Press.

Zaman, A. (2020). *The effects of stress on working memory, inhibitory gating, and motor symptoms in Parkinson's disease.* (Publication No. 28027147) [Doctoral dissertation, Iowa State University]. ProQuest Dissertations & Theses Global.

Zar, J.H. (1994). Candidate for a Pullet Surprise. *Journal of Irreproducible Results, 39,* 13.

Zasa, M. (2015). The accuracy of references in five sport science journals. *Science & Sports, 30*(1), E31-E33. 10.1016/j.scispo.2014.10.007.

Zelaznik, H.N. (1993) Ethical issues in conducting and reporting research: A reaction to Kroll, Matt, and Safrit. *Quest, 45,* 62-68.

Zhu, W., Fox, C., Park, Y., Fisette, J. L., Dyson, B., Graber, K. C., . . . & Raynes, D. (2011). Development and calibration of an item bank for PE metrics assessments: Standard 1. *Measurement in Physical Education and Exercise Science, 15*(2), 119-137.

Zhu, W., Rink, J., Placek, J.H., Graber, K.C., Fox, C., Fisette, J.L., . . . & Raynes, D. (2011). PE metrics: Background, testing theory, and methods. *Measurement in Physical Education and Exercise Science, 15*(2), 87-99.

Zhu, W., & Yang, Y. (2016). Item response theory and its applications in kinesiology. In Ntounamis, N., & Myers, N.D., *An introduction to intermediate and advance statistical analyses for sport and exercise scientists.* Wiley & Sons.

Ziv, A. (1988). Teaching and learning with humor: Experiment and replication. *Journal of Experimental Education, 57,* 5-15.

SUBJECT INDEX

Note: The italicized *f* and *t* following page numbers refer to figures and tables, respectively.

Jerry Thomas, EdD, retired in 2016 from the University of North Texas, where he served as a professor and dean of the College of Education. He has authored more than 200 publications, 120 of which are in refereed journals. In 1999 SHAPE America named him the C.H. McCloy Lecturer for his production of research throughout his career. Thomas has been editor in chief of *Research Quarterly for Exercise and Sport* and a reviewer for most major research journals in kinesiology and numerous journals in psychology. He has also served as president of the National Academy of Kinesiology (NAK), founding president of American Kinesiology Association (AKA), AAHPERD Research Consortium (now SHAPE America), and North American Society for Psychology of Sport and Physical Activity (NASPSPA). In 1990 he was named an AAHPERD Alliance Scholar and in 2003 was named a NASPSPA Distinguished Scholar based on lifetime achievement in research. Thomas received an honorary doctorate of science from his undergraduate institution, Furman University, in the spring of 2015.

Courtesy of College of Human Sciences at Iowa State University.

Philip Martin, PhD, is a professor emeritus in the department of kinesiology at Iowa State University. Previously, he served as the head of the department of kinesiology at Penn State and as chair of exercise science and physical education at Arizona State University in Tempe. Martin's research activities have addressed mechanical factors influencing the economy and efficiency of walking, running, and cycling. He has authored 80 research articles and book chapters and presented at regional, national, and international professional meetings. Martin is an active member and fellow of the American Society of Biomechanics (ASB), the American College of Sports Medicine (ACSM), and the National Academy of Kinesiology (NAK). He serves on the board of directors and the executive board of the American Kinesiology Association (AKA). He is a former president of the NAK and the ABS and former vice president of the AKA. He has also served on the editorial board of the *International Journal of Sport Biomechanics*, as an associate editor for the *Journal of Applied Biomechanics*, and as a biomechanics section editor for *Research Quarterly for Exercise and Sport*. Additionally, he served on the scientific advisory committee for the U.S. Olympic Committee Sports Medicine Council.

Courtesy of Susan Calkins Photography.

Jennifer L. Etnier, PhD, is a Julia Taylor Morton Distinguished Professor and chair of the department of kinesiology at the University of North Carolina at Greensboro, where she has received the Health and Human Performance Teaching Award and UNCG Alumni Teaching Excellence Award. Etnier's research focuses on the cognitive benefits of physical activity. She has authored over 90 peer-reviewed research articles and contributed to over 20 book chapters and editorials. She is also the author of two books for the lay public focused on the youth sport experience: *Bring Your "A" Game* and *Coaching for the Love of the Game*. Etnier is a past president of the North American Society for the Psychology of Sport and Physical Activity (NASPSPA), where she had previously served on the executive board. She is a fellow of the American College of Sports Medicine (ACSM) and the National Academy of Kinesiology (NAK), where she also served as member-at-large. Etnier serves on the editorial board for the *Journal of Sport and Exercise Psychology* and the *Journal of Aging and Physical Activity*, where she previously served as editor in chief.

Steve Silverman, EdD, is a professor of education and senior advisor to the provost for research preparation at Teachers College at Columbia University. He has taught and written about research methods for more than 30 years and has conducted research on teaching in physical education focusing on how children learn motor skill and develop attitudes. He has published more than 80 research articles in addition to many books and book chapters. Silverman is a fellow and past president of the National Academy of Kinesiology (NAK) and the AAHPERD Research Consortium (now SHAPE America) and a fellow of the American Educational Research Association (AERA). A former coeditor of the *Journal of Teaching in Physical Education* and former editor in chief of the *Research Quarterly for Exercise and Sport*, Silverman was an AERA Physical Education Scholar Lecturer, a Research Consortium Scholar Lecturer, and Weiss Lecturer and Alliance Scholar for AAHPERD. In 2010, he was inducted into the Kinesiology and Health Education Hall of Honor at the University of Texas at Austin.

ISBN 978-1-7182-1304-3

Human Kinetics